MONTY AND ROMMEL

Dr Peter Caddick-Adams has been a professional military historian for over twenty years. He lectures at the UK Defence Academy and staff colleges around the world on military history, war studies and media operations. Specialising in battlefield tours, doctrine and leadership, he worked with the inspirational Richard Holmes for twelve years and has led visits to more than fifty battlefields worldwide. As a regular and reserve soldier he has experience of three major war zones: Bosnia, Iraq and Afghanistan. In 2010 he was elected a Fellow of the Royal Historical Society and is a frequent TV and radio broadcaster on military and security issues.

'A quite brilliant piece of writing. Here in a single volume we have a first-rate expose of two of the war's best known commanders . . . If Caddick-Adams were a landscape painter his book would be the equivalent of Monet, full of rich intriguing colours and patterns. The resultant effect is spectacular'
BBC History Magazine

'Caddick-Adams writes with authority and a deep knowledge of both his subjects and the two world wars in which they fought. He has produced an utterly absorbing and hugely entertaining book that will provide a new benchmark on how we view these two divisive generals'   James Holland

'Peter Caddick-Adam's comparison of two entwined careers is full of penetrating new insights, illuminated by a clear understanding of the world wars . . . one of the best military histories of the year'   Richard Holmes

'Caddick-Adams has achieved a first by entwining biographies of two WW2 adversaries . . . an exciting character-driven read and an excellent example of how the personal experience of war can create great and yet humane generals. First-class history'   *Military Illustrated*

'[Caddick-Adams] succeeds in contextualising these two commanders and in analysing the contributions each made to the art of military leadership. The use of the comparative method enables us to see not only how different these two men were, but also how difficult it remains to define what successful military leadership is all about'   *Tablet*

'A wonderful mine of information for fans of either general . . . [Caddick-Adams] is a military historian of great industry who shows an impressive grasp of his materials . . . The author's central purpose is achieved with impressive and cumulative success as his book progresses'   *New Republic*

MONTY AND ROMMEL

PARALLEL LIVES

Peter Caddick-Adams

arrow books

Published by Arrow 2012

10 9 8 7 6 5 4 3 2 1

Copyright © Peter Caddick-Adams 2011, 2012

Peter Caddick-Adams has asserted his right to be identified as the author of this work
under the Copyright, Designs and Patents Act 1988

First published in Great Britain in 2011 by Preface Publishing

20 Vauxhall Bridge Road
London, SW1V 2SA

An imprint of The Random House Group Limited

www.randomhouse.co.uk

Addresses for companies within The Random House Group Limited can be found at
www.randomhouse.co.uk

The Random House Group Limited Reg. No. 954009

A CIP catalogue record for this book is available from the British Library

ISBN 978 1 84809 154 2

The Random House Group Limited supports The Forest Stewardship Council
(FSC®), the leading international forest certification organisation. Our books
carrying the FSC label are printed on FSC® certified paper. FSC is the only forest
certification scheme endorsed by the leading environmental organisations,
including Greenpeace. Our paper procurement policy can be found at
www.randomhouse.co.uk/environment

Typeset in Minion by Palimpsest Book Production Limited, Falkirk, Stirlingshire

Printed and bound in Great Britain by CPI Group (UK) Ltd, Croydon CR0 4YY

To Stefania and Emmanuelle for their
love, support and tolerance in allowing me to
spend so much time with the field marshals.

Contents

List of Maps

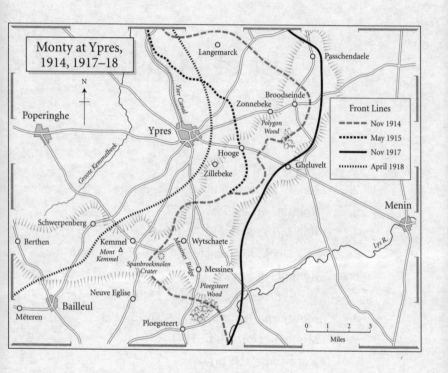

Monty at Ypres, 1914, 1917–18

Front Lines
- - - Nov 1914
····· May 1915
— Nov 1917
····· April 1918

There was rarely a quiet moment in the Ypres area, which witnessed near-continuous fighting, from October 1914 to October 1918. Monty was dangerously wounded in Méteren in 1914, and from 1917–18 he planned operations as a staff officer with IX Corps in the battlefields east of Ypres. Within the Ypres salient are the villages of Gheluvelt and Messines, where Adolf Hitler fought, and Ploegsteert, where Lt.-Colonel Winston Churchill's battalion (6/Royal Scots Fusiliers) was stationed between January and May 1916.

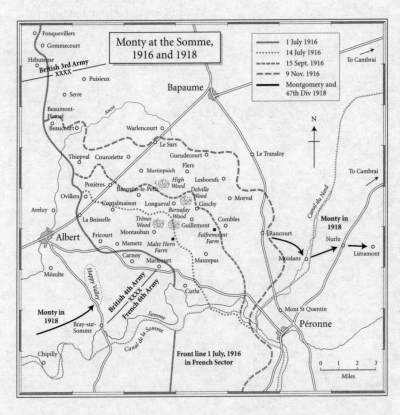

Monty at the Somme, 1916 and 1918

————	1 July 1916
··········	14 July 1916
– – – –	15 Sept. 1916
— — —	9 Nov. 1916
━━━━	Montgomery and 47th Div 1918

N

Named after a river that most British troops never got to see, the Somme battlefields witnessed Monty's presence twice. Initially during the 'Great Push' of 1916, whose slow progress is charted here, Bernard's 104th Brigade fought in the area of Trônes Wood, Malz Horn Farm and Guillemont. His advance with 47th Division in 1918 is also shown, including Happy Valley, where his brigade paraded in 1916, but which his division attacked in 1918.

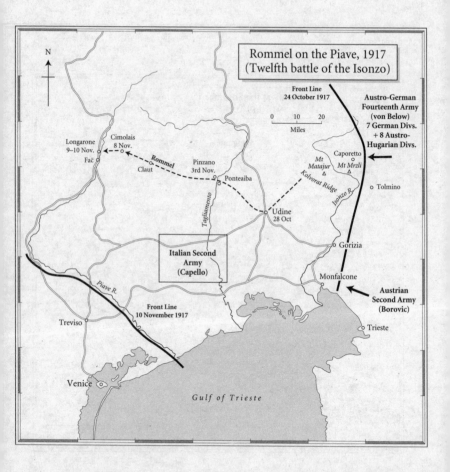

Rommel on the Piave, 1917
(Twelfth battle of the Isonzo)

Front Line
24 October 1917

0   10   20
Miles

N

Austro-German
Fourteenth Army
(von Below)
7 German Divs.
+ 8 Austro-
Hungarian Divs.

Longarone
9–10 Nov.
Fač

Cimolais
8 Nov.

Rommel
3rd Nov.

Claut

Pinzano
3rd Nov.

Ponteaiba

Mt
Matajur

Caporetto

Mt Mrzli

Tolmino

Kolvorat Ridge

Isonzo R.

Udine
28 Oct

Tagliamento

Italian Second
Army
(Capello)

Gorizia

Piave R.

Monfalcone

Austrian
Second Army
(Borovic)

Front Line
10 November 1917

Treviso

Trieste

Venice

Gulf of Trieste

Rommel's 1917 war in Italy was one characterised by speed and movement, in contrast to the static nature of the Western Front. Erwin advanced 150 miles over twenty-two days, ending spectacularly on the River Piave, at Longarone, on 10 November. His achievement is all the more impressive when one remembers that most of it was conducted in contact with the enemy, over tiring mountainous terrain and on foot.

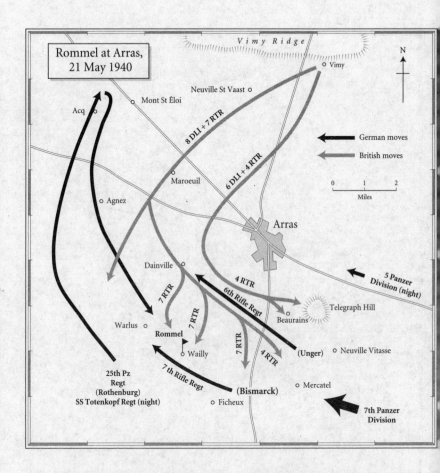

The armoured counter-attack against 7th Panzer Division at Arras was the first time Rommel encountered British troops. Deploying from Vimy, the Royal Tank Regiment's *Matildas* completely surprised disorganised German units; only the speedy use of artillery at Telegraph Hill and Wailly saved 7th Panzer division from annihilation. Caught in the midst of the fracas, Rommel was left with an indelible impression of British resourcefulness.

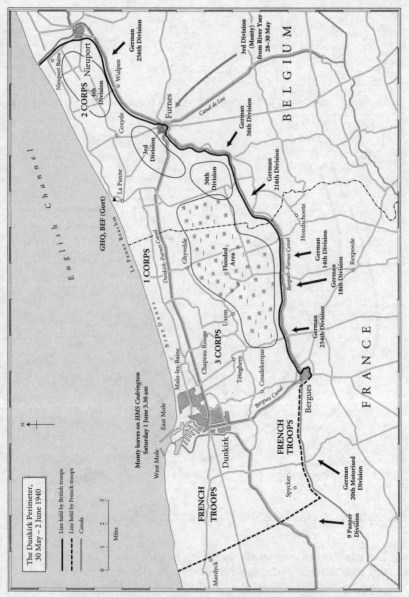

The Dunkirk Perimeter,
30 May – 2 June 1940

Line held by British troops
Line held by French troops
Canals

Miles
0 1 2 3

N

English Channel

BELGIUM

FRANCE

German 256th Division

3rd Division (Monty) from River Yser 28–30 May

Nieuport Bains
Nieuport
2 CORPS
4th Division
Wulpen
Coxyde
Furnes
Canal de Loo
3rd Division
German 56th Division
La Panne
50th Division
German 216th Division
GHQ, BEF (Gort)
La Panne Beaches
Dunkirk-Furnes Canal
1 CORPS
Ghyvelde
Flooded Area
Hondschoote
Bergues-Furnes Canal
German 14th Division
Rexpoede
German 18th Division
Bray Dunes
Uxem
German 254th Division
Chapeau Rouge
3 CORPS
Téteghem
Coudekerque
Malo-les-Bains
Monty leaves on *HMS Codrington* Saturday 1 June 3.30 am
East Mole
West Mole
Dunkirk
Bergues
Bergues Canal
FRENCH TROOPS
FRENCH TROOPS
Spycker
German 20th Motorised Division
9 Panzer Division
Mardyck

Historians agree that of all British formations which fought in 1940, Monty's 3rd Division performed best. It trickled into the Dunkirk perimeter whilst fighting rearguard actions to slow the German advance, before rescue from the beaches or port. Dunkirk also witnessed Monty's elevation from divisional to corps command, when his boss (Brooke) departed for England.

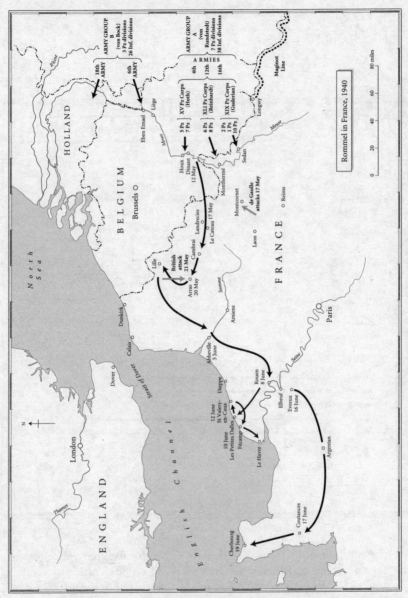

It was 7th Panzer Division's dramatic progress through northern France in 1940 that brought Rommel national acclaim throughout Germany, a Knight's Cross and led directly to his North African posting. He covered over 500 miles in forty days – with battles on the River Meuse and at Arras and the capture of Cherbourg on 19 June.

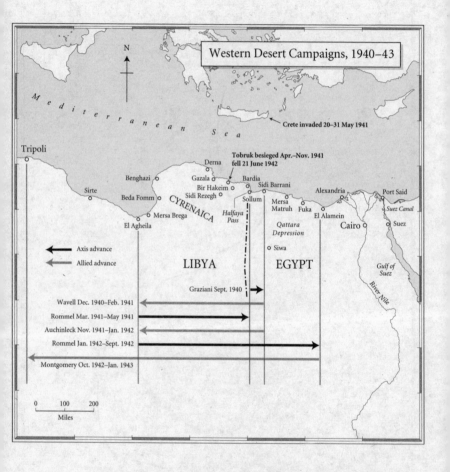

Western Desert Campaigns, 1940–43

Crete invaded 20–31 May 1941

Mediterranean Sea

Tripoli

Tobruk besieged Apr.–Nov. 1941
fell 21 June 1942

Derna

Benghazi   Gazala   Bardia
                Bir Hakeim   Sidi Barrani
Sirte   Alexandria   Port Said
      Beda Fomm   Sidi Rezegh   Sollum
                                Mersa   Fuka   Suez Canal
            CYRENAICA   Matruh   El Alamein
      Mersa Brega   Halfaya   Qattara   Cairo   Suez
El Agheila   Pass   Depression

Axis advance
Allied advance

LIBYA   EGYPT   Gulf of Suez

Siwa

River Nile

Graziani Sept. 1940

Wavell Dec. 1940–Feb. 1941

Rommel Mar. 1941–May 1941

Auchinleck Nov. 1941–Jan. 1942

Rommel Jan. 1942–Sept. 1942

Montgomery Oct. 1942–Jan. 1943

0   100   200
        Miles

The advances and retreats of both sides in North Africa took place over huge distances, underlining how success rested on efficient logistics. All supplies had to be imported by air or sea and trucked to the battlefields. The campaign had a see-saw nature at the operational level, and Rommel came very close to his goals of Cairo and the Suez Canal.

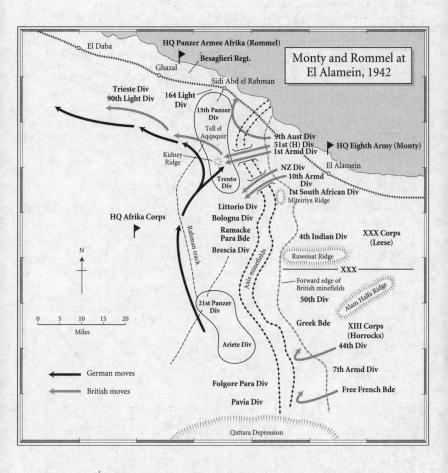

As well as numerical superiority, some of Monty's success at El Alamein was due to his ability to deceive his opponents into believing that Eighth Army's strike would come from the south. At this stage the Afrika Korps were 700 miles from their nearest port, Benghazi, and 1,400 miles from Tripoli, whilst Cairo for the British was a mere 150 miles east by road or rail.

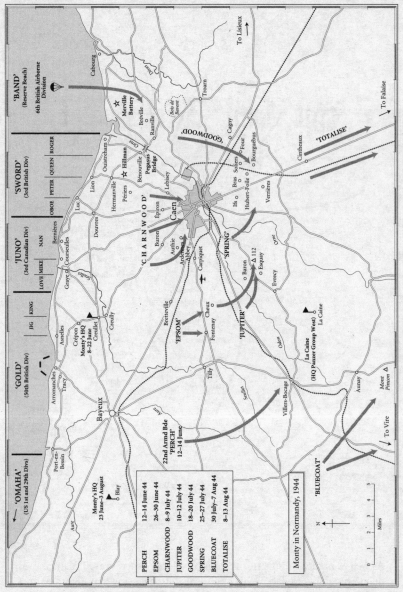

Caen dominated the eastern sector of the Normandy battlefield. Failure to seize the city on D-Day necessitated a series of frustrating and attritional operations, each of increasing size, in an effort to fight in more favourable terrain and break out of the bridgehead. Note the proximity of Monty's headquarters to the battle zone, whereas Rommel's headquarters was a hundred miles east, at La Roche Guyon.

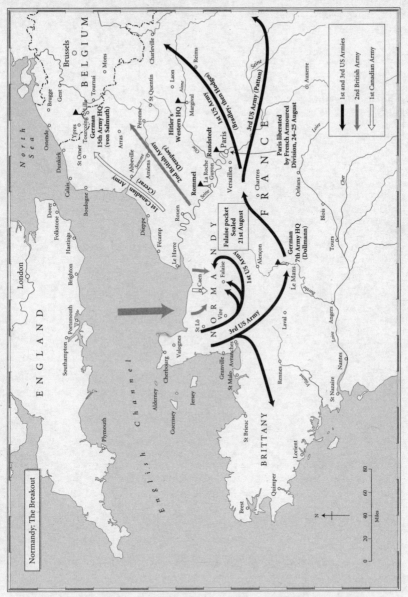

Normandy: The Breakout

Paris liberated
by French Armoured
Division, 24–25 August

Falaise pocket
Sealed
21st August

1st and 3rd US Armies
2nd British Army
1st Canadian Army

Considering that the bulk of Allied forces were still around Falaise on
21 August, their subsequent breakout was conducted with great speed.
Rheims was liberated on 30 August, Arras and Mons on 2 September, Brussels
the following day, Charleville on 4 September and Ypres two days later.
Once the Normandy front collapsed there were few German reserves to stem
the extremely rapid advance, sometimes dubbed 'Monty's Blitzkrieg'.

Prologue

*At every crossway on the road to the future, each progressive spirit
is opposed by a thousand men appointed to guard the past.*
Count Maurice de Maeterlinck

IT IS MID-AFTERNOON ON 21 May 1940, a warm Tuesday, near Arras
in northern France. Last year, the scaffolding came down, ending years
of reconstruction. The Great War had seen Arras shelled mercilessly.
Now it looks like happening all over again.

The same grey-clad invaders are once more at the gates. One of them,
an officer who had done well in the previous conflict, is now a divisional
commander. The sun catches the unique bauble *Generalmajor* Erwin
Rommel wears at his throat, the *Pour le Mérite*. His business is reducing
attrition. Not in avoiding combat, but by superior tactics which will limit
the fighting. Using tanks at high speed, and lots of them, he aims to slice
through his opponents' lines before they know what's happened. So far
he has succeeded beyond his wildest expectations. The forty-eight-year-
old took over his command just two months earlier, when his men were
still in training. Now they are further into France than any of their fellow
invaders, and have become known as 'the Ghost Division' for their wraith-
like ability to materialise anywhere on the battlefield.

The weather is considerably better than it had been back in 1917,
when last a major battle erupted around Arras. Warm sun has turned
the chalky topsoil to fine powder; it is hot. The dips and mounds in the
fields around signify the old trenches and dug-outs of earlier battles,
mellowed by time. These were once occupied by the headquarters of the

British 33rd Division, where his future Second World War rival served as a staff officer. Rusting strands of old barbed wire remain, as if handing over the fighting from one generation to the next.

Rommel can now see dust clouds trundling towards him: this can only mean one thing. Tanks. He curses his luck. Normally he would be riding in the tank of his friend, Colonel Rothenburg, commanding his 25th Panzer Regiment, but today he has let them continue their lightning advance without him. His infantry – the 7th Rifle Regiment – though in trucks, are too slow for his liking and he has dropped back to chivvy them along. And now the British are attacking his unprotected flanks.

Fifty miles away to the north-east, on the outskirts of another big city, Lille, another divisional commander is facing *his* traditional foe. It is a warm day here too. He sports khaki battledress and wears the special ribbon of the Distinguished Service Order. Since the Germans invaded over a week earlier, his troops have been rushed from place to place in Belgium, achieving little. There's no overall strategy and absolutely zero cooperation with his French and Belgian allies. His own warriors are conveyed in a laughable mixture of camouflaged trucks and brightly coloured baker's vans, because of a lack of investment by politicians over the years. The soldiers of his formation have cursed him ever since the war began for their harsh training – tougher than any other division – but now the results are beginning to pay off. They are coping with the endless marches and little sleep better than the rest of Britain's Expeditionary Force.

Deployed along a ten-mile stretch of the lazy River Escaut, between Pecq and Avelgem in Belgium, his men are dug in and have been under intense shellfire since 2 a.m. Peering through the smoke, Bernard Montgomery can see that the factories and homes lining the water's edge are crumbling under the tornado of German shells. He was woken earlier by a series of distinctive blasts signifying the end of the many bridges across the river, blown by his engineers to block the German advance. His own guns speak again, but the weight of fire is coming from the invaders.

Meanwhile, in the eleven days since Generalmajor Rommel's advance began, the French have barely attacked his panzers with more than a shotgun. Yet, the General is annoyed. His young troops have tasted victory and are off their guard. They have paused, thirsty, and are resting their tired feet in scuffed jackboots. Grimy-eyed, dozing in the sun, their limp fingers hold cigarettes trailing smoke. They are overconfident. The Luftwaffe own the skies, their bombers drone lazily overhead, seeking prey. In the hamlets surrounding Arras, Rommel's vehicles are backed up in the narrow lanes, unable to move. Exactly what he warned them about in training. Quick-tempered, the General is about to shatter the peace of the day.

A shell whines over; then another. Under fire, a half-track explodes, showering the cobbled street with burning fuel and rubber: confusion reigns. A soldier, badly burned, runs screaming. The General is angry at the chaos. Yet he knows exactly what to do: he has seen it all before. The smell of high explosive takes him back, instinctively, to other battles.

He urges his driver through the village and up a hill. Towards the firing. Always *towards* the trouble. They turn right, past the cemetery. Under the leaves of a small copse, he spies more gunners milling around in panic, their hot meals abandoned. He quits his command car and snaps out a hurricane of orders. Then an inner calm descends. His young aide, Leutnant Most, scuttles after him, maps and notebook at the ready, as he has been taught. With astonishing energy the General moves and thinks supremely fast:

'You men, unhook those guns, NOW!'

'You, the ammunition.' He starts to form a gun line. Anything will do.

'*Herr General*, these are anti-aircraft guns.' No problem. As long as it has a barrel and will fire.

'*Herr General*, we have only anti-aircraft ammunition, not suitable against tanks.' Always problems, not solutions.

'The Tommies will not know the difference. Open Fire!'

He dashes about. Seconds matter. Here and there, he lends a hand to push gun wheels through the dirt, pulls on a barrel to swing this or that cannon into position. A quick squint through his binoculars. The leading tanks are no more than two hundred metres away. It is going to be close.

His artillerymen sweat to feed their guns quickly.

'Make every shot count,' he orders, and picks out targets for his men.

The heavy British tanks, 'Matildas', are now close enough for the squeal of their tracks to be audible above the gunfire; they halt and shudder as the rounds strike home. Sparks fly and a metallic 'ping' sounds, as shells bounce off the armour plate.

Curses! Then some penetrate. The monsters erupt in flames. Loose caterpillar tracks writhe like snakes. The nearest tank pauses, a turret hatch opens with a resounding clang and a grim-faced officer climbs down, cap askew, and – arms raised – walks unsteadily towards the guns that have just killed his driver.

It is over. The chief's quick thinking has saved the moment.

He rests his binoculars and rolls a spent brass shell case, still hot, with his boot. He turns to his aide, a sparkle of triumph in his eyes that he has not known since 1917. But the expression is wiped from his face and replaced with one of horror as Most, so young and keen, falls towards him and collapses in his arms. Blood gushes from his mouth, he is mortally wounded.

Rommel not only mourns his friend, but ponders, as we could, on what might have been – had the British soldier adjusted his sights and aimed just a fraction to the left.

In the same hour, back on the Escaut, Major General Montgomery watches as the shelling subsides. Suddenly he sees the muddy river is full of little specs – rubber dinghies, manned by German assault troops, paddling furiously. Their machine guns spit fire from the far banks to cover them. One boat flips over, a lucky mortar round shatters another. A sniper takes care of the NCO urging his men on in a third craft. Binoculars reveal that the splashing menace is still swarming across the water. A few have reached the near bank but are met by screaming men in khaki, wielding bayonets. Softened almost into melody by the distance, the mosquito whine of German machine guns alternates with the slower, rhythmic *thump-thump-thump* of British Brens. Quickly, his artillery find the range and a pattern of water plumes engulfs the picture. As the spray and smoke subside, the menace has become a mass of broken boats and twisted life, which ebb slowly downstream.

Although the first attack has failed, there will be others; the British general's life story is one of having prepared for moments such as these; he knows that now is not the moment for sentiment over the dead. Earlier wars and battles have told him what to expect. His formation, who also have a nickname – the Iron Division – must now live up to their sobriquet and hold on the rest of the day and into tomorrow, before orders permit them to withdraw. Much against his expectations, within days he will find himself promoted to command a corps of several divisions, from sand dunes around a little port called Dunkirk. Bernard

Montgomery vows that if he ever visits the continent as a general again, it will not be under such ignominious circumstances.

I have the letter still. It is addressed to my maternal grandfather, who left it to me. The paper is still crisp; the blue ink of the unmistakeable, slightly immature handwriting remains sharp. My grandfather happened to be a bishop, as was the letter writer's father, and in handing me a keepsake dated 6 July 1952, Clifford Arthur Martin, 4th Bishop of Liverpool, triggered a sense of curiosity which never departed. The letter's contents are unremarkable, but I always wondered who this busy Field Marshal Montgomery was, who excused himself as too busy to come and preach in my grandfather's cathedral. Then – after one Sunday afternoon black-and-white feature film – I discovered Monty's nemesis, Erwin Rommel. Gradually the similarities and interrelations between the two commanders dawned on me.

Born four years apart, with birthdays separated by two days, Montgomery and Rommel were wounded and decorated within days of each other in 1914. They were 'outsiders' in several ways: neither came from families with a military tradition, and they originated in provinces distant from their capitals: Ulster and Swabia. Their lives would be closely intertwined, fighting as opposing divisional commanders in 1940, then leading their respective armies during major duels in North Africa and Normandy, which saw both elevated to the dizzy rank of field marshal. Wiry and slight of stature, they came from large families, but produced single sons in the same year, 1928. Both were notoriously thrifty in their domestic milieu, neither was particularly sophisticated, their lives revolved around work; families came second.

The two capitalised on their experience of the First World War through writing, produced tactical textbooks, and kept voluminous notes and diaries in the Second World War with an eye to post-war publication. The way they would in the future meet and inspire front-line soldiers, plan operations, hire and fire subordinates, deal with logistics and take account of casualties – both estimated and actual – was forged in the fury of combat of the trenches and, by 1918, their respective ideas on leadership and command had already become well established.

The future field marshals reacted to the distance of their superiors in the First World War and became beacons of hope for their respective nations and iconic leaders for their troops, with whom they communicated in person where possible and if not, via newspapers, radio and propaganda. The conflict we commonly associate with the two is the Desert War in North Africa, yet they served in an extensive, and arguably far more important, range of campaigns.

Both men conducted their battles from captured armoured caravans, taken as booty. The pair never met, but Rommel later paid warm tribute to Monty's skill.[1] Whilst interrogating a captured British officer over tea in May 1944, Rommel asked after his 'old friend General Montgomery'.[2] For his part, Montgomery had Rommel's portrait hung in his own battle caravan 'to understand what made him tick', and named one of his dogs and a horse after him. Monty observed later in life that 'I would have liked to discuss the battle [of El Alamein] with him. But he is dead and we cannot tell the story together.'[3]

Montgomery and Rommel often found themselves tested and frequently triumphed over adverse circumstances that would defeat most of us just reading about them, from Montgomery's attempts to shore up his brigade's morale on the Somme in 1916, to Rommel's juggling of diminishing resources to meet the next thrust of the Eighth Army in late 1942. Were they similar in character, or did they just travel the same path through war and life together? Both were highly controversial during their lifetimes, were often at war as much with their superiors as with their opponents, and yet still beat the drum from afar today. It is perhaps also important to understand both commanders in the context of modern military leadership, especially whilst we are again at war, so as not to perpetuate their faults – hidden as they may be, by the dazzle of the reputation we have given them. They commanded in an era where decisions were made in the war rooms of distant capitals, which they were obliged to enact. They were products of different political systems: Rommel grew up under a militaristic autocracy and served a dictatorship; Monty was the product of a democracy, with all that implies; neither was particularly scholarly or politically aware (to the detriment of their careers) so this may not have occurred to them.

In their different ways they made *themselves* the centre of the decision-making process, wrenching back the initiative from their superiors. Both marshals may have been over-promoted: we shall see. They were certainly stubborn, poor team players and notoriously intolerant of allies. Montgomery refused to accept Churchill's directives as to when to fight at Alamein; the price Monty paid for his squabbles with superiors and colleagues was eternal damage to his reputation. Rommel, meanwhile, frequently disobeyed Hitler. Eventually this cost him his life.

PART ONE

BAPTISM
AND FIRE

The past is a foreign country: they do things differently there.
L. P. Hartley, *The Go Between*

1

The Irishman

BERNARD MONTGOMERY and Erwin Rommel shared a gift for communication. This is how we remember them: boosting their troops' morale in the dark days, inspiring their countrymen and befriending the world's media when this was uncommon amongst professional soldiers; leaving a mark for posterity through photographs, film and the books they wrote. Both came from long lines of worthy, middle-class stock – communicators by trade: one was the son of a vicar, the other of a school teacher. Curiously, neither had any significant military tradition in their ancestry. Such solid lineage provided each with continuity and self-assurance, a firm bedrock on which to found any career. The two were born into empires now long gone, in an era of confidence and growth. In Britain and Germany, each then the centre of an imperial web, society was enjoying the fruits of hard-won wealth after the nation-building social upheavals of the early nineteenth century. Arthur Conan Doyle, Henry Rider Haggard, Robert Louis Stevenson and Thomas Hardy were at the height of their literary fame. Renoir, Toulouse-Lautrec, Gauguin, Monet and Van Gogh were producing some of their finest canvases, whilst Brahms, Delius, Grieg and Verdi were delighting audiences with their music. Richard Wagner had just died and Edward Elgar was a decade away from finding fame.

Bernard Law Montgomery, the fourth child of an eventual nine (six boys and three girls), was born in the fiftieth year of Queen Victoria's reign and would serve six monarchs. He arrived on 17 November 1887 in the middle of the long Victorian afternoon of imperial achievement. 1887 was a relatively quiet year; there were no major wars or upheavals. Indeed, the year epitomised stability and success. At the summer solstice, 21 June, Victoria celebrated fifty years on the imperial throne with a banquet to which fifty European kings and princes were invited; an impressive review of her Royal Navy at Spithead followed.* Western Europe and the United States were brimming with prosperity, represented in Germany (then a mere sixteen years old) by the unveiling of the first Daimler automobile and the patenting of the gramophone. This was an era when much of the world looked to Britain for leadership, guidance or moderation; a time which often conferred on the Empire's sons a degree of confidence in their place in the world, and sometimes a touch of arrogance or xenophobia.

The year of Monty's arrival witnessed the birth of another enduring facet of British life: the resident of 221b Baker Street. Mr Sherlock Holmes appeared in the same year as Thomas Hardy's *The Woodlanders* and Ryder Haggard's *She*. His creator, Arthur Conan Doyle, would later chronicle the history of the First World War. Lord Salisbury was Prime Minister and had just lost his talented Chancellor of the Exchequer in a surprise resignation. Monty would spend his most influential years working with the ex-Chancellor's son, Winston Churchill. 1887 also saw the arrival of the future war poet, Rupert Brooke, and several future generals, including Alan Cunningham, an Irish gunner and the Eighth Army's first commander (whose elder brother, Andrew, would command the Mediterranean Fleet, 1939–43); Henry Pownall, Chief of Staff to the British Expeditionary Force (BEF) in 1939; and 'Pug' Ismay, deputy secretary to Churchill's war cabinet. Amongst Monty's future adversaries born during this time was Wilhelm Canaris, head of German intelligence in the Second World War, and two important armour specialists: Geyr von Schweppenberg (born 1886) and Heinz Guderian (1888). Three days before Montgomery's birth, on 14 November, a future President of the United States was delivered of a son, whom he christened Theodore – Teddy – after himself; Teddy would later lead the US 4th Division ashore on D-Day.

It is often said that Mars favours the month of November for some of its greatest sons and daughters, and whilst we are now aware of Monty, Rommel (15 November) and Teddy Roosevelt, we ought to remind ourselves that George Smith Patton was born in 1885 on 11 November, whilst Charles de Gaulle arrived in 1890 on the 22nd day of the same month. All of these warriors fall under the sign of Scorpio, which to

* The 1887 Spithead Review included 126 vessels, mostly from the Royal Navy – then the largest military fleet in the world, with 179 major warships.

astrologers (if you follow such things) is one of the four fixed signs of the Zodiac. These signs are thought to represent determination, power, natural leadership, purpose, reliability, loyalty and self-confidence – but also stubbornness and immovability, with a tendency to get stuck in ruts. Those born under a fixed sign supposedly pursue their goals with dogged persistence and have great powers of concentration. Scorpio is known also as a water sign, representing passion, intuition and imagination. In the language of astrology, Scorpios like activity, mysteries, secrets, winning, being acknowledged – and strategy. Apparently the best occupations for such enquiring, searching and calculating types are said to include medicine, science, research, insurance, financial analysis, politics – and soldiering.

Bernard (Law was the surname of his great, great, great grandmother, which subsequently became a family Christian name) Montgomery, the third son and fourth child, was born in Kennington, a modest south London suburb, to a Church of England clergyman, the Reverend Henry Hutchinson Montgomery, and his wife Maud (née Farrar). Henry was the vicar of St Mark's, Kennington, officiating from a fine old neo-classical church and residing in a red-brick vicarage overlooking Kennington Oval. The church still stands – despite being bombed and burnt out in 1940 – and remains the centre of a thriving evangelical parish. Both of Bernard's parents were products of colonial rule and Anglican evangelism, two of the solid platforms of Victorian Britain. His mother, Maud, was the daughter of a prolific author, Dean Farrar. Born Frederic W. Farrar in Bombay, the son of a Church Missionary Society chaplain, Farrar was a bright cleric who became headmaster of Marlborough in 1871 and Chaplain to the Queen in 1873. Three years later Prime Minister Benjamin Disraeli urged him to accept the post of Canon of Westminster and Rector of St Margaret's – Parliament's own church, adjacent to the Abbey.

Dean Farrar was also vice president of the Temperance League and thus pushed all the right buttons for austere, prim, middle-class Britain – perhaps a hint here of the teetotalism that was to rule Monty's later life. Through a huge number of books (he wrote seventy-two, the best-known being *Eric, or Little by Little*, the tale of a public schoolboy's descent into immorality) and his oratory (he was regarded as a gifted speaker), Farrar's influence was as wide as his congregations were full, often to overflowing, so that it became necessary to reserve seats. (Sadly, when paring down our family library some years ago, I consigned our copy of *Eric, or Little by Little*, to the local charity shop, before I realised its significance, along with a whole pile of unread quasi-religious novels and strident evangelical missionary tomes of the late nineteenth century; for my forebears were not dissimilar and, I suspect, also caught up in Farrar-fervour.)

In his lifetime Dean Farrar's brand of revivalist Anglicanism works was as popular as Thomas Hughes's *Tom Brown's School Days*. Hughes put the fictional Tom Brown at Rugby where he himself had been educated, under the reforming headmaster Thomas Arnold. It was Arnold's potent and fashionable late-nineteenth-century brew of muscular, evangelical Anglicanism and a Christian-Socialist sense of 'fair play' that provided exactly the sort of pulpit-thumping ideas promoted by Hughes and Farrar. As his extensive duties kept him away from his parochial responsibilities, Dean Farrar took on curates to help ease the load. One of these was Henry Montgomery. Connections helped: Farrar had been housemaster to Henry at Harrow.

Though gentler in temperament, less of a 'mover and shaker' than Dean Farrar, Henry Montgomery came from a not dissimilar background. He was the eldest son of Robert Montgomery, a humble, religious and self-made colonial administrator who rose to become Lieutenant Governor of the Punjab and was rewarded with a knighthood. To the biographer Alun Chalfont, Sir Robert represented, 'for better or for worse, Victorian values at their most admirable, and he strove quite explicitly to pass them on'. Henry Montgomery was a Trinity College Cambridge man, ordained in 1871. Five years later, after an extensive tour of the Middle East, he took up the curacy under Dean Farrar at St Margaret's, Westminster. That Farrar was an influential figure in his day is reinforced by the callers to his house, amongst them key literary figures, educators and artists including Arnold, Tennyson, Browning, Wordsworth and Millais.[1]

Henry Montgomery, the bachelor curate, was one such caller who drifted to the house for intellectual refreshment, and it was there that he fell under the spell of one of the dean's daughters. He first met Maud when she was eleven. He proposed three years later – when he was thirty-two and she fourteen – and married her two years later with the dean's blessing.* Henry was given a parish of his own in 1879 – St Mark's, Kennington – and it was to here that he brought his young bride. He

---

* Her age on marriage and the gap in years between husband and wife were less remarkable then, even for a clergyman. In the latter half of the nineteenth century, British society was marked by the collision of Victorian urbanisation and public health and hygiene; unregulated, neither co-existed happily until well into the twentieth century. The solution was to marry young and produce as many offspring as possible, in the hope that some would survive.

had a practical side and ran his parish like a military unit. Perhaps he needed to, for his responsibilities included 16,000 parishioners, 250 church workers and 125 Sunday school teachers. Henry was bent on his own calling, and became emotionally and spiritually detached from his young wife. Maud, formerly the centre of attention at her father's soirées, threw herself into her husband's work, and took control domestically of the home and of their children.

The parish was soon regarded as highly successful and a template for others, so successful indeed that in 1889 Henry was appointed Bishop of Tasmania, necessitating the removal of the family overseas. The appointment and commensurate rise in salary brought some welcome financial relief to the family. Sir Robert Montgomery had died just after Bernard's birth, leaving Henry the family estate of New Park at Moville, County Donegal, but also a £13,000 mortgage outstanding. Despite selling off some outlying farms and juggling finances in order to run New Park and St Mark's vicarage, it had been a struggle to raise the family on a vicar's wage, and this had brought a degree of austerity and tension to the lives of the young Montgomerys.

The family estate at Moville, on the west bank of Lough Foyle and a boarding point for emigrants to the United States and Canada, was a special place for the Montgomerys. The family, who had settled in the area in the seventeenth century (Ulster remains home to many distant and distinguished strains of the Montgomery family), were descended from Hugh Montgomerie, 5th Earl of Eglington, whose nephew, Colonel Alexander Montgomery, crossed to Ireland in about 1640. His descendant, Samuel Montgomery, a Sheriff of Londonderry, built the family home in 1776, ironically – in view of Bernard's later abstinence – having made his fortune as a wine merchant. The Reverend Henry honeymooned at Moville with Maud in 1881 and they and the children – Bernard, Sibyl (who died aged seven in 1889), Harold, Donald, Una, Maud Winifred (known as Winsome), Desmond (who died aged thirteen in 1909), Colin, and Brian – would spend a six-week holiday there each year, where the family rode and shot, sailed and fished. Henry and Maud later retired here.[2] *Life Magazine* 'dropped in' on Lady Maud Montgomery at Moville in a photo feature of their 28 June 1943 edition: the impression is conveyed of a frugal old lady in threadbare woollen cardigans struggling with a big, damp old house full of family mementos – an image familiar to generations of British readers, but appearing amusingly eccentric to *Life's* predominantly American readership.

The Montgomery family could trace their lineage back to Roger de Montgomeri, lord of a small Normandy town, who was a kinsman of William the Conqueror, and one of his most loyal supporters. After the latter's victory at Hastings on 14 October 1066, Montgomeri was rewarded with land that eventually became Montgomeryshire and a title, Earl of Shrewsbury. On his death in 1095 he was buried in Shrewsbury Abbey. Bernard and his siblings were very aware of their family's ancestry and the estate at Moville that had housed their family for over a century and a half. The irony of Bernard reconquering his distant ancestor's homeland was not lost on him during the Normandy campaign of 1944.

In 1889 the Montgomerys left England for their new home in Bishopscourt, Hobart with the five children. Bernard was two and his earliest memories were of Australia. There is less Outback in Tasmania, making it is as close to England in appearance as you can find in the Southern Hemisphere. In the 1890s it was still Anglican Middle England, with an accent. This was a world of big houses, of high ceilings and rooms made into chapels; of lessons at home with the governess; of family prayers twice a day and grace before all meals; of stained-glass and Gothic arched windows. It was an orderly, predictable world, where a Dickensian missionary zeal prevailed, run like clockwork to Maud's daily routine. On Sundays there was church – morning and evening – plus Sunday school. Teetotal and non-smoking, there was an overarching sense of moderation in all things – wartime austerity without the war. I was brought up with something not dissimilar, and can still remember the sense of eternal damnation just around the corner, mixed with the whiff of furniture polish. You could either embrace the regimen, or become the rebel. Becoming a rebel was easy, but you had to be prepared to take the consequences: so Monty and Maud clashed – in all things.

Bishop Henry was away for long periods, visiting every remote corner of Tasmania, so Maud's rule was absolute. She would beat or berate her children for any infringement of her rules, which included daily prayers, scrupulous honesty, strict time-keeping, starchy cleanliness, instinctive politeness, good table manners, and neat and accurate schoolwork. She was determined too that the children would not adopt an accent. The strictness – even cruelty – in Maud may have come from her not having had much of a childhood herself; she was married at fourteen. Without necessarily malign intent she withheld from her own children what she had been denied: affection. The eldest children, Sibyl (who died shortly after arrival in Hobart), Harold and Donald, stuck together and submitted, as did the later, younger ones – Una, Winsome, Desmond, Colin and Brian. But Bernard, always a loner, fought back. Apparently at one children's party, when Maud was trying to make herself heard, Bernard jumped on a table and yelled: 'Silence in the pig market, the old sow speaks first!'[3]

Though harsh and uncompromising, it proved an effective regime. Bernard and his siblings were brought up well by the standards of the day; as Monty later wrote: 'We have all kept on the rails. There have been no scandals in the family.' Eventually three would emigrate (driven away? – who knows): Harold fought in the Boer War then lived in Kenya, Donald in Vancouver and Colin in South Africa. Bernard's own coping mechanism was emotional detachment. He appears to have persuaded himself that he was indifferent to his mother and thus impervious to her withholding of affection. Indeed, the loveless environment gave Bernard's character a hint of a bully. He later described himself as 'a dreadful little boy', the 'bad boy of the family, the rebellious one, and as a result I learnt early to stand or fall on my own.'[4]

The family returned to London once – for the Lambeth Bishops' Conference of 1897 – and left Tasmania for good in 1901, when Bernard was thirteen. His father had been asked to take on the post of Secretary to the Society for the Propagation of the Gospel (SPG), and the family returned to the capital reluctantly. Bernard followed his elder brothers to St Paul's School, a daytime fee-paying college in Hammersmith, where the Montgomerys had settled, to which Bernard would cycle each day. Money was the chief reason that prevented Bishop Henry, an Old Harrovian, from sending Bernard to his alma mater, but St Paul's was in many ways an obvious choice for an impoverished cleric with many sons. Founded in the City of London in 1509 as St Paul's Cathedral School, the high-achieving institution had relocated to Hammersmith in 1884, and was able to offer an important churchman like the bishop a substantial discount on school fees.[5] Between 1902–21, five of the six Montgomery brothers attended St Paul's. Sadly, Bernard's younger brother, Desmond, died of a medical complaint whilst a pupil there in 1909.

Monty (or 'Monkey', his nickname at St Paul's, a reference maybe to his agility and his sense of mischief) found the experience of a large public school unsettling at first. He was put in an 'Army Class' of ten boys, that is, for the least academically inclined (Churchill, too, was the product of an 'Army Class', at Harrow), but took advantage of St Pauls' expansive playing fields. By nature short and slight, 'Monkey' was fit and wiry. Rebelling against his mother in Hobart and at Moville had usually meant messing around out-of-doors and he found he was well suited to games, playing for the school Cricket XI and captaining the Rugby XV. To no one's great surprise he also became a prefect.

Fears of aerial bombing caused St Paul's to be evacuated to Berkshire in 1939 and the buildings were requisitioned by the War Department. In an odd *Brideshead*-esque quirk of fate, St Paul's School would become the headquarters of 21st Army Group in the early months of 1944, and its commander, one B. L. Montgomery, would take great delight in demanding the same study his headmaster had occupied nearly forty years before. Neither set of buildings, Monty's Chiswick home at 'Bishopsbourne', 19, Bolton Road, W4, or St Paul's School, exists today: the former fell victim to the Luftwaffe, the latter was demolished in an act of 1960s architectural vandalism when the school was rehoused in Barnes.

Meanwhile, in the autumn of 1907, 'Monkey' passed the entrance exam to Sandhurst (72nd out of 177 candidates) and in January 1907, at the age of nineteen, began his military career at the Royal Military College. He apparently decided on the army to spite his parents who were pushing hard for the Church as his career.[6] An Army career was probably, however, not a conscious choice for Bernard (the nickname seems not to have accompanied him), but recognition that he was no academic and enjoyed sports and the outdoors life. In Edwardian Britain, public school education coupled with a lack of private income invariably pointed towards a career in the Church or Empire; the Church of England favoured bookish, brainy types, but the Empire required fit young men who were brave and resourceful. If Bernard had developed a coping mechanism in Tasmania that drew him away from his family, particularly his mother, it was accentuated by his departure for Sandhurst. From here on he would be a stranger, only intruding occasionally into the lives of the younger members of the Montgomery clan, usually Una, Winsome or Brian.

The Royal Military College, Sandhurst had been producing officers since 1802 and by Montgomery's day it was the main route to gaining a commission – though gunner and engineer officers had been educated separately at the Royal Military Academy in Woolwich since 1741. Most of Monty's superiors and Regular contemporaries were Sandhurst-educated (Douglas Haig had passed out from the college in 1883 and Winston Churchill in December 1894). Montgomery's prowess on the playing fields soon elevated him to the Rugby XV. St Paul's had prepared him admirably and within the first few weeks – the most testing of times, when a cadet needs quickly to find his feet – Monty ascended the first rung of gentleman cadet promotion, to lance corporal. Sandhurst then was very much a finishing school for young, moneyed swells and Bernard ascribed his progress not necessarily to zeal, but to the stinginess of his mother's financial allowance which prevented him from engaging in the traditional outside attractions of late-night carousing in London, fine wines and fast women, and forced him to knuckle down and work. Monty soon emerged as the leader of a Flashman-like gang of hearties in his cadet company, the 'Bloody B', who eventually attacked an 'A' Company cadet changing in his room. This was ragging of the public school variety (from where nearly every cadet had arrived), which not infrequently drew blood. Whilst others held the cadet, Montgomery set the poor unfortunate's shirt tails alight, with the result that the cadet was hospitalised and Montgomery held culpable for this excess.

What to do with Gentleman Cadet Montgomery? On the one hand, he was doing well at Sandhurst and his infantry regiment, the 6th of Foot, the Royal Warwickshires, were eager to have him; on the other, he had broken the rules and offended the authorities' sense of fair play, everything that Thomas Hughes, Matthew Arnold and his own father-in-law, Dean Farrar, had stood for. At this juncture Maud took a hand in her son's future (the only recorded time when she proactively intervened on his behalf) and came down from London to plead Bernard's case. Bishop Henry's then heady status as Prelate of the Knightly Order of St Michael and St George* meant that Bernard's behaviour threatened a public scandal, and there was a very real possibility of him being sent down and thrown out of the army. In the event he lost his lance corporal's stripe and was held back a term, to graduate in a sea of martial music, flashing swords, spiked helmets, scarlet tunics and gold braid, in the summer of 1908.

In his memoirs, Monty singles out a Major Forbes of the Royal Scots Fusiliers (rather than Maud) as the person who saved his bacon; or it may have been Captain George Crossman of the West Yorkshire Regiment, CO of the Company of Gentleman Cadets. The Commandant, Colonel William Capper (whose high-flying brother, Thompson Capper, would die as a divisional commander on the Western Front in 1915), would also

* Henry Montgomery would receive a knighthood, a KCMG, in 1928.

have been involved. At any rate, someone perceived that Bernard possessed the 'right stuff', but needed taking down a peg or two. In his memoirs Monty was quite open about the crime and consequent punishment, perhaps with an unconscious sense of in-built religious self-flagellation.

Achievement at Sandhurst was measured by the final examination results, and for Monty especially the prize was India. At this time, a private income (in addition to salary) of £100 a year for infantry officers or £400 for those in the cavalry was a basic requirement for junior officers based in England (a multiplication factor of at least sixty brings these figures into the twenty-first century). Monty was desperate for a posting to India, where an officer could live well on relatively modest pay. Making sure that one's pay plus any family allowance at least equalled one's bill for fine claret and cigars at the end of each month was a huge concern to those like the impecunious Monty.

Accordingly he set his sights whilst at Sandhurst on getting into a British-officered Indian Army regiment, which, funded by the Indian War Office, paid more than a prohibitively expensive regiment in England. He was quite specific about the money issue: he could not afford mess life except in India. Money went further there and officers could ride government-supplied mounts (riding was obligatory for all officers in all armies in this era), though buying one's own, or several, was infinitely preferable. In this much-parodied era there were regimentally sponsored race meetings, shooting parties and the whole business of dressing for dinner each evening in the mess, when the silver came out and the regiment's band played stirring melodies through the meal. In cavalry regiments, the best Indian soldier in the troop or platoon had the honour of unravelling his regimentally striped turban around his officers' midriff before dinner, hence the origin of the cummerbund.

Montgomery just missed out on one of these Indian Army vacancies, but was then placed high enough (36th out of 150) to be sure of acceptance for his next choice, a British Army regiment, with one of its two battalions stationed in India. His choice fell on the old Sixth of Foot, the Royal Warwickshire Regiment. With no personal connections to the regiment (the various regiments still prefer some kind of personal connectivity today), he is alleged to have made his choice based on their antelope cap badge. But the real reason was the location. Thus, in December 1908, he joined the Royal Warwicks' 1st Battalion at Peshawar on India's North West Frontier, with two other newly minted second lieutenants.

Monty later claimed that his first taste of alcohol had been on arrival in his regiment's mess at Peshawar: 'a young officer was soon taught to drink. I have always disliked alcohol since', but it seems incredible that, the high jinks of Sandhurst considered, drink hadn't surfaced earlier in his life. This, however, was Monty speaking in his seventies, conscious of his teetotal reputation and omitting to recall, as we shall discover, that he later found alcohol extremely welcome in the trenches of the Western Front.

Monty appears to have made little impression on his new battalion. Hardly challenged mentally, he embraced inter-regimental sports, excelled at hockey and cricket, and hunted with the Peshawar Vale Hounds on an Indian cavalry charger named Probyn. Athleticism was (and remains) one way of gaining rapid acceptance in an all-male, insular society, and this was where Monty could achieve. Though his capacity for joining in the merry japes of his colleagues was, as at Sandhurst, limited by finances, he once more offset this latter disappointment by study. He learnt Urdu and Pushtu and organised mountain warfare training in the nearby hill station of Cherat. It was at this stage that his talents for teaching, training and communication (no doubt inherited from his father bishop) came to the fore. To his fellow officers he may have also come across as a bit of a swot, or as Alun Chalfont put it, having him in the mess 'was a little like finding a player in the Gentleman's XI'. From this time, to the end of his days, Monty never appeared wealthy, with his preference for old sweaters and corduroy trousers. He was certainly serious at a time when seriousness was at best unusual and at worst discouraged, though in Monty's case, this may simply have been a way of disguising his financial embarrassment. In choosing to stand apart, we might also divine hints of his natural aloofness and arrogance. Whatever the underlying causes, the seriousness paid off. The army always needs good organisers and he was made assistant adjutant on promotion to lieutenant in 1910.

What India seemed to bring out in Monty was an aversion to a certain type of officer:

Soldiering in India seemed to me at that time to lack something. I saw a good deal of the Indian Army. The men were splendid; they were natural soldiers and as good a material as anyone could want. The British officers were not all so good ... An expression heard frequently was that so-and-so was a *good mixer*. A good mixer of drinks, I came to believe, for it soon appeared to me that a good mixer was a man who had never been known to refuse a drink.[7]

There is prejudice aplenty seeping out of this passage as surely as the slice of lemon flavouring a good gin and tonic. Monty was never a good mixer, being always slightly shy and awkward in social situations. He refers warmly to the British soldiery in India, though they were certainly no better than their officers, and frequently worse in a drink-fuelled hot climate, where they still managed to enjoy, and abuse, a status considerably higher than that which soldiering in Blighty would have offered. Further evidence of Monty's attempts to integrate with his men are the several tattoos, rarely caught in photographs, he had applied to both his forearms. A butterfly amongst other designs adorned his right arm and an antelope (inspired by the regimental cap badge) was applied to his left. Discreet tattooing of the limbs or torso remains a tradition amongst British sailors and soldiers to this day, but is more closely associated with other ranks than officers. Perhaps embarrassed by this early adventure into attention-seeking, as he rose in seniority, Monty went to great lengths to conceal his tattoos by keeping his sleeves rolled down in the desert and elsewhere. They are only visible in photographs of Bernard in old age, when, frankly, he didn't care. But Monty's real problem with British India was that he believed he would never be taken seriously there: that with little money and few apparent social graces, he could never compete. And he came to despise that of which he could not be part. Indeed it was with some relief, as he later observed, that his battalion, after a spell of duty in Bombay, returned to England in January 1913.[8]

It was whilst in India, in the insular world of early-twentieth-century society, that Monty first encountered Germans. In December 1910, the cruiser *Gneisenau* put into Bombay with the Crown Prince on board. Montgomery was told to organise a football match against the crew – and to make sure the *Gneisenau's* team won. Monty's earliest brush with diplomacy resulted in the Germans being thrashed 'by forty to nothing I think – I wasn't taking any chances with those bastards', he later recalled.

2

The Swabian

JOHANNES ERWIN EUGEN ROMMEL was born in Heidenheim-an-der-Brenz on 15 November 1891 and baptised two days later. The provincial town, then of about 15,000 inhabitants, lies deep in the state of Württemberg, one of the twenty-six kingdoms, duchies and independent states that merged to form the German nation in 1871. Thus Rommel had allegiance to two kings: the Kaiser, Wilhelm II, in Berlin, and another Wilhelm II, hereditary King of Württemberg, who had acceded to his throne only some forty days before Erwin's birth. In 1888, Kaiser Wilhelm I died in Berlin and was succeeded by his son, the liberal Frederick III, who spent all ninety-nine days of his reign battling throat cancer before succumbing on 15 June. His successor was his twenty-nine-year-old son, Wilhelm II, who would prove to be the architect of so many of Europe's twentieth-century woes.

1891 seems to have been as unremarkable as 1887: a quiet, but promising year, which witnessed the start of building the Trans-Siberian railway in Russia and the opening of Scotland Yard in London as the capital's police headquarters. Reigning as chancellor since the young German state had been founded, Otto von Bismarck had resigned the year before at the insistence of the new Kaiser, and the Fatherland as a whole mourned the passing of *Generalfeldmarschall* Helmuth von Moltke, architect of the 1870 victory over France. The peacetime German army then numbered 506,000 officers and men (far outstripping Britain's regular force which totalled 144,000 in the same year), whilst the Imperial navy included seventy-seven major warships.

Rommel was accompanied into the world by an impressive clutch of future generals. Amongst his contemporaries was a second field marshal-in-waiting, Walther Model, who, like Erwin, was a schoolmaster's son. Ritter Wilhelm von Thoma, who served with him in both world wars, and Frido von Senger und Etterlin, later a Rhodes Scholar and the defender of Monte Cassino (who was a subordinate of Erwin's in 1940), were also born in 1891; as was Erich Marcks, the one-legged commander of LXXXIV Corps in Normandy in 1944. Three other important future subordinates were very close to Erwin in age, belonging to the 1892 vintage: Ludwig Crüwell and Walther Nehring, who both commanded the Afrika Korps at different times, and the Bavarian former sergeant major, Sepp Dietrich, Hitler's one-time bodyguard and by 1944 an SS corps commander in Normandy and unlikely friend, labelled by some as 'decent, but stupid'.

The year also ushered in two future British field marshals: Harold Alexander, the third son of an Irish earl, who would be every bit as responsible for Rommel's downfall as Montgomery. The other was Bill Slim, born to lower-middle-class parents in urban Birmingham, who in due course would be commissioned into the same regiment as Monty. Bracketing Rommel in age were two important future RAF adversaries: Arthur Tedder, who would become Eisenhower's deputy in 1944, born in 1890, and Trafford Leigh-Mallory, born 1892, future supremo of the Allied air forces on D-Day. Indeed the four-year bracket 1887–91 would see the birth of the lion's share of commanders who would rise to prominence in the Second World War, including some 404 British and colonial wartime generals. These years also witnessed the arrival of 399 future US army generals, including Eisenhower, Brereton, Devers, Gerow, Middleton, Patch, Simpson and Spaatz. With de Gaulle, we must add the names of the Frenchmen Juin and de Lattre de Tassigny, 169 future Russian generals and Chiang Kai-shek of China. Born right amidst this galaxy of future stars on 20 April 1889 was an altogether more sinister entity, christened with the single forename Adolf.

The nation Bismarck had forged (it was never the Kaiser's in the way it was Bismarck's) was an erratic assembly of many German-speaking states. Prussia was dominant militarily, but one of the largest regions that made up the south-west of the country was Swabia. A mediaeval duchy, since incorporated into the larger kingdoms of Bavaria and Württemberg, Swabia still had a culture and dialect all of its own, which Erwin Rommel enjoyed speaking. It was claimed that a typical Swabian was tight with money, crafty, reserved, level headed – and loyal. In those far-off days Swabians were hardly regarded as cosmopolitan, but in Rommel's era they were esteemed fighters, like the Irish. But Swabia had long since ceased to exist as a political entity, so the wily, cunning south-western Germans found their voice as Württembergers. Their monarch retained a degree of political autonomy, ran his own railway and postal services, and appointed officers to his own army corps, albeit incorporated into the larger Prussian army.

These contexts are important in understanding Montgomery and Rommel: both regarded themselves as outsiders; both were scions of regions renowned for producing generations of warriors – Ireland and Swabia. This sense of being on the periphery of the military establishment gives us one of many true parallels for both leaders.

Rommel is a surname that crops up all over central Europe in every Germanic state and countries ranging between Denmark, Sweden and Poland.* The Inspector General of the Polish Army in 1939 was a Major General Juliusz Rómmel, who went on to command the Lódz Army during the German invasion and survived five years' subsequent captivity. One genealogist claims to have traced the family's ancestry back as far as 1175, to the Drakenburg estate in Sweden. Some historians have even suggested a French branch of the family, the Romelles, and have attempted to link it to Normandy, the family perhaps removing themselves to Germany and elsewhere as practising Protestants following the massacres of Hugenots in 1572 and Louis XIV's Edict of Fontainebleau, declaring Protestantism illegal. This is difficult to substantiate and may be a post-war confection, driven by the undoubted connections of the Montgomery clan to Normandy.

The German-born Catherine II (the Great), Empress of Russia, certainly attracted over 100,000 German settlers in the late eighteenth century, including several impoverished Rommel clans, to European Russia. A century later, anti-German sentiment encouraged their onward migration to Canada and the United States. The Rommel family today is not small and forebears from every state, duchy and principality in central Europe were caught up in the great Germanic migration to America of the eighteenth and nineteenth centuries.[1] Some Rommels are recorded as migrating to Louisiana as early as 1720, but many arrived

* In origin, Rommel is fifteenth-century German or Dutch slang for an obstreperous person, or one who creates a disturbance – quite appropriate for the future Desert Fox.

in Pennsylvania one hundred years after. Of Rommel's direct ancestry, at least one of his father's siblings emigrated to America. Many of the millions of descendants of European émigrés were eventually drafted in both world wars, producing the scenario of Corporal Bobbie Jack Rommel, from California, of 506th Parachute Infantry Regiment, landing in Normandy on D-Day, to do battle with his third cousin, Erwin.

Rommel's father, also Erwin, was born in 1860, the son of a schoolmaster, a vocation he in due course followed in the attractive little provincial town of Heidenheim, thirty miles north of Ulm. Erwin (senior) was one of four – Georg, Johanne and a sister, Greta. Whilst some of his siblings migrated to the United States, 'Herr Professor' Erwin seemed content to stay in Heidenheim teaching mathematics, as had his father. In 1886 he married Helene, daughter of a local civic dignitary, Regierungs-Präsident Karl von Luz. They enjoyed a happy marriage which resulted in five children: a daughter, Helene, Manfred who died young, and three further sons, Erwin, Karl and Gerhard.[2] Although, like all Germans, the professor was obliged to undergo compulsory military service (in 1880 he was a lieutenant in the Württemberg artillery), there was no military tradition in the family. According to Brigadier Desmond Young, the British author responsible for the first and bestselling biography of Rommel, the professions of young Erwin's siblings indicated 'a respectable Swabian family of moderate means, far removed in education and environment from the Prussian officer class'. Helene never married and became, like her father and grandfather, a teacher in Stuttgart, the capital of Württemberg, where her nephew was eventually mayor. Karl became an army pilot and was awarded both classes of the Iron Cross and the Knight's Cross of the Württemberg Military Merit Order in September 1917, but caught malaria whilst serving with the air force in Turkey and Mesopotamia and was invalided out of military service. He produced two sons, Jürgen, Wilhelm and a daughter, Ilse. Gerhard became an opera singer in nearby Ulm. Much later, during the Second World War, one of Erwin's nephews joined the SS, and in 2001 a British newspaper revealed that documents in Britain's National Archives pointed to the complicity of a Leutnant Rommel in the execution of three British prisoners of war in Italy in 1943. Erwin at least would have the foresight to forbid his own son, Manfred, from joining the organisation.[3]

In 1898, Herr Professor Erwin was made headmaster of the secondary school in nearby Aalen, about ten miles north of Heidenheim, where young Erwin attended school. These are attractive small towns, lost in the surrounding hills, boasting baroque architecture that emphasises their pedigree, their growth very much the result of the unifying and industrialising of Germany. The professor may have provided a solid education for his family (he could afford the extra tuition that Erwin needed to enter his first school), but he provided less in the way of direction. School-mastering was in his blood, but not, it seems, in that of his offspring, and their diverging career paths suggest this. Notions of civic service clearly ran strong in Helene Rommel's veins, and she was probably more ambitious for her children than her husband, hoping that they would at least match her father's achievement and status, possibly beyond Swabia itself*. Certainly in 1890s Germany, a country developing rapidly, with a newly discovered world footprint and small empire of its own, the opportunities for advancement and achievement seemed limitless. But what to do with young Erwin? Attracted by mathematics, but no theoretician like his father, Erwin's interests always had a practical application; his other strong subject was science. There evolved in the young Rommel an instinctive mechanical interest in new technology, which stayed with him all his life; aged fourteen he and a friend had built a full-size glider, which apparently flew. In due course he would earn his pilot's wings, and later he would take apart and successfully reassemble a motorcycle. Wireless sets (and later on, guns and tanks) fascinated him and this interest in mechanics led him for a while to consider approaching the vast new Zeppelin works at Friedrichshafen, on the shores of Lake Constance, for an apprenticeship as an engineer.

Rommel was, unlike his teacher father, an outdoors person: extremely fit (as was Monty – both emphasised the importance of personal fitness throughout their army careers, when this was considered innovative, even radical), he would ski in the winter – the surrounding terrain is made for it – and cycle in the summer. Both were relatively new middle-class sports, requiring some money to invest in the necessary equipment and clothing. They grew quickly in popularity across Europe as they offered, among other things, the opportunity of meeting the opposite sex, unchaparoned. Just like Monty, though a good sports all-rounder, the young Rommel was no intellectual. Recalling his own army service it seems the possibility of a military career first came from Rommel's father, and was taken up by the son. Some historians claim that the Herr Professor was against a military career for his son, but the opposite was true: Erwin would go with his father's blessing; he even paid for his son's first uniform.

* Both Maud Montgomery and Helene Rommel appear to have been the dominant parent in the early lives of the future marshals.

Erwin initially approached his father's local Württemberg artillery regiment, which was quite scientific and technical in its leanings, but it was fully recruited. Gunnery had made enormous strides since the Franco-Prussian war, with the introduction of rifled breech-loaders replacing the centuries-old muzzle-loading cannon. The construction of an ocean-going war fleet for the Kaiser further encouraged this expertise. Germany, through the Krupp steel works and other manufacturers, was a world leader in the art and science of gunnery, and this attracted many enquiring minds. The heavy reliance on horses to tow artillery about also gave those who yearned for the cavalry, but could not afford the life, a second chance for a smart uniform on horseback. Erwin also considered the other technical arm, the engineers, but the small Württemberg army corps maintained only a single battalion, which again was full. Thus in March 1910 Erwin applied to the 124th Württemberg Infantry Regiment, based in the very old southern garrison town of Weingarten, for a vacancy as an officer cadet. Accepted for service, Rommel arrived on 19 July to start a military career that would span thirty-four years.

It was Frederick II (the Great) who really established the reputation of the Prussian army in the early 1700s, and used it to expand his domain. That force, built on traditions of unquestioning obedience, harsh discipline and leadership, had stagnated by the time of Napoleon, who crushed and humiliated it in the Jena campaign of 1806. Thereafter the Prussian army was rebuilt from scratch by reformers who included Scharnhorst, Gneisenau and Clausewitz. They concluded that one of the reasons for defeat had been the poor example of many junior leaders who by dint of birth were too remote from their soldiers. Amongst their solutions, some of which endured until 1945, was the founding of a *Landwehr* (National Reserve Army) based on conscription, a *Kriegsakademie* (the German war school, or Staff College) to train a General Staff, and the establishment of a new officer corps. To ensure that junior officers bonded with their men, they required every candidate for a Regular commission to serve, as Rommel would, as a *Fahnenjunker* (an officer cadet, perhaps more similar to a midshipman of the Royal Navy) in the ranks for around six months. During this time, they would be taught the basics of an NCO's duties and serve as a private, corporal and sergeant.

It was a challenging commitment. The cadets ate with the officers and slept in barracks for the first six weeks, and throughout this period they were watched and mentored closely by the other officers. If all went well they would be then despatched to one of several officer schools, *Kriegsschulen*, which prepared them to pass examinations that would result in a commission. This was a process more egalitarian in nature than the British Army's solution. The system Montgomery passed through was born out of an all-volunteer army led by a ruling class, which was to maintain the gulf between officers and men by educating potential officers at Sandhurst and commissioning them first, before letting them loose on soldiers.

As the historian Hew Strachan has pointed out, the German army then (as now) drew a social distinction between NCOs and privates, whereas the British lumped them together (as today: officers and *men*), and this elevation of the NCO engendered an ethos of initiative and individual responsibility not present in their British counterparts. A comparison of responsibilities illustrates this point. The backbone of the British Army in 1914, the infantry battalion, numbered thirty officers and 977 men, and was commanded by a lieutenant colonel. Majors commanded the German equivalent, of twenty-three officers and 1,050 men. Thus, German officers of at least a rank lower than their British equivalent were doing the same job. With an officer to men ratio of one to thirty-three in British battalions and one to forty-six in German, the German army, devolving far more responsibility onto NCOs, thereby practised a more efficient use of their officers.[4]

Rommel spent his obligatory first few months in the ranks of the 124th Regiment, being promoted to corporal in October and sergeant in December. He was assigned to the Danzig *Kriegsschule* in March 1911. Whilst there were many officer schools located throughout Germany – at Potsdam, Glogau, Neisse, Engers, Hannover, Kassel, Anklam, Danzig and Munich – the posting of a Württemberger from the south-west to an officer school in the north-east of Germany was part of a deliberate policy of fostering a pan-German ethos, in the process breaking down the limited horizons, mindsets and even accents of candidates from the provinces and imposing something of Prussian traditions and approaches to soldiering. It was another way of removing some of the insularity that had been the downfall of the old Prussian army in 1806. Officer schools educated cadets in a variety of military subjects: drill, engineering, tactics, fitness, administration, riding and in the social graces – for the army at this time occupied a high place in the nation's social hierarchy, just as it did in Britain. Rommel did well in the eight-month course, passing out with higher than average marks. His commandant rated him as 'good' in leadership, and competent at all else: he was 'firm in character, with immense willpower and a keen enthusiasm'. He gained his commission on 27 January 1910.

The young cadet Rommel was especially keen to foster the social graces on account of one of his dancing partners. Lucie Mollin was the daughter of an east Prussian landowner, where her family, Italian Catholics in origin, had settled in the thirteenth century. Her father had died when she was young and she had migrated, not far, to the big cosmopolitan port of Danzig, to learn languages and make her own way in the world. Invitations to balls at the Königliche Kriegsschule were much sought after and rarely refused, and it was at one of these splendid affairs that Erwin and Lucie first met. Although his studies taxed him – Rommel not being a natural academic – they found time to fall in love during that summer of 1911. When Rommel left Danzig at the end of the year, they began an almost daily correspondence that would persist for much of the rest of their lives. They were not at that stage formally engaged, but their friends assumed a formal announcement was not far off.[5] Junior officers were discouraged from marrying (until even quite recently) and required their commanding officer's permission to do so. It was hoped that they would spend all their waking hours applying themselves to their profession. In Rommel's day that encouragement was more by way of a prohibition. Although the Kaiserreich made great efforts to ensure that German officers were socially acceptable, made 'of the right stuff', there was no corresponding way of vetting their wives. Two brakes were therefore applied: one being an age limit of around twenty-five, below which junior officers were unlikely to be granted permission to marry, in the hope that they would be less inclined to make rash and unsuitable choices in their brides; the second was a requirement for the bride to provide a dowry. Known colloquially as the *kaution* (literally, bail), this, in theory, was meant to maintain the appropriate lifestyle – horses, servants and so on – but in practice it was in place to deter the unsuitable. With her father dead, this was probably beyond Lucie, which may also have dissuaded the young, impetuous Rommel from tying the knot during this time.

Erwin returned to his parent unit, the 124th Regiment at Weingarten, housed in its secularised eleventh-century abbey (and which until the 1990s was home to a series of *Bundeswehr* and NATO units). His job for the next couple of years was to train the Kaiser's conscript army – a role, with teaching and communication in the blood, for which he was practically and temperamentally extremely well suited. All German males were liable for military service between the ages of seventeen and forty-five. They were usually called up at twenty, and would spend two years on active service (three for the cavalry or artillery) and four or five years on the Regular Reserve. This reserve liability still meant being mobilised for military service (usually for the annual *Kaisermaneuver*, mock engagements held on a vast scale, which were an international showcase for German military prowess), for a maximum of six weeks. Soldiers then passed into a series of National Guard-type organisations. First, trained men formed the *Landwehr* (roughly approximating to the British Territorial Army) until they turned forty – still with compulsory training – followed by *Landstürm* (more akin to the Home Guard, or National Militia of poorly trained, armed citizens) liability until reaching forty-five. Such planning meant that the 1913 Army List contained the names of 23,000 Reserve officers and 11,000 *Landwehr* officers, all trained and capable of quick recall if required. Others not called up for family or economic reasons formed the *Ersatz* (Supplementary Reserve), which would, nevertheless, be called up in time of war.

Back in Weingarten, Rommel quickly settled into the routine of training his recruits. The pretty Swabian garrison town was (and remains) more pleasant than many a north German urban blot on the landscape. The history and architecture absorb the military element leaving one unaware there are soldiers living in the vicinity. The town is small and on leaving one is soon lost in the hills and forests that terminate to the south abruptly with the delight of Bodensee, Lake Constance, an easy place in which to fall in love, as Leutnant Rommel discovered in 1912 when he met Walburga Stemmer, the twenty-year-old daughter of a seamstress. Lucie was in Danzig pursuing her studies; besides, there was no formal arrangement between them, just an 'understanding'. A year later, on 8 December 1913, with Rommel still in Weingarten, Walburga gave birth to a daughter, Gertrud, to Rommel's delight. He wrote, calling their baby girl his 'little mouse', and talked of setting up home together: 'it's got to be perfect, this little nest of ours'. But the talk came to nothing, and, though he maintained contact with Walburga and Gertrud, he returned to Lucie, having confessed the relationship and the child. He and Lucie would marry three years later.*

---

* Erwin had apparently been quite prepared to terminate his career in the army, marry and settle down with Walburga in Weingarten. Although Rommel's father died three days before Gertrud's birth, his strict and ambitious mother was doubly horrified by the child born out of wedlock and social gulf between her twenty-one year old son, an officer, and Walburga, a fruit seller eight months older, and managed to persuade young Erwin to remain in uniform. Within a few months, the onset of war swept away much of the scandal.

Walburga Stemmer died in October 1928 aged thirty-six. Though the public explanation was that she had died of pneumonia, her grandson, Josef, later disclosed that she had taken an overdose just before Lucie gave birth to Rommel's son, Manfred: 'the family doctor told me she had taken her own life'.[6] A collection of 150 letters and photographs written to Gertrud was recently discovered. It says a great deal for the Rommels, particularly Lucie, that they stayed in touch with Gertrud. She exchanged hundreds of letters with her father and was a frequent visitor to the family, where she was known by her half-brother, Manfred, as 'cousin Gertrud'. She knitted her father a scarf, which he often wore at the front. This is possibly the hallmark tartan scarf he was often photographed wearing in the desert.[7]

The Imperial army that Rommel joined had been forged and shaped by three wars, against Denmark in 1864, Austria in 1866 and finally France in 1870. This last conflict laid the foundations for future strife, as France sought a military/diplomatic formula for the return of the provinces of Alsace and Lorraine, which were incorporated into the new Reich on the proclamation of Wilhelm I of Prussia as German emperor. (Ironically this event took place not in Berlin, but near Paris, in the celebrated Hall of Mirrors at Versailles on 18 January 1871.)

It was their 1870 victory against France that made the world sit up and take notice of the Prussian-German army and its training methods. This victory was in part attributed to the training of staff officers, as devised by Helmuth von Moltke during his lengthy time as Chief of the General Staff (1857–87). By 1870, most brigade and divisional commanders had studied under Moltke, as had all their Chiefs of Staff. The result was a hitherto unprecedented uniformity of doctrine within the Prussian senior command that went to war in 1870. The Franco-Prussian War of 1870–71 was the military watershed between 1815 and 1914; although the earlier American Civil War of 1861–64 had hinted at the direction of future military technology, 1870 was the first conflict where infantry of both sides carried breech-loading rifles, where shells fired from breech-loading Krupp field guns replaced Napoleonic round-shot, the first machine gun (the French *Mitrailleuse*) made its appearance, and unprecedented numbers of troops could be concentrated and conveyed, thanks to railways. This was also a confrontation that inspired and obsessed the young Hitler. In *Mein Kampf* he wrote of that war becoming his favourite reading material: 'it was not long before the great heroic struggle had become my greatest spiritual experience. From then on, I became more and more enthusiastic about everything that was in any way connected with war or, for that matter, with soldiering.'[8]

For the next forty years European diplomacy was played out against a backcloth of the resultant Franco-German arms race. The two sides fine-tuned plans devised by their respective General Staffs; the French offensive plan (inevitably) was geared around the need to recover the two provinces lost to Germany, whilst their opponents came up with the Schlieffen Plan.

Modern historians still debate whether there was ever a set war plan to which Count von Schlieffen, Chief of the German General Staff 1891–1905, put his name, but the essential point is that following the Franco-Russian entente of 1894, the Count had to suggest a way of fighting France and Russia, at the same time if need be. A series of war plans evolved, developed by his successor, Helmuth von Moltke (by chance the namesake and nephew of the architect of the 1870 victory). These envisaged a quick, knockout war with France, after which Germany could turn its attention to Russia. The German General Staff fed on the paranoia most Germans felt of being surrounded by powerful, militaristic neighbours – not unlike the anxiety shared by many in Israel today – a sense of never letting one's guard down, and always having a powerful military force and a plan ready for all contingencies. The essential result of the Schlieffen-Moltke planning was a scheme that envisaged a rapid mobilisation followed by three armies, First, Second and Third, sweeping into northern France, passing through Belgian territory to do so. The Fourth Army would turn north of Sedan, whilst the Fifth, Sixth and Seventh would defend Alsace-Lorraine. Schlieffen (who always doubted the strength and logistics of the German army to undertake such a huge venture in the time available) emphasised that First Army must swing wide, passing beyond Paris, before turning south to encircle French forces in the Rheims-Champagne area.

In a modern democracy it is difficult to picture the authority and importance of the 650-strong German General Staff. It was much more than an old and venerable institution that organised the German military machine. It was both defence ministry and foreign affairs department. It had an authority over Germany's railway system, it was responsible for some internal security and its topographical department controlled much of the country's mapping. In the Kaiser's day it virtually dictated foreign policy and with a heavy investment in conscript soldiers, shaped social and educational policy, too. As Germany was less reliant on her navy (although Kaiser Wilhelm loved his battleships) than Britain or America, the army and its General Staff totally eclipsed the maritime arm and were unrivalled in their exercise of power within Wilhelmine Germany.

The Imperial German Army of Rommel's day, then, was a unique organisation, a coalition-based army made up from several formerly independent nations. The Kaiserreich consisted of twenty-six states that included free cities, principalities, duchies and kingdoms, with a combined population of 68 million in 1914, where the five largest states contributed their own forces. The greatest of these was the Prussian army, a large and long-established force which provided the backbone of the overall German war effort, fielding 158 infantry and eleven guards regiments – seventy-five percent of the army. In 1914 nearly two-thirds of Prussian army officers were aristocrats. Regiments as a rule numbered 3-5,000 men and recruited from a specific locale. Bavaria with the second largest army provided twenty-four infantry regiments (including the 16th, which 'adopted' Hitler); the Kingdom of Saxony contributed seventeen regiments of infantry. The fourth army, with whom Rommel served, comprised ten infantry regiments recruited throughout the Kingdom of Württemberg (population, 2.5 million), whilst the duchy of Baden provided its own army corps of eight regiments. These totalled a standing army of around 800,000 men. The mobilisation of conscripts would inflate this figure to nearly 4 million. Junior leaders like Rommel and his NCOs accordingly spent most of their time training the 200,000 new recruits the army received each year.

Thus, Rommel and, as we shall see Montgomery, spent their pre-war days commanding platoons, the building blocks of an infantry battalion. German platoons comprised eighty men, larger than their British counter-parts, where fifty was more the norm; platoons formed companies, and usually four companies made a battalion. In Rommel's case, his first infantry company, commanded by a captain, numbered 250 men in varying states of training, plus five officers: the commander, his second in command and three platoon (*zug*) commanders. In reality, NCOs often ran the training platoons, as directed by their officers, and German NCOs thus had more responsibility on their shoulders than their British opposite numbers. As all German officers started in the ranks, and had to spend a while as an NCO, the tactical direction of the German army rested far more with NCOs, than with their British counterparts, as both officers and NCOs came out of the same pool. In combat, during the big, attritional battles of both world wars, as German company strengths varied wildly, it would not be unusual to find officerless companies run very ably by a sergeant major. It was at this level – the company – that Rommel served for most of the First World War, and where Montgomery began.

Rommel was taught to be able to perform the job of the next two ranks up as well as well as his own, as were his NCOs. The Germans accepted that war was chaotic by nature (as Clausewitz advocated) and taught junior leaders to take advantage of the chaos, and always to take some form of effective military action. They believed that the man on the spot was the best judge of his immediate circumstances, a concept that leant itself to decentralised command. This expressed itself in their method of issuing orders, which was handed down to the Second World War *Wehrmacht*, and which Western armies know today and practise as 'mission command'. By contrast, the British were reluctant to sanction such individual initiative and required junior officers such as Bernard Montgomery to react to the orders of a (sometimes very long) chain of command.

The highest aspiration for professional officers such as Montgomery and Rommel would have been to join the General Staffs of their respective countries, achieved by attending the Staff College at Camberley, or Kriegsakademie in Berlin, though in 1914 this would have been a distant ambition for both men. Entrance to the Kriegsakademie for the three-year course was by competitive examination. By contrast, up to the First World War, many candidates for the Camberley Staff College two-year course were nominated for selection, rather than examined.

Montgomery and Rommel became officers in an era when promotion in both armies was slow, and since neither was recognised initially as being a particularly gifted leader who would rise to the peak of his profession, they might have anticipated a plodding career that might lead as far as the rank of colonel – not much further – bringing some status and financial security that neither possessed. The First World War changed all that. It gave able youngsters access to vastly accelerated promotion, enormous responsibility, a taste of staff experience, and of working with allies in wide-ranging terrain. It forced on young officers of both sides the ability to accept barely trained civilians into their ranks and integrate them quickly. All this, with awards for bravery and leadership that would gild their careers for life, in exchange for risking life and limb by leading men in conditions of extreme danger.

The Great War – the very term itself reveals how the generation who passed through the conflagration regarded it – the first total war, when the whole resources of a nation were geared towards war-making and national survival, was fought on such a scale, and in a variety of locations, that its participants gained incredible experience in a remarkably short space of time. Any commander, however junior, who survived his first combat, was bound to be leading larger bodies of men later on in the fighting. Although today we are encouraged to look back on the events of 1914–18 in terms of futility, despair, or 'impending doom', memoirs published in the 1920s and other later recollections suggest that for some it was a positive experience. That generation of 1914, fresh out of school or university with no work experience, who found they could manage the daily adrenalin rush of danger, did not necessarily look forward to a quick return to peace, with the prospects it brought of financial uncertainty, reduced status and unemployment. The youthful Charles Carrington, a Territorial officer in Montgomery's Royal Warwickshire Regiment, was one of a number who even believed that war did his men a favour:

> The skinny, sallow, shambling frightened victims of our industrial system . . . who were given into our hands, were unrecognisable after six months of good food, fresh air and physical training. They looked twice the size and as we weighed and measured them, I am able to say that they put on an average of one inch in height and one stone in weight during their time with us. The effect on me is to make me a violent socialist when I see how underdeveloped capitalism has kept them, and a Prussian militarist when I see what soldiering makes of them.[9]

Rommel's first biographer, Desmond Young, a young staff officer attached to Fourth Army HQ in 1917, recalled meeting Roland Bradford, who joined the British Territorial Army just before the war. From lieutenant in 1914, Bradford demonstrated such powers of leadership that he was promoted acting brigadier general by the age of twenty-five, winning both a Victoria Cross and Military Cross. He was killed in 1917. Young remembered the twenty-five-year-old brigadier:

> In all armies there is a small minority of professional soldiers (and a few amateurs) who find in war the one occupation to which they are perfectly adapted . . . I remember riding over, unduly conspicuous, I felt, on a white horse, to his brigade headquarters in front of Bourlon

Wood and thinking, as I talked to him, that here was someone at last who knew his trade and was equal to any demands that war might make. I remember too, A. N. S. Jackson, the Olympic runner, my contemporary at Oxford, and in the regiment, whom I saw married in 1918 on Paris leave, wearing one ribbon only, the DSO with three bars. There were others like them but not many. Of this small company of exceptional young men was Rommel, on the wrong side.[10]

A whole generation rose to this leadership challenge and many perished in the act. The First World War prompted many young men to assess themselves as never before, bringing self-knowledge and a confidence that bordered on arrogance in their ability to lead and direct. This extended way beyond the purely professional. The war coloured their approach to civilians and to personal relationships. Wives and sweethearts remained often on the fringes of their lives; other friendships were hard won, whilst loyalty to soldiers, 'the regiment', the profession of arms, remained paramount and without question. The war hardened hearts to loss, yet created reluctance in commanders to 'spend' more lives than they had to. Both Montgomery and Rommel were part of the generation that was sucked into the trenches, who had a sense of fortune (or was it guilt) that they had survived and a deep sense of loss that so many good friends had not. Both proved keen, when the opportunity presented itself, to bridge the gap that they personally witnessed between those in combat at the front and senior commanders and their planning staffs who were usually further to the rear. Both would place themselves consciously closer to their troops in reaction to their own Great War experience and very deliberately share, and be seen to share, the discomfort of their men.

3

First Combats

THE ASSASSINATION OF the heir to the Austro-Hungarian throne in faraway Sarajevo on 28 June cast a shadow over the summer of 1914; but the First World War really began its irrevocable path with the German declaration of war on Russia of 1 August, which would leave an estimated 16 million dead by the Armistice of 11 November 1918, some 1,564 days later. An average of 10,230 deaths for each day; 426 per hour, or seven per minute.

Montgomery, Rommel – and Hitler – were all products of this war waged on an unprecedented scale; it is fundamental to understanding the leaders they later became. Hitler was a mere two and a half years older than Rommel, came from a non-military, relatively humble family and was a geographical 'outsider' – he was not even German, but born in the north-west Austrian border town of Braunau am Inn. The young Adolf had been a failure at whatever he had attempted, but welcomed joyously the coming of war in 1914. Despite having been screened as unfit for military service by the Austrian army that February, he rushed to volunteer with a German unit, joining the 16th Bavarian Reserve Regiment (also named the Infantry Regiment List after its commander). In a letter of February 1915, Hitler recalled of his first combat experience:

> Alongside me were Württembergers, under my feet dead and wounded English soldiers. The Württembergers had stormed the trench before we arrived. And now I knew why I had landed so softly . . .[1]

Perhaps these positive encounters with the Württembergers at Ypres (possibly his first contact with *any* Württemberger) and the perceived similarities in family backgrounds helped draw the Führer, in the 1930s, to Rommel, the diminutive but outspoken native of that ancient, south-western German kingdom.

Rommel and Hitler both left memoirs that reflect a similar idealism and the patriotic naivety of young men heading to combat and death or glory for the first time. In one of the few readable passages in *Mein Kampf*, Hitler recorded his experience of the fighting at Gheluvelt, near Ypres, on 29 October 1914:

> We marched in silence throughout the night and as the morning sun came through the mist an iron greeting suddenly burst above our heads. Shrapnel exploded in our midst and spluttered in the damp ground. But before the smoke of the explosion disappeared a wild *Hurrah* was shouted from two hundred throats, in response to this first greeting of Death. Then began the whistling of bullets and the booming of cannons, the shouting and singing of the combatants. With eyes straining fever-ishly, we pressed forward, quicker and quicker, until we finally came to close-quarter fighting, there beyond the beet-fields and the meadows. Soon the strains of a song reached us from afar. Nearer and nearer from company to company, it came. And while Death began to make havoc in our ranks, we passed the song on to those beside us: *Deutschland, Deutschland über Alles, über Alles in der Welt!*[2]*

Rommel's war diary, published as *Infantry Attacks* in 1937, would become a highly regarded military textbook and a minor bestseller in his life-time. More significantly, it would attract the attention of Hitler, which directly altered Rommel's career path and future prospects in the post-ings Hitler showered on the young colonel. Hitler's patronage, because of Rommel's First World War experience, would 'make' Rommel in a way that ordinary promotion and military achievement was unlikely to have ever done.

As a battalion runner Hitler's task was to take despatches to and from the front. This role had an unusually short life expectancy, even by First

---

* The singing-into-battle episode seems to have been recalled only by National Socialist writers; neighbouring units and the British never heard it, and in all probability it is Nazi hogwash.

World War standards: three of the eight runners in Hitler's battalion were killed in a single day in November 1914. At the front, Hitler found he had no anxiety of death and for the first time in his life discovered a calling at which he could excel; he was in action after just six weeks of training and his battalion suffered around seventy percent casualties in its first battle, but soon after he won the Iron Cross, Second Class. By 1918, the drifter and failure, who found personal relationships difficult, had been recognised for bravery on several occasions (by his Jewish company officer Leutnant Hugo Gutmann), had received another Iron Cross, First Class, the Bavarian Military Medal Third Class with bar, and the Cross of Military Merit. By the war's end, the twenty-nine-year-old Hitler had risen to lance corporal, a modest promotion, but beyond the wildest dreams of the directionless, pre-war refugee from Austria. Importantly, he had found fellow souls and revelled in the comradeship and respect of his front-line comrades. As Ian Kershaw observes: 'The First World War made Hitler possible.'[3] Hitler never forgot the war that made him and shaped his future. He would *always* wear his Iron Cross on his Führer's uniform. German historian Joachim C. Fest argues that the knowledge of having survived extreme danger gave the future Führer a sense of self-belief, of destiny that brought with it a sense of invincibility.[4] Both Rommel and Hitler possessed the Imperial Wound Badge in black, indicating a combat injury that was not crippling. Hitler was wounded in the thigh by a shell splinter in October 1916 near Bapaume on the Somme (some of his comrades were killed in the same blast), and gassed exactly two years later near Ypres. Rommel was wounded in combat on 24 September 1914 in the Ardennes. Three weeks later, on 13 October, Montgomery suffered multiple bullet wounds at Ypres. Facing danger and expressing no fear also had an immense effect on Hitler and Rommel's colleagues, who were actively inspired by their lucky friends.[5]

That all had survived a wounding in the unforgiving conditions of the Western Front perhaps furthered a personal sense of divine intervention and provided fortitude for the dark moments of the future world war. Indeed, many of the twentieth century's later decision-makers were also wounded when passing through the meat-grinder of the First World War, including Clement Attlee, Harold Macmillan, Benito Mussolini and Charles de Gaulle, as well as many of Hitler's entourage, including his deputy, Rudolf Hess, his bodyguard commander, Sepp Dietrich, and the brownshirt leader, Ernst Röhm.

*

Neither Montgomery nor Rommel had the typical Western Front experience. For a start, both belonged to that tiny group of professional warriors, the Regular Army. Although nearly 6 million Britons and Commonwealth troops served in the British Army during 1914–18, the pre-war Regular Army numbered just 250,000, of whom 13,000 were officers.[6] By contrast, the pre-mobilization 1914 German Army mustered some 800,000 men, including 30,000 commissioned officers.

The outbreak of hostilities found both men in training camps. In January 1913 the 1/Royal Warwicks returned from Bombay to Napier Barracks at Shorncliffe, near Folkestone and twenty-five-year-old Lieutenant Montgomery attended a musketry course at Hythe, 'the infantry subaltern's professional nursery'. After some pretty slothful years in India, where he felt he had learnt nothing but acquired some strong prejudices, Montgomery felt clearly there was more to soldiering. At the first opportunity – Hythe – he applied himself, and excelled, passing out top. Monty was now finding something he could get his teeth into besides the metaphorical slice of lemon in the mess gin.

The German invasion of Belgium on 4 August 1914 triggered Britain's military mobilisation and the following day Monty's battalion moved from Shorncliffe. Whilst Montgomery had been learning his trade at Hythe, Leutnant Rommel had been doing the same, via an attachment to a nearby artillery unit. It was usual in the German army to cross-post infantry and artillery officers to effect better cooperation in time of war, so that commanders understood the nature of support that artillery could offer the infantry, and the way those guns operated. This says much for the professionalism of the German armed forces of this era, for the British did not undertake such military musical chairs. The result was that Rommel gained an early understanding of the power of artillery, whereas Monty's knowledge of the army was limited in 1914 to the infantry.

Although young Erwin had been commissioned into the 124th Württemberger Infantry Regiment, which traced its origins back to 1673, he had been posted temporarily in March 1914 to Ulm, where he commanded a section of horse-drawn 77mm field guns in an artillery regiment that would support the Württembergers in war. The 77mm *Feldkanone* field gun, which could hurl a 19-pound shell up to 9,180 yards (8,394m), was pulled by teams of six horses, and when formed into a section of two guns, with extra horse teams pulling ammunition wagons, was quite a proposition.

For Rommel, the first hint of trouble had been on Friday 31 July, when France called up her conscripts and it became clear that Germany was about to do the same. Erwin was with his artillery regiment and desperate to return to his parent infantry unit. He recalled leading the guns back to barracks 'accompanied by an enthusiastic crowd whose numbers ran into thousands'.[7] News of war was greeted enthusiastically throughout Germany, by 1914 a deeply militaristic country. As a military band played the usual 'um-pah' strident marches in Munich's Odeonsplatz, local cameraman Heinrich Hoffmann caught the mood on that first Saturday morning of August; only later when he had become Hitler's court photographer did he re-examine his images of that day and, amazingly, was able to identify the twenty-five-year-old Adolf Hitler cheering with the rest.

The coming of war was greeted with no less enthusiasm in the streets of Britain, by a country that had had no experience of a major war since Waterloo in 1815. The nation's youth was understandably curious, having been brought up on a Victorian diet of horses, swords, bright uniforms and acts of derring-do. Modern scholars now suggest that the keenness with which the crowds had embraced the idea of war in early August soon subsided, as the reality of reporting for duty, squeezing first into ill-fitting uniforms, then into railway cars (usually reserved for horses – every nation seems to have sent its men off in 1914 in wagons marked '40 men or 8 horses'), hit home. Society for the first time was torn apart as literally millions disappeared in the smoke of railway locomotives conveying them away to the unfamiliar. The pre-arranged Schlieffen plan ground into motion – rather like a giant industrial machine, with an 'on' button, but no 'off' switch, it envisaged the German First, Second and Third Armies invading France, swinging around through Belgium, adhering to the requirement to 'let the last man on the right brush the Channel with his sleeve'.

Allocated to the 27th (2nd Royal Württemberg) Division, Rommel's regiment was not part of the huge flanking sweep into Belgium and France; it had none of the blister-making marches of the first three armies which struggled to keep to the punishing Schlieffen-Moltke timetable. As part of XIII Corps assigned to the Crown Prince of Prussia's Fifth Army, Rommel and his fellow soldiers acted as the hub around which the Belgo-French invasion force pivoted. With Fourth Army, Fifth's job was a plodding advance through the heavily forested terrain of the Ardennes (a perpetual sore in France's military history), where no amount of rehearsal and pre-war manoeuvres could prepare conscript battalions for the difficulties of deploying in dense woodland. This put excessive emphasis on the leadership and command talents of junior leaders like Rommel, at war not only with the French, but with the terrain itself.

With less distance to march, 124th Württembergers crossed the German–Luxembourg frontier only on 18 August, long after their colleagues further north, and during three days of hard marching they closed with their opponents, the sound of artillery fire echoing through the trees in the distance. On 22 August, Rommel's battalion found themselves advancing to make contact with their opponents, French troops who had deployed forward, in the Belgian Ardennes, just west of the small village of Bleid. It had been exactly three weeks from the date of mobilisation to first combat.

Within that short, three-week window, Rommel's unit, like everyone's, had to collect and integrate its reservists (until that moment civilians with other responsibilities, who had often forgotten what military skills they had ever acquired), kit them out and deploy. The Württembergers now consisted of three 1,000-man battalions, each of four companies. Rommel, who commanded a platoon in II Battalion, had been up all night, first on a reconnaissance mission, then taking messages between various command posts, and he started the day hungry. Here then was a handy young officer who was used and used again whenever a crisis loomed.

At 5.00 a.m. on Saturday, 22 August, the same day the BEF traded their first shots with German troops at Mons, Rommel led his men across potato fields in early morning fog. French rifle volleys intermittently flew through the mist towards them (as is usual when riflemen cannot see their targets, they overcompensate and aim high) whilst he went on ahead, feeling his way through the murk. Moving west along the road from Mussy, as he reached the outskirts of Bleid, he spied a group of French troops occupying some farm buildings:

I withdrew quickly behind the building. Was I to bring up the platoon? No! Four of us would be able to handle this situation . . . We quietly released our safety catches; jumped out from behind the building; and standing erect, opened fire on the enemy nearby . . . I stood taking aim alongside a pile of wood. My adversary was twenty yards ahead of me . . . We both aimed and fired almost at the same time and missed. His shot just missed my ear. I had to load fast, aim calmly and quickly and hold my aim . . . my rifle cracked; the enemy's head fell forward on the step . . . there were still about ten Frenchmen against us . . . I signalled to my men to rush them. With a yell we dashed down the village street. At this moment Frenchmen suddenly appeared at all doors and windows and opened fire. Their superiority was too much; we withdrew as fast as we had advanced and arrived without loss at the hedge where our platoon was getting ready to come to our aid.[8]

Rommel later returned, leading a second, successful assault into Bleid, where there was another moment of: 'Should I wait until other forces come up or storm the entrance of Bleid with my platoon? The latter course of action seemed proper.'[9] He carried the village by house-to-house clearance; its successful capture surely a tribute to Rommel's pre-war training of his conscripts, for he observed: 'the formations became intermingled. Rifle fire came from all directions and casualties mounted.'[10] He then led his men beyond Bleid in attacks to the north-west: 'We rushed forward in groups, each being mutually supported by the others, a manoeuvre we had practised frequently during peacetime . . . Again I found myself well in advance of my own line with my platoon.'[11]

Then, Rommel blacked out. One of the early effects of the fighting on Rommel had been a stomach complaint, partly a medical condition that would recur throughout his life, and partly a psychological reaction to the stress of being under hostile fire for the first time. Soldiers from all conflicts and ages recount how there is always some kind of reaction to the 'baptism' of enemy fire, whether it be excessive sweating, soiling one's trousers, vomiting. In Rommel's case,

suddenly . . . everything went black before my eyes and I passed out. The exertions of the previous day and night; the battle for Bleid and for the hill to the north; and last but not least, the terrible condition of my stomach had sapped the last ounce of my strength . . . I must have been unconscious for some time.[12]

On coming to, he heard the bugle sounding the recall. As the regiment assembled he saw that

there were many gaps in their ranks. In its first fight the regiment had lost twenty-five percent of its officers and fifteen percent of its men in dead, wounded and missing. I was deeply grieved to learn that two of my best friends had been killed . . . Night fell. Nearly dead from fatigue, we finally reached the village of Ruette, which was already more than filled with our own troops. We bivouacked in the open. No straw could be found, and our men were much too tired to search for it. The damp, cold ground kept us from getting a refreshing sleep.[13]

The combat at Bleid was significant because it was Rommel's first action: on a misty morning in August, a young, fearless platoon commander had tried to capture a village virtually on his own. Early combat encounters are important too because they shape the attitude towards one's opponent, as Rommel would later observe: 'After the first exchange the German rifleman became imbued with a feeling of superiority vis-à-vis his French counterpart.'[14]

The ground around Bleid reveals how the tough young commander kept pushing ahead to the detriment of his foes, and how in little time he became a 'force-multiplier' all by himself. The route of Rommel's first attack makes for a pleasant, if tiring, cross-country walk today. Bleid, barely a kilometre over the French border into Belgium, is now little more than the hamlet it was then: a small collection of cottages and houses, surrounded by farmland, where six local roads meet, lost between several other small settlements, all of which Rommel fought over in 1914. There is no trace of that conflict around the settlement, no trenches or craters, for the land reverted to its valuable role of agriculture as soon as the fighting ceased. The only reminders are a series of unobtrusive First World War cemeteries containing French and German dead. This is typical Ardennes country: large forests and small copses abound where herds of wild boar roam and game birds shriek. Small hills and ridges, which loom over Bleid to the north, channel all vehicular movement; ground that was similar in nature to the area of Rommel's break-in battles. This challenging terrain is why the French assumed that German armour could not operate quickly or effectively in the Ardennes area in 1940, and why former Sergeant André Maginot's line of surreal concrete fortresses, built in the 1930s, extended only as far as Longwy, a few miles to the south-west, leaving Bleid and the area west towards Sedan largely undefended.

In 1914, to counter any German aggression against France, Britain had determined to send its Expeditionary Force to join the large, conscript French army. Commanded by the sixty-two-year-old Sir John French, who had made his name in South Africa, the BEF comprised initially two corps, each of two divisions, and a cavalry division. The names of the commanders (Douglas Haig of I Corps, James Grierson of II Corps and Edmund Allenby of the cavalry division) were in all probability unknown to the twenty-six-year-old Montgomery, for whom his own Royal Warwickshire battalion would have formed the very edge of his military awareness. Haig and Allenby eventually became field marshals but in the event Grierson, II Corps commander, never reached the front: en route he expired of a heart attack and was replaced by Sir Horace Smith-Dorrien.

As the BEF expanded, most of the brigade commanders of 1914 were, by 1916, commanding divisions and corps, despite the lack of any kind of training course or qualification to better enable them to handle such large numbers of men and equipment. The question of the competence of British generalship still bedevils any rational debate over the First World War, but the issue of rapid promotion was not unlike elevating a competent primary school headteacher to the dizzy heights of a university vice chancellor. The profession is the same, but the skills required and numbers involved are totally different; some could manage the transition and others would find themselves (and be regarded by their subordinates as) 'over-promoted'. The same issue arose less within the French or German armies, which were already large through conscription with pools of well-trained officers at every level.

By 31 August a third BEF corps was formed of 4th and 6th Divisions, of which Lieutenant General Thomas D'Oyly Snow's 4th Division, including Lieutenant Montgomery, was hurried across the Channel to join Smith-Dorrien's II Corps. Monty's battalion, 1/Royal Warwicks, was brigaded with three other Regular battalions into the 10th Brigade, which totalled about 4,000 men. This in turn was assigned to Britain's 4th Infantry Division, comprising some 18,000 soldiers. 10th Brigade's commander was the well-connected James Aylmer Lowthorpe Haldane, a member of an industrious Scottish clan from Gleneagles that produced writers, scientists and the politician who founded the Territorial Army, R. B. Haldane. Monty is unlikely to have known his brigade commander, who fortunately left us with two useful memoirs, but his commanding officer was altogether a different man.

On 22 August 1914, as he gazed over the stern rail of the SS *Caledonia* taking his battalion out of Southampton on the overnight crossing to Boulogne, Monty, now a senior subaltern (unofficial mentor to junior officers) would have mused on the wisdom of his commanding officer. Aged forty-eight, Lieutenant Colonel John Ford Elkington had taken command only that February. He was the eldest son of a previous 1/Royal Warwicks commanding officer who had risen to the rank of Lieutenant General; his four brothers were also Regular officers. In Blighty, Monty had sought out Elkington's advice. Having sharpened his sword for combat, Monty was concerned as to whether he would need any money in the forthcoming war. The CO said no, but Monty nevertheless bought ten pound's worth of gold sovereigns. He was surprised to receive further avuncular wisdom that in war a very short haircut is best, being easiest to keep clean. Whilst Elkington had 'all his hair removed with clippers by the regimental barber and looked an amazing sight', Monty recollected that he just had a short haircut. An attitude of disapproval towards his CO seeps through the pages of Monty's memoirs, as does his self-advertised independence, in that the young subaltern heeded neither piece of advice from Elkington.

After landing in France, Monty and the rest of 4th Division marched to the small market town of Le Cateau, a few miles east of Cambrai, along an old Roman road. Coping with the French cobbled roads in the August sun would have been particularly challenging for the old sweats who had left the colours perhaps years before and were now hastily recalled to Elkington's battalion. The 1/Royal Warwicks war diary records 'between 500–600 reservists collected from all over the country' – butchers, bakers and candlestick-makers only days earlier, now perspiring along poplar-lined roads, cursing blisters and new, tight boots.[15] Reservists were men like Frank Richards, who after eight years with the Royal Welch Fusiliers, had spent five and a half years a civilian, in his case a timberman's assistant in a South Wales coalmine. On 5 August he was in a pub drinking with his fellow old soldier reservists 'when someone happened to come in with a bit of news. He said that war had broken out with Germany and that the Sergeant of Police was hanging up a notice by the post office, recalling all reservists to the colours.'[16] The transition from civilian to soldier was often bewilderingly quick. By 7 August, Frank Richards had reported for duty, been medically examined, issued kit and volunteered to join a draft heading to France. By the 10th he had married up with 2/Royal Welch Fusiliers, then in Rouen to protect the BEF's lines of communication.

The woollen khaki service dress worn by Monty's men had been introduced only after the Boer War, with high-buttoned collars for the men and a shirt-and-tie for officers, who carried their rank on the cuff in 1914–17 (it was the mud of the trenches that necessitated the return of rank insignia back onto a man's shoulders by 1917–18). Contrary to the 'Captain Edmund Blackadder' myth, by 1916 officers in the front lines would routinely discard their traditional peaked cap, Sam Browne belt, collar and tie when leading an attack, dressing as similarly to their men as possible, in an effort to dodge the sniper's bullet. There were initially no helmets, but a peaked khaki cap, adorned with regimental badge, which was routinely worn by all ranks, and matched their khaki two-piece woollen uniform – adorned with 1908-pattern khaki canvas pouches and equipment and made of wide webbing belts with brass buckles (holding 150 rounds of .303-inch ammunition in five-round clips) – knee-length puttees and hobnailed ammunition boots. The switch from traditional leather to canvas webbing equipment was made because canvas did not freeze in wintry conditions, whereas leather adopted the consistency of stiff cardboard, making the swift removal of ammunition impossible.

After a day's combat at the Belgian city of Mons, on 23 August, the first Anglo-German battle of the war, British casualties stood at 1,642 men lost and German at around 5,000. The twelve divisions of General Alexander von Kluck's First Army, the outermost 'wheel' of the German juggernaut rolled through Belgium and France, pushing back the four-division BEF, who withdrew slowly at first. After two long days, the BEF was tired and ragged with von Kluck on its heels. 4th Division's original orders were for it to cover British II Corps' withdrawal and initially it was to do this from Le Cateau, where II Corps had paused during their retreat. Smith-Dorrien learned at midnight on 25 August that Allenby's cavalry had already relinquished the ridge forward of Le Cateau, where his own rearguards were still arriving, dead beat. Unless II Corps was clear by first light, von Kluck would be on him. Smith-Dorrien thereupon determined he should, in fact, stop his retreat, turn and give the pursuing Germans a 'bloody nose' (not that he really had much option). Both Allenby and Snow, with his newly arrived, but still incomplete, 4th Division, neither technically under II Corps' command, agreed to fight under his orders.

Smith-Dorrien had fought in many of the late-nineteenth-century's wars of Empire and was an unlikely survivor of the Zulu massacre of British troops at Isandlwana in January 1879. His rich experience had led him to fits of temper and an independent mind. Despite orders from Sir John French (with whom he had experienced several pre-war differences of opinion) to continue the retreat, but with fresh troops in the form of his flank guard (Snow's 4th Division), he felt he had to stand and fight as the only way of managing a clean break. His force would have to fight on the ground it occupied, and no doubt he was too pre-occupied to recall that 26 August was the anniversary of Crécy (1346), when the rapid fire of English archers had defeated a vastly superior enemy. At Le Cateau, the 10-round magazines of their .303-inch Short Magazine Lee Enfield rifles would have to do the same. He placed 4th Division on his left, around Haucourt, just south of the Cambrai-Le Cateau Roman road; 3rd Division occupied his centre, based on Caudry, whilst 5th Division occupied his right, on high ground south of Le Cateau itself. There was precious little time to dig more than shell scrapes out of the heavy, clayey soil with the pathetically tiny issue entrenching-tools, which better resembled a gardener's trowel than an infantryman's spade – a defenders' nightmare. On the right flank some farmhands had been enlisted to make a start at trench digging, though the results were not very military.

Frank Richards, meanwhile, with the rest of 2/Royal Welch Fusiliers, was hastening through Le Cateau, whilst II Corps covered their back:

> We arrived in Le Cateau about midnight, dead-beat to the world. I don't believe any one of us at this time realized we were retiring, though it was clear we were not going in the direction of Germany . . . Le Cateau that evening presented a strange sight. Everyone was in a panic, packing up their stuff on carts and barrows . . . [we] camped on the square in the centre of town . . . I slept the sleep of the just that night, for three hours. I could have done with forty-three, but we were roused at 4 a.m. and ordered to leave our packs and greatcoats on the square. Everyone was glad when that order was issued; the only things we had to carry now, besides rifle and ammunition, were an extra pair of socks and our iron rations which consisted of four army biscuits, a pound tin of bully-beef, and a small quantity of tea and sugar. At dawn we marched out of Le Cateau with fixed bayonets . . . We were all fed up with the marching and would have welcomed a scrap to relieve the monotony.[17]

Snow's 4th Division only received word that Smith-Dorrien intended to stand and fight at about 5.30 a.m. As staff officers set about issuing new orders, word came back from the battalions (as is often the way of these things) that the fight was *already under way*. What happened was this. On Smith-Dorrien's extreme left 1/King's Own (Royal Lancaster) Regiment of 12th Brigade had marched through the previous day and arrived, exhausted, in the small hours. They drew up on high ground, beyond a ravine on the northern edge of Haucourt, where they piled arms and paraded in preparation for breakfast. In the absence of 4th Division's cavalry, 1/King's Own had been assured that French cavalry was screening their front. Although horsemen were seen to the north, these were dismissed as friendly – until at about 6.00 a.m. devastatingly accurate machine-gun fire scythed through the King's Own's neatly drawn-up ranks, skylined on high ground. Although they eventually beat off their attackers they suffered 400 casualties for their pains.[18]

In his memoirs, Montgomery remembered bivouacking in a cornfield after a long night march, then watching his neighbours on an adjacent hill eating their breakfast, with their rifles piled.

> That battalion was suddenly surprised . . . and fire opened on it at short range; it withdrew rapidly down the hill towards us, in great disorder. Our battalion were deployed in two lines, my company and one other were forward . . . The CO galloped up to us . . . and shouted to us to attack the enemy at once. This was the only order; there was no reconnaissance, no plan, no covering fire. We rushed up the hill, came under heavy fire, my Company Commander was wounded and there were many casualties. Nobody knew what to do . . . if this was real war it struck me as most curious and did not seem to make any sense against the background of what I had been reading.[19]

'C' Company, including Monty's platoon, hastened uphill from the Warnelle Ravine to help 1/King's Own reoccupy its former position, reached a hedge at the top, but were unable to hold it in the face of German machine-gun fire, and Captain Day, Monty's company commander, was badly wounded. Montgomery and two soldiers tried to rescue Day (his company commander in India also) from the exposed hilltop, but were unable to recover and evacuate him and were forced to leave him to his fate as a prisoner of war. In 1973, Monty's youngest brother Brian, to whom he was relatively close, brought out his own memoirs, *A Field Marshal in the Family*, which help illuminate aspects of Monty's life. And it is from Brian we learn that Monty's first taste of battle was as comic as it was tragic:

Waving my sword I ran forward in front of my platoon, but unfortunately I had gone only six paces when I tripped over my scabbard, the
sword fell from my hand (I hadn't wound that sword strap round my
wrist in the approved fashion!) and I fell flat on my face on very hard
ground. By the time I had picked myself up and rushed after my men
I found that most of them had been killed.[20]

The scabbard-tripping tale, perhaps unsurprisingly, makes no appearance
in Monty's own *Memoirs*. (Bernard's autobiography is based on a
scrapbook he put together much later in life containing sixty-one pages
of letters he'd written home, almost always to his mother, only very
occasionally to his bishop father. Bernard annotated many of the
scrapbook pages, in his later handwriting, and his recollections are therefore a mixture of the contemporary letters and later scrapbook musings.[21])
Indeed the tale of the scabbard at Le Cateau may have been apocryphal,
for he should also have possessed a service revolver, attached to his person
by a lanyard. The latest Webley & Scott model would have cost him
around £6 from the Army and Navy Stores in Victoria. Officers could
privately purchase (with their kit allowance) any sidearm, providing it
chambered government-issue .455-inch calibre ammunition. The same
emporium would have charged him £3 15s for his regulation sword and
another £1 5s.6d for the leather-bound scabbard (though one 2nd
Lieutenant Robert Churchill-Longman of 2/Royal Sussex kitted himself out
at Moss Bros in Covent Garden, where his sword cost him 30 shillings).[22]
Waving a sword – sharpened or not – whilst charging up the sides of
the Warnelle Ravine made no sense in 1914, just as surely as it would
have made little sense to an earlier generation in South Africa, during
the 1899–1902 Second Boer War (where Winston Churchill, for example,
used his own Mauser pistol resisting capture during a Boer ambush on
his armoured train). So the suspicion lingers that Monty may have been
exaggerating, to underline the point that the BEF in August 1914 was
already dangerously old-fashioned.

Stalled opposite 4th Division, von Kluck contented himself with
chipping away at them with a tornado of shells and machine-gun fire.
In the centre, he made little progress against 3rd Division, but soon after
midday – when there was a lull – the ceaseless hammer of guns signalled
a renewed assault on the other flank, the Germans dragging Maxim
machine guns forward (putting one in Le Cateau's church tower) to
decimate 5th Division on the extreme right. By then Smith-Dorrien

knew he had precious little time left if he were to enact the clean break he intended. 4th Division had to hold for longest and pull out in the dark; Elkington's 1/Royal Warwicks had kept their ground throughout the day, exposed to shrapnel and machine-gun fire, but Brigadier Haldane at 10th Brigade was, according to the official history, only informed of the withdrawal plans at 5 p.m., his brigade being detailed as the rearguard.

Pressure on 10th Brigade mounted; the weather turned and rain set in. The brigade, deployed across several villages, and having taken steady casualties all day, especially amongst its signallers, started to disintegrate. Under fire, without precise orders and in conditions of relative chaos, the Warwicks split into several groups and intermingling with the other withdrawing 10th Brigade battalions, made their way independently, without any direction, southwards, sometime after dark at around 10 p.m. Lieutenant Colonel Elkington led a group as far as St Quentin (fifteen miles away), where he met Lieutenant Colonel Mainwaring, CO of the 2/Royal Dublin Fusiliers. There, the two colonels decided to capitulate with their exhausted men and prepared a surrender document at the insistence of the mayor, who was concerned to preserve his town. Before they could give up, however, a patrol of 4th (Royal Irish) Dragoon Guards under Major Tom Bridges entered the town, collected the stragglers and restored morale. They also retrieved the incriminating paper which resulted in the two COs being cashiered. Both narrowly escaped a firing squad and Elkington, as Montgomery observed, 'joined the Foreign Legion, where he made good in a magnificent manner'.[23] Elkington would go on to win a Croix de Guerre and Médaille Militaire, rejoin the British Army and win a DSO.

Monty's group of about 300 was led by the second in command, Major Poole, across country for two days and nights with little food or water, surrounded by Germans. They covered a very creditable fifty miles in forty-four hours, moving mostly by night and hiding from the advancing German cavalry screen in the daylight hours. His memoirs observe with understatement that 'the subsequent days were very unpleasant', though Monty's fuller, more explicit account to his father of 27 September gives a better picture:

> The whole air seemed full of bullets and bursting shells. I have several friends amongst the list of killed and wounded . . . Our men fell out by the dozens & we had to leave them; lots were probably captured by the

Germans, some have since rejoined. The chocolate Mother had sent me was invaluable, as of course we had no rations all the time and there was no food in the villages. The villagers were all fleeing before the advancing Germans. All our kit was burnt to make room in the wagons for the wounded etc., so we had only what we stood up in. We had passed our kit burning on the wayside; altogether the outlook was black and we were in low spirits ... At 10 p.m. on the 28th we caught up with our Division and they put us on the motor lorries of a supply column.[24]

According to Brian Montgomery, this is when his brother used his gold sovereigns to buy food (so Monty's account of 'no food' is not quite right, but he was clearly very hungry and tired). His description is supported by the account of a fellow 1/Warwicks officer in Monty's group, Captain Clement Tomes, who wrote that during the retreat

the flames of the houses on fire in Ligny and Haucourt only seemed to accentuate the darkness. It seemed an interminable march, the men were dead weary, and many, falling asleep as they moved, pitched forward into the man in front and fell with a clatter to the ground.

Tomes especially recalled

the look of misery on the faces of the few people about, and near one village a group of women were on their knees by a crucifix; the generosity with which the inhabitants brought out all they had to feed us, although, mind you, we were retreating and leaving them to their fate; pathetic small parties of refugees with all their portable household goods in bundles, hand-carts or perambulators and their children being hurried along in panic and fear. Poor souls, they had not a hope of escape.[25]

Tomes's account has the air of Dunkirk about it. The effect of this first encounter with battle, the famine of orders, subsequent chaos of retreat, and his CO's behaviour cannot have been lost on Montgomery throughout his subsequent career. Few armies get close to experiencing the misery of the retreat from Mons, but Monty would have to lead his 3rd Division through not dissimilar circumstances in May 1940 and counter a sense of rampant defeatism before El Alamein. Monty also acknowledges how lucky he was to be collected into Major Poole's group ('a first-class regimental officer'), for with the retreat over and a consequent rise in morale, Poole took command of the remnants of the battalion on 5 September, promoting Lieutenant Montgomery to temporary captain and putting him in charge of a composite company of 350 men, usually a major's appointment.

Though Monty later criticised the chaos and the lack of a plan at Le Cateau, he saw the positive qualities of Poole's leadership: his personal belief in attending to his troops' morale and keeping soldiers informed of the situation at all times (Monty initially had no clue of what was happening at Le Cateau, nor did his immediate superiors, let alone his men); of the need for personal fitness (both the march from Boulogne and retreat had tested his personal stamina); for good reconnaissance and intelligence about the enemy (he knew nothing of his opponents and witnessed the surprise arrival of an enemy force which nearly destroyed a neighbouring battalion); and the requirement for a plan, with clear, concise orders (again, lacking at Le Cateau). All of these future hallmarks of Montgomery the General date back to this period, just six years into his military career.

The passage of time has been kind to Le Cateau, which has changed remarkably little over the years. It is possible to walk over the windswept corn and beet fields with an eye for the ground Montgomery would have recognised. Le Cateau itself, untouched in both world wars, nestles in a hollow and was soon occupied by Germans as battle commenced. Only St Martin's church spire and that of the *mairie* peep up over the surrounding fields. The terrain is very open, with no tree cover whatsoever and little 'dead ground' from which to resist a combined infantry and cavalry advance, supported by artillery. The future official historian, Colonel James Edmonds, described 4th Division's ground as 'open fields under cultivation, with some of the crops, notably beetroot and clover, still ungathered, soaked by the rain of the previous night, and in many places churned into deep mud by the passage of men, horses, guns and vehicles'.[26]

The BEF made use of the network of sunken farm tracks that still crisscross the area, where occasionally an old brass shell case peeps out from the chalky loam, a reminder that this was where the 13-pounders and 4.7-inch howitzers of the Royal Field Artillery did their business side-by-side with the battalions. Here, too, three VCs were won as a detachment of gunners, Captain Reynolds and Drivers Luke and Drain, retrieved their howitzer from under the noses of the Germans, when 5th

Division withdrew on Smith-Dorrien's order. Monty's 1/Royal Warwicks have no memorial to their twenty-six killed and many more wounded. The ploughing season each year still turns up the detritus of 1914: rusting shells, corroded buttons and five-round clips of .303-inch bullets that Smith-Dorrien's infantry no longer had time to load. It is easy to see why the BEF lost 7,812 casualties and thirty-eight guns at Le Cateau, a loss rate comparable with Waterloo, fought ninety-nine years previously.

Many of the II Corps dead lie in several Commonwealth war cemeteries around the town, including a thought-provoking, multinational military graveyard by the junction of the two Roman roads, where the fallen soldiers of several nations were buried by the Germans during their occupation. The cemetery lies astride a Roman road which was the main German axis of advance to the south-west, and looks onto a slight ridge, which Allenby's cavalry had previously occupied, where on 26 August the Germans placed machine guns and artillery to pepper Monty's battalion, amongst others, all day. Le Cateau was the birthplace of the painter Matisse; within his lifetime (1869–1954) three generations of German invaders, in 1870, 1914 and 1940, passed through his hometown. Montgomery returned briefly to Le Cateau during a bicycle-mounted battlefield tour in April 1927. But of far greater significance was the arrival, twenty-six years later, of the next cohort of field-grey invaders, preceded in person by a panzer bearing Generalmajor Erwin Rommel.

4

Cheating Death

THURSDAY, 24 SEPTEMBER 1914 saw Rommel and his Württemberg battalion attacking the French west of Verdun. Promoted from platoon leader to battalion adjutant, he had been in the area for several days, on the edge of woods with thick undergrowth, attacking dug-in French positions to the south and east of Montblainville. According to his own account, in *Infantry Attacks*, Rommel made it his business to know the ground better than anyone else. On several occasions he suggested plans to his battalion commander which were adopted and proved successful. Fighting in woods is like fighting in urban terrain: the environment encourages formations to split down into several smaller units, which are consequently more difficult to control. Noise is also easy to hear but extraordinarily difficult to locate, which further increases the command and control challenges.

Rommel often acted as a guide, or took command of machine-gun platoons or infantry sections as the occasion demanded, and demonstrated that he was a very capable infanteer. 'During the last few years of peace the junior officers of the 124th Infantry had been given intensive training in the use of the compass at night, and this training now reaped its just reward,' he observed during a night march.[1] In this kind of terrain, casualties amongst the regiment's leaders were high, and Rommel was forever going forward on his own to see the situation for himself. During the afternoon of the 24th, he had worked out a way of outflanking the French defenders by an attack east along a road towards the village of Varennes. The attack inevitably bogged down in the thick woodland and, after reporting to his battalion commander Major Salzmann, ever needing to be at the scene of action, he

> took a rifle and ammunition from a wounded man and took command of a couple of squads. It was impossible to handle a larger unit in those woods . . . The calls for aid men told us that our casualties were increasing . . . It was becoming harder to get the men to move forward; consequently we gained ground slowly . . . Once again we rushed the enemy in the bushes ahead of us. A little group of my former recruits came with me through the underbrush.[2]

Here again Rommel refers to the loyalty he inspired, with his recruits following him. In this leafy environment the field grey of German uniforms would have given Rommel a distinct advantage over the bright red trousers and dark blue tunics worn by French troops at the beginning of the war, who would have stood out like rare birds of paradise. Unlike Monty, Rommel had not taken his sword to war. He usually carried an issue 9mm automatic pistol in a holster on his left hip. But on this occasion had borrowed a rifle.[3]

> Again the enemy fired madly. Finally, scarcely twenty paces ahead I saw five Frenchmen firing from the standing position. Instantly my gun was at my shoulder. Two Frenchmen, standing one behind the other, dropped to the ground as my rifle cracked. I was still faced by three of them. Apparently my men sought shelter behind me and couldn't help me. I fired again. The rifle missed fire. I quickly opened the magazine and found it empty. The nearness of the enemy left no time for reloading, nor was any shelter close at hand. There was no use in thinking of escape. The bayonet was my only hope. I had been an enthusiastic bayonet fighter in time of peace and had acquired considerable proficiency. Even with odds of three to one against me, I had complete confidence in my weapon and in my ability. As I rushed forward, the enemy fired. Struck, I went head over heels and wound up a few paces in front of the enemy. A bullet, entering sideways, had shattered my upper left leg; and blood spurted from a wound as large as my fist. At any moment I expected a bullet or a bayonet thrust. I tried to close the wound with my right hand and, at the same time, to roll behind an oak.

'Count your rounds' is the cry every weapons instructor gives to his recruits during range practice and this passage is a good example of why. That dreadful click of an empty magazine would have been a heart-stopping moment for the twenty-three-year-old officer, who must have cursed his own stupidity. (In his 'Observations' section in *Infantry Attacks*, he comments: 'In a man-to-man fight, the winner is he who has one more round in his magazine.'[4]) Given this is an autobiographical account of being wounded in action, this reads as almost coldly impersonal, without adjectives or metaphors. It is, nevertheless, absurdly positive in tone: whatever Rommel's ability, the bayonet is not a weapon of choice in any three-to-one encounter. His account continues:

> For many minutes I lay there between the two fronts. Finally my men broke through the bushes and the enemy retreated. *Gefreiter* Rauch and *Einjährig-freiwillige* [one-year volunteer private][5] Rutschmann took care of me. A coat belt served as tourniquet and they bandaged my wound . . . As the sun set, two men carried me back to Montblainville in a shelter half-attached to two poles. I felt little pain, but fainted from a loss of blood.[6]

After passing through his battalion's aid post,

> my wound was dressed again, and I was loaded into an ambulance beside three wounded, groaning comrades. We left for the field hospital, the horses trotting over the shell-torn road; and the jolting which resulted caused me great pain. When we arrived around midnight one of the men beside me was already dead. The field hospital was overcrowded. Blanketed men lay in rows along the highway. Two doctors worked feverishly. They re-examined me and gave me a place on some straw in a room. At daylight an ambulance took me to the base hospital at Stenay, where a few days later, I was decorated with the Iron Cross, Second Class.[7]

There are hallmarks of the future Rommel here: the chancer, the risk-taker. Coming across five opponents when he was alone in dense wood-land and, not needing to engage, Rommel's immediate action was to attack, and surprise initially helped him. Killing or disabling *some* of the opposition is never enough: it summons up the blood in the remainder to kill the perpetrator and avenge their comrades. Rommel was extremely lucky that his men arrived when they did, or the young Erwin would have likely ended his days as a pincushion for French bayonets. Although this piece was written before Rommel had commanded anything more than a battalion, the stamp of his boldness, and rashness, too, are already apparent.

Rommel's Iron Cross, his first decoration, also begs a moment's thought. Although the medal in Second Class was given in huge numbers (an alleged 5.2 million during 1914–18), his was one of the very early ones, so was more of a novelty. It was a very visible award: the cross itself was only worn with parade dress, but the black-and-white striped ribbon was worn looped through the second buttonhole of the combat tunic. It was a spectacular launch to Rommel's soldierly career, especially when coupled with another visible trinket, the Black Wound Badge (the colour signifying the grade, or severity of injury), which he was also awarded at this time. By January 1915 he had acquired the Iron Cross, First Class, a far more exclusive decoration, to be worn on the right breast at all times, even in combat. Fewer than 200,000 were awarded, but the recipient had to have already won the Second Class medal. Such were the ways Rommel's grey tunic began to be decorated: a portable curriculum vitae.

The wood fighting on that day had cost Rommel's battalion dear: thirty dead, including two officers and eight wounded. The high number of officer casualties (six in total) represented twenty-five percent of a battalion's full complement. One is struck, too, by the (unexplained) ratio of dead to wounded, the inverse of the usual proportion of three wounded for every fatality. This perhaps reflects a high number who did not survive medical evacuation: wood fighting is particularly deadly, where wood splinters caused by artillery fire and bullets can be every bit as lethal as high-velocity hot metal.

It is easy to miss the tiny village of Varennes-en-Argonne today, hurrying on to Verdun, but the area repays a visit. To the French, Varennes is significant chiefly because this was where Louis XVI and Marie Antoinette were arrested by Revolutionaries in June 1791, en route to safety. They were then taken to Paris, and ultimately became the play-things of Madame la Guillotine on 21 January 1793. The town was levelled in the war, so the inn where the detention occurred no longer stands, but a plaque today commemorates where the building stood. Rommel was wounded in the woods a few metres north of the D38, about 1km out of Varennes heading west to the Four-de-Paris crossroads, although the exact location is impossible to determine from his narrative; he was attacking towards Varennes at the time. The ground, subsequently fought over in 1918 by Lieutenant Colonels George Patton and Harry Truman, was so torn by shrapnel during the war that the woods have been encouraged to grow within their old boundaries; a walk through the undergrowth hints at the remains of moss-covered, crumbling concrete shelters, trenches and shell craters, softened by time – even old barbed wire still reaches out to tear at one's limbs and clothing – and the whole Verdun area remains one vast battleground.[8]

The British retreat from Mons officially ended on 5 September. On 10 September the 1/Royal Warwicks crossed the Marne where they contributed to a BEF counterattack which halted von Kluck's advance and obliged the Germans in their turn to withdraw and ultimately dig in to consolidate their gains. Monty wrote home laconically, explaining Le Cateau and the aftermath:

> We have fought a battle with the Germans but I suffered no injury; I am awfully fit . . . we had a bit of marching with no food and very little sleep . . . I have everything I want except tobacco, if you could send me some; I smoke Capstan Navy Cut Medium. 1 oz tins are the most convenient size.[9]

The battalion subsequently advanced to the Aisne, crossing under hostile artillery fire on 13 September, where on 30 September Major Poole (now confirmed in command of 1/Royal Warwicks) noted in the daily war diary that discipline was 'noticeably worse than in South Africa, probably due to socialistic ideas imbibed by reservists'.[10] This was a reminder, if one was needed, that the BEF relied heavily on civilians to make up its numbers and such men needed sensitive handling. Most of the soldiers that Bernard Montgomery would lead in this war, and the next, were civilians; but Monty and his fellow officers had no experience of any troops other than Regulars. In this sense, Poole's war diary observation is an acknowledgement that he and his colleagues had yet to find the right formula for training these temporary warriors, for to suggest that they were all 'socialists' (and by implication, 'anti-establishment') is clearly nonsense.

On 14 September, Bernard's parents were warned by telegram that he was 'missing in action [at Le Cateau]' – which Monty's later scrapbook

annotation observes 'was technically correct!'[11] The muddle was soon
sorted out and shortly afterwards he asked them to send '25 cigarettes
every three days. The cigarettes must be in a tin box.'[12]

On 20 September Monty wrote to his mother:

> Good luck pursues me & I am quite safe. My clothes are in an awful
> state of mud & of course wet through. But it doesn't seem to matter
> much as I haven't even a cold after it; I came straight in, in my wet
> clothes, threw myself down and slept as I was, without taking off
> anything. I find that rum is a great standby; they give us some every
> day.[13]

Two days later he wrote: 'Will you send me a pipe; a plain shilling one.
My present one is all but finished. Tell anyone how much we out here
appreciate letters from home; and cigarettes too . . . It is no small responsibility being in command of 250 men on active service and within 600
yards of the German Army!'[14] A week later, he wrote a long letter home
with an account of the previous fortnight, mostly vignettes of the advance
past abandoned German kit, and signed off with the observation:

> I have a big beard. I have not washed my face or hands for 10 days; there
> is no means of washing & no time for it. The necessary things are sleep
> & food; washing is unnecessary. Don't forget the cigarettes, will you.[15]

Curiously, there is no evidence in the war diary of 1/Royal Warwicks
that they were too busy to wash or shave at this time, and beards of any
size would surely have challenged the very notion of regimental pride
and discipline. The boasting of his poor personal hygiene may have been
Monty's way of retaliating against his iron-willed mother: now free from
her apron strings, he was demonstrating, albeit rather juvenilely, that he
could now behave as he wished. More significantly, from these various
letters it is clear that the future teetotal general and obsessive persecutor
of smokers drank alcohol and smoked both cigarettes and a pipe. A
collection of photographs published by Brian Montgomery in 1985[16]
betrays the fact that Bernard's pipe smoking went back to his days in
India. Monty's rather prim annotation at the top of the scrapbook page
reads: 'It will be noticed that I ask for tobacco. I cannot recall that I
smoked in those days, but obviously I did.'[17] A later annotation on a
different scrapbook page reads: 'I still ask for cigarettes and matches;

I think they must have been for the soldiers in my company.'[18] Even in old age, Bernard was trying to recast history as he would like it to have been. Monty mused on 29 September:

> We are very short of officers as we lost eight in our first fight [Le Cateau] and we have had two out from home since then. I had a youngster join my company yesterday,* he only applied for a commission the day war broke out, so he has no experience . . . It looks to me as if the war will last over Xmas. I hope it won't . . . I think it is too scientific a war for it to last too long; Europe would not stand for it; and where is the money going to come from?[19]

Strategically, German troops further north were approaching Antwerp, threatening the BEF's line of communication via the Channel ports. On 6 October the whole of 4th Division was withdrawn from the Aisne and sent by train to St Omer, where they arrived early on the 12th to bolster the northern flank of the Allied armies. Haldane's 10th Brigade was ordered to attack Méteren (south of Ypres), which had just fallen to the enemy, and intelligence on the German dispositions was this time sought and plans were made. The battalion was now back up to strength after the losses at Le Cateau and the arrival of a replacement captain obliged Monty to relinquish his company and assume once more the role of platoon commander.

The thirteenth of October 1914, the day that Montgomery nearly lost his life, was a Tuesday – not that days of the week tend to matter to soldiers, they tend to merge into a meaningless jumble of twenty-four-hour segments. It witnessed III Corps mounting the first formal British offensive operation of the war (up until now the BEF had been on the back foot) to regain the hill line, south of Ypres, running up to Kemmel. Major Poole's 1/Royal Warwicks fronted 10th Brigade's attack on Méteren, which lies on a slight ridge, deploying all four companies but without any artillery support. 'A' and 'C' companies led, preceded by the divisional cyclist company and a detachment of cavalry, followed by 'B' and 'D'. Besides heavy mist, a drizzle was falling, which drastically reduced visibility. Captain Clement Tomes recalled the day:

---

*Bernard had taken a shine to the youngster, called Briscoe, known as 'tin-eye' on account of his monocle. According to Bernard's scrapbook annotations, he was killed at Ypres in 1915.

The ground was very difficult to what we had left; instead of open rolling downs we found ourselves in a slightly undulating country much cut up by hedges, wet ditches, gardens and hop fields, in which the poles and wire made progress difficult. The *pavé* road ran straight as an arrow from Caëstre to Flétre then to Méteren and Bailleul and was obviously impossible to use if the enemy held these places.[20]

As 1/Royal Warwicks approached Méteren, the Germans withdrew back into the town and at about 11 a.m., as 'A' Company crossed the Méteren Becque (one of the 'wet ditches' Tomes referred to), they came under fire. With the forward pair of companies fixed by enemy fire, the reserve two charged forward, including Monty, at the head of his platoon. This was another sword-waving moment (again, no word of·a revolver). Montgomery recalled: 'When zero hour approached, I drew my recently sharpened sword and shouted to my platoon to follow me, which it did.'[21] Here was the importance of Major Poole's recent concern over discipline and 'socialism' in the ranks. Unless a platoon commander, such as Monty, had the absolute confidence and trust of his men, he could not hope to get them to follow him into danger and battle, as at Méteren, sword or no sword. He continued:

We charged forward towards the village; there was considerable fire directed at us and some of my men became casualties, but we continued on our way. As we neared the objective I suddenly saw in front of me a trench full of Germans, one of whom was aiming his rifle at me. In my training as a young officer I had received much instruction in how to kill my enemy with a bayonet fixed to a rifle. I knew all about the various movements – right parry, left parry, forward lunge. I had been taught how to put the left foot on the corpse and extract the bayonet, giving at the same time a loud grunt. Indeed, I had been considered good on the bayonet-fighting course against sacks filled with straw, and had won prizes in man-to-man contests in the gymnasium. But now I had no rifle and no bayonet; I only had a sharp sword, and I was confronted by a large German who was about to shoot me. In all my short career in the Army no one had taught me how to kill a German with a sword. The only sword exercise I knew was saluting drill, learnt under the sergeant major on the barrack square. An immediate decision was clearly vital. I hurled myself through the air at the German and kicked him as hard as I could in the lower part of the stomach; the blow was well aimed at a tender spot. I had read much about the value of surprise in war. There is no doubt that the German was surprised and it must have seemed to him a new form of war; he fell to the ground in great pain and I took my first prisoner![22]

I have read this passage many times and in trying to picture the event there seem to me several things wrong with Monty's account, which reads more like a good after-dinner anecdote. Bernard is running, waving his sword in his right hand (as a right-handed man would do). Presumably he has tied the leather sword strap round his wrist (having omitted to do at Le Cateau), whilst his regulation pigskin-bound sword scabbard flaps against his left leg. Heaven knows why he hadn't discarded his sword already as impractical; Le Cateau should have told him that. The weather was poor and would have made the fields slippery underfoot also, so Monty's charge would have been more of a ragged stumble. He sees the German (Brian Montgomery records that the man was a corporal) with the rifle and is close enough to launch himself off the ground and kick him hard in the groin, in the manner (presumably) of a studs-up football tackle. Monty's 'launch' would have been despite the mud, which the battalion's historian describes as 'extremely trouble-some as it worked its way into rifle mechanisms with the consistency of cement'.[23]

Photographs in the regimental museum confirm the toffee-like nature and state of the ground at this time. Bernard's kick would have to have been powerful enough to penetrate the German's thick woollen tunic and leather equipment to cause him discomfort. If the German was standing in a trench, his groin might well have been inaccessible. Kicking his German in the groin implies that Monty ended up on the trench floor, or on the ground, at the mercy of other German defenders milling around him: not a very advantageous position. The German trenches were on high ground, and surely a German rifleman would have had time to spot the slowly advancing officer (recognisable by elevated sword, scabbard, brown leather Sam Browne belt, tie and cap), track him and fire. We know also that there were sharpshooters in the vicinity from what happened to Monty later in the day. It is also astonishing that Montgomery's German had not fixed his issue bayonet onto his rifle (in his memoirs he makes no mention of it, though he had just discussed bayonets) – for they had seen the Royal Warwicks coming, having opened fire on them earlier – and did not attempt to ward off the flying Monty.

German *soldaten* in those days still employed the older M1898 long bayonet with a blade length of just over 20 inches, which when fixed onto the standard 49-inch long Mauser rifle produced a weapon nearly six feet long, not unlike a mediaeval pike or lance: more than dangerous to a plodding man waving a 32½-inch sword blade about.

Monty had left Sandhurst in an era when edged weapons still counted, when just ten years earlier his army had mounted its last (as it turned out) sword-and-lance charge at Omdurman (well reported by one of its participants, the young Lieutenant Winston Churchill of the 4th Hussars). Coming, as he had, straight from India, by no means a 'quiet' posting, where sword-making remained a village craft and most rebels possessed a fearsome example, it is therefore difficult to accept the claim in Monty's memoirs that the only sword exercise he knew was saluting drill. And where were Monty's sword and scabbard whilst he was flying through the air? They would surely have disturbed his equilibrium (you may try it for yourself, as did I in the garden – the result is impossibly painful). At a later date, Monty's 1897-pattern infantry officer's sword, its edges since re-blunted for ceremonial purposes, was donated to the Imperial War Museum. On inspection, the challenge of waving the 1lb 12oz (800g) weapon about for any length of time is immediately apparent!

Whether or not Monty's famous and oft-repeated tale is true (and I suggest that it is not), it appears that he was using an event to highlight two issues: the pointlessness of swords in modern war, and the value of surprise. In a wider sense, this is an old army that Monty portrays here, very different to the one he believed he had forged in the Second World War. To talk (truthfully or not) of sword-waving was to exaggerate the difference between the BEF of 1914 and his own very modern British Army of 1944–45.[24]

What is not in question is what happened next. The capture of the town had stalled, so after 2 p.m., Méteren again became the focus of an assault, this time by the eight battalions of 10th and 12th Brigades, instead of the lone 1/Royal Warwicks attack of the morning. Monty had withdrawn his platoon to a ditch about a hundred yards away:

> I went out myself to see what the positions looked like from the enemy point of view – in accordance with the book! It was then that I was shot by a sniper; the bullet entered at the back, which was towards the enemy, and came out in front, having gone through my right lung, but broken no bones.

I collapsed, bleeding profusely. A soldier from my platoon ran forward and plugged the wound with my field dressing; whilst doing so, the sniper shot him through the head and he collapsed on top of me. I managed to shout to my platoon that no more men were to come to me until dark. It was then 3 p.m. and raining. I lay there all afternoon; the sniper kept firing at me and I received one more bullet in the left knee. The man lying on me took all the bullets and saved my life.[25]

Montgomery would have stood out as an officer (that sword again) and thus was an obvious target for a sniper, who was close enough to discern he was still alive, so kept firing at him. (E. S. Turner makes the point that the sword was a foolish weapon to flourish in 1914, and may well have been the cause of many unnecessary officer deaths.[26]) But here was evidence, if any were needed, of the loyalty Montgomery had generated already within his platoon: one of his 'boys' risked his own life to attend to his wounds and was killed; yet Monty still had to caution the rest not to attempt another rescue until darkness had fallen (the affection went both ways). In the meantime he had to lie in the mud and rain, shrouded by a dead comrade, the life slowly seeping out from him. In another account, Monty recalled: 'When dark, a party from my platoon soon came to me; they had no stretcher, so four of them carried me in an overcoat to the road, and down it they met some stretcher-bearers.'[27]

The barely conscious Monty may not have been entirely aware of the correct details, for in the Imperial War Museum's archives, Private H. W. Jackson left an account recording that he and Captain Phillips of 21st Field Ambulance crawled out to rescue the young officer; as the ground was wet, they manhandled him onto a ladder and laboriously dragged him back.[28] However retrieved, he was then evacuated, still conscious, via the well-proven military medical route of the regimental aid post to an advanced dressing station run by the Royal Army Medical Corps. Monty continued:

the doctors reckoned I could not live and, as the station was shortly to move, a grave was dug for me. But when the time came to move I was still alive . . . I survived . . . I think because I was very fit and healthy after two months of active service in the field.[29]

Here might be the origin (if not derived from the Le Cateau retreat) of Monty's obsession with the physical fitness of his men, one that would often make him unpopular with headquarters staff.

The marksman who caused Monty's great discomfort was one of a growing cadre of men trained to spot 'high-value targets' (to use modern terminology) and kill or disable them, the word 'sniper' deriving from the fact that snipe were the most difficult of all game birds to shoot, and therefore required an accomplished shot. That he was a sniper and Monty's demise was not a lucky shot from a German rifleman is evidenced by the fact that he continued to use Montgomery as a target throughout the afternoon, scoring several more hits on him or the unlucky private now draped over him. Snipers were to cause a disproportionately high number of senior officer casualties, including in February 1915 Brigadier General Johnnie Gough, a VC-winner and Douglas Haig's Chief of Staff, who had just been selected to command a division; Major General Sir Thompson Capper (brother of the Sandhurst Commandant who had saved Monty's bacon back in 1907) was sniped and killed in 1915 and at least twelve other senior officers (all brigadiers) were killed by snipers and scores more wounded.[30]

Monty returned to the spot on 22 February 1916 with his brother officer Captain Clement Tomes, shortly after returning to France with his brigade:

> Went with Tomes to see the place where I was wounded, Méteren. It was most interesting going over the ground again and I remembered it all quite well. The trench I charged and captured has been used as a grave to bury the Germans we killed. I saw the exact spot where I lay for over two hours while they shot at me. The place is just the same . . . All the fields about there are full of graves of our men.[31]

Passing through field hospitals in France, Monty was then admitted to the military hospital at Woolwich on 18 October, where his condition was assessed as suffering from gunshot wounds (at least two bullets had passed through) to the right side of his chest, and another in his knee. In contrast to Le Cateau, Monty's parents seem at no stage to have been informed that he was wounded until he wrote to them from hospital. Brian Montgomery makes the point that the bullets passing through his brother indicated maximum muzzle velocity and that consequently the sniper was close by. Had the range been greater and the bullets towards the end of their trajectory, they might, with less power, have lodged in his body, necessitating a life-threatening operation to extract them, which in all probability Montgomery would not have survived.[32]

With his right lung pierced, he would take time to recover; indeed the recovery was never quite complete: the injuries left him slightly short of breath for the rest of his life and this may well be the source of his legendary hostility towards tobacco, as he was obliged to relinquish his favourite cigarettes and pipe. His left knee was also damaged, so all his future attempts at fitness were a challenge against discomfort. Did Monty take this life-altering chain of events personally? Difficult to say, although late in life, no doubt with a twinkle in his eye, he wrote to Trumbull Warren: 'the Germans nearly killed me in October 1914. But I got my own back on them later on!'[33]

Aware that he was still alive, but assuming his soldiering days were over, Major Poole confirmed Monty's promotion to captain on 14 October, as a parting gift, and recommended him for a gallantry award at the same time. Montgomery would spend his twenty-seventh birthday in hospital recovering – staying, in fact, until 5 December, by which time Poole's recommendation for bravery had been approved. This materialised as a DSO, one of the earliest of the 8,981 awarded during the First World War, which was gazetted on 1 December 1914. Traditionally awarded to majors and above, the award to a captain (as Montgomery now was) was rare. (Nigel Hamilton makes the astute observation that Bernard's award might have been seen at the time as helping to restore the 1/Royal Warwicks' reputation, so recently tarnished by its former CO, Lieutenant Colonel Elkington.) Monty was incredibly lucky with the timing of this decoration, for on 28 December the Military Cross was instituted for junior officers, with precisely Monty's kind of gallantry in mind. Bernard annotated in his scrapbook: 'The MC had not been invented, if it had been, being only a lieutenant, I would have got that.'[34] Monty was also 'Mentioned in Despatches', the first of six 'Mentions' he would accumulate by 1918.

What of Méteren? The Warwicks were shortly afterwards withdrawn, but the rest of 10th and 12th Brigades captured it finally in the late afternoon of 13 October. 1/Royal Warwicks lost forty-five killed and eight-five wounded in the battle, including a company commander and a lieutenant. They remained in the area digging trenches and witnessing the Christmas truce, and led by Lieutenant Colonel Poole, as he had become (he would eventually reach brigadier general).

In the German spring offensives of 1918, war returned to Méteren and on 16 April the town was retaken by the Germans. It was literally levelled by Allied artillery over a fortnight, prior to a successful counterattack on

19 July. Postcard photographs taken immediately after the war portray it simply as a smudge of brick dust and timber in a torn landscape, littered with unexploded shells; the townsfolk who returned lived initially in windswept Nissen huts. Its Commonwealth War Cemetery reminds one not only of 1914 (all the forty-five from Monty's battalion killed on the 13th are buried there, with twenty-three 'unknowns') and the later fighting, but that a second British Expeditionary Force tangled here with another German army, hastening on to Dunkirk in May 1940. Méteren's second liberation in 1944 is celebrated by the town each 6 September. Still surrounded – as it was in 1914 – by hop fields and reconstructed of stark Flemish brick, with prominent gables and steep-pitched roofs, it is nevertheless a Méteren that Monty would have recognised, a settlement where he nearly lost his life, but gained a DSO.

5

Learning the Trade

Monty, though he wouldn't have known it, was extremely lucky to sit out the First Battle of Ypres, which began a few days after his wounding. In the final battle of the year, the BEF prevented the Germans from outflanking the allies and continuing their advance westwards, but the fighting sucked in men and inflated casualty lists. It effectively sounded the death knell for the old Regular Army (less those battalions in India), who by the end of the year had lost 90,000 men, including 58,000 at Ypres, and was unrecognisable. The cost to the Germans at First Ypres was even higher, at 20,000 killed and 80,000 wounded.

In 1917 the War Office announced that those who had fought up to the end of First Ypres, 22 November, would qualify for the first campaign medal issued, the Mons Star, which Montgomery received, but by the time the medal was announced very few were left alive to wear it. Witnessing the Christmas Truce, 1/Royal Warwicks stayed holding the line on the southern edge of the Ypres Salient and when at the end of the year, the old BEF was divided into the First (Haig) and Second (Smith-Dorrien) Armies, as troops from India and the Territorials arrived to help, the Warwicks found themselves part of Second Army, whilst Kitchener's New Armies trained for victory back in Blighty.

Across the Channel, whilst Montgomery had been fighting at Le Cateau and Méteren, the Secretary of War, Lord Kitchener, had taken steps to boost the Army's size and called for 100,000 volunteers, aided by a very successful poster campaign. Kitchener was amongst a tiny minority who refused to believe the war would be over by Christmas and consequently Britain needed a much larger force to defend her interests. Britain's management of her manpower during 1914–18 was chaotic in the extreme, in that it put four separate and distinct armies into the field in the First World War.

First of all there was the Regular Army of professional soldiers of which Montgomery was a part – the only volunteer army in Europe. It numbered 246,000 in 1914 (of which two corps totalling 154,000 men, one cavalry and six large infantry divisions, formed the early British Expeditionary Force). Whilst the Regulars deployed and fought in the opening battles, the part-time soldiers of the Territorial Force mobilised and trained. These were civilian volunteers who for one evening a week and one weekend a month learned the rudiments of soldiering, which they perfected on an annual two-week camp. Mirroring the Regular Army in structure, rank and (usually) equipment, from 5 August 1914 they mobilised and left their civilian jobs to fight for King and Country. Founded only six years earlier on 1 April 1908 out of the pre-existing hotchpotch of Rifle Volunteer battalions and yeomanry regiments by Secretary of State for War Haldane (a cousin of the Haldane of 10th Brigade), they supplied initially fourteen infantry and five cavalry divisions, with a strength of 268,777 in July 1914. (My paternal grandfather was one of them, joining his battalion via the OTC of his public school in 1912 when aged twenty-three; it was very much a social club, for he found he knew all his fellow officers and his family business employed many of his part-time privates.)

A few Territorial formations deployed abroad in 1914, but most started to arrive in France and Belgium during March–April 1915, bolstering the numbers of the depleted Regulars, who had been bled almost dry. The Territorials – who would be first blooded en masse at the battle of Loos in September–October 1915 – fought in their own brigades and divisions, staffed by Territorial officers, but commanded by Regulars. During the war nearly 50,000 officers and a million men passed through the ranks of the Territorial Force, comprising 568 battalions and equivalent formations, of whom 6,000 officers and 106,000 other ranks would be killed, and they were awarded seventy-one Victoria Crosses.

On the outbreak of hostilities in August 1914, Field Marshal Lord Kitchener, hero of the Sudan and Egypt, was appointed Secretary of State for War in a move designed to reassure the nation, and took steps to

divert recruiting away from the Territorials. Kitchener had seen the need
for a large army when others did not, but had a low regard for the
Territorial Force, who had been promised that they would be expanded
to support the Regular Army in time of crisis. For his own reasons,
instead of recruiting through the Territorial County Associations, he
chose to raise a third force, his 'New Army' (often referred to as
'Kitchener's Army'), through the Adjutant General's Department of his
War Office. His cabinet colleague Winston Churchill, himself a Territorial
officer, put it down to an experience in 1870, when Kitchener had
witnessed a French Territorial division give ground at the battle of Le
Mans, and was unaware – or unwilling – to accept that British Territorial
formations were completely different in all but name to their French
counterparts.[1] Peter Simpkins has observed that patriotic fervour was
high in 1914, and volunteers did not particularly care whom they joined,
as long as they got to the front before it was all over; 298,923 enlisted
into the three competing Armies (Regulars, Territorials and Kitchener's)
in August, 462,901 in September and an average of 113,660 *per month*
over the successive fifteen months of 1914–15.

Had Kitchener been true to the spirit of Haldane's army reforms, they
would have enlisted in the Territorial Force, not in the New Armies –
of which there were eventually five, totalling twenty divisions, fed by 556
service and Reserve battalions. Many of these were 'Pals' or 'Chums' units
– groups of friends or work colleagues who enlisted together, and died
together. The Kitchener men were a unique phenomenon in British
history, and millions of Britons today are descended from them.
Approximately one in three of the UK population in 2009 had an ancestor
who had served in uniform during the Great War, the vast majority of
whom would have been soldiers on the Western Front and the largest
proportion of these Kitchener men.[2] My maternal grandfather – the
future bishop, then aged nineteen – was one of these, who fled the tedium
of his clerical job in London for excitement overseas; he agreed with
Harold Macmillan, the future PM, whom he knew well, that 'the general
view was that it would be over by Christmas. Our major anxiety was by
hook or crook not to miss it'.[3]

Like the Territorial Divisions, New Army formations were based on
local recruiting areas; the north of England produced a disproportionate
number of them, Lancashire, Yorkshire, Northumberland and Scotland
providing over one third of the total recruits. They were not necessarily
the seasonally unemployed urban workforce, but men with steady jobs,

comfortable backgrounds and even families.[4] Their motivations for joining were various, but must have included boredom and curiosity as well as the usual calls of patriotism, duty and employer and peer pressure. The encouragement to enlist was intense: music hall songs, Kitchener's beady eyes and pointing finger following men down the street, marching bands, and posters with discomforting catch phrases like 'Daddy, what did <u>YOU</u> do in the Great War?', all persuaded young men to volunteer. That particular brand of late-Victorian, empire-building Anglican religion – 'muscular Christianity' – so familiar to Monty, also had a hand in this: church youth groups such as the Church Lad's Brigade and the Boys' Brigade joined en masse, whilst from the pulpit clergy preached martial spirit, were granted army ranks and wore uniform. In Welsh Nonconformist chapels, the approach was the same, with ministers producing more recruits than any other method. Organisations like the National Rifle Association and the Boy Scouts (as well as the Rifle Volunteer movement, founded in 1860) taught generations of young men useful military skills and when the call came, channelled them in the direction of the recruiting sergeants. It took nearly a year to train Kitchener's New Armies before they deployed to France to be bloodied on the Somme in 1916; there, seven of the thirteen assault divisions were based on New Army battalions.

Britain's final army in the Great War dated from after the formation of the New Armies, when Britain's reserves of voluntary manpower were trickling away, and was another innovation. In January 1916, the Military Service Act was introduced, conscripting single (later married) men between eighteen and forty-one. Thereafter it was conscripts who arrived at the front, though not as formed units, but as drafts of reinforcements. There had been many calls before the war from the influential National Service League for conscription, which was only avoided by a reluctant government through the 1908 Act which created the Territorial Force instead. The four armies were symptomatic of a failure to coordinate the nation's pool of labour, both at home and at war.[5] Territorials suffered at the expense of the Kitchener Armies, and after conscription had been introduced, reinforcements were no longer guaranteed to come from parent regiments, or the same part of the country. Even by the end of the Somme campaign, it is arguable whether there was any difference in the social make-up or performance of a Regular, Territorial Force or New Army division. All tended to contain a mix of professional, volunteer and conscript.[6] One thing all these formations pointed towards was that the British Army was going through a growth spurt of unprecedented proportions; it wasn't merely getting bigger, it was becoming huge – and it needed a competent staff to run it.

There was nothing in Montgomery's career to date to suggest that he was even remotely qualified for his next job. Lucky to be alive, it was assumed (not without reason) that his wounds would prevent him from pursuing any further military career. He had two consolation prizes in the form of his captaincy, important for any eventual pension entitlements, and the coveted DSO. After four weeks in hospital, his first medical board in November concluded he would be out of action for at least five months, and then only fit for home service. But they had not reckoned with the power of self-improvement in one as single-minded as Montgomery. By the time of his second medical in December, the board members felt his recovery had progressed sufficiently to let him home, to his parents in Chiswick, with three months leave. His picture would also appear in *The Graphic*, a long-established illustrated weekly magazine; in this case on 24 July 1915, Bernard was featured in the 'Roll of Honour – British Officers Who Have Suffered'.

Impatience was already surfacing as a Monty trait and within two months he had lobbied the Army Board sufficiently to be graded by a third board in February as fit for home service, but not active service – a remarkable about-turn in the health of a soldier for whom a grave had been dug less than four months previously. This was not just the effects of hospital and nurses; this was the individual willing and exercising himself towards recovery, rather in the manner that British service personnel wounded today push themselves to physical and mental fitness at the Headley Court rehabilitation centre in Surrey. In Monty's day there was no set programme of rehabilitation, so recovery was down to the soldier concerned. The bizarre aspect of the story is not so much the recovery as what happened next, for having persuaded the War Office he was fit enough to wear uniform again, he was posted as brigade major (Chief of Staff) to the headquarters of the New Army's 112th Infantry Brigade in Manchester.

Brigade major is a rite of passage for the ambitious in the British Army to this day. Nearly all of Monty's future contemporaries at the top served in this role during the First World War (including Field Marshals Wavell, Alanbrooke and 'Jumbo' Wilson). Monty, given his experiences to date as a platoon commander, could have made as much or as little of the posting as he chose. His Kitchener volunteer charges knew nothing of military life, and initially there were not even enough uniforms to go around, no rifles or webbing; and his lone superior, Brigadier General G. M. MacKenzie, was, according to Monty's scrapbook annotation, 'an old retired officer – a very nice person but quite useless'.[7] As the New Army expanded, there were too few qualified officers in England to run it; everyone was busy making their name at the front.

In all some 247,000 would be commissioned as British Army officers during 1914–18. Accordingly the War Office fell back on 'dug-outs', retired officers with some military experience 'dug-out' of other careers, and even those wounded, like Monty, who remained mentally alert but would never fight again, to train the Kitchener men. In most cases, any kind of education beyond secondary school was deemed enough to make the applicant an officer (there was no officer selection as such). This was a world where Second Lieutenant Desmond Young (Rommel's first biographer) could talk himself into a commission in the King's Royal Rifle Corps merely by doorstepping the Military Secretary to Lord Kitchener.[8] Such was the expansion that invariably New Army battalions contained only one officer with any Regular service. Promotion also came quickly to those with ability: my bishop grandfather rose from private via company sergeant major to second lieutenant in four years. Others leapt from second lieutenant to lieutenant colonel within two.

After much War Office prevarication, Monty's brigade was redesignated the 104th; its battalions eventually comprised 17th, 18th and 20/Lancashire Fusiliers (the 20th were also known as the Fourth Salford Pals) and 23/Manchesters. A former Territorial CO, Lieutenant Colonel G. E. Wilke, had raised both the 17th and 18/Lancashire Fusilier battalions in December 1914 and January 1915; the 17th was over-recruited (1,350 men) within five days, whilst the six Salford Pals battalions, formed between November 1914 and March 1915, were the brainchild of the South Salford MP, Montague Barlow. The 23rd Manchesters were raised by that city's mayor and corporation. A local newspaper observed of the latter: 'Many collier lads were forthcoming last evening. Some of them had gone straight to the Town Hall from the pit,' reporting that they were 'a sturdy set of fellows, and men who will make capital fighters'.[9]

As volunteer enthusiasm waned after the initial surges of August–September 1914, entry restrictions were eased and eventually the height requirement stood at just five feet. Diminutive soldiers, those under five feet, four inches, were collected together into 'Bantam' units. Montgomery's brigade comprised such a formation. The vast majority of Bantam units were fit miners escaping pit life in the Lancashire coalfields. All the infantry battalions in Major General Reginald Pinney's 35th Division were Bantams (originally the divisional badge was a bantam cock, but this was altered in December 1916 as not appearing sufficiently martial, to a circle of seven 'fives'). The artillery, engineers, signals, field ambulance and service corps (transport) were New Army volunteers of average height. A December 1914 photograph shows the 17/Lancashire Fusiliers in snowy Salford, parading as a somewhat bedraggled battalion in their cloth caps and civilian clothing, underlining the acute shortage of uniforms and rifles: just one of the many frustrations with which Monty would have had to deal.[10]

Montgomery and his brigade collected around Masham, North Yorkshire, in June 1915, by then uniformed, and in August moved to Salisbury Plain, fully weaponed, for a series of pre-departure manoeuvres. His twenty-eighth birthday – anticipating a return to the front line in an important post – was in complete contrast to the previous, when he was recovering from a life-threatening wound, his military career assumed to be over. Before the end of 1915 the brigade received orders to prepare for service in Egypt, to defend the Suez Canal against the Turks in the aftermath of Gallipoli; rumours abounded, too, of Mesopotamia, and tropical uniforms and pith helmets were issued. In the event they would leave for France in January 1916. Monty threw himself into his new role with enthusiasm and for nearly a year oversaw the training of his four battalions. A later, typical Monty, annotation in Bernard's scrapbook of his war letters home reads: 'it would be true to say that I really ran the Brigade and they all knew it'.

Rommel's thigh wound sustained at Varennes in September 1914 was serious enough to put him in hospital, where he passed his twenty-third birthday as the fighting in the west bogged down. Clausewitzian friction, in the form of the BEF and then the weather, had long since taken hold of the Schlieffen-inspired invasion plan, and shredded it. Through the winter months – apart from a Christmas Truce, informally arranged in some sectors only – both sides dug in to consolidate gains and find protection from hostile fire and the weather. The trenches also acted as a force multiplier, enabling both sides to hold a continuous front lightly and rush reinforcements to wherever they were required. In such a period of adaptation, casualties amongst the leaders, officers and NCOs were bound to be particularly high and it was Rommel's good fortune to miss this particular attritional period.

When Rommel left hospital and returned to the II Battalion of his 124th Regiment, he found it engaged in trench warfare near Binarville, about five miles west through the Argonne Forest from Varennes where he had been wounded. Trenches: another skill to master. Try digging a trench in forest and you'll understand that it's worse than in open terrain: tree roots are forever in the way, the ground is invariably full of shale or clay, and water has nowhere to drain. Even the onset of warmer, drier weather does not alleviate the unpleasantness of operations in woods; cold and the endless drip of water in winter is replaced by the incessant drone of biting mosquitoes in summer. In the wet months, bone-chilling damp penetrates leather marching boots quickly and no amount of foot-stamping in the wet clay at the bottom of a trench can shift the feeling of intense cold. Everywhere would be a clinging chronic aroma of moist earth.[11]

Though not fully recovered, Rommel limped back to his old unit early, on 13 January 1915, rather than be posted to strangers. Here was another emerging trait: impatience, a characteristic we have also discovered in Monty. Appointed commander of 9th Company, which needed some 'ginger' injecting into it, the twenty-three-year-old immediately set about improving the 400-yard stretch of trenches his company occupied. In some cases his men had only managed to dig down three feet, to their detriment whenever the French started shelling; shelter was often nothing more than a coffin-shaped hovel scooped out of the forward edge of the trench, closest to the French. Coffin was the appropriate word, for near misses caused his men to be buried and sometimes Rommel dug for hours to rescue unfortunates trapped in this way.

Leutnant Rommel was now responsible for 'about two hundred bearded warriors' in a position under intermittent shellfire and unable to see his French opponents through the underbrush. On 29 January he was invited to join a raid and led 9th Company against French positions through the woods; rushing the enemy trenches, he found his opponents taking to their heels: 'red trousers flashed through the underbrush and blue coat tails were flying'. Following hard behind, he led his troops on a mad gallop through several lines of French outposts, his quarry always just ahead. They ended up half a mile downhill and held up by wire entanglements and under fire. Crawling through, Rommel found himself alone:

> The commander of my leading platoon lost his nerve and did nothing, and the rest of the company imitated him and lay down behind the wire. Shouting and waving at them proved useless . . . I crept back . . . [and] informed the platoon leader that he could either obey my orders or be shot on the spot. He elected the former, and in spite of intense small-arms fire . . . we all crept through the obstacle and reached the hostile position.[12]

Rommel occupied a disused position deep in the French defences and sent for ammunition and reinforcements, whilst his opponents recovered and started to counterattack from both flanks. Running low on ammunition, he eventually received an order to withdraw.

> Now for a decision! Should we break off the engagement and run back through the narrow passage in the wire entanglement under a heavy cross fire? Such a manoeuvre would, at minimum, cost fifty percent in casualties . . . I had one other line of action: namely, to attack the enemy, disorganise him, and then withdraw. Therein lay our only possible salvation. To be sure, the enemy was far superior in numbers, but French infantry had yet to withstand an attack by my riflemen.'[13]

Which is exactly what he ordered. Stunning his opponents with an immediate counterattack of his own, he took advantage of the confusion and made it back to his own lines with minimal casualties. He had to leave five seriously wounded men behind – small change by First World War standards. He later observed in *Infantry Attacks* that: 'breaking off combat is most easily accomplished after successful offensive manoeuvre'. Whether or not he would have pulled the trigger on one of his subordinates will never be known (perhaps it would not have been wise to test the Rommel resolve on that occasion) but this merely underlined the utter dedication and foresight he possessed: he threatened one life to save a couple of hundred; he had total confidence in his mission and complete disdain for shells and bullets – and this from someone barely recovered from a serious wound who had rejoined the fighting just two weeks previously. We should not be surprised that this action brought Rommel the Iron Cross, First Class, the first awarded in his unit.

As casualties mounted among Rommel's comrades, reinforcements were sent from a reserve training unit in Württemberg, dedicated to keeping 27th Division fully manned, though numbers were all too soon depleted by combat and the chronically poor weather. Fighting was often a relief from the endless hours of shivering in the alternate snow and sleet that comes to the Argonne each winter. Much of the action Rommel experienced consisted of raids, exchanges of hand grenades and mining. The geology for the latter was challenging, as was the discovery of corpses of French soldiers in the trench walls. But shelter could only be obtained by making dugouts. 'The officers pitched in,' Rommel observed; 'we found that sharing the work helped morale.'

Rommel remained in this sector into May 1915, and, as invariably happens in all institutions, military or civilian, a new arrival to the regiment was posted into the 9th Company over Rommel's head, though the newcomer had no combat experience. Rommel was offered another company but elected to remain with his men and reverted to platoon command instead; it is an echo of Monty gaining a company after Le Cateau, only to lose it again before Méteren. The instance here is different, because Rommel chose to revert to his platoon, rather than being demoted forcibly. This suggests flexibility in the German system not present in the BEF, which was more rank- and position-conscious.

It also demonstrates how contented Rommel was with his war and his men. It has been suggested that there would have been a 'degree of Schadenfreude as his green replacement coped with having a highly decorated [as he was by 1915 standards] regimental hero looking over his shoulder'.[14] If so, he dealt with the situation remarkably well. After a period of well-earned leave, the irrepressible Erwin regained a neighbouring company, and became the regimental commander's fireman, nursing a unit here, counterattacking there, and leading his men in rebuilding positions. One of the youngsters he came across at this stage was a very junior platoon commander of the sister Württemberg Guard Grenadier Regiment, named Hans Speidel, destined one day to become his Chief of Staff and, ultimately, a senior NATO general.

Rommel's unit remained static, rotating through the same sector for most of 1915. There is a sense of diminishing challenge here: he knew the terrain, he had the measure of his opponents, he understood his men; the novelty of combat, Argonne-style, was wearing off. Promoted to *Oberleutnant* (lieutenant) in September, he was posted away from his beloved Württembergers at the end of the month. The regiment had been his pre-war home and he had spent almost a year with them in action. He was the regimental hero and was going places. Rommel had also come to the attention of the senior Württemberg commanders.

By the time Montgomery and his brigade finally set foot in Le Havre on Sunday 30 January 1916, the British Army in France had grown from the tiny Expeditionary Force of Regular troops in August 1914, to a huge field force of thirty-eight divisions (1.5 million men) by 1 August 1916, and was still increasing in size. A Third Army had been formed in July 1915, to which would be added Fourth and Fifth Armies in 1916; eventually this field strength would peak at almost 2 million.

If he had been reluctant to use his officer's revolver before Méteren, Monty was certainly better equipped now, for at some stage before his return to France he bought a diminutive .380-inch Colt M1908 automatic pistol with a seven-round magazine,* an investment which would, according to the relevant Army & Navy Stores catalogue, have cost him £6. 10s. At the same time, he may have also purchased the steel helmet which appears in all the early trench photographs of him. Its shape is wrong for an issue helmet, which only made their way to the BEF in late December 1915 and January 1916, on a limited scale of fifty per battalion, initially for frontline troops only, to be handed over from sentry to sentry. Monty's tin helmet looks more like a bowler hat made of steel than the iconic soup-bowl headgear that Tommies wore through two world wars. It is clearly a private purchase helmet, an odd accoutrement. One wonders what Monty is doing with such an item that would have marked him apart from his men. Perhaps it was a well-meant present? But, then again, soldiers have always bought items of kit that create identity in an organisation that seeks to rob its personnel of exactly that same sense of individualism.

Initially Monty's brigade was attached to the New Army 38th (Welsh) Division to learn the lore of trench warfare which had set in after Bernard's departure, over the winter of 1914–15. This presented some difficulties to the Bantam-height battalions who found the trenches too deep and the firesteps in them too low. The 35th Divisional History tells how each man was issued with two sandbags each time he rotated through front-line trenches to fill, then stand on so as to be able to see over the parapet. In these early days, Monty would have been cheered by the fact that his 1/Royal Warwicks colleague from Le Cateau days, Captain

---

* On display at the Imperial War Museum, London.

Clement Tomes, was brigade major of the neighbouring 106th Brigade and the two would have been able to learn much by consulting each other. Eventually they undertook their first front-line duty at Richebourg, fifteen miles due south of Méteren and close to Neuve Chapelle, site of a British attack in March 1915 by other Regular and Indian Army battalions. Although troops had penetrated the lines on 10 March 1915, German counterattacks sealed the breach and prevented any breakout. Major General Pinney had served there commanding 23rd Brigade during the battle and it was his performance then that marked him out for a division. This was the first extensive effect of shelling the brigade had seen, which Monty's letter home of 26 April caught:

> Neuve Chapelle is really a most extraordinary sight. At one time it must have been quite a pretty little village; now there is nothing to be seen there at all. The whole village has been razed to the ground and is nothing but a heap of rubble; not even the walls of the houses are left standing. Our line goes about 200 yards in front of what was the village.[15]

The vision of such military vandalism to a town cannot have been lost on Monty (sights like this were not seen in India) or his troops, most of whom came from the towns and cities of industrial Lancashire.

During this time Montgomery lost two good friends: his cousin Valentine Farrar was killed during a trench raid by 17/Lancashire Fusiliers, and Brigadier MacKenzie, who was sent home on 14 April, basically to make way for a younger commander. That Monty liked his first brigade commander is evidenced by an earlier letter, in which he noted that MacKenzie 'is too old to command a brigade out here . . . He is fifty-six and is old-fashioned and out of date in most things he does; a younger, more modern man is really wanted'; he 'is a very nice man, quite charming, but that of course has nothing to do with it!'[16] MacKenzie was Douglas Haig's near contemporary in years, and the age Monty would be when he commanded in Sicily as a full general. Here, Monty witnessed his first sacking.

His letter illustrates a hardness, remarkable in a twenty-six-year-old, that soon became routine. Monty became obsessed with age and youth in his commanders and saw the ruthless, but necessary, side of the military profession in MacKenzie's departure. By the end of August 1916, Monty would be the last surviving member of the brigade staff who had deployed from England, for his immediate subordinate, the brigade staff captain, the Hon. J. M. Balfour, was also weeded out as too 'old' (he was thirty-eight), whilst the brigade signals officer was wounded.[17] As a general, Montgomery became a legendary 'hirer and firer' of senior officers and, unlike many contemporaries who avoided removing the loyal, Monty went out of his way to dispose of the 'belly-achers', the worn-out, and the inefficient. In 1944, during the Normandy campaign, he would remove a corps commander (Bucknall) and three divisional generals (Erskine, Verney and Bullen-Smith).

MacKenzie was replaced by the forty-two-year-old newly promoted Brigadier General J. W. Sandilands, 'a first-class officer from whom I learnt a great deal', Monty later observed. For his part Monty's new superior consciously took his protégé under his wing and taught him all he knew. Sandilands had been newly promoted from lieutenant colonel with 2/Cameron Highlanders and Mentioned in Despatches. He had also been Mentioned in Despatches twice in South Africa, where he won a DSO and was wounded. He had attended Staff College in 1909–10, so had much to impart to Monty, and would command the brigade for the rest of the war. Montgomery was conscious that he owed a lot to this brigadier, under whom he served for nine months, and in 1945 he penned a private letter giving Sandilands the supreme accolade as 'the best general I ever served under'.[18]

Most troops never saw the feature which gave the British summer campaign of 1916 its name, the River Somme. The wide, sluggish river was the inter-army boundary between British Fourth Army to its north and the French Sixth Army on its southern bank, and a natural choice as the springboard for a joint Allied attack, the original intention being to balance Anglo-French participation. Due to the German assault on Verdun (begun on 21 February), the summer campaign was launched early to take pressure off the French, and became largely a British affair. Brian Montgomery (too young to serve in the Great War) observed that his elder brother's almost daily visits to the front-line trenches at this time were fraught with great danger from German artillery, about which Bernard remained tight-lipped. In June 1916, the 104th Brigade war diary noted the visit of Brigadier Sandilands and Captain Montgomery to the 17/Lancashire Fusiliers:

At 5 a.m. The enemy were extremely nervous and evidently feared another attack. Intermittent barrages of great intensity were kept up by the enemy . . . this made progress very difficult and exciting. However the outward and return journeys were successfully accomplished through the barrages.[19]

Monty's 104th Brigade were disappointed to have missed out on the 'big push' of Saturday 1 July 1916, but were committed to the later stages of the Somme campaign, passing through Albert – the main town behind the British lines, but within German artillery range – as Bernard reported home on 12 July:

> Albert is a very pretty little place. There is a gilt figure of the Virgin on top of the church tower there; its supports have been shot away and the figure is leaning out at right angles to the tower. The French say that when the figure falls completely down the war will end and the Germans be beaten.[20]

Hundreds of thousands of Tommies who served on the Somme remembered the leaning Virgin statue, which also appeared on wartime picture postcards. In fact the Virgin's demise was caused by British gunners on Albert's occupation by the Germans during their March 1918 offensive. Initially 35th Division was housed in inherited French trenches and Monty was taxed with hygiene problems. One company officer recorded: 'Trenches full of liquid mud. Smell horrible. Floors and walls contain many dead Frenchmen, too bad to touch. Our men quite nauseated.'[21] The division fought its first major engagement along the Bazentin Ridge from 15 July, when the sister brigades 105th and 106th were detached to fight with neighbouring divisions. It was then the turn of Monty's 104th Brigade to join in the attack, between 19–27 July. In the midst, Monty wrote home:

> The fighting here has been very fierce; we have lost heavily in the brigade, particularly in officers. However you can't take part in a show of this sort without losing. We have been unlucky in losing rather a lot of officers in proportion to men . . . The whole country round here is a perfect shambles; everything absolutely shelled to bits. We have an enormous amount of artillery here and we shell the Germans all day and all night. His artillery fire is also very intense as he too has amassed a lot of guns against us. I have come through untouched so far though I had a lucky escape this morning. I was out on an important reconnaissance with another officer and four big shells (8-inch) burst quite close to us; he was hit in the head and I was not touched at all.[22]

Monty's matter-of-fact, almost detached tone is striking here, given that his brigade was to lose nearly a thousand casualties in the battle of which he wrote. But what was eventually called the Somme campaign had been 'sold' to all concerned as the breakthrough needed to end the war, and there was optimism, despite the losses – particularly the oft-cited casualties of 57,000 on the opening day – that to keep on pushing would soon wear the Germans down. Historians of the First World War now accept that the Germans were more worried about the attrition they were suffering on the Somme than at Verdun, and after the Somme senior German commanders observed their army 'was never the same again'. Rather than a callous disregard for human life, Monty's reaction during the fighting shows an understanding of the inevitable lot of a military commander, for he would find himself going through the same thought process at least twice more: before Alamein in 1942 and Normandy in 1944. In both cases there would be an attritional element to the attacks in which the assault waves would suffer disproportionately heavy casualties. It seems from this letter home of 23 July that Monty, with maturity beyond his young years, had already grasped this essential military truth.

This battle was still raging as Bernard penned his next letter home on 27 July, again with none of the restraint he might have employed to keep his family from worrying:

> We have had an absolutely hellish time the last few days and yesterday morning I went up to Trônes Wood to help extricate one of our battalions which had had a particularly bad time. They were scattered about all over the place and very much shaken up. To collect them, I had to run about the place in the open between Trônes Wood and Guillemont . . . the latter place is strongly held by the Germans & we have so far failed to take it. Heavy shelling was going on and I was sniped at incessantly by the Bosches. However I escaped with the exception of a bit of high explosive shell which grazed the palm of my hand. It is quite alright and I am at work just the same.[23]

This part of his letter had actually begun: 'I was returned as wounded yesterday, but am still doing duty', so it sounded more than a mere 'graze'. Ernst Jünger, then a stormtrooper, whose writing Hitler would admire, recalled of nearby Guillemont, just east of Trônes Wood, in August 1916: 'The village . . . was distinguished from the landscape around it only because the shell-holes there were of a whiter colour by reason of the houses which had been ground to powder. Guillemont railway station lay in front of us. It was smashed like a child's plaything.'[24] Trônes Wood,* situated on a slight ridge – as all the Somme woods were – and between the German first and second line trenches, acted as a magnet for shells, because the Germans tended to locate their artillery, supplies and reserve troops there, under cover from Allied reconnaissance aircraft. Trônes was linked by trenches to Bernafay Wood, a few yards to its west, and beyond to the key village of Montauban.

This village lay on the same ridge and represented the extreme right wing of the British assault on 1 July. Further east (on the British right) were the French Sixth Army, whom British historians tend to overlook in narrating the Somme campaigns. Forward of Trônes, Bernafay and Montauban lay two features on a little ridge whose capture was also vital to success – the Montauban brickworks and Maltz Horn Farm. In contrast to much of the story of 1 July, Montauban had been captured on time and as planned by New Army Pals battalions from Liverpool and Manchester, with the fewest casualties suffered by any of the assault units. Many of these Pals would have been close friends of Monty's 104th Brigade, recruited in the same area. With the fall of Montauban a window of opportunity existed during the afternoon of 1 July to roll up the German positions from the south, but it was not exploited, for the rigid instructions issued to the two assaulting divisions ordered them to 'bite' and 'hold', but not 'exploit'.

Better communications might have enabled messages to be got back and the plan altered, but the one area of technological advancement that failed all armies in the First World War was that of signalling equipment. Compared with other battlefield innovations (amongst them, tanks, gas, flamethrowers, aircraft, trench mortars, rifle grenades) wirelesses were almost as primitive in 1918 as they had been in 1914. The realistic alternatives were pigeon, or runner, both of which had absurdly low life expectancies. In his memoirs, Montgomery recounted the story of another brigade being issued with a pigeon to convey news of an important advance. It was given to a soldier going forward:

Time was slipping by and no pigeon arrived; the brigadier walked feverishly about outside his HQ dugout. The soldiers anxiously searched the skies; but there was no sign of any pigeon. It last the cry went up: 'The

---

* Originally appearing on British maps as Bois des Troncs (wood of tree trunks), a wartime cartographical misspelling gave the feature its present name.

pigeon,' and sure enough back it came and alighted safely in the loft. Soldiers rushed to get the news and the brigade commander roared out: 'Give me the message.' It read: 'I am absolutely fed up with carrying this bloody bird about France.'[25]

German units were still hanging on to the dominating features of Trônes, Bernafay and Maltz Horn Farm when Monty's brigade arrived to renew the attack on 20 July. Particularly hard hit were 23/Manchesters, who in their first assault on machine guns in Maltz Horn Farm lost nine officers (including the CO) and 162 other ranks. Meanwhile, 17/Lancashire Fusiliers attacked Trônes Wood itself, after which Fusilier Heath, recalled: 'When we went to help at Trônes Wood, we found the front line was just a jumble of shell holes under heavy German gunfire. What a sight! Mangled bodies, arms and legs and bits and pieces everywhere, and about a hundred and fifty Royal West Kents lying dead in no-man's land.'[26] This is exactly where the steely demeanour and presence of the Brigadier and his Brigade Major made all the difference: the daily visits to as many of their front-line troops as they could reach (recorded in the brigade war diary) were instrumental in shoring up morale and set an inspiring example.

In contrast to the two woods and Montauban, the brickworks and Maltz Horn Farm have disappeared – the odd mound covered with nettles in a ploughed field and a Calvary cross is all that is left to remind the visitor of these battalion-stopping bastions over that summer of 1916. Montgomery also wrote of the next village east, Guillemont, which was much contested. Nestling alongside the road between Trônes Wood and Guillemont is a British war cemetery, where the Prime Minister's son, Raymond Asquith, lies; the gap in wealth and privilege between him, the Hon. Edward Tennant – a fellow Grenadier Guardsman buried nearby – and some of Monty's backstreet Salford Pals, also within the same graveyard, symbolises the all-encompassing national effort and sacrifice of the Somme. War returned to the area in May 1940, when the sleek, slate-grey painted tanks of Generalmajor Rudolf Veiel's 2nd Panzer Division rumbled through Guillemont and past Trônes Wood, speeding on their way to defeat another generation of khaki-clad, tin-hatted Britons.

Monty's 104th Brigade left the area depleted in numbers and exhausted on 30 July. The Bantams' historian, Sidney Allinson, recorded how they reformed in 'Happy Valley', near Bray-sur-Somme on 1 August, with other units of the Lancashire Fusiliers for a combined Minden Day parade, a regimental battle honour dating back to 1759. It is a day when Fusiliers traditionally wear wild rose briars in their headgear, and in 1916, 'someone saw to it that enough roses were available to enable three battalions to wear them on their steel helmets'.[27] Brigadier Sandilands reviewed the troops with his Brigade Major and thanked them for their efforts in Trônes Wood. By an odd twist of fate, Monty would be drawn back to Happy Valley exactly two years later, when the division of which he was Chief of Staff was obliged to assault exactly the same area on 22 August 1918. For Monty, this was the worst fighting he had seen to date. On 6 August 1916, he wrote home:

> I don't suppose anyone can realise the awfulness of the show generally in this Somme push unless they've been through it. The French, who are on our immediate right, say that Verdun at its worst was as nothing compared to this.'[28]

He was working long hours too, routinely waking at 5 a.m. and working until 11 p.m.; he admitted to his father in letters home later in 1917 of feeling 'very tired' and 'beginning to feel the strain a bit'.[29] Later, too, in August 1916, 104th Brigade returned to the Somme fighting, this time being ordered to attack Falfemont Farm, a formidable complex of buildings on high ground, south-east of Guillemont on the road to Maurepas. It was on the Franco-British operational boundary and after several inter-Allied misunderstandings (many officers, Montgomery included, never learnt to speak French) and because of the dominating nature of the position, Sandilands eventually persuaded his divisional commander to cancel the attack scheduled for 24 August, anticipating crippling – and pointless – losses. It is noteworthy that senior commanders felt morally courageous enough to object to plans by this stage in the campaign, and lobby for change or cancellation, in stark contrast to the pre-Somme, pre-1 July air of expectation. By mid-August, it seems, there was more lassitude. Importantly, Montgomery would have witnessed this exchange of views prior to the proposed Falfremont Farm operation and seen how to brief succinctly and accurately in order to win over a superior. All 35th Division's troops left the Somme area on 26 August, leaving Falfremont in German hands. One officer who died attacking the area on 9 September was Major Cedric Charles Dickens, grandson of the writer, for whom a private memorial was erected which still stands.

Compared with what we know of other British infantry brigades serving on the Somme in 1916, Monty's 104th seem to have got off relatively lightly. Their first tour of 20–30 July was costly – nearly a thousand casualties, though less than many other units – but brief; their second, a week during August, was tiring and by no means casualty-free; their third of 1916, in early September, was in the quieter sector of Arras, immediately to the north of the Somme battle area. On moving up, by bus and train, they found,

> they would be fighting a different style of warfare. At their back was the ancient city, resting on a system of catacombs hollowed out in medieval times and which now provided safe, dry quarters during rest periods. The trenches also promised to be a great improvement over the wet ditches and precarious above-ground embrasures of the Somme. Miner-like, the Bantams set to digging, shoring and revetting their lines, while sending out reconnaissance patrols to get to know the territory.[30]

In December 1916, Lieutenant General Sir Aylmer Haldane, VI Corps commander (fast-risen after being Monty's brigadier at Le Cateau) inspected a Bantam battalion, and failing to understand their unique nature, started personally weeding out the most undersized, 'deformed' or sickly-looking of them. The 35th Division had in any case experienced problems recruiting enough Bantams to sustain the special diminutive appearance of its battalions, and Haldane's behaviour drew attention to the issue. Five hundred men were removed from Monty's 104th Brigade and sent home; it sounded the death-knell to the Bantam concept. 35th Division's bantam badge was dropped before the New Year. Monty neatly interpreted this as another demonstration of the Army's need to be cruel in the name of efficiency.

In September 1915, Oberleutnant Erwin Rommel, with his growing collection of medals, was posted to a unit of the new *Alpenkorps* ('Alpine Corps'), which had been founded that May, mostly from Prussian and Bavarian units, to bolster Austria-Hungary's military operations, many of which were being conducted in mountainous terrain. The harsh training of the *Alpenkorps* developed unusual powers of resilience and initiative amongst its junior leaders, many of whom would rise to prominence in the coming years – becoming senior Nazi figures.[31] One officer assigned to the staff of the Bavarian *Jäger* Regiment, a founding unit of the original *Alpenkorps*, was Leutnant Friedrich Paulus. A few months older than Erwin, Paulus and Rommel had much in common and their respective careers would intertwine several times before the former ended up, famously, at Stalingrad.

Rommel's formation, the *Königliche Württemberg Gebirgs-Battalion* ('the Royal Württemberg Mountain Battalion') was known throughout the German military system as the 'WGB'. Whilst not exactly a regiment, it was certainly larger than the battalion it advertised itself to be. At 2,000 strong, its structure was unique and reflected the requirement to operate independently. Furnished with six (rather than the usual four) rifle companies, it had a generous allocation of machine guns, mortars and signallers. Most of the WGB's initial strength came from Württemberg itself, many motivated to volunteer out of boredom, but about ten percent (a very high proportion for any unit at this stage of the war), like Rommel, possessed bravery decorations. In command was the man who would do more to mould Rommel's future than anyone else, a forty-five-year-old fellow Württemberger, Major Theodor Sprösser, who was determined to make something of Württemberg's own élite force.[32]

Rommel called Sprösser 'a martinet', whilst the major was already aware of Rommel's reputation (Sprösser had come from the sister 125th Infantry Regiment). He was so impressed with his new subordinate that he immediately gave the twenty-four-year-old Rommel his 2nd Rifle Company. Everyone fully expected that once fully established, the WGB would be sent to the southern front, where Italy, encouraged by the Allies with promises of territorial expansion, had declared war on Austria-Hungary in April 1915. It was thus the cause of some disappointment to find that they were sent instead to the Vosges, arriving 'during a rainy, howling New Year's Eve'. The regiment took over the 10,000-foot Hilsen Ridge, ten miles south-west of Colmar, just north of another ridge that had witnessed most of the previous fighting – Hartmannswillerkopf (to the French, Le Vieil Armand).

Most of young Rommel's activities in the Vosges were conducted with the support of 'a terrific concentration of fire [of] all types and calibres of shell', most usually from the 77mm field guns with which he had trained in August 1914. One French soldier recorded his experiences of being on the receiving end:

> a stench of sulphur and black powder, of burned and calcified earth
> which roams in sheets about the country . . . The country is bodily lifted

in places and falls back again. From one end of the horizon to the other it seems to us that the earth itself is raging with storm and tempest . . . We strain our eyes, and one of us has thrown himself flat on the ground; others look instinctively and frowning towards the shelter that we have not time to reach, and during those two seconds each one bends his head. It is a grating noise as of huge scissors which comes nearer and nearer to us, and ends at last with a ringing crash of unloaded iron. That one fell not far from us . . . We crouch in the bottom of the trench and remain doubled up while the place where we are is lashed by a shower of little fragments.[33]

One of Rommel's younger contemporaries, the future author Ernst Jünger, has left us with a vivid metaphor of himself being under fire:

I believe I have found a comparison that conveys what I, in common with all the rest who went through the war, experienced in situations such as this. It is as if one were tied to a post and threatened by a fellow swinging a sledgehammer. Now the hammer is swung back for the blow, now it whirls forward, till, just missing your skull, it sends the splinters flying from the post once more. That is exactly what it feels like to be exposed to heavy shelling without cover.[34]

Although Rommel seems to have mastered his fear of shellfire, in July he was wounded a second time by a piece of shrapnel, which damaged his shin. It is curious that he omits the event altogether from *Infantry Attacks*, in a culture where wounds were considered an honourable occurrence rather than carelessness or bad luck. Rommel put his Argonne experience to good use in a night raid he made on French positions in the Vosges in early October 1916. He inserted an assault team of over twenty men into the opposing trenches, having first cut through several wire obstacles, covered by the weather of a filthy storm. He personally conducted a thorough reconnaissance and rehearsed the raid exhaustively beforehand – Rommel trademarks – and succeeded brilliantly.

German assault teams were another development of the Western Front, and these eventually evolved into stormtroop detachments. They were initially small groups of hardy men with plenty of initiative, trained purely to break into enemy defences. German officer and NCO training provided the leadership to make such a concept successful, and German weapons technology gave it the means. Heavy M1908 Maxim machine guns were stripped down to make lighter M1915 models – the 1915 still weighed in at forty-three pounds; flamethrowers and small mortars were issued and assault teams also carried large quantities of stick grenades. Their tactics were of stealth by infiltration, followed by a very hard punch, then a rapid exit. By their very nature they risked being surrounded and cut off, so they trained to coordinate their activities closely with supporting artillery, but were utterly reliant on their own commanders' initiative, rather than a cumbersome chain of command. Initially called assault pioneers, when commanded by Hauptmann Willy Röhr from August 1915, they gained a higher profile. The stormtrooping name was devised, which in modern terms became a 'brand', along with practical clothing, based to a certain extent on that of the *Alpenkorps*: mountain boots, added patches on knees and elbows to facilitate crawling, carbines instead of rifles. In a masterstroke of self-advertisement, Röhr ensured they were the first German army unit to adopt the new coal-scuttle-shaped steel helmet in early 1916. Röhr's stormtroop detachments operated in the Vosges with great tactical success from October 1915 (when the WGB was being formed) and later on Röhr demonstrated and taught other units his tactics, including the Bavarian Life Guards Regiment of the *Alpenkorps*. There is no conclusive evidence that Röhr encountered Rommel, though it is perfectly plausible that Erwin did in fact meet this highly influential and colourful character, but they were certainly operating in the Vosges at the same time. Nevertheless, Röhr's influence was widespread at this time, and Rommel demonstrably experimented with his ideas during 1916. It is easy to see why the principles that lay behind stormtrooping appealed to him: one can imagining Erwin's innovative, magpie mind reaching out for the new, to enhance and make it his own.

Rommel made his mark quickly with the WGB, justifying Sprösser's faith in him. The qualities that Rommel brought to his 2nd Rifle Company were regarded as unique by those under him, as one of his platoon commanders later recalled:

> He was slightly built, almost schoolboyish, inspired by a holy zeal, always eager and anxious to act. In some curious way his spirit permeated the entire regiment right from the start, at first barely perceptibly to most, but then increasingly dramatically until everybody was inspired by his initiative, his courage and his dazzling acts of gallantry.[35]

Although Theodor Werner was writing here with an element of retrospective hagiography, it is clear that the young Rommel possessed a forceful, personable, character.

Both sides dug or chiselled fire positions into the sides of the massif, connected with elaborate communications trenches, and tunnelled deep bunkers into the rock. It tells us much about the mentality of trench warfare that the same tactics were imported even into mountainous regions, for one might have assumed that in hilly regions, the easiest course of action would have been to garrison the hilltops and dominate areas of interest by fire. Not a bit of it; the sophisticated, work-intensive set-up of the trenches, more appropriate to Flanders or the Somme, was imported into the mountains by the French who were determined not to give up more than one square inch more of France than they had to, even if it meant taking up disadvantageous positions.

The front barely moved during the entire 1914–18 period and the two opposing lines grew very elaborate. Rommel is not specific enough for one to be able to find his base or the positions he raided, but one can sit amidst the pine cones on the lip of an old trench, now shallow with the passing of years, and ponder his activities here, as I did so recently in the Vosges. Quite by chance amongst the pine needles, as though dropped only yesterday, my fingers alighted on a soldier's brass belt buckle of the era, bearing the Bavarian crown and a motto, *In Treue Fest* – 'In Steadfast Loyalty'. Although it may not have belonged to one of Rommel's men, it sums up the bond that kept them together.

Rommel's raid of October 1916 concluded with the capture of eleven prisoners, but the regiment didn't hang around to celebrate. For, by the end of the month, they had been posted to another country. After two years of fighting in France, Rommel was about to witness war in an altogether different theatre – Romania.

Throughout the Great War, the vast majority of British troops fought in France and Belgium: 64 out of 101 British and Empire divisions were to be found there in November 1918, whilst 5 were in the UK and the rest scattered in the Middle East (Palestine, Salonika, Mesopotamia), Italy, India and Africa.[36] For Germany there was an Eastern Front as well: the dominant state of Prussia particularly had always felt exposed and threatened by nearby European Russia, with her endless armies. The East beckoned for several reasons: as the Western Front ground to stalemate at the end of 1914, German commanders – Hindenburg and Ludendorff chief amongst them – considered that the Eastern Front, twice the length of the Western and less urbanised, offered the greatest opportunities for deep, operational-level encirclements, recreating the dreams of Clausewitz and Schlieffen. Such envelopments were taught as concepts to the German General Staff (though were considered far too ambitious and large scale for the British) and based historically on the battle of Cannæ, a battle that not merely defeated one side, but virtually annihilated it. Hannibal's 216 BC victory over the Romans was so perfectly executed that it became, and remains, a model for military strategy in its purest, textbook form – a knock-out blow against an opponent on such a scale, so as to become 'a battle without a tomorrow'.

Thus, German forces were consistently deployed in the East, as much aspiring for the perfect battle, as to shore-up Austria-Hungary. Germany soon came to regard supporting its Axis partner as being 'shackled to a corpse' – such was the poor state of Austrian industrialisation, and the political will, military leadership and strategic vision of its commanders. Romania's treaty with the Allies of 17 August 1916, which was encouraged by French and British military missions to Bucharest, had been a direct challenge to the Central Powers, but in many ways was premature and excited an immediate German-Austrian response. Lured by the siren voices of territorial expansion and initially reinforced by three Russian divisions, Romania's invasion over its western border into Austria-Hungary of 27 August 1916 had gone well. But by October German counterattacks had pushed all Romanian forces back to its borders.[37]

Naturally suited to the mountainous terrain, the WGB was posted to Romania in October 1916, and Rommel deployed with an advance guard to the front line, which lay at an altitude of 6,000 feet. Although trained to fight in winter conditions, Rommel's pack animals had yet to arrive and his troops were soaked with icy rain on the way up, which froze on reaching the summit. Ill equipped and prematurely deployed, Rommel was 'threatened with court-martial proceedings if we yielded one foot of ground'. The already poor weather deteriorated into blizzard conditions, where fires were impossible, as was warmth; 'a horrible night began ... A long, long night! When the day broke the doctor had to evacuate forty men to the hospital.' However, as Rommel later observed, 'we saw what the soldier can endure in the presence of the enemy'. Within days conditions had improved: 'Fog covered the plain far below us and broke like ocean waves against the sunlit peaks of the Transylvanian Alps. A wonderful sight!'[38]

By November 1916, the Württembergers had left the high mountains and were advancing further into western Romania. On the 11th, Rommel led an attack against an opposing garrison on the important Mount Lesului, and saw the value of humping his so-called 'light' machine guns through the foothills: suppressive automatic firepower kept Romanian heads down whilst his men broke into the enemy trenches and secured the important hill. Rommel's assault of the 11th developed into a bigger battle on 12th–13th around the village of Kurpenul-Valarii, where he saw off Romanian troops he estimated at ten times his own force in conditions of thick fog and darkness.

It is an agile mind that copes with leaving this sort of intense camaraderie and high-tempo combat, to go home and get married, but that is exactly what Rommel did – ten days later. He returned to Danzig, Lucie's home, where they married just after his twenty-fifth birthday, on Monday 27 November. Leave for any purpose is an extremely rare commodity in wartime, with a high premium on transport, principally reserved for reinforcements, the wounded and logistics. So a wartime marriage can be difficult to arrange with little foreknowledge of exactly when leave might be granted. This is what happened to my paternal grandfather in May 1915; although he had forewarned his local church, guests and his future in-laws, he had no firm idea of when he would be able to get home; so too with Rommel. In his case it was more complex because he was Lutheran, and Lucie, whom he had known since 1911, was Catholic; her father had died, and interdenominational marriages were less common in that era – and routinely opposed by the Catholic Church, unless the husband was prepared to convert and any children were brought up as Catholics.

As with Erwin's professional life, these were merely obstacles to be overcome. He regarded his time away as a brief window, an opportunity to be seized (though most of this time away would have been taken up with long tedious train journeys). After a necessarily brief honeymoon, Rommel was by mid-December back with his WGB comrades on the Romanian front.

6

Bloody Red Tabs

ENJOYING A WEEK's leave in London during January 1917, Monty discovered that his efforts as Brigade Major had been recognised sufficiently to warrant a posting as GSO2 (General Staff Officer, Grade 2) to the staff of 33rd Division. This was definitely promotion, although the majority he might have expected with this (or his last job, where he was, indeed, succeeded by a major) still eluded him. Accordingly, he packed his bags and took up his new post on 16 January.

Why did Montgomery move at this particular stage? His former divisional commander had himself moved on in September 1916 and been replaced by Major General Herman Landon, who had been commissioned into the old 6th of Foot (shortly to be renamed the Royal Warwicks) in 1879. He was much older than Brigadier Sandilands, but the two had served at Omdurman in 1898, where both had been Mentioned in Despatches. Landon had commanded 1/Royal Warwicks back in 1906, and had actually left regimental duty before Second Lieutenant Montgomery arrived. Nevertheless, he would no doubt have heard from Brigadier Sandilands the good work that young Captain Montgomery was doing and would certainly have gone out of his way to foster the career of a Regular officer from his own regiment. It seems certain that, not having a vacancy in his own divisional HQ, Major General Landon pushed Monty with a high recommendation towards the first suitable vacancy in a neighbouring division. Landon's predecessor as GOC of the Bantam Division, Major General Reginald Pinney, had not been sacked, but had taken over command of 33rd Division and almost certainly requested Monty's services when he heard from General Landon that he was due for a staff promotion. Efficient and personable Regular staff officers were rare acquisitions and the young Royal Warwicks captain would have been already well known to Pinney, especially if Monty's later claim – that he really ran 104th Brigade for Brigadier MacKenzie – was even a half-truth.[1]

Montgomery did not move very far, for 33rd Division was already based in the Arras sector. Although plans were afoot for a big assault to be launched in April, Monty was packed off almost immediately to the Staff School at Hesdin, about thirty miles due west of Arras, and about ten miles away from Sir Douglas Haig's HQ at Montreuil-sur-Mer, to learn more about what had become his new 'trade'. He reported home that 'there was no exam at the end . . . nor at any time during the course. We just went there to learn and we all helped each other and acquired knowledge from each other.' In peacetime, the natural next step would have been for Monty to have attended Staff College at Camberley; the next best thing was the wartime abbreviated staff course. The course began as soon as Monty moved to 33rd Division, and it is likely that his nomination was submitted some time in advance, while he was still at 104th Brigade, from the pen of Sandilands, and endorsed by a clinching recommendation from Landon. Both these instances – of Monty's next posting and his staff course – are examples of the Regular Army 'mafia' at work. In a vastly inflated force, the network of the 'old army' was still determined to protect and advance the careers of its own, excluding often deserving and talented Territorials and New Army men.

The Senior Staff Course answered the need for staff officers, but carried no qualification on completion, whereas graduates from the army's two-year Staff College course at Camberley were entitled to the post-nominal suffix 'psc' (passed Staff College) after their names in the *Army List*, which carried enormous weight in terms of promotion potential. However, in a bizarre War Office move, Camberley had shut its doors on the outbreak of war (maybe expecting everything to be over by Christmas), and many of its small but important cadre of officers died on regimental duty in the first months of the war.[2] As the BEF expanded rapidly, the shortage of junior officers even suitable to take up staff positions was felt at every level – brigade, division, corps and army – prompting two BEF staff courses to be run at Hesdin from 1915. Hesdin would shut down during the Arras and Passchendaele operations of 1917, and reopen at Cambridge University in October the same year, where Lieutenant Colonel Alexander of the Irish Guards found it 'excellent value'. As a brigade major, Montgomery would have been aware that the earlier training and credibility of staff officers had suffered due to the constant promotion and movement of officers as the BEF expanded rapidly over 1915–16, which prevented any continuity of thought or leadership within the staff. The creation of the Hesdin and Cambridge Staff Schools sought to correct this. Montgomery would have been one of twenty students on the six-week Senior Staff Course. All qualified staff officers, regardless of rank, were required to wear the red cap band and red collar tabs of the general staff, and a coloured brassard on the right arm, which immediately marked them as distinct, separate and unpopular. Hence the derogatory term, 'Bloody Red Tabs'.

Whilst Montgomery was at Hesdin, the real cost to the German army of the 1916 Somme campaign was acknowledged in their withdrawal back to a shorter, pre-prepared fortress system, known generally as the Hindenburg Line. At around this time, Martin Kitchen reminds us the traditional German patriotic marching song of 1840, '*Die Wacht am Rhein*' ('The Guard on the Rhine') had been rewritten as '*Die Wacht an der Somme*'. Ernst Jünger recalled:

> The spirit and tempo of the fighting altered, and after the battle of the Somme the war had its own peculiar impress that distinguished it from all other wars. After this battle the German soldier wore the steel helmet, and in his features there were chiselled the lines of an energy stretched to the utmost pitch.[3]

The Germans pulled back on 17 March leaving booby traps and thoroughly destroyed territory – the forerunner of Stalin's 1941 'scorched earth' policy in miniature: towns demolished, trees uprooted, crops trampled, livestock looted or slaughtered, wells poisoned and roads cratered. For nearly a month there was a spell of mobile warfare and troops on both sides operated in open country ('just like pre-war manoeuvres', some remembered) until the BEF closed up to the new German lines and trench warfare recommenced. Monty may have felt more than a twinge of frustration and envy during his course, for whilst he was learning more of staff duties, both 33rd Division and his old 104th Brigade (with its sister formations, 105th and 106th Brigades) were pursuing the Germans up to the Hindenburg Line.

Montgomery's course finished just in time for the April attack at Arras: 'I would have liked a few days leave after the strenuous brain work at the Staff School, but with the Division moving up to fight it was out of the question to ask for it.'[4] His new post was a key one in the hierarchy of Pinney's divisional headquarters, his duties he explained in a letter to his father:

My job is an interesting one, and not such hard work as when I was a Brigade Major. Then I had to be an expert on every subject . . . now I have to worry about nothing but G work, operations, training etc . . . There are 3 General Staff Officers in a Division, GSO1, GSO2, and GSO3. The latter does all the intelligence work, maps, training, etc. Ours is a Territorial and has to be told a good deal. The next up from GSO3 is Brigade Major [the job he had just finished]; then GSO2. I have no fixed work like GSO3; the GSO1 and myself divide the work between us, and I am responsible that it is done and nothing is forgotten. There is a lot to be done, but one has more time to do it than a brigade has. The GSO1 and I take it in turn to go out and visit the trenches: he goes out one day and I the next. I use a motor car more than my horses as it is quicker . . . Divisional headquarters are in dugouts; there is no accommodation in this part of the country as everything was battered in as we advanced.[5]

Montgomery's charges were based on a New Army division recruited in London, which had arrived in France during November 1915. It too had attacked the Bazentin Ridge in mid-July alongside the 35th Division, returned to tackle High Wood at the end of July and revisited the Somme in the dreadful weather of late October to early November. It had sustained high casualties on the Somme and by 1917 the division's order of battle had altered sufficiently to well illustrate the 'four armies' concept that bedevilled the BEF at this stage in the war. Its 98th Brigade included two Territorial and two Regular battalions, 100th Brigade had two Regular, one Territorial and one New Army battalion, and its third brigade, 19th (a refugee formation from another division altogether), had the same make-up as 100th Brigade: 5/6th Cameronians – Territorials; 3rd Public Schools Battalion (20/Royal Fusiliers) – New Army; 1/Cameronians and 2/Royal Welch Fusiliers – both Regular. The latter included Frank Richards, last heard of trudging out of Le Cateau in August 1914. This Royal Welch Fusiliers battalion has passed into history as being the most literary of all on the Western Front, for it also contained two men of letters, Robert Graves and Siegfried Sassoon, and Frank Richards, who wrote his own memoirs, as did the battalion's medical officer, Captain James Dunn. Into all of these battalions trickled conscripts, who might have come from anywhere; certainly the Cameronians were obliged to welcome Cockneys from London's East End and the 2/Royal Welch were compelled to take conscripts from Edinburgh or Glasgow. As the United States had entered the war on 6 April 1917, the first American soldiers were also appearing: not infantrymen, but young doctors and medical students, who volunteered to serve as medical officers in British infantry battalions, with all the risks that entailed, so great had the call been on qualified British medical staff.

The fight that 33rd Division was moving towards was the battle of Arras, which opened well with the attack and seizure of Vimy Ridge on Easter Sunday, 9 April 1917. It was planned in conjunction with the French high command, who intended to attack simultaneously under General Robert Nivelle on the Chemin des Dames, much further south. The Germans, it was thought, would only have reserves sufficient to stem one assault, but not both. In the event, Nivelle attacked late, on 16 April, but his plan was already hopelessly compromised and the Germans were waiting. The British effort, which was repaid initially with success at Vimy, involved an attack by the four divisions of the Canadian Expeditionary Force, fighting together as a Canadian Corps for the first time.

Vimy, often seen as the birth of Canada as a nation, succeeded partly because of the extensive preparatory tunnelling into the ridge itself. The tunnels offered concealment and shelter from German eyes and artillery, and also the weather, which was unusually nasty: 9 April was a day of driving snow and freezing temperatures. The other factor influencing Vimy's undoubted success was the fire plan, which included a creeping barrage using new No. 106 proximity fuses, which detonated shells on the slightest impact (proving effective against barbed wire), and counter-battery fire, whereby German guns were identified by flash-spotting (pinpointing muzzle flashes) and sound-ranging, using microphones. Both techniques enabled German artillery positions to be accurately plotted, and destroyed at H-Hour on the day of assault. One staff officer responsible for this effective and innovative fire plan was an artilleryman on the Canadian Corps staff, Major Alan Brooke, future GIGS and Monty's patron and protector in 1942–45, whilst practitioners of Brooke's scheme included the Canadian gunners Lieutenant Colonel Andrew MacNaughton and Major Harry Crerar, who were successive commanders of the First Canadian Army in 1942–45. Monty may not have been privy beforehand to the artillery plans, but he certainly witnessed them work and understood the scale of preparations that enabled such success. Having started spectacularly with a strategic gain for few casualties, the rest of the battle at Arras failed to produce a break-through or seize any more significant ground. The twenty-five Victoria Crosses awarded testify to the ferocity of the campaign, but the cost (around 150,000 casualties for *each* side) did not justify any of the gains.

One of the authors of this attack was Edmund Allenby, commanding Third Army, who shortly afterwards was sent to Palestine to coordinate attacks against the Turks. A fellow cavalryman, Allenby was Haig's direct contemporary at Staff College and a rival; there had always been tension between the two, which is why his departure for the Middle East in June came as no surprise to anyone in the know. It is instructive that when taken away from the physical (and mental) constraints of the Western Front, Allenby achieved stunning victories over the Ottoman forces, employing surprise, irregular troops under T. E. Lawrence and great flanking movements, yet he fought the Arras campaign after Vimy with little of the flair and vision he demonstrated later in the Middle East. This suggests that terrain itself can create a mindset difficult to break – an affliction that was surely shared by the New Zealand Corps Commander, Bernard Freyberg, when battling before Monte Cassino in early 1944. The nine-times wounded, triple-DSO and VC-winner Freyberg (whom the other Bernard would also encounter in the desert and became a close friend) was himself an Arras warrior, where he had commanded 173rd Brigade as a captain-acting-brigadier. His order of the day before attacking at Arras had concluded:

> The word 'retire' does not exist. Any man making use of this word in action will be treated as an enemy and shot.
>   By order of Brig.-Gen. commanding 173rd Brigade.

The order was instantly the stuff of legend: copies were circulated around adjacent commands, including 33rd Division. Apart from the terrain, another factor over which the BEF had no control was weather, and the unseasonally late snow and rain conspired to make a quagmire out of the ground that defeated man and tank alike.

Of Monty's charges, 19th Brigade was committed first on 14 April; it was the turn of the whole division on 24 April, during the sub-combat known as the Second Battle of the Scarpe. Working with tanks, the division pushed east towards Gavrelle, but were unable to make headway and by 27 April the division's casualties stood at nearly 3,000. Out of the line for a rest, Pinney put on a divisional horse show to raise morale, in which Captain James Dunn witnessed Montgomery's talent for horsemanship:

> May 4th. – Sweltering. Greenery is appearing everywhere. In a four-furlong race between Yates [a fellow officer] on *Girlie* and Montgomery, the GSO2, who rode a fine gelding, *François*, for the ASC [Army Service Corps], Yates was giving away over four stones and could not do it on the day, so the battalion dropped a few hundred francs.[6]

Although the Arras campaign formally ended on Wednesday 16 May, the 33rd Division was involved in a final *hurrah* against the German front lines at Bullecourt on 20 May, without result. It was either when on the staff course or during this period at Arras, as Montgomery later confided to his brother Brian, that he encountered an instruction issued from his higher formation, entitled 'Offensive Policy of VII Corps', which concluded:

> The value of patrol work, [night-time barbed-] wiring, and raids, in raising the confidence of our infantry, and giving them a feeling of superiority over the enemy cannot be too strongly emphasised; casualties in carrying out these duties must be accepted.[7]

According to Brian Montgomery, the last phrase of this instruction seemed to Monty an encouragement to take unnecessary risks and damn the casualties. Brian later reflected that this instruction was for Montgomery the beginning of a sense that he must do all he could to *minimise* battlefield losses. The corps commander responsible was none other than Snow ('Snowball' to his friends, 'Slush' to his detractors), Monty's old GOC of 4th Division at Le Cateau. In 1938, when Monty was in correspondence with Captain Cyril Falls, compiling the Arras volume of the official history, he observed that 'the real people who were tired were the commanders behind; Corps Commanders were getting pretty old by 1917 and few of them knew what went on up in front'. He considered Snow to have been 'quite useless' at the Battle of Arras: 'he merely told his divisions to get on with it' and provided 'no coordinated artillery plan'.[8]

Monty had worked hard through the Arras campaign – drafting orders, coordinating the different elements of the army, anticipating the logistical needs of his units and visiting the troops. As the battle subsided, he was posted again. This too was a promotion and confirmation of his talent for staff work. Major General Pinney had been well served, though Frank Richards of 2/Royal Welch Fusiliers was less complimentary about his commanding general. In his memoirs he referred to Pinney as a 'bun-punching crank and more fitted to be in command of a Church Mission hut at base than a division of troops'.[9] The reason? Pinney, a teetotaller, had banned the rum ration for the division and substituted tea; if Montgomery was still as partial to his rum ration as he had been back in 1914, then during his stay at 33rd Division he would have been working in an enforced alcohol-free zone, possibly contributing to his own later insistence on abstinence. It has even been suggested that it was Pinney who inspired the enduring and well-known ditty penned by Siegfried Sassoon, 'The General'.[10]

> 'Good-morning; good-morning!' the General said
> When we met him last week on our way to the line.
> Now the soldiers he smiles at are most of 'em dead,
> And we're cursing his staff for incompetent swine.
> 'He's a cheery old card,' grunted Harry to Jack
> As they slogged up to Arras with rifle and pack.
>
> ... But he did for them both by his plan of attack.[11]

If this is correct and the general is Pinney – which is certainly possible as he was Sassoon's GOC – then Monty as his GSO2 also features anonymously as one of the 'incompetent swine', putting Monty as close to the unfeeling, stereotypical staff officer as it was possible to be.

7

The Mountain Lion

JANUARY 1917 SAW Rommel and the WGB continuing their campaign in Romania. They had now been active for a year since they first deployed to the Vosges in January 1916. For most of the first month of 1917, Rommel, commanding his company of a couple of hundred warriors, harried his Romanian opponents, attacking hills like Magura Odobesti (3,300 feet) in four inches of snow and exploiting his success ahead of the rest of the WGB into the valley beyond to take the village of Gagesti. This would have been unremarkable had it not been for the fact that through a mixture of bluff and violent action, he managed to capture the garrison there of over 400 whilst leading a force one quarter of that size. On the morning following that victory, Rommel wrote:

Futtermeister [Mess Sergeant] Pfäffle and I took a morning ride through the lower village in the direction of Odobesti . . . I let Sultan step out briskly and paid more attention to the horse than to my surroundings . . . Pfäffle rode about ten yards behind me. We were about 1,100 yards from Gagesti when something moved on the road ahead of my horse. I looked up and was surprised to see a Romanian scout squad of about fifteen men with fixed bayonets right in front of us. It was too late to turn and gallop away, for any indication of intended flight would have brought me a couple of bullets. I quickly made up my mind, trotted up to the scout squad without changing pace, greeted them in a friendly way, gave them to understand that they must disarm, that they were prisoners, and were to march towards the church . . . I doubt very much whether any of the Romanians understood my words. But my demeanour and my calm, friendly tone of voice had a convincing effect. The fifteen men left their weapons on the road and moved off across the fields.[1]

Here was the resourceful Rommel at his unflappable best; he was a fit, tough soldier, happy operating in the unforgiving conditions of the Carpathians. He was also at times commanding as many as two rifle companies and a machine-gun detachment, exceeding the remit of a usual company commander. On the Western Front this would have been unlikely, but in Rumania – and in some of the other disjointed theatres in which he would serve – he became his own boss and no stranger to much increased responsibility, which he shouldered lightly. Rommel was in a very real sense a product of the sectors in which he served. On the Western Front, contrary to popular perception, soldiers generally served at the sharp end for short periods of a week or less, and were then rotated back. Rommel's theatres required more time at the front; they were more self-contained and all-involving. In this sense, he owed his military development and fortune to places such as the Vosges, Romania, and later Italy, and to people like Theodor Sprösser, his enlightened commander, who clearly understood the special qualities of his talented subordinate, for whom no task seemed too onerous.

Sprösser, who had led the WGB from its inception in 1915, encouraged a very modern system of delegated command. Being the high-profile, sole mountain battalion of the Kingdom of Württemberg, he seems to have been allowed to create his own order of battle, devising a large self-contained unit of 1,500–2,000 personnel. His WGB was self-sufficient in terms of transport (horses, wagons and mules) and had its own veterinary officers and a doctor on the strength of each company. Instead of the usual reporting chain, each of his nine company commanders (six rifle companies and three of machine guns, plus transport, medical and signals platoons) reported direct to him. It was a 'band of brothers' horizontal model of command, rather than the traditional vertical hierarchy. In November 1917, Sprösser formally published his tactical doctrine, enunciating seven points:

- Maintain pressure at the head of the attack
- Install and maintain communications day and night (by field telephone, cyclists or runners)

- Send written communications with simple sketches or prompt messages carried by reliable runners
- The staff (adjutant or night duty officer or telephonist) should remain especially vigilant at night
- Mark the route for the next detachments
- No unnecessary captured material to be carried
- All must be inspired by the need to press on – pursuit without rest saves casualties.[2]

The principles of war that Sprösser taught the young Rommel, especially through his delegated command concept, can be seen as the forerunners of the way many armies aspire to think and fight today. Much of the Kaiser's army's doctrinal thinking permeated through to Hitler's *Wehrmacht* too, inevitably perhaps, since it featured the same cast of characters. After 1945, many of the senior German military thinkers assisted the US Army in devising lessons learned from various theatres of the Second World War, especially with a Russian threat in mind. Elements of this in turn were adopted by NATO, the more so once Germany joined NATO in 1955 and created the new *Bundeswehr*, whose senior staff, of course, were former *Wehrmacht* officers.

It was clear that Rommel had gained an unusual amount of experience in the previous year, working in challenging terrain, but within a flexible organisation and a command chain that exploited his best leadership qualities. Once the war on the Western Front had grown static, dominated by the trenches, there was little room for innovative thinking or independent action, but in the sectors where Rommel fought subsequently, he was able to maximise his tactical talents, both physical and mental. According to his colleague, Theodor Werner:

Anybody who once came under the spell of his personality turned into a real soldier. However tough the strain, he seemed inexhaustible. He seemed to know just what the enemy were like and how they would probably react. His plans were often startling, instinctive, spontaneous and not infrequently obscure. He had an exceptional imagination, and it enabled him to hit on the most unexpected solutions to tough situations. When there was danger, he was always out in front calling on us to follow. He seemed to know no fear whatever. His men idolised him and had boundless faith in him.[3]

Following the arrival of the WGB and the rest of the Austro-German forces in late 1916, much of Romania was subdued, but by the middle of January 1917 the campaign had stalled. This was clearly no place to keep an elite shock force like the WGB (as they were now regarded), hanging around and underutilised. So in mid-January, Sprösser, Rommel and their men were conveyed back to France over ten days in the bitter cold in railway wagons, and found themselves in their old sector of the Vosges mountains. This was a quiet time in the Vosges and Rommel was able to take companies out of the line and train them; though there was little actual fighting, there were many tests of the WGB's alpine skills. In May, Rommel was back on the Hilsen Ridge, being 'hammered' by the French for a couple of days. In the summer of 1917, the worsening political situation in Russia made continued Russian military assistance to Romania unlikely: this was a good moment for the Central Powers to strike and complete their occupation of the country. Hence, another long train journey across Europe, this time in boiling weather, back to Romania, where the WGB arrived on 7 August 1917.

The unit deployed straight away with four days' rations and was thrown into the fierce struggle for Mont Cosna.* By this stage Rommel was acting very much as Sprösser's second in command, and took tactical command of three rifle companies and sometimes all six. In *Infantry Attacks* he personalised his command, calling it '*Die Abteilung Rommel*' (the Rommel unit), though it was never, formally, identified as such. He, of course, led the way:

> This type of combat under a burning August sun called for tremendous exertions on the part of the troops who had to contend with their heavy packs as well as with the steep slope. Several men collapsed from exhaustion. We drove the enemy from five successive positions each one stronger than its predecessor until Leutnant Hausser and I together with ten or twelve men were the only ones left in pursuit of the enemy.[4]

Appreciating the supreme value of communications (as in Sprösser's doctrinal points), Rommel payed out field telephone wire whenever he advanced, rather than risk the time-consuming method of runners. There is evidence of stormtrooper doctrine in his orders to assault groups, who were to be armed with spades, sharpened daggers, bags of stick grenades and wire cutters – no rifles, only pistols; and they would use captured

* Also called Hill 788; modern surveying techniques have caused it to be retitled Hill 789 today.

weapons and ammunition where needed. Rommel was also usually close to danger; later on, having surrounded a key Romanian position and pinned down by machine-gun fire, he noted: 'Anyone who showed himself drew an immediate burst of rifle and machine-gun fire; in this manner Vizefeldwebel Büttler received an abdominal wound whilst observing close beside me.'* As with Montgomery at Ypres, the unit loyalty and trust that Erwin Rommel inspired meant that his medics carried their wounded sergeant the full eight miles over mountainous terrain back to the medical aid post for a life-saving operation. Again, a few hours later, 'one of my combat orderlies killed a Romanian who was aiming at me from a distance of about fifteen yards'. In combat amongst the foothills of Mount Cosna, early on 10 August 1917, Rommel was wounded again, in the arm:

> I intended to dash on ... when I was suddenly shot in the forearm from the rear and blood spurted out ... In spite of severe pain and exhaustion through loss of blood, I did not give up command on the unit ... the tightly bandaged arm and overcoat thrown over my shoulders hampered every movement. I was considering giving up the command but the detachment's difficult position prompted me to remain at my post for the time being.[5]

Throughout 11 August, Rommel soldiered on, realising that he was the driving force behind his unit's success: 'we proceeded slowly as the terrain caused great difficulties ... My wounded arm hampered my climbing very much and my combat orderly had to help me over the more difficult spots.' At one stage, his attack stalled in front of an unusually determined Romanian position on the slopes. In *Infantry Attacks*, Rommel recalled his tactical appreciation of the situation (the act of weighing-up options that soldiers are still taught today) and he takes us back, musing out loud:

> Should I commit my reserve? Would its fire power turn the tide of the battle in our favour? No! Should I order a withdrawal? No! For then our dead and wounded would have to be left in enemy hands and we would have been driven from this position down into the ravine where the Romanians would have annihilated us with ease. The situation seemed desperate, but we had to master it or ... remain on the spot [i.e. perish].[6]

---

* Curiously, Rommel's unit had just taken over the sector from the Bavarian 18th Reserve Regiment, where *Vizefeldwebel* (Sergeant-Major) Rudolf Hess – the Nazi Party's future deputy leader – had also been seriously wounded by a bullet which pierced his lung and exited below his shoulder blade: another example of the 'shared war experience' of senior Nazis and Erwin Rommel.

He led the attack that swept his opponents off Mount Cosna later that day; Rommel had a great feel for terrain and observed that this and other successes were due to his personal reconnaissances, often conducted in the pre-daylight early hours. When counterattacks threatened his positions, he explained: 'In spite of the unfavourable situation, I decided to attack the Romanian positions without artillery support. I knew my men could do it and it was better to be a hammer than an anvil!' He was certainly the hammer – to himself, also – for several times he makes the point that: 'sleep was out of the question'. Eventually, after two days and a night, when Rommel felt 'completely exhausted and did not feel able to continue in command', he sought permission to head to a dressing station.

Before he could do so, Sprösser summoned all his officers to a conference on the night of the 12/13 August, to discuss ways of defending Mount Cosna against the overwhelming Romanian counter-assault that intelligence suggested was now massing against them. This act of inviting discussion to weigh up options (also used by Horatio Nelson) may be a facet of modern leadership techniques, but it was rare in armies of 1914–18, and it underlined the enlightened leadership of Sprösser, which in turn influenced Rommel, and enabled the WGB to flourish in comparison to other combat units.

It was Rommel's proposals that Sprösser accepted, with the request that Rommel, in spite of his wound, take command of the defence, knowing the ground better than anyone. With direct control of six companies (and more would arrive during the course of the battle), his detachment now exceeded that of a battalion commander. Furthermore, it was a true Axis coalition effort, for it included Württemberger, Bavarian and Hungarian units, all happy to fight under the dynamic young lieutenant. Rommel's tactics here were the future hallmarks of his success in defence: light front-line garrisons, with strong counterattack groups held in reserve and under cover at weak points; machine guns sited back, in depth positions; field telephone communications between all units; detachments dug in and trenches dug to facilitate the movement of reserves under cover; plenty of artillery on call (how little Rommel could have known that his 1914 attachment to the 49th Württemberg Artillery Regiment would have proved so useful).

Throughout 13 August, waves of Romanians attacked the WGB. Rommel lost control of the summit, but not the slopes, and hung on.

During the two-day lull that followed he ordered a couple of draughtsmen to reproduce and grid a sketch map of the Mount Cosna terrain he had drawn (Rommel was a talented artist and often used pictures to explain his intentions to his men). The maps were then distributed to every artillery unit in the vicinity, to enable swift defensive fire – another instance of the able tactician multiplying his efficiency by using whatever resources he had to hand. By the night of 15 August, with fresh reinforcements, Rommel found himself as a lieutenant commanding sixteen and a half companies 'more strength than an entire regiment'. Yet, no one thought of replacing him; neither did he (still bandaged and untreated) ask to be relieved. A second Romanian attack developed on 19 August and this time it was seen off with a swift counterattack, so swift that his pursuit stayed on the Romanians' heels, allowing him to take several successive positions. Even so, the unforgiving terrain was so steep that Rommel's hand grenades rolled past their attackers before exploding. Artillery support was weak through shortage of ammunition. On the same day, the 19th, Mount Cosna returned to German hands, and Rommel, on his feet 'until midnight taking care of provisions for the troops, replenishing ammunition and preparing my combat report', managed to get some sleep. It had seemed an endless kind of action: he commented that he hadn't managed to take his boots off for five days. By 20 August, he wrote,

I was so exhausted, probably because of the exciting activities of the past days, that I could give orders only from a lying position. In the afternoon because of a high fever, I began to babble the silliest nonsense . . . in the evening I turned over command . . . and walked down.[7]

The WGB followed him a few days later. The campaign had been a supreme personal test for Rommel, with enormous responsibility, and he had passed with flying colours. It had been a fluid battle, fought over ten days in unforgiving terrain riddled with ravines and ankle-twisting scree slopes devoid of cover; the common characteristic for all concerned was exhaustion. Of all his early battles and campaigns, the Cosna battles were some of Rommel's most significant. The key lesson he drew, as recorded in *Infantry Attacks*, was that one can always dig deep and find the extra resources to carry on, despite the odds. His life in combat would always be one of driving himself to the limits – and beyond.*

* October 2009 saw the release of *Hill 789: The Last Stronghold*, a documentary by the US-based Kogainon Films examining Rommel's 1917 battle. Much footage was shot on location and the importance of the hill is immediately obvious. Over ninety years later, the slopes of Mount Cosna are littered with military debris and the outline of German and Romanian trenches.

Naturally a commander behaving like this will inspire his men to do likewise, and it was perhaps on the Cosna battleground Rommel first found he could do this. He also gained unprecedented battle experience – certainly for a company commander – at times commanding the equivalent of a regiment (in British terms, a brigade). The WGB, too, collectively gained extensive experience though suffered their highest losses at Cosna of all the theatres in which they fought (88 dead, 299 wounded and 6 missing). On the climbs, Rommel observed that some carried 110lb loads whilst clambering uphill; others collapsed in the assault, pushed to exhaustion under a merciless August sun.

In retrospect it is surprising that Rommel received no promotion or decoration as a direct result, but that was not necessarily the German way. German army leaders (NCOs as well as officers) were encouraged, even then, to think of commanding 'two-up' if necessary. Thus a platoon commander was taught to be comfortable temporarily commanding not just a company, but a battalion; or, as in Rommel's case, a company commander could direct not just his battalion, but a regiment or brigade. Time after time in both world wars, German combat reports relate how sergeants commanded companies and battalions in moments of crisis. To the German army this was less incredible than it would seem to British or Americans: this is what the German soldier was trained to do. There was far less inclination to follow this pattern of leadership within the British Army, where the strictures of class were stronger and citizen-soldiers were a strange and unknown quantity – and in a country where conscription was introduced for the first time ever only in 1916.

Promotion was unlikely for Oberleutnant Rommel, the sometime regimental commander, because the German military system only promoted when a vacancy in the army became available: rank reflected seniority and service, and was often divorced from the responsibility the individual had. The British were more comfortable with attaching rank to responsibility, which is why the two world wars witnessed many cases of acting and temporary rank. Thus Rommel remained a lieutenant until 1918, and by the war's end he was still only a captain. Nevertheless, even without an inflated rank, he would observe in 1937 that he looked 'back on my days as a commander of such troops with intense pride and joy': the Desert Fox he would one day become; but mountain lion he had proved himself to be. His reward: a few weeks leave with Lucie on the Baltic coast.

8

Mud and Mountains

Back in France, Montgomery was posted on 6 July 1917 from 33rd Division to the busy staff of IX Corps (another rung up the staff ladder), which had just executed a spectacular assault on the German lines south of Ypres along the high ground of the Messines Ridge. The attack was a prelude to a much larger campaign, the Third Battle of Ypres, due to start on 31 July, although Second Army would be playing second fiddle to Sir Hubert Gough's Fifth Army in it. Montgomery's new boss was the mercurial General Sir Alexander Hamilton-Gordon – an artilleryman and a pre-war Director of Military Operations in India, where he had reported to Douglas Haig, then Chief of the Indian General Staff. Though Second Lieutenant Montgomery had also been in India at the same time, the gulf in rank makes it unlikely that they ever met.

In 1916 Haig appointed Hamilton-Gordon to command IX Corps, though not being a Staff College man he required a good staff around him. He would have seen that artillery had grown in importance throughout the war: guns and gunnery had evolved greatly since Le Cateau; they had increased in number, calibre, destructive power and the quantity of ammunition at their disposal. In 1915, the BEF fielded 801 artillery pieces on the Western Front, by July 1917 this figure had risen to 5,277.[1] Whilst sixty percent of the BEF were infantry some twenty percent were artillerymen by 1917. Guns of most calibres could fire high explosive, shrapnel, smoke and gas rounds with ever-increasing accuracy, at longer ranges, in huge quantities: 'artillery was the battle winner, artillery was what caused the greatest loss of life, the most dreadful wounds, and the deepest fear'.[2] The steady improvement of artillery technique also owed much to the development of artillery staffs at division, corps and army level, as well as the growing expertise of artillery observers, who were harder to train than gun commanders. All of this Monty had begun to observe at 33rd Division, but the lessons would be reinforced by his experiences with IX Corps.

The most innovative senior officer on the Western Front was Hamilton-Gordon's boss, the Second Army commander, sixty-year-old Sir Herbert Plumer. Montgomery's proximity to the professionalism of 'Plumer of Messines' (as he styled himself with his post-war viscountcy) was to have a lasting effect on Monty's military thought thereafter. An infantryman commissioned into the York and Lancaster Regiment in 1876, Herbert Plumer was the caricature of a bluff old Great War general; with his red cheeks and a bushy white moustache he was often taken to be the model for David Low's 'Colonel Blimp' character to whom there was, indeed, a close resemblance. Appearances can be deceptive and 'Daddy' (soldiers have always nicknamed their commanders and this sobriquet indicated affection rather than derision) Plumer possessed a fine brain and is now considered to have been one of the ablest First World War commanders. He understood that the success of any campaign rested on the handling of artillery – and any other innovation he could deploy. Monty later wrote of him: 'Plumer was one of the very few commanders in the war who was a soldier's soldier, held in trust and respect by his men – also being a high-class professional soldier, with all that implies.'[3]

Sir Douglas Haig had long planned to fight a return match in the Ypres area (where he knew the ground, having made his name there commanding I Corps in October–November 1914), in conjunction with an amphibious landing on the Belgian coast. Plumer was the driving force behind the 7 June Messines Ridge operation; a hallmark of his meticulous preparation was the fact that the tunnelling activities, which enabled the explosion of nineteen mines under German lines, were commenced up to eighteen months in advance. Second Army cracked open the German positions at Messines; IX Corps was in the centre of this advance and benefited from Plumer's planning and rehearsals. All its objectives were seized within twelve hours with 7,000 prisoners taken. All this was recounted to Monty as he arrived at his new post, and shortly afterwards he was taken over the ground to see the results for himself.

At the Corps HQ, the senior staff officer (GSO1) was a brigadier and the next grade down (GSO2) could be held by a lieutenant colonel or a major, yet such was the faith in Monty's staff work that he fulfilled this latter role as a captain. He was hugely aware of being centre stage as 'big events' unfolded and looked forward to watching the exercise of battle-field command at the highest levels. On 11 July, he wrote home:

> I like my new job very much. I am lucky to get it. There are 3 GSO2s in a Corps and when the senior appointment fell vacant it would have been only natural if one of the other two had moved up. But it was given to me, although both the others are senior in rank to me, being Majors ... I am easily the youngest holder of the appointment, and am also the only Captain holding such a job. In nearly all other Corps it is held by a Staff College man.[4]

Haig was keen to exploit his Messines success, but was unable to take advantage of the immediate time window presented: London was reluc-tant to sanction an attack without French support, which would not be forthcoming because French troops had started to mutiny against their casualties and conditions, expressed in a refusal to obey orders, or move to the front. Pétain never revealed to Haig the true extent of the indis-cipline (49 out of 113 infantry divisions were affected to varying degrees) and Berlin never realised what had happened. After persistent lobbying, Haig was authorised to proceed only on 25 July, six weeks after Messines. Success depended largely on the artillery fire plan. New targets had to be pinpointed from scratch. With the ground so destroyed and all roads impassable to vehicular traffic, the only practicable way to move Gough's 2,174 artillery pieces and sufficient ammunition forward was by horse teams or mule train. This task was gargantuan and required huge numbers of beasts (who themselves also required fodder and water); mules for example only carried eight 18-pounder shells each, and Gough fired 4.2 million shells of all calibres between 14–31 July. This kind of staff calcu-lation was precisely the sort of work occupying Monty at Corps HQ; indeed many of his greatest lessons in planning were learned at Third Ypres rather than elsewhere. He would have understood exactly why Fifth Army could not have followed on with a sequential attack imme-diately after Second Army had finished their work, let alone mounted a simultaneous one.

All had gone well with an effective preliminary bombardment but on Tuesday 31 July, the day of the assault, the weather broke. Instead of the usual dry spell Flanders experienced each August, it started raining ceaselessly, with five times the norm falling in August–September, converting the Gheluvelt Plateau into a giant marsh. The fields flooded immediately, rendering the ground impassable for tanks, the going slow for infantry and nullifying the effect of shelling. It was the worst weather in Flanders for seventy-five years, and made the loss of six weeks of fine weather after Messines particularly galling. After three weeks of battling against the formidable Hindenburg Line and the weather, Gough's attack slithered to a halt at the end of August.

Amongst the Fifth Army divisions pulled out of the line to recover was Monty's old 33rd, still commanded by the abstemious Major General Pinney (who sadly still failed to realise the efficacy of administering a rum ration to wet soldiers). Advised by intelligence that the Germans were being worn down quicker than the BEF and likely to crack, Haig determined to carry on, but switched the effort back to Plumer's Second Army, in whom he had more confidence. Before attacking, Plumer requested and got a three-week pause whilst his staff tackled the problems of how to penetrate the three separate lines of German defences (wire, trenches and concrete pillboxes) and defeat their inevitable counter-attacks; devising cleverer artillery tactics and improved logistics support, his troops rehearsed and (as the rain had stopped) the ground dried out. Monty was at the heart of this problem-solving exercise. The pause enabled light railways to be laid to bring Plumer's guns and their ammunition forward. His bombardment began on 31 August and the infantry attacked in the early hours of 20 September.

Second Army's initial 'bite-and-hold' successes, in contrast to Fifth's embarrassing failure, can be put down to the exhaustive staff work undertaken beforehand. Although the guiding hand was Second Army, Monty at IX Corps made a significant contribution. In the first of the many training documents he would issue in his career, he wrote the IX Corps' 'Instructions for the Training of Divisions for Offensive Action', issued to its two divisions on 3 September. This was not simply Monty's enunciation of Second Army doctrine, but a distillation of all that he had learned on the Western Front to date and included doctrine for every asset under IX Corps command, including infantry, artillery techniques, engineers, aircraft, aerial photography and intelligence, communications and logistics. The only omission was for the use of armour, and this was because the ground was too soft for tanks. It was a remarkably clear document for the era, setting out exactly what was expected of the battalions, brigades and divisions within IX Corps in sixty pages of instructions, forty pages of appendices and twelve of maps.

Before the attacks went in at 5.40 a.m. on Thursday 20 September, IX Corps units – predominantly 19th (Western) and 37th Divisions – had tested, perfected and rehearsed their techniques on the Berthen Training Area (about three miles north of Méteren, where Monty was wounded in 1914). Orders went out to the six infantry brigades within IX Corps on 7 September that each infantry brigade was to 'stage a full rehearsal of the attack in every detail . . . The services of the GSO2 Training, IX Corps are placed at the disposal of GOC 19th Division for the purpose of assisting in the preparation of the ground.'[5] Such meticulous attention to preparation was a Plumer hallmark and it is no surprise that it became a Montgomery one also; this instruction was as close as Monty would get to having his name personally on corps-level orders.

IX Corps' 19th (Western) Division attacked first with two of its three brigades in what became the battle of the Menin Ridge. The fighting here lasted for an exhausting five days until 25 September. On the 26th, Plumer delivered the second of his assaults, on a hot, dry day (not the usual image of mud-coated Passchendaele), supported by Fifth Army units, amongst them Monty's old 33rd Division, who spent 26 September to 3 October battling for possession of the high ground of Polygon Wood. For these two assaults, Plumer upped the number of guns to 3,125 and for each attack fired 3.5 million rounds in preparation and another 3.5 million in the opening assault. Despite the negative images and associations that have been transmitted down to us today of Passchendaele, for Monty the September attacks were a success. On 9 October he wrote home to his mother, even eschewing censorship to name his battlefields:

> Things are going very well here and we are entirely successful every time we attack. We are of course learning many very valuable lessons and there are few, if any, mistakes. The attacks delivered on 20th Sept (Polygon Wood), 26th Sept (Menin Road), 4th Oct (Broodseinde), and today, are really masterpieces and could not have been done better.[6]

The weather broke again on the afternoon of Thursday 4 October, and at the 5th of October 1917 conference, when Haig determined to carry on, most historians today agree that he should have taken Gough and Plumer's collective advice offered that grey Friday, to bring the campaign to a close. Both Hamilton-Gordon and Plumer particularly understood that their field guns' platforms would become quickly unstable in the poor weather, thus affecting accuracy. However, as Plumer (mostly) and Gough seemed to have discovered the correct formula for victory in these earlier assaults, it is possible to see why the arguments to carry on prevailed. Second and Fifth Armies renewed the assault, but IX Corps was not involved in any more offensives. Monty would have kept an eye on the progress of his old 104th Brigade as they put in a major attack with the rest of 35th Division on 22 October (most of their Bantam-sized troops had now been replaced with average-height conscripts and returning convalescents). Third Ypres thus carried on for a further month yielding very little reward for greater effort.

It was mostly the weather that brought the campaign to a close once the village of Passchendaele (tellingly, one of Gough's 31 July objectives) had been seized in early November. Mud is the chief factor blamed for the failure of the campaign, for it was a failure in operational terms, even if in strategic terms it degraded the German army beyond recovery. Mud at Ypres was not quite the show-stopper it was traditionally thought to have been; Royal Engineer after-action reports written at the time concluded that success or failure at Ypres rested on the ability of engineers to repair roads or build new ones, and in 1917 too few engineers and equipment were allocated to the task. These kinds of invaluable lessons reverberated around the corps and army headquarters. The IX Corps war diary is full of reports of meetings and conferences, which Monty attended, picking over the impact of Third Ypres. Important conclusions such as these were, in any case, soon forgotten in the imme-diate post-war blame culture, led by Lloyd-George, when the full impact of the casualty bill hit home. It is, indeed, difficult to justify any battle that cost Britain and her Empire around 300,000 dead. But modern historians now accept that of the eighty-eight German divisions on the Western Front in 1917, over half were drawn into the Third Ypres battle and thoroughly chewed up. Germans killed amounted to around 260,000 and thus the revisionist view is that had there been no 1917 Ypres campaign, the eventual autumn victory of 1918 would have been most unlikely, with an even higher casualty bill to follow.[7]

On 24 July 1927, Field Marshal Plumer of Messines returned to Ypres to unveil the Menin Gate memorial to the 'missing' of the Ypres battles (54,896 names are now chiselled into its Portland stone) and lay the cornerstone of the new St George's church, built for the town's sizeable British community. Across the town square a plaque on the impressive cloth hall – an exact interwar copy of the building destroyed by First World War fighting and funded by German war reparations – reminds us that Ypres was liberated in the Second World War, on 6 September 1944, by the 1st Polish Armoured Division fighting in Field Marshal Bernard Montgomery's 21st Army Group.

Remnants of mine craters pockmark the landscape still. At Spanbroekmolen the largest surviving crater of 1917 has been preserved as the Pool of Peace. Some mines, however, remain unexploded and lost. In the 1950s a lightning strike on a tree growing over the site of a lost mine triggered an enormous explosion, resulting in dead cows and a vast unwanted crater; another remains undiscovered. Up on the ridge, the observant might spot lurking in the fields the crumbling concrete of British and German bunkers, overlooked by the rebuilt Church of St Nicholas in Messines, where in November 1914 the headquarters of Hitler's Infantry Regiment List was based. As a regimental signals runner, Corporal Hitler frequently sheltered here in the crypt of the ruined church; later he painted a watercolour of its ruins. Monty walked the ridge in June 1917 and returned in April 1927 (as did Hitler in 1940). On both occasions Monty observed, but did not visit, the remains of Ploegsteert village, once adorned with Bruce Bairnsfather's 'Old Bill' cartoons. On both occasions he would have been unaware of the identity, or the significance, of the officer commanding the 6/Royal Scots Fusiliers, whose headquarters had been nearby: one Lieutenant Colonel Winston Churchill.

If there had been any will to carry on for much longer at Ypres, despite the weather, then it was removed by the collapse of the Italians at Caporetto in October 1917 and the decision to send five British divisions (which the Western Front could ill-afford to lose) led by Herbert Plumer to that theatre. Amongst the seven divisions of General Otto von Below's German Fourteenth Army spearheading the Austro-Hungarian attack was a mountain battalion containing Erwin Rommel.

After a few weeks leave with Lucie on the Baltic (in reality their postponed honeymoon) Rommel rejoined the WGB in Carinthia in preparation for the Fourteenth Army's strike at Italy. Here the WGB was brought up to strength after the losses of Mount Cosna, given a permanent battalion of mountain artillery and trained in readiness for a harsh mountain campaign alongside the smart Bavarian Life Guard Regiment and Jäger battalions.

Italy had declared war on the Central Powers in May 1915. For two and a half years it had been attacking Austrian territory along its northern borders with the intention of reaching Trieste, though without much success. With stalemate on the Western Front in late 1917, previous pleas to open up a new Austro-German front in Italy had found some willing listeners in the German high command. When the requests were renewed in September, Generalleutnant Konrad Krafft von Dellmensingen ('Krafft' hereon), the Bavarian founder of the *Alpenkorps*, was despatched to assess the military options for Germany. He concluded that German troops could meet the challenge, that the Italians were disorganised and had few depth positions and, crucially, that something needed to be done to shore up the Austrian army.

It was in reaction to Krafft's advice that the German Fourteenth Army was established under von Below, with Krafft as his Chief of Staff. Generals Below and Krafft sought to assist their attack by using surprise (assembly of troops and guns in hours of darkness only, no pre-assault bombardment until the final minutes and denial of airspace to the Italian air force) and deception (overt displays by the *Alpenkorps* in other sectors of the front and decoy radio broadcasts).

Their opponents, the Italian army, were already suffering from low morale, which reflected the war weariness of the nation as a whole. The recent eleventh battle of the Isonzo had cost over 330,000 casualties for little gain; domestic food rationing was widespread and unpopular, and there was much anti-war feeling; Corporal Benito Mussolini, who had been discharged from the army in August 1917 after suffering shrapnel wounds, would soon seek to capitalise on this disillusionment with the war. Soldiers were executed at random for disciplinary offences, and others – including officers – disciplined or transferred: 170,000 sentences were imposed by Italian courts martial, resulting in 4,028 death sentences during the course of the war. Many soldiers deserted to their homes and some to the Austrians. The net result was an army that was poorly motivated, led by an officer class that was distant and had little notion, or care, of how to motivate peasant soldiers just wanting to get back to their families. However the army was reasonably well equipped, and supplied with good quality British (artillery and ammunition), French (helmets, uniforms, rifles) and later American, military surplus. Although the Italians had built three defensive lines through their northern foothills, it was the first – and weakest – that was most strongly garrisoned and there was confusion amongst some Italian subordinate commanders as to whether they should adopt an offensive or defensive posture.

The approach march before the assault, Rommel noted, was exhausting – sixty-three miles at night in the pouring rain, over five nights. Sprösser had been deliberately assigned some of the most challenging terrain facing the Fourteenth Army, dominated by Hill 1114, the first peak of a row forming the daunting Kolovrat Ridge, and ultimately by the highest peak, and prize, Mount Matajur. Wednesday 24 October 1917 dawned with a mixture of mist and rain, perfect cover for the advance of Rommel and his mountain troops.

> A few shells struck on both sides of the long column of files without doing any damage. The column halted close to the front line. We were frozen and soaked to the skin and everyone hoped the jump-off would not be delayed. But the minutes passed slowly.[8]

The Italians had been drenched with high explosive and gas from 2,000 guns for five-and-a-half hours before Rommel set off, leading six companies, at 8 a.m.. Toiling up the hillsides silently, his men labouring under the weight of their machine guns and ammunition, he sneaked through the Italian first lines, following camouflaged paths the defenders used for resupply and surprising troops sheltering from the elements in their dugouts. Rommel, as ever very much his own boss, had been given an overarching mission that allowed him, the man on the spot, a great deal of lassitude to: 'without limiting the day's activities in space or time, continue the advance to the west, knowing that we have strong reserves with and behind us'.[9] For such orders to have been issued in any other theatre of the First World War would have been highly unusual. Normally commanders of this era were inclined to be far more proscriptive; the British on the Somme in 1916 were reminded how many hand grenades and field dressings to carry in which pocket in divisional, brigade and battalion orders with no reference to exploiting any successes. Such wording, partly influenced by the impossibility of good communications over difficult terrain, has a curiously modern feel – a form of delegated authority or 'mission command' – the closest translation of the more precise German *Auftragstaktik* (literally 'order, or mission, tactics'). This was initiative at grass-roots level.

Rommel's WGB unit went on to capture seventeen field guns, and, more importantly, a hot meal, prepared for some Italian officers who were captured before they could eat it. By nightfall they had met up with a flanking Bavarian Life Guards battalion. Its commander, Major Graf von Bothmer, being senior, tried to commandeer the WGB units for his own assault on the Italian third lines, but Rommel demurred and was backed by Sprösser when the latter arrived. Meanwhile Rommel had come up with his own plan to capture the next feature along the Kolovrat Ridge line. Before dawn on Thursday 25 October, with Sprösser's blessing, he set off, deploying two companies and machine-gun support, the rattle of Italian small arms soon giving him the satisfaction of knowing his Bavarian rivals had run into stiff opposition. Shortly afterwards he breached other barbed-wire defences and slipped through a few more off-guard positions (some were asleep, others washing). Rommel was trading on a potent mixture of reputation and surprise:

> One mountain soldier was enough to supervise the evacuation, disarming and falling in of a hostile dugout garrison. In the sentry posts the sentries were still watching the valley where they had a beautiful early morning picture of the Isonzo and its twin 6,500-foot peaks gleaming in the sunlight. The sudden apparition of a mountain soldier behind a sentry was enough to paralyse them with fear.[10]

Rommel's luck could not last indefinitely and his 2nd Company soon encountered some well-sited opposition on the lower slopes of the Kolovrat Ridge, which outnumbered and threatened to overwhelm them.

> My estimate of the situation was that the 2nd Company could be relieved only by a surprise attack by the entire detachment, this attack to be against the enemy flank and rear. Under such conditions I believed that the superior combat capabilities of the mountain soldier would prevail.[11]

As predicted, Rommel managed to creep up and attack that battalion-sized position in the flank, whereupon his pinned-down point company launched itself at their attackers. Caught off guard, the Italian battalion of eight officers and 500 men surrendered, thus yielding the peak known as Kuk to him. Italian officers who demurred and drew pistols were overwhelmed by their men.

By about 9.15 a.m., and with over 1,500 prisoners already in the bag, Rommel embarked on one of the boldest strokes of his career. He led those of his detachment he could muster downhill at a trot, taking a hidden path he had spied that Italian military engineers had blasted into the hillside. Steaming into the mountain village of Ravna he then headed for Luico, a logistics hub for that sector, astride a lateral road running *behind* the Italian lines and down to the key town of Caporetto, six miles north. In Luico more troops were completely taken unawares and captured, along with vehicles and many supplies.

> As we hurried by, we snatched eggs and grapes from the baskets of the captured pack animals. We moved on at the double! . . . I chose the way through clumps of bushes and small woods, for we had to move down-hill unobserved . . . [The] apparition of the leading soldiers . . . who suddenly rose from the bushes a hundred yards east of the road, petri-fied a scared group of Italian soldiers . . . they were totally unprepared to encounter an enemy two miles behind the front . . . Soon we had more than a hundred prisoners and fifty vehicles. Business was booming. The contents of the various vehicles offered us starved warriors unex-pected delicacies . . . Soon all efforts and battles of the past hours were forgotten. Morale two miles behind the enemy front was wonderful![12]

With some of his command left uphill, Rommel was busy building a roadblock to halt the arrival of any reinforcements when an endless column of marching soldiers hove into view, easily outnumbering Rommel's small gang (about 150 at this stage). Audaciously, Rommel initially attempted to persuade them to surrender, but the gamble failed; however a quick firefight in which his well-sited machine guns domin-ated the encounter ensured that the better part of the IV Bersaglieri Brigade laid down their weapons – 2,000 men and fifty officers, most of whom had not had the chance to unsling their weapons from their shoulders. Rommel conferred with Sprösser, who had caught up with him in Luico, regained the heights and unified his command once more, whilst making plans for the morrow, 26 October.

By moonlight, Rommel had occupied Jevscek, a small settlement west of Luico, where the friendly Slovene inhabitants fed his men coffee and dried fruit. The following morning he surprised three Italian companies at Jevscek and conducted a frontal assault against Italian positions on Mount Cragonza – Hill 1096 – which overlooked him.* There was no

---

* Patient detective work in mountain boots by John and Eileen Wilkes has revealed that Rommel actually attacked the Italian-occupied Hill 1096 (which he incorrectly identified as Mount Cragonza in his narrative). The heights in metres of the peaks and villages give a hint as to how much he was climbing – and descending – to reach each objective.

guile to this attack on Hill 1096, for there was no cover, flank to use or element of surprise to be had. It was just a question of the fighting spirit of Rommel's men, and tempo. By 7.15 a.m. they were in possession of the peak. Regrouping, his urged his men on to the next height, Mount Mrzli (1,348 metres). Having roughed the defenders up with artillery fire, summoned by his signal flares, and after some minor skirmishes, Rommel approached the defenders slowly, on foot and alone, bearing a fluttering white handkerchief.

> With the feeling of being forced to act before the adversary decided to do something, I left the edge of the forest and, walking steadily forward, demanded, by calling and waving my handkerchief, that the enemy surrender and lay down his weapons. The mass of men stared at me and did not move. I was about a hundred yards from the edge of the woods, and a retreat under enemy fire was impossible. I had the impression that I must not stand still or we were lost.[13]

An evocative final sentence if ever there was one, and one that would sum up his entire career – and life: '*I had the impression that I must not stand still or we were lost . . .*'

His opponents, the 89th Regiment of the Salerno Brigade, suddenly caved in and surrendered in droves, hoisting him on their shoulders, shouting '*Viva Germania!*'; one Italian officer who resisted was shot by his own men. Concerned about the growing number of prisoners, which he had no means of policing or securing, Sprösser issued an order to withdraw. From the acreage of prisoners he encountered, Sprösser not unreasonably assumed the battle was over and the prize of the dominating height, Mount Matajur, was safe in Rommel's hands. Nothing could have been further from the truth and one can imagine that the air turned blue with Rommel's oaths on receiving these instructions, with no means of correcting them, the sunlight glinting on Mount Matajur in the near distance.

> The battalion order to withdraw resulted in all units of the Rommel detachment marching back to Mount Cragonza [actually Hill 1096], except for the hundred riflemen and six heavy machine-gun crews who remained with me. I debated breaking off the engagement and returning . . . No! The battalion order was given without knowledge of the situation on the south slopes of the Matajur. Unfinished business remained

> ... the terrain favoured the plan of attack and – every Württemberg
> trooper was in my opinion the equal of twenty Italians. We ventured to
> attack in our ridiculously small numbers.[14]

Most of Rommel's *Abteilung* were directed back, in strict obedience,
whilst Rommel craftily kept back a few men in the immediate vicinity:
these he deployed for an attack on the beckoning mountain, and opened
fire. Fragmentation of the rocks from ricochets magnified the effect and
some Italians were seen to retire. On an off-chance, he deployed the
white handkerchief again and was astonished to see the 90th Regiment
of the Salerno Brigade (whose colleagues had earlier surrendered to him
on Mount Mrzli) lay down their arms.

> Deeply moved, the regimental commander sat at the roadside,
> surrounded by his officers and wept with rage and shame over the
> insubordination of the soldiers of his once proud Regiment. Quickly,
> before the Italians saw my small numbers, I separated the 35 officers
> from the 1,200 men ... the captured colonel fumed with rage when he
> saw that we were only a handful of German soldiers.[15]

Racing up to the summit, 200 metres higher, a few more bursts of fire
persuaded the remaining 120 defenders to throw in the towel, and just
before midday on 26 October, Rommel was able to fire one white and
three green signal flares to show that the 1,642-metre twin peaks of
Mount Matajur – the pre-war Austro-Italian border – had fallen to him.
He had captured 81 guns, 150 officers and 9,000 men (almost five
complete regiments) for the loss of six killed and thirty wounded in
fifty-two hours of combat. Most of this achievement had been due to
his personal efforts. Once news arrived of the fall of Mount Matajur, a
defeat of operational proportions, the Italian high command was forced
to relent and recognise the inevitable: the rest of the day was taken
up with its commanders issuing orders for the previously unthinkable
retreat.

Alas for Rommel, the prize promised by General von Below – an
award of the coveted *Pour le Mérite* to the first of his men to stand on
the summit of Mount Matajur – did not come to Rommel himself,
though it was undoubtedly his. Leutnant Walther Schnieber, a company
commander in the 63rd Regiment, reported in 'mistaken good faith' that
he had reached Matajur's summit much earlier, at 7 a.m. and therefore

claimed his reward. Rommel only realised after notification of the award was published on 27 October, and could not be retracted. Worse was to follow, for a second medal was awarded to the soldier who first gained Hill 1114; Rommel had been key in opening up the defences to this feature by taking the Kolovrat Ridge; this second decoration, he felt sure, was his. Again, the air turned blue when he discovered it had gone to the Bavarian Life Guard Leutnant Ferdinand Schörner, who had indeed stood on the summit, albeit thanks to Rommel's exertions. Rommel was outraged.

What happened next gives us as much an insight into Rommel's character as his storming over the mountain passes. As far as Rommel was concerned, the decoration, founded in 1667, which was the personal gift of the Kaiser, and had been hitherto mostly awarded to great generals and commanders, and the odd air ace, was his. So he protested, he lobbied, he sent reports up through the command chain via Theodor Sprösser, bombarded the *Alpenkorps* commander, Generalmajor Ludwig Ritter von Tutschek, protested through the corps commander, Generalleutnant Hermann Freiherr von Stein, to the Chief of Staff, Generalleutnant Konrad Krafft von Dellmensingen to General von Below himself. This was unheard of – but it was Rommel all over. Another emerging trait: he would not meekly submit to higher authority if he thought it was wrong. According to one account, General von Below was so impressed by the evidence that he requested another decoration for Rommel from Generalleutnant Moriz Freiherr von Lyncker, personal military secretary to the Kaiser, but this was refused.

This is all part of the Rommel legend, and we may want it to be true, but have to remember the only accounts that have surfaced to authenticate the episode are those of Erwin himself and Theodor Sprösser. It didn't stop there: for several years after the war Rommel mounted a 'determined (and, eventually, successful) campaign . . . to persuade the official historian to amend the record by crediting him with the lion's share of the praise'.[16] Rommel even made sure that in the official history Schörner's rank was changed from *Oberleutnant* to the correct *Leutnant*.[17] Perhaps, too, the dice were loaded, for the only non-aristocrats in the 1917 command chain (without a 'von' or a title like 'Freiherr' or 'Ritter') were Rommel and Sprösser. It seems likely that Schörner and Schnieber won their decorations because, as Dennis Showalter shrewdly observes, both 'officers belonged to established organisations, able to make ongoing cases for their candidates. The WGB was a wartime creation, having nothing but its deeds to speak for it to senior officers at higher headquarters.'[18] Desmond Young, Rommel's later biographer, observed of his own experiences of 'gongs' in the Great War, 'everything depended on ... recommendations being written up in a flamboyant and often fanciful style which does not come naturally to the good frontline commander'.[19]

What was remarkable about Rommel's capture of Mount Matajur was his ability to quickly assess a tactical situation and exploit it: he could sense when military units might be about to crack, or were determined to resist hard. 'He showed shining courage,' one of his fellow officers, Kurt Hesse, reflected; 'despite the greatest strains, he possessed apparently inexhaustible strength and freshness, an ability to put himself in his opponent's mind and anticipate his reactions. His planning was often surprising, intuitive, spontaneous and not immediately transparent ... Danger did not seem to exist for him.'[20] Another emerging facet: the military psychologist. Rommel started to look for behaviours in his opponents which would dictate, in turn, his next move. The speed with which he logged this information, decided and acted upon it was extraordinary. In the 1950s, US Air Force Colonel John Boyd dubbed this process 'the decision-making cycle', which he had based on observations and analysis of air combat encounters between MiGs and US pilots over Korea. The 'Boyd Loop' recognises that time is the dominant characteristic in outsmarting any opponent, by making appropriate decisions quicker than an adversary, a concept that favours mental agility over brawn. The speed with which the process can be undertaken often marks out the better commander, or command team. Understood by leaders around the world today, Rommel was displaying aspects of the Boyd Loop long before the process was formally devised.

With their defence lines swept away, the vast Italian Second Army, covering the sector where the break-in occurred, started to collapse. To a certain extent the Italians had made a rod for their own backs with their Second Army: it was unmanageably vast, containing twenty-five divisions in nine corps, but no attempt had been made to streamline it into a smaller, more efficient organisation. It was really too cumbersome to administer in the normal course of events, and its commander, Generale Luigi Capello, was further handicapped by a kidney complaint requiring long spells away in clinics. Inexplicably, he had refused to give up command of his forces and he was personally at odds with his commander, Generale Luigi Cadorna. The net result was disaster: the Austro-German armies pierced the initial Italian defence lines and started advancing beyond the outer crust of the Julian Alps, crossed first the Isonzo, then made for the riverine plains of north-east Italy.

On 27 October, the German penetration bypassed Cividale and had reached the Torre river, but the weather was changing again. Further to the north, mountain troops were battling with fresh snowfalls of up to a foot deep; all troops were soaked and the river waters were rising rapidly, but the advance continued, spurred on by its own success. On the 28th the Torre was crossed, Udine had fallen and the WGB was approaching the Tagliamento river, over thirty miles beyond their start line. Here the Italians had managed to destroy most of the crossings behind them, and as the attackers drew up to this new obstacle, they paused. Originally, von Below's attack was envisaged as the spearhead of a limited operation to regain the initiative on the Isonzo front. Once it seemed to be delivering beyond the planners' wildest dreams, the Austrian high command cautiously authorised an advance as far as the Tagliamento, but none of the necessary logistics were in place beforehand to sustain the new objective.

On the 30th, the goalposts moved again as an offensive across the Tagliamento was authorised, with a view to capturing Venice, although the appropriate bridging equipment was far behind in the logistics chain. By the 31st all the Austro-German attacking forces had drawn up to the line of the Tagliamento, in order from north to south, Tenth (Austrian), von Below's Fourteenth (German), Second (Austrian) and First (Austrian) Armies: opposing them were Capello's Second and the Duke of Aosta's Third Italian Armies. The Italian defenders had not managed to adequately destroy all the Tagliamento's bridges and on 2 November some were seized and in other places a lull in the weather allowed the freezing river to be forded. Cadorna had no choice but to order a withdrawal back to the Piave, which all units had crossed by 9 November.

Rommel had to bury his annoyance at missing out on the *Pour le Mérite*; there was work to be done. By 30 October, commandeering abandoned vehicles, horses and cycles, he led his detachment as far as the swollen Tagliamento. Even before reaching the main river, in fording one of the tributaries, Rommel recounted:

An Italian prisoner carrying a large medical kit on his back was torn from the rope by the strong current and, lying on his back, floated downstream. The man could not swim. Besides, the heavy knapsack dragged him down. I felt sorry for the poor fellow. Spurring my horse, I galloped after the Italian, and succeeded in getting near him in the stream. In his deadly fear the Italian seized the stirrup and the good horse brought us both safely to land.[21]

They soon crossed the main river, but whilst there, the whole of the WGB was ordered to make for Longarone on the Piave with all haste, and in order to facilitate as speedy an advance as possible, were transferred to the command of 22nd Austrian Division. Along the route, on 7–8 November, they clashed continually with Italian rearguards fighting hard from prepared positions. The fight beyond Cimolais was particularly tough, the Italians using the rocky terrain very much to their advantage; Rommel had to rely on the clever siting of his heavy machine guns to break up the Italian defence. With the main resistance broken, he sent cyclists and horsemen to overtake the retreating Italians, and hurtled down through tunnels to bridges beyond that he desperately needed to seize before they were demolished. Italian engineers blew one, but incompletely, and Rommel and his men raced across a second, pulling out fuses as they went. Emerging from a last tunnel into the dazzling sunlight – the weather had picked up again – at about 11 a.m. on 9 November 1917, Rommel was presented with

a beautiful sight. The Piave valley lay before us in the brilliant light of the midday sun. Five hundred feet below us, the bright green mountain stream rushed over its . . . stony bed. On the far side was Longarone, a long and narrow town; behind it 6,000-foot crags soared up to the heavens . . . An endlessly long hostile column of all arms was marching on the main valley road on the west bank . . . Longarone [was] jammed with troops and stalled columns.[22]

At Longarone, the Piave flows south, rushing through a narrow gorge north of the town, and then broadens to a network of shallows, over 500 metres wide, just south of the settlement. It is a small route centre and consequently in 1917 attracted many retreating soldiers, all of whom were blissfully unaware of Rommel's presence. His detachment had now covered about eighty miles as the crow flies in fifteen days – perhaps triple that distance in weaving, twisting, climbing and descending, fighting all the way.

Immediately Rommel engaged the Italians he could see with the ten riflemen at his disposal, but as more of his own troops caught up, they watched helplessly as the bridge spanning the Piave gorge was blown before midday. Meanwhile an enterprising platoon of eighteen men, led by one of Rommel's junior officers, had managed to ford and swim the Piave downriver, where it was much wider and shallower, and reached the west bank at Fae, where, despite their small numbers, they had established a road block, cut telephone wires and ambushed and took prisoner several groups of Italians heading south out of Longarone. Back on the east bank, Sprösser arrived at around midday and set up his HQ high on the slopes, overlooking the scene. As darkness fell, Rommel then forded the icy Piave, joining his gallant band at Fae with two companies, whilst the haul of compliant prisoners – now amounting to an incredible 780 men – splashed across into captivity, prodded by the odd bayonet. Fifty Italian officers, kept separate from their men, were locked up in Fae castle.

At about 9 p.m., reluctant to wait and determined to maintain the offensive spirit, Rommel led half his force up the road, north to Longarone, scouting ahead with the vanguard. He got to within 100 metres of the town before being challenged and fired upon by an Italian roadblock; he was also machine-gunned by his own troops from across the river, who were unaware of his activities, despite messages back.

Bullets scythed into Rommel's group from two directions and felled his company commander. Everyone else took cover and the very dark night aided their withdrawal. Rommel then ran back down the road, pursued by what sounded like a vast mob of Italians. At Pirago, 1,000 metres south of Longarone, he collected enough men to form a roadblock and faced the Italians, who kept advancing. Rommel was uncertain of their intentions, but at ten metres they opened fire and stormed his position, killing or capturing most of the Württembergers. Rommel was one of the few to avoid this fate (though several – misinformed – accounts have it that he was initially taken prisoner, but escaped), hiding behind a roadside wall. He then took off at high speed, cross-country, hell bent on warning his remaining troops in Fae of the Italian attack about to descend on them. He reached Fae just in time to organise a defence from the castle, with the remnants of two companies and a few machine guns. Images of a P. C. Wren-type French Foreign Legion detachment, outnumbered and besieged in a remote desert fort, spring to mind as Rommel mounted a defence, and was obliged to set fire to nearby buildings to illuminate the scene. He did, however, have plentiful supplies of Italian weapons and ammunition impounded from the prisoners taken that afternoon and stored in the castle courtyard, and turned them against their former owners.

From about midnight, Sprösser, whom Rommel had kept in the picture by having his runner swim the Piave, started sending reinforcements across the river. The Italians grew increasingly desperate, as Rommel blocked their only route to safety. Aided by artillery, they kept up the pressure, launching fresh attacks at about three in the morning. Rommel was now more confident of success, as the Italian pressure slackened and reinforcements trickled in. He awaited daylight with some captured bottles of Chianti. Meanwhile a howitzer battery of eight guns and two Austrian infantry battalions arrived at Sprösser's command post, and one of these units he sent over to join Rommel in Fae. Sprösser also sent a message via a prisoner to the Italian commander in Longarone, claiming that the town was surrounded by an entire Austro-German division and he should capitulate to such superior numbers.

However, early on the morning of Saturday 10 November, Rommel appeared to know nothing of this, and advanced with a very small detachment cautiously up the road he had used before, picking over all the signs of the previous night's encounter – spent shells, abandoned weapons and the recumbent corpses of his own men and Italian soldiers. As he did so, another crowd of Italians appeared; Rommel drew back and at first couldn't make out the figure advancing at their head, but it turned out to be one of his own missing officers – Leutnant Schöffel, captured during the night – leading a vast column of surrendering Italians who had responded to Sprösser's surrender summons. Most of Rommel's missing men, and 10,000 troops of the 1st Italian Division inside Longarone (with 200 machine guns, three mountain batteries, 600 pack animals, 250 loaded vehicles, ten trucks and two ambulances) surrendered to Rommel's small detachment. Rommel's losses were an astonishing six killed, twenty-one wounded and one missing.

Even the Italian official history accepted the figure of 10,000 captured at Longarone, going so far as to quote an official Austrian account. The Austrians, however, ascribed the victory to the activities of their own 22nd and 94th Divisions rounding up the Italians, as on a pheasant drive – Longarone was merely where they were trapped and forced to surrender – and failed to mention the role of the WGB at all. The Austrians had a point, but the main determining factor was the pace set by Rommel and Sprösser. The speed with which they arrived at Longarone and then surprised the garrison, showing up in an unexpected quarter, no doubt induced many more Italians to surrender than might have been the case in a normal set-piece battle. Had the advance been slower, without the WGB's presence, certainly some Italians would have been caught, but many more would have escaped. It was the high point of their war.

On 11 November, the day after the triumph at Longarone, the WGB received a personal message from von Below's Chief of Staff, General Krafft, the first commander of the *Alpenkorps*, which read: 'The Battalion has achieved a brilliant performance of the first order, which as an old mountaineer I well appreciate.' Neighbouring German formations also sent in congratulatory messages, but importantly, so too did the Württemberg War Ministry and King Wilhelm II of Württemberg. His Majesty would no doubt have been heartened by the performance of his mountain battalion, as against those of his rival, King Ludwig III of the large old kingdom of Bavaria (two of whose junior officers, Schnieber and Schörner, had already picked up a *Pour le Mérite* apiece in the campaign).

Today we would term Sprösser's ploy of lying about his strength and challenging the Italians in Longarone to surrender 'psychological warfare', but even in 1917, the more scholarly of officers might have recognised that this was very much in keeping with one of the major lessons taught in Sun Tzu's *Art of War*, already widely translated into German, French and English, that 'he who defeats his enemy without fighting demonstrates the epitome of skill'. It was also fortunate that Sprösser deployed this psychological device, his surrender summons, because at Longarone Rommel had overstretched himself. He was at the end of a long logistics tail, both operationally and tactically, and was on the wrong side of the Piave from the bulk of his colleagues. He had attacked the Italians in Longarone prematurely when he should have waited overnight for reinforcements. Had Sprösser not issued his surrender summons when he did or sent reinforcements the following morning, his talented subordinate would, indeed, have been overrun. At Longarone Rommel was lucky, not clever. Nevertheless, on other occasions, he tried the same ploy and got away with it. The truth of Longarone might have harmed his reputation, but perhaps it was already getting too strong to damage.

On 24 October 1917 the Italian army in north-eastern Italy comprised sixty-nine divisions (685,000 men); sixteen days later it dropped to a recorded strength of just thirty-three fighting divisions and had lost over half its artillery (3,000 guns). The Italian retreat was hampered by nearly half a million refugees, who now took flight, carrying whatever they could. Because of the snow and rain only movement by road was possible; thus the few lateral east–west routes became clogged with over a million people – soldiers and civilians – on the move. Law and order broke down, and command appeared to depend now on trust rather than rank. Some notable Britons were caught up in the catastrophe, including the distinguished Cambridge Whig historian and Italophile, George Macaulay Trevelyan, in charge of a British Red Cross detachment working on the Italian front, whilst nearby, in a British Royal Garrison Artillery siege battery, was thirty-year-old Lieutenant Hugh Dalton, the future Labour politician and Chancellor of the Exchequer. Dalton, who several times recorded German aircraft attacking civilian traffic on the roads, was awarded an Italian military decoration 'for contempt of danger' during the retreat. In 1919, Dalton (who was Monty's exact contemporary in age) recorded his experiences as *With British Guns in Italy*, the same year as Trevelyan published his *Scenes From Italy's War*. This was also the setting for Ernest Hemingway's 1929 novel *A Farewell To Arms*, based on his experiences as a Red Cross driver in Italy.

The Italian collapse was on a strategic scale and one which had not been anticipated by anyone. General von Below had frankly not expected such a breakthrough in two weeks and his troops began to outrun their transport. In the nature of horse-trading between coalition partners, Germany had contributed combat troops to the Caporetto offensive, looking to Austria to supply the logistics. The 70-mile advance had caught both by surprise and the latter's logistical arrangements – the creaky bureaucracy of the old-fashioned dual monarchy with everything in three languages and arranged in triplicate – simply wasn't up to the job. Whilst Rommel and the WGB were more used than most to living off the land, scrounging abandoned transport and turning captured weapons on their former owners, the Fourteenth Army as a whole was less proficient at sustaining a long advance: there were simply not the resources available to pour into northern Italy and exploit this success. Having to bridge the fast-flowing rivers in mid-winter did not help and hitting the hills beyond the Piave proved the final buffer. Clausewitzian friction had set in. The advance had run out of steam and was called off in late November.

Rommel buried his fallen comrades at Longarone on 11 November and, four days later, spent his twenty-sixth birthday on the march with the WGB, heading for more battles in the mountains above the Piave, where they stayed until being relieved on 1 January 1918. On 18 December they had withdrawn to rest in a small valley, where mail awaited them. Amongst it were two small packages for Oberleutnant Rommel and Major Sprösser: both had been awarded the ultimate accolade for a soldier of the Kaiserreich, the *Pour le Mérite*. News of their efforts at Longarone (and almost certainly Rommel's earlier complaints over the lack of recognition at Mount Matajur) had travelled all the way up the command chain, to Berlin. On 13 December it was announced that both officers would receive Germany's highest award; as Rommel later observed: 'two awards was a hitherto unheard-of honour for one battalion'. The act of posting the awards to the recipients may well have been an acknowledgement that neither Sprösser nor Rommel was expected to survive and that, rather than wait for an official investiture by the Kaiser, both should enjoy possession of the medal whilst still alive.

The awards were not only recognition of the extraordinary soldier Rommel had become but a real tribute to his insightful commander. Under Sprösser's leadership the WGB had achieved great things in France, Romania and Italy, causing tactical ripples in the campaign pond far beyond that expected from a force of 2,000 men. They were expertly trained, superbly motivated and outstandingly well led. Sprösser may have been lucky in his subordinates, but as their founder, he was the ultimate source of their leadership, direction and management, and forged the WGB into the outstanding organisation it became. Photographs show him during the Italian campaign, short, stocky, with a goatee beard and bristling moustache, stomping around the hillsides in his greatcoat in foul weather, a man twenty years Rommel's senior. Rommel constantly pushed himself to his physical limits, but so too did Sprösser, who was never far behind him. Moreover, Sprösser had the good grace to indulge his pushy lieutenant. It is a rare gift for a chief to be able to handle a subordinate whose talents are prodigious, without hindering their development or cramping their style. His inspired conferences in formulating plans were rare for the mindset of the day. Sprösser seems to have played Rommel just right, without putting him down or curbing his ability to think outside the box.

Today, we would rate the WGB as a sort of special forces unit, and Rommel, force-multiplier that he was – by his toughness and tactical excellence – would be regarded as the equivalent of an officer in the British SAS. Theodor Sprösser was the best teacher Rommel could have possibly encountered, and more than anyone moulded Erwin into the outstanding general he became.

In 1977 the historian David Irving wrote of discovering Rommel's *Pour le Mérite* in a dusty cardboard box, 'its enamel slightly chipped where it once struck an asphalt road', alongside his peaked cap and a pair of the signature Perspex goggles, then yellowing with age. I was unaware that *Pour le Mérite* recipients also received a gold pocket watch from the Kaiser; Rommel's came up for auction in October 2005, along with other mementos relating to his time with the WGB. The watch, from Eppner of Berlin, bears the Wilhelmine cipher – an enamelled 'W' beneath the Imperial crown – on the exterior. Inside the engraved inscription reads: '*Dem Oblt. Rommel vom Württemb. Gebirgsbtl. Als kaiserlichen Dank. 10.12.1917*' – with a facsimile of Wilhelm's signature, Wilhelm IR. ('To Oberleutnant Rommel of the Wurttemberg Mountain Battalion as a Gesture of Imperial Thanks, 10 December 1917'). Somewhat beyond my pocket, it made 6,500 Euros.

Sprösser acknowledged the award of the *Pour le Mérite* in his 13 December Order of the Day, and concluded with the humility of a true leader:

> Comrades, the Imperial Kaiser's thanks apply to yourselves, the officers, under-officers and men, whose incomparable energy attained all that was humanly possible. It is an honour to belong to the Battalion. The highest honour is to be its Commander.[23]

*

On Monday, 7 January 1918, Rommel left his beloved WGB proudly wearing at the neck (as he would for the rest of his life), every German soldier's dream, the highly distinctive light blue enamelled cross of the *Pour le Mérite**. After a few days leave with Lucie, he was posted to a staff appointment on the Western Front.

On the face of it, this was a bizarre appointment for an accomplished combat commander. But alternating between combat and staff work assignments was a routine policy within the German army, and even the British. The logistics tail of all armies is always much larger than the combat element at its head. Enabling troops to get, say, from Ulm to

---

* Whereas few First World War winners of the *Pour le Merité* continued to wear their decoration in the years of the Third Reich, Rommel never wore a uniform without it, and when later awarded the Knight's Cross of the Iron's Cross in various grades, commenced wearing *both* at his throat – a unique occurrence, surely underlining the importance to him of events at Longarone in 1917.

Belgium, required knowledge of logistics. This is the world of hand-books and timetables, of understanding how to calculate how many tons of rations for the men, fodder for horses, water for both, fuel for trucks and ammunition would facilitate a major advance along five roads for seven days. It requires all sorts of detailed knowledge of other military arms and services – not just the infantry – how many trucks a transport company possesses; how many beds, surgeons and ambulances in field hospitals should be stood-to in readiness; in what order should the army move and what training should be undertaken in preparation, when and where; where will the wooden crosses for the casualties be stored, and so on.

Staff work requires the ability to write clear orders and memoranda to a set template, and more than a nodding acquaintanceship with railway timetables and the speed with which ports could unload ships. At the very least, staff officers needed to be able to turn the whims and orders of a commander into functional reality. It has been calculated that to send just one of Germany's army corps (of the forty eventually deployed) into war in August 1914 required 1,135 railway carriages for the infantry, 2,960 for cavalry units whilst artillery regiments required 1,915 – some 6,010 in all, grouped into 140 trains – and an equal number again for their supplies; 'from the moment the order was given, everything was to move at fixed times according to a schedule precise down to the number of train axles that would pass over a given bridge within a given time'.[24] This was partly the world that Montgomery had been applying himself to since 1915, blending the ability to command and lead men in combat with the art of what is logistically possible.[25]

Rommel then, against his will, was transferred away from his battalion (which was at time expanding into a larger formation, the Württemberg *Gebirgsjäger* Regiment) to the sedentary, safe and predictable world of a junior staff officer at the headquarters of LXIV (Württemberg) Corps, far behind the lines. It must have been a huge mental leap, in the midst of total war, to go from leading men to glory over the Italian Alps to completing forms and filing in returns. But assignments to the staff also fulfilled another, important, function: it rested talented leaders whose luck might be about to run out and whose mental health might be altering from the stress of continuous combat, and preserved them for the future. Captain Charles McMoran Wilson (writing later, when Lord Moran, Churchill's physician) suggested in his seminal *Anatomy of Courage*, published in 1945, that everyone has reserves of courage on which they can draw, but everyone also has a breaking point, which cannot be anticipated until met. It would have been as much in anticipation of talented officers suddenly meeting their breaking point as to acquire knowledge of staff work, that Rommel was posted away from the front.

The Württemberg Corps headquarters, at Colmar in Alsace, in the relative quiet of the Vosges mountains sector, was far removed from Italy or the main drama of the Western Front. This was a world of order, where Rommel's decoration would have mattered less than his rank and the correct press of his uniform jacket (a consideration hardly on Rommel's radar until that point), and where the private soldiers he loved to command were outnumbered by officers, who, in turn, were frequently petty and obstructive and, in private, jealous of the bauble around Erwin's neck.

The Württemberg Corps had also sent one of its divisions, the 26th, to the October–November 1917 Italian campaign, led by the distinguished Generalleutnant Eberhard von Hofacker, who would later write their official history. He had already won the *Pour le Mérite* in April 1917, and would be one of only 111 to win the very rare supplementary award of *Eichenlaub* (Oakleaves) in November 1917, for his part in the Caporetto campaign. Hofacker was then promoted to command LI Corps, which would be on the receiving end of the British counterattack at Amiens in August 1918, which included 47th London Division, whose Chief of Staff was one Bernard Montgomery. After Amiens the much-decorated Hofacker worked in the Württemberg War Ministry, which is how he and Rommel first met. Eberhard, with his war record in Italy and decorations, became something of a hero to Rommel, and probably his first patron. In due course the future Desert Fox would become friends with his son, Caesar. As Oberstleutnant von Hofacker of the Luftwaffe, Caesar would play a pivotal role in the events that led to 20 July 1944, and Rommel's eventual death.

9

The Last Year

MONTY SAW IN the New Year of 1918 still a captain with IX Corps, which happily now included his old 33rd Division, and remained busy drafting meticulous instructions, mostly for defence. The BEF now manned a huge proportion of the Western Front, whilst the French rebuilt their army after the 1917 mutiny and Pershing's Americans began to trickle to the front. The Allies' problems of defence were multiplied by Britain's shortage of manpower (the last Conscription Act extended the call-up to married men, aged forty-one to fifty), provoking the reduction of all infantry brigades from four battalions to three, the balance of manpower being redistributed to bring units up to strength. This meant that divisions could call on the services of just nine rather than the previous twelve infantry battalions. At a stroke, 141 New Army battalions were disbanded, creating much tedious staff work for Monty and his colleagues playing musical battalions, just when clear heads were required to devise defensive strategies. There was an attempt to offset the manpower shortage by an increase in firepower; lavish quantities of Vickers and Lewis machine guns were distributed to battalions, but many troops were new and poorly trained in their use.

With the Treaty of Brest-Litovsk (3 March 1918) Russia formally exited the war. This enabled the Germans to transfer their entire military effort in the east – nearly a million men, or fifty divisions – to the Western Front. It was obvious that a big German offensive would soon break, but no one knew where or how. Haig and GHQ expected the attack in the north, where a penetration along the coast towards the Channel ports made strategic sense. Accordingly the defences there were strengthened, but remained incomplete.

Collectively known as the *Kaiserschlacht* ('Kaiser's battle') what actually manifested were several, sequential attacks. The main assault fell on Gough's Fifth Army, before Amiens, which had the least-prepared defences, manned by the weakest formations with fewest men: when 66th Division's engineers had arrived at the front on 1 March in Fifth Army's sector, they encountered a 'pleasant golf course, an officers' mess with a garden, but the defences were difficult to find'.

The main blow arrived finally at 4.40 a.m. on Thursday, 21 March 1918 with a hurricane of hot steel and gas falling on the battle zone, whilst the forward zone was overwhelmed by stormtroopers – experts in infiltration – who emerged, wraith-like out of the fog and smoke in small groups, invariably armed with sub-machine guns and stick grenades. Stormtrooping doctrine (with which Rommel had been experimenting) had been perfected in 1917 at Cambrai, as had artillery tactics: advancing soldiers could accelerate or slow a creeping barrage by means of Very lights. The result was a series of concentrated punches to break through the Allied lines once and for all. Gough's Fifth Army, the victims of the ferocious campaign, 'Operation Michael', were pushed back, in some cases up to eighty miles in a few days – an unprecedented movement on a hitherto static front. A breakthrough at Amiens would split the Allies, for the French would fall back to defend Paris and the British, with Haig needing to defend the British line of communications, would retreat to the Channel ports.

So effective was the initial German attack that by 25 March General Hamilton-Gordon and his staff at IX Corps (then in reserve) was tasked to mark out a new defensive line *fifty* miles to the rear. It was at this point that Montgomery was promoted to field rank (long overdue) as a temporary major. On 3 April, IX Corps was ordered to relieve another corps at Ypres and in so doing arrived just before the Germans launched a further assault, this one orientated towards the southern edge of the Ypres sector, in the vicinity of Méteren, and aimed, in fact, right at IX Corps. This second tornado of violence, known to the Allies as the Battle of the Lys (to the Germans it was code-named Georgette), struck in thick fog on 9 April and initially showed the same promise as the initial attack, targeting a weak Portuguese division holding the line at Neuve

Chapelle, then sweeping away all the ground gained in the previous autumn's Passchendaele offensive in just a few days.

Despite all the German assaults and early spectacular gains, inertia soon set in. This was partly because the German break-in forces were élite *sturmtruppen*, trained in infiltration tactics only, who then handed over to Regular infantry once their can-opening job was done. By 1918, the bulk of the German infantry divisions were less well equipped and less mobile than the *sturmtruppen* – and less well fed; stormtroops did, indeed, have special rations. The Imperial Army therefore tended to stop and feed itself when ration dumps were seized (a great many were discovered and looted deep in the British rear) and drink itself silly when the quartermasters' rum rations were captured. The sheer numbers involved, and the distances they had to cover, also contributed to a slowing down of each assault after the initial gains, just as in 1914, allowing the British time to gather their wits, build new defence lines and man them.

The period was a shocking one for the BEF at every level. Gough, commander of Fifth Army, was sacked on 28 March, together with his staff ('there but for the grace of God go I', many staff officers in neighbouring formations must have mused), and was replaced by Sir Henry Rawlinson. In Monty's sector the neighbouring XV Corps commander, Lieutenant General Sir John du Cane, was replaced on 11 April with the more dynamic cavalryman, Beauvoir de Lisle. Hamilton-Gordon's IX Corps was pushed back as far as the dominating Kemmel Heights, where they held on for ten days, while the six divisions it loosely commanded at different times suffered 27,000 casualties. Amongst the German troops fighting for the Kemmel Heights between 10–29 April were the Bavarian units of the *Alpenkorps*, including Oberst Ritter von Epp, Leutnants Friedrich Paulus, Ferdinand Schörner and Rittter von Thoma. Writing after the war to Captain Cyril Falls (the official historian), Monty recalled that his IX Corps commander 'appeared to be in his dotage, in a state of gloom and depression but was not sent home until September 1918 to make room for younger men'.[1]

It remains difficult to follow the IX Corps battle of these days because the reality was that it was fought by a variety of detached brigades and battalions who were drawn in literally from anywhere to block the German thrust; Australians and cyclist troops fought alongside bandsmen, cooks and returning convalescents in composite battalions. Such improvisation would have taxed a busy staff officer like Monty to the extreme; his job would have evolved into one more akin to running a fire brigade, in contrast to the luxury of weeks to prepare for big operations. At one stage, 100th Brigade, of Monty's old 33rd Division, mounted a near-suicidal but 'spirited defence' of Neuve Eglise for several days, but it was one which almost destroyed them as a formation. With the crisis appearing to worsen, on 11 April Douglas Haig issued an Order of the Day (the only one of his anybody remembered), which concluded: 'Every position must be held to the last man; there must be no retirement. With our backs to the wall, and believing in the justice of our cause, each one of us must fight to the end.' This reads more like a stirring and articulate 1940s diktat from Hitler or Goebbels than the product of a man whose critics have always condemned as an appalling communicator.

The Operation Michael attacks (technically there were three of them, Michaels I, II and III) eventually ran out of steam in late April, ending with the German capture of Villers-Bretonneux on the Roman road leading to Amiens, a beckoning spire in the distance. In taking the village, the Germans used captured British armour and some of their own A7-V tanks which happened on some defending British armour. Thus ensued history's first tank-versus-tank engagement which ended in a tactical draw. During the 24–25 April battle, several British and Australian brigades wrestled Villers-Bretonneux back from its German occupiers and the *Kaiserschlacht* ground to a halt. Operationally, the German breakthrough had failed.

Montgomery had first witnessed British tanks at Arras the previous year; on 14 April 1917 he had 'watched two Tanks walking down the Hindenburg Line ... The Germans shelled the Tanks hard for several hours but did not get a direct hit on them and nothing else will knock them out.'[2] Early tanks were often less impressive than Monty's letter implied. Nine Corps did not use them at Passchendaele, but later on in November, Julian Byng's Third Army had attacked Cambrai, throwing nearly 500 armoured vehicles of the Tank Corps at section of the Hindenburg Line in what was termed a 'raid'. It had been spectacularly successful and broke through, but the BEF was so tired after Third Ypres and, with five divisions in Italy, possessed neither the reserves nor the energy to exploit the initial success. An inevitable German counterattack had retaken most of the gains, but Cambrai had proved what tanks could achieve – and would inspire military theorists such as Colonel J. F. C. Fuller (Chief of Staff of the Tank Corps) and practitioners like Captain Heinz Guderian and Sergeant Major Sepp Dietrich. Monty at IX Corps had read and pondered over the post-action reports that were swiftly circulated, though in early 1918 he was more concerned with defensive infantry tactics.

Georgette subsided in late April at a cost of 120,000 casualties to the BEF and around 100,000 to the Germans. At this stage, Montgomery was fortunate to get a short spell of leave in London, whilst the whole of IX Corps was rotated away from Kemmel (the incoming French formation subsequently lost it and the Schwerpenberg beyond to the German *Alpenkorps* on 25 April) and travelled south by train to the quieter sector of the Chemin des Dames and along the Aisne river, where they found themselves under General Auguste Duchêne of the French Sixth Army. By 12 May, IX Corps had completed its move, but it led them straight into the jaws of the dragon: Ludendorff's final attack, code-named Blücher-Yorck, began exactly in this area on Monday 27 May. Armed with a well-argued briefing paper, probably written by Montgomery, Hamilton-Gordon warned his French superior that German tactics invariably drenched forward troops with high explosive and gas before assaulting and that the ideal solution would be to hold the majority of troops back. Duchêne wouldn't listen to the new doctrine (despite being ordered also by his own superior, Pétain, *not* to reinforce the front-line positions) and packed the forward areas of IX Corps' sector. Hamilton-Gordon reluctantly deployed his three available divisions forward.

Monty was back for the attack, which was heralded by a violent bombardment from over 3,700 guns along a thirty-eight-mile front. It penetrated the Allied lines by some twelve miles on the first day, striking at the boundary between British and French formations. Again the Germans achieved tactical success, but operational failure: they ran eventually into a wall of Pershing's Americans at Château-Thierry and Belleau Wood, the first blooding of US forces in the First World War on any scale. Suffering 29,000 casualties in nine days (in addition to the Georgette losses), many of IX Corps' formations had ceased to exist or were too decimated to be tactically viable. Monty's papers do not reveal whether any distrust of the French Army lingered after Duchêne's tactics (Duchêne was sacked on 9 June), but it cannot have helped his attitude towards future Anglo-French relations. More importantly, on 26 March the Allies had taken the long-overdue step of appointing a Supreme Allied Commander to coordinate strategy – in the short term, to devise a military response to the *Kaiserschlacht*. This was to be the French commander,

General (soon to be Marshal) Ferdinand Foch, underlining that for Britain the war was, after all, a coalition effort.

In overall terms, the *Kaiserschlacht* had not only tested the bravery and guts of the BEF, but the efficiency of its headquarters. Staff officers particularly had to drop their myopic trench warfare mentality and suddenly become flexible and responsive. This ability to take or devolve control rapidly and coordinate the activities of many arms – and Monty was at the very centre of this – bore fruit in what happened subsequently in the war. On 8 August the BEF took the offensive and launched a series of attacks which concluded only on 11 November. Hitherto there had been very long lead times for big offensive actions: the 1 July 1916 Somme attack had taken 115 days to plan; the Vimy assault was conceived over ninety-five days and Third Ypres was born over sixty days. The *Kaiserschlacht* forced on the BEF's staff a flexibility and responsiveness that allowed the subsequent battle of Amiens (8 August) to be planned in twenty-six days from scratch, the attack to retake Albert (21 August) was put together in eight days and the crossing of the St Quentin Canal (29 September) was achieved after eleven days of planning. These were all successful, formation-level attacks fought as much with conscripts as volunteers, incorporating arms such as the Royal Air Force (created only in April 1918), which had hitherto played little role in offensive warfare. Arguably these triumphs were due to the experience of the army, the standardisation of its doctrine and trust in its staff. Though casualty numbers remained huge, this new, enlightened approach by a flexible staff was to prove the most effective and successful way of conducting war.

On 3 June, the now very experienced GSO2 at IX Corps was promoted to brevet major, bringing him seniority within the army, but not within his regiment. As the battered IX Corps went into reserve, on Tuesday 16 July, Monty was posted for a final time to become both an acting lieutenant colonel and the GSO1 (Chief of Staff) of 47th London Division.

Monty probably learned far more at IX Corps than he did with 47th Division, which was more his own 'train set', on which he would experiment with what he had discovered elsewhere. He had been with IX Corps from July 1917 to July 1918 – a full year and ten days at exactly the time when the size, influence and professionalism of the staff was expanding. Within that period he had overseen and helped with the set-piece arrangements for the big assault of Third Ypres, but also had to react very quickly to the unexpected shocks of the *Kaiserschlacht*, of which the unfortunate IX Corps suffered two doses – Operations Georgette and

Blücher-Yorck. Arguably, coping with the unexpected is a far better test of skill, rather than launching a campaign where everything begins according to a predetermined plan. The way the Germans achieved surprise in their March 1918 attacks would hold valuable lessons for Monty in the future; the fact that no battle zone (middle-line defences) had been broken since trench warfare commenced in 1915 led to a false sense of security, which made the surprise all the more complete when the German penetrations of 1918 took place in thick fog.

During the winter of 1944–45, many British commanders would see parallels between the *Kaiserschlacht* and the German Ardennes attack of 17 December, likewise launched in poor weather and fog. These were links the US high commanders in 1944 were unable to understand because very few had direct experience of March 1918, and fewer still had studied it. There is evidence that the British in December 1944 were less concerned by the sudden German break-ins than their US counterparts, because they had seen it all before. Monty's Second Army commander in 1944–45, Sir Miles Dempsey, would later write of the 1944 Battle of the Bulge: 'The Americans of course had not experienced this sort of thing before. March 1918 and Dunkirk were part of our lives. I really do believe that this explains most of what happened in the Ardennes, and all the dramatic stories which have been written.'[3]

While remaining outwardly calm Monty, too, was changing. According to Alun Chalfont, who had the opportunity to interview many of Monty's contemporaries, 'Those who knew him then say that there were one or two outward changes. His face became warier and tauter than ever; the eyes more piercing; the voice sharper and more authoritative. Beneath the surface more profound transformations were taking place. The basic simplicity of his attitudes was hardening.'[4]

Monty's new charges were a Territorial division whose battalions were part of the London Regiment. The Londons were a unique organisation of twenty-six battalions, each recruited in a different area of the metropolis, or from a distinct profession. Monty's new boss was the irascible Major General George Gorringe, who was in his fiftieth year and the only divisional commander at the Armistice who had been a major general in 1914. Apparently, Gorringe's nickname, 'Bloody Orange', was rhyming slang which partly acknowledged the speech patterns of the division he commanded (London cockneys), but also reflected his 'rude and unpleasant' personality. Though 'a large, arrogant, tactless, officious man ... often loathed and distrusted', Monty got on with him very well, due no doubt in part to the fact that both were bachelors, utterly dedicated to their chosen profession.[5]

Being one of the first Territorial divisions to land in France in March 1915, the 47th had seen continuous fighting on the Western Front, and when Monty arrived they were near Amiens as part of Lieutenant General Sir Richard Butler's III Corps. Immediately, the new GSO1 started issuing instructions, mostly relating to defensive arrangements should the Germans try again. With Operation Michael having ground to a halt just east of Villers-Bretonneux, the BEF maintained an offensive spirit through raids and the capture of prisoners; intelligence analysis started to conclude that the Germans had lost the operational initiative. Henry Rawlinson, Gough's replacement and now commanding the renamed Fourth (ex-Fifth) Army, had been worried about the continued threat to Amiens, but raids, notably by the Australians (a tough bunch to run into on a dark night in no-man's land) demonstrated the German defences were also weak. They showed no signs of trying to resume the advance; furthermore the front-line troops captured seemed to be of poor quality with low morale. By June, Rawlinson had begun to ponder going over to the offensive just south of the line of the Somme river, and was accumulating men and equipment. Using the Amiens–St Quentin road as an axis, he envisaged three corps advancing in tandem, from north to south; Butler's III Corps with four divisions; all five Australian divisions fighting together for the first time in John Monash's ANZAC (Australian & New Zealand Army Corps) in the centre, and Arthur Currie's Canadian Corps of four divisions taking the south. By this time the colonial forces, Australians and Canadians, had acquired a fearsome reputation as the Allies' shock troops.

In reserve, Rawlinson retained three divisions, including 47th London and the American 33rd Division, who were acclimatising to the Western Front. Loitering for the right moment were the three horsed divisions of Sir Charles Kavanagh's Cavalry Corps. The attack would incorporate all the technology and lessons learned over the previous year, and Rawlinson was particularly concerned to retain the elements of surprise (never a principle of war associated with the Western Front) and operational security. With great care he managed to achieve both, particularly by doing away with the traditional pre-assault bombardment.

The first thing the defenders knew was at 4.20 a.m. on Thursday 8 August when Allied guns, having pre-registered every hostile battery, began destroying their opponents' gun lines, then rolling forward in a realistically paced, controlled creeping barrage. Rawlinson had built up an artillery density of one gun every twenty-five yards, nearly 400 armoured vehicles, and could call on 2,000 Allied aircraft. The new RAF as a single service had started thinking about doctrine and using aircraft innovatively, devising tactics for strafing ground troops and disrupting their lines of retreat or slowing down the arrival of reinforcements, as well as maintaining contact with ground forces. In this sense, Amiens is regarded as a very modern battle, where the combined arms of today's battlespace – infantry, armour, artillery and air – came together for the first time. The result was catastrophic for the German army, an Allied advance of eight miles, inflicting 27,000 casualties on the defenders; 'our blackest day', recorded Eric von Ludendorff. Military historians regard 8 August as the beginning of the 'Hundred Days' (there are actually only ninety-five of them) which led inexorably to the Armistice of 11 November, for the BEF's story is of an almost continuous advance after Amiens.

Monty might have understood his transatlantic cousins better during the Second World War had he any experience of Pershing's forces in 1918. But it seems this opportunity passed him by (despite the fact that many British battalions in 1917–18 had American medical officers, including in Monty's division). It is quite possible Montgomery never met any of them – or worse, confused them with Canadians.[6] When on 6 April 1917 America had declared war on Germany, Montgomery had written home: 'I am glad America has finally declared war; she ought to send us large supplies of food and money.'[7] From the very beginning, then, Monty's world view (perhaps not uncommon at the time) was how other nations, such as America, could assist Britain, rather than building a workable coalition to defeat Germany – tragically, a view he seems to have retained all his life.

By the end of 1918, there were significant examples of American military achievement on the Western Front. Pershing's policy had been to concentrate and train all American troops until they could deploy and fight as a single US force. He knew the political connotations of trickling his soldiers into the Allied line would not be well received in Washington, and even the Allies were not keen to let the Americans loose before they were ready, though by the time of the Battle of Amiens on 8 August, scattered US units had built up some considerable experience of war fighting.[8] This no doubt induced Lieutenant General Butler, III

Corps commander, to unleash the US 131st Regiment of 33rd Division on 9 August, who first seized the Chipilly spur – on the north bank of the Somme, west of Bray – then carried on the advance with the British 58th and Australian 4th Divisions until 20 August. Although the tactical significance of the Chipilly action was small (albeit at a cost of 1,400 casualties), the political ramifications of the highly successful Anglo-American–Australian–Canadian assault were huge.

This was underlined by the well-timed visit of George V to the sector on 12 August, where he scattered decorations on troops of all the assaulting nations and knighted the excellent John Monash, the Australian Corps commander. At the same time, between 8–14 August, in a comparatively minor operation, battalions of the US 88th Division retook the towns of Serre and Puisieux across the crater-strewn wilderness of the old 1916 Somme battlefields, as German units, startled by the Amiens offensive, started to retire. In early September, 27th and 30th U.S. Divisions also helped to re-secure the Ypres sector after the losses to Operation Georgette. Thus, whilst the main American efforts would always be associated with the St Mihiel salient (south of Verdun) on 12–15 September and the Meuse-Argonne campaign, which began on 26 September, American involvement with other sectors of the Western Front was both significant and visible, and it is unfortunate that Montgomery was not more aware of this.

Although 47th London took a back seat in the Amiens enterprise, Monty was fully briefed with the other Chiefs of Staff and integrated into the planning cycle as soon as he arrived to take up his post. His first serious test came on 22 August when 47th Division attacked along a feature just north of Bray-sur-Somme, Happy Valley, ironically where Monty and 104th Brigade had paraded for Minden Day two years before. It is a barely perceptible mile-long dip in the ground, running north, towards the airfield at Méaulte, now known as the Vallée du Bois Ricourt, and lies between the two roads running north out of Bray-sur-Somme. Here, Butler's III Corps was clearing the north bank of the Somme and the Morlancourt Ridge south of Albert on a four-division front, with (left to right) 18th, 12th, 47th and 3rd Australian Divisions. The Corps planned to advance 3,000 yards from a line running 1,000 yards west of Bray to Méaulte. The attack was orientated north-east, along a four-mile front between Méaulte and Bray and set off behind a creeping barrage; the corps cavalry regiment (Northumberland Hussars) and tanks were on hand to exploit any gains.

The assault went in at 1 a.m. on 22 August and, noting that the division contained many inexperienced conscripts who were in action for the first time, the official history observed that 141st Brigade suffered.

> Owing to bad staff work and the insufficient training of the young troops in movements in darkness, smoke and mist, the two leading battalions, 1/20th and 1/19th London, lost count of distance and though the Germans surrendered freely, the battalions halted considerably short of the intermediate objective.[9]

As planned, 142nd Brigade leapfrogged through their colleagues, but in the confusion two squadrons of the Northumberland Hussars entered the southern end of Happy Valley and charged forward unaware that the Germans still held the high ground beyond, to the north-east; as the leading squadron cantered uphill it encountered a barbed wire obstacle, and being skylined was chewed up by machine guns and artillery and withdrew. Although, as David Kenyon's painstaking work shows, it was still possible for horsed cavalrymen to make an effective contribution to offensive operations on the Western Front, launching them was a matter of fine judgement, and intact wire remained a show-stopper. Although the two forward brigades from 47th Division then held their ground against a strong counterattack, the operation was effectively over by 8 a.m. The assault had failed to achieve its objective, but in withstanding the counterattack, some of Monty's preparatory training had been vindicated. The warm day (described in the divisional history as 'the hottest of all the war') ended with the capture of Albert (the town contained the church with the leaning virgin statue Monty had remembered back in early 1916), which fell to 18th Division and, psychologically, because of its associations with the 1916 battles, for many Britons this represented real progress.

The reference to 'insufficient training of the young troops' points up, in the London Division at least, the large numbers of recently arrived conscripts in their ranks, such was the turnover due to casualties in the March offensive. This was something the whole of Haig's BEF felt and another of Monty's challenges. Monty's Londoners paused to take stock and renewed their assault in moonlight at 1 a.m. on 24 August, this time with total success. Robert Angel of the Civil Service Rifles (15/Londons) recalled of the later attack,

> Around midnight a terrific barrage opened up. We lay in our shelter shivering with a mixture of excitement and fright until we were called outside and formed into small parties. The parties moved forward and down the slight slope into Happy Valley ... As it got lighter we could see the whole valley, the sides honeycombed with shelters and the wide bottom full of debris, corpses, dead horses, abandoned machine guns and overturned wagons.[10]

The division put in another attack the following evening, renewing the assault, but they encountered more problems from the thick fog than the Germans; most casualties in fact came from the British barrage, where a slightly worn gun barrel or faulty shell could result in a 'drop shot', a round landing short, amongst friendly troops.

The objective of the 22 August attack was taken two days later. The official history specifically identified 'poor staff work' as a contributory cause for the initial failure, yet there is no evidence that this accusation rebounded on Montgomery, the divisional GSO1, who must have been livid at the slur. This was a big attack to organise. Lieutenant-Colonel Montgomery (his job would be done in a British division today by a full colonel) was coordinating not only his own nine infantry battalions, each with an average strength of about 700 and equipped with a total of 324 Lewis light machine guns, but also the division's artillery, forty-eight 18-pounders and 4.5-inch howitzers; two heavy trench mortar batteries; a divisional machine-gun battalion, and cavalry squadrons armed with sword and lance. The division and all other BEF formations of late 1918 looked very different in terms of structure, size and fire-power to how they had appeared when they landed in France in 1915. And now the BEF was about to fight across the old Somme battlefields. Its very difference in 1918 was what would make it so successful: its German opponents had not collapsed; if anything they were more profes-sional. But by 1918 Haig's army was better.

Monty had to coordinate the 47th Division's assets where necessary with the support of the Tank Corps and RAF. His operations orders (and they were *his* – he was the facilitator of General Gorringe's wishes) also included plans for anti-aircraft defence. Within the divisional HQ Monty had two junior staff officers: a GSO2 (a major) and GSO3 (a captain) and an intelligence officer (a lieutenant) to assist him. Three other staff officers looked after the division's personnel and quartermaster issues. Although each divisional commander had his own style, in 47th Division,

'Gorringe laid down the policy, and made whatever major decisions were necessary; the implementation he left to his Chief of Staff – as was done in the German army.'[11]

In a letter home to his mother of 3 September, Monty's self-confidence, verging on boastful arrogance, shines through in every sentence, remarkable in one who had held his job for eight weeks and three days, though during which time he had organised six divisional attacks:

> One is gaining wonderful experience in this advance. As Chief of Staff I have to work out plans in detail for the operations, and see that all the branches of the Staff, and administrative arrangements are working with my plans. The day generally commences with an organised attack at dawn, after which we continue to work slowly forward all day; then another organised attack is arranged for the next morning to carry us forward again, and so on. It means little sleep and continuous work; at night guns have to be moved forward, communications arranged, food and ammunition got up . . . The general and I work out the next day's plan and he tells me in outline what he wants. I then work out the detail and issue the orders. Then I send for the heads of each branch of the staff, tell them the plan and explain the orders. They tell me what they propose to do to fit in with the scheme. If I think it is bad I say so and tell them what I think is a better way to do it; there is often no time to refer to the general and I take the responsibility on myself. I know his ideas and thoughts very well now and so far he has always backed me up and approved of everything. The heads of the various branches are all Lt.-Colonels, older and senior in rank to me and one is a Brigadier-General. But we all work very well together and they do what I tell them like lambs.[12]

*

By a bizarre twist of fate, the German defence in this sector had been given to what was left of 124th Württemberg Regiment of 27th Division, the formation with whom Rommel had started the war. They had already sustained heavy losses on the Chemin des Dames, where the long-suffering Theodor Sprösser had been wounded. Had Rommel returned to the old unit, as he wanted to do in 1918, he would have found himself fighting Montgomery, perhaps as a battalion or regimental commander, in what would have constituted an opening, perhaps one-off, bout between the two. For, in making their counterattack, the official German history described how 'in the attempted advance of the 27th Division, the 124th Regiment was completely shattered' – and Rommel would almost certainly have been killed.[13] The local German corps commander was Rommel's model hero from Italy, Generalleutnant Eberhard von Hofaker.

Indeed, the timing of Rommel's posting to the staff on 11 January 1918 was highly fortuitous, for the March–May 1918 attacks cost the Germans nearly 700,000 irreplaceable casualties (and a million men by July 1918), a disproportionately high percentage for the number of attackers, and many talented commanders who had as much by luck as judgement managed to survive until then, were killed in this last great offensive. Rommel, with his insistence of always being at the front to lead his men would have stood a high chance of injury if not death. One can sense his natural frustration, mentally still with his men, reading the reports of their advances and the casualty returns, literally hating his desk and comfortable surroundings, wishing to be back at the front doing the only job he knew how to do, where he had found his métier.

1918 would be Rommel's only experience of working on the staff and it is useful to contrast his experience with that of another near contemporary and panzer-exemplar, Heinz Guderian, who was the last head of the German General Staff. Born three years before Erwin, in June 1888, Guderian joined the army in 1907, was commissioned as a signals officer, attended the Kriegsakademie in 1913 and rose to command a battalion, whilst alternating between General Staff appointments. After a brief General Staff officer's course in the captured fortress town of Sedan (which he was to assault and capture in May 1940), Guderian transferred permanently to the General Staff Corps in February 1918 and remained with it thereafter. Had Rommel been so motivated, he could have followed suit: whatever intellectual capacity he lacked (much has been made of this, but he wrote some excellent, almost scholarly books and memoirs), he could have made up for with hard work – and of course the *Pour le Mérite* would have helped.

Not surprisingly, Rommel remained tight-lipped about his year on the staff in 1918, although he enjoyed two short breaks away, instructing on his Italian experiences to Landstürm battalions, from 20 July– 8 September. His major biographers have thus emphasised his combat experience in Italy, whilst downplaying his career during 1918 in a sentence or two. Rommel's attitude towards staff work was forged in this January–December 1918 period, an attitude which was to dog him for the rest of his life and, it can be argued, indirectly contribute towards his death. Impulsive, task-orientated and impatient to an extreme, the risk-averse staff world of caution and planning was not for him. All of Hitler's other field marshals would be trained in the German General Staff, which would give them a feel for logistics and a wide network of colleagues who furnished the senior echelons of the army.[14] In the Second World War, Rommel was definitely the odd man out of the marshals, but then Hitler (who, like Rommel, was not only ignorant but scathing of staff work) always regarded himself as an outsider, and was drawn to others similarly disposed. It was a personal failing and need not have been so. In a formal group photograph of Erwin with the staff of his HQ, thirteen of the thirty-one officers in the picture appear to wear decorations of some kind, implying that Rommel was not the only front-line warrior working there – though perhaps the only one there against his will

Meanwhile, in late August 1918, Monty and the 47th London Division pulled back from the Somme front to rest and refit for three days before rejoining what had become a constantly moving battle. 'We have been through it all, hammer and tongs, the last few days . . . Things have gone well . . . We are killing a lot of Germans . . . [and] capturing large numbers of prisoners. I am with a splendid crowd. They are like little lions – these London men,' observed Lieutenant Colonel Rowland Fielding, commanding officer of the Civil Service Rifles, writing home to his wife.[15] H. S. Moore of Monty's 47th London Division later recalled that his war 'passed in a kind of romantic haze of hoping to be a hero. Sometimes in France there were three or four days of great danger when you thought there wasn't a chance of getting through, and then all one felt was sadness at having taken so much trouble to no purpose; but on the whole I enjoyed the Army.' Moore's 'trouble' paid off, for he emerged as the sculptor Sir Henry Moore.[16]

The Germans were now falling back continually, the division pursuing them, the brigades leapfrogging forward. On 1–3 September 1918, 47th Division went into the attack again, crossing the old Somme battlefields and assaulting the (mostly dry) Canal du Nord, which snakes north–south across the area, linking the Scarpe river at Arras with the Somme river at Peronne. Initially the division's attack fell on German positions between Rancourt and Moislains, forward of the canal line and just north of Peronne. The first attack went well with all objectives seized by

7.30 a.m. and 200 prisoners taken. On the 2nd, the division followed up with a second assault with the aim of clearing the ground as far as Nurlu, two miles further east. This was a bigger attack, launched jointly with the 74th (Dismounted Yeomanry) Division on its left, to the north. The 15/Londons, however, lost half their number to artillery and machine-gun fire in open ground. Lieutenant Colonel Fielding described the day in a later letter home:

> Most of the men are very young – in fact, quite boys. They wear khaki shorts [it was early September and still very hot], with grey hose-tops turned down over their puttees. On their sleeves they have canary yellow hearts as a distinguishing badge ... Almost immediately the enemy opened with a heavy artillery barrage ... soon supplemented with machine-gun fire ... [which] grew heavier and heavier. It came from the front and from both flanks. With their khaki shorts showing about 4 inches of bare knee the men went forward ... It was a truly wonderful sight: each man with his shoulders squared to the objective, walking with bayonet fixed, apparently unconcerned, through the deadly fire; many dropping; the remainder carrying on ... The last hundred-yard lap was the worst, and had it not been that the ground was pitted with shell-holes, not one of us could have got across it alive.[17]

Though this passage has definite overtones of 1916, the outcome of these attacks, over pretty much the same ground as two years before, was very different. The change lay in the professionalism of the commanders, who now truly understood how to wage war at the senior tiers of command, and the equipment – and the doctrine devised for its use – with which the BEF was now fighting. By the start of the Hundred Days campaign on 8 August these factors had dovetailed together to make the BEF the most professional army Britain has ever fielded, and present it with some of its finest hours.

The weather broke that night with thunderstorms, and there was little subsequent fighting before 6 September, when 47th Division was pulled back into reserve. Their most serious challenge was that none of the casualties since 22 August had been replaced, nor would they be (another headache for Monty), such were the strains on manpower across the BEF. The 15/Londons had, for example, lost 389 in that period, about half the combat strength of a 1918 infantry battalion. Although there were more offensive operations to come, the divisional history regarded this as their last battlefield, for their subsequent activities 'were of secondary importance and out of the area of the decisive battle'.[18] It is quite precise as to the achievements of the 22 August–6 September period, which amounted to 1,500 troops and 150 machine guns and cannon captured. The distance from Bray, where they started this last campaign, to Peronne is about twenty-five miles by road; the last time most British units had battled across this area was in that bloodletting summer of 1916 when an advance of one mile was considered a major achievement and many never managed that, instead being forced to measure gains in yards. These figures underline that the Somme battles of 1916 so injured the German army that it never recovered and by August 1918 was a shadow of its former self. Meanwhile, Monty's professionalism typified the learning curve of the BEF as a whole, which had grown and matured hugely in terms of leadership, staff work, personnel and tactics: it is generally accepted that by 1918, half of the BEF's platoon commanders had worked their way up through the ranks, being granted battlefield commissions, also making it the most egalitarian army Britain has ever had. Such was the institution Monty witnessed winning in 1918, which he strove to recreate in the Second World War.

The advance grew easier as the Germans evacuated their positions in front of the division during the night of 3–4 September, and by the 6th Gorringe's men had reached Liéramont; that evening they were relieved by 58th Division and departed their battlefield by bus – a sure sign of the times (in 1914 Monty's battalion had marched to and from the war). The divisional war diary recorded violent thunderstorms at this time, but on 3 September, Monty wrote home: 'At present I live in a tent which I have carried on my motor car. I prefer a tent pitched on clean ground rather than some *bosche* hut or dugout full of *bosche* straw, and smells and bugs.'[19] Here we see the beginning of Monty's declared preference for sleeping out of doors, culminating from 1942 with his favourite caravans – a product more of Monty the boy scout, rather than any clear reaction to 'château generalship', which in any case had no place in the fast-moving advance of late 1918.

As the division was pulled back into reserve on 6 September, it was posted to XIII Corps, Fifth Army, and Monty's staff-trained mind had to get his formation up to the coast by train and bus, with the aim of preparing to exchange with a British division in Italy. Arrangements were almost complete and an advance party had left when, as is ever the nature of things with large military organisations, the plans were cancelled suddenly. Italy was off and on 30 September 47th Division was assigned to XI Corps of Fifth Army, now commanded by Sir William Birdwood, concentrating at Lestrem, about thirty miles west of German-held Lille. On 3 October the division resumed combat operations; this time the aim was to stay in touch with the retreating Germans, maintaining pressure with two brigades forward.

The divisional history noted that the reinforcement problem had become acute, with some battalions managing to field 'a trench strength of hardly more than 300'. On most days there were raids, engagements and artillery exchanges, but there was also wariness, with the realisation that the Germans were definitely on the run. The pursuit was maintained until 17 October when they were pulled back and the neighbouring 57th Division passed through them, liberating Lille – a key garrison town and logistics hub – which the Germans had just evacuated. The occupation came, without doubt, as a sudden slap in the face to Gorringe and Monty, whose division had almost surrounded the place and started to engage the defenders. But a consolation prize came on Monday 28 October in the form of 47th Division being selected to stage a formal victory march through the city. The Armistice was exactly two weeks away, but as yet they still had no idea. The parade was another challenge for Monty, who later recounted to his mother:

All the arrangements had to be very carefully worked out and I have been busy for several days over it . . . I rode at the head with the General at the head of the Division till we reached the Grande Place, then we got off our horses and stood with the Mayor while the troops marched past; it took 2½ hours for them to go by . . . We reduced the column to eight miles [from fourteen] by leaving out several units and most of the transport.[20]

This was as formal an occasion as the British Army on campaign could manage – colours flying, bands playing, and several days of parade drill beforehand. Captain Leslie Walkingon, adjutant of the 47th Divisional Machine-gun Battalion (who would find himself fighting in France again in 1940) recalled how

we had specially generous issues of new uniforms and boots . . . [and had] to paint all our shrapnel helmets the standard khaki colour, all horses and mules had to be freshly shod and the metal parts of their harness and the hubs of the limbers burnished. When we finally did our march past we must have looked pretty good . . . Everywhere flags and bunting brightened up the scene and the pavements were packed with people eight or ten deep. The girls in the crowds kept rushing out to hug and kiss the soldiers, who responded enthusiastically . . . At the saluting base we gave 'Eyes right' to an assembly of big-wigs . . .[21]

Taking the salute with the Mayor of Lille was the Fifth Army commander, Sir (soon-to-be Lord) William Birdwood, Sir Richard Haking of XI Corps, Gorringe the GOC and Minister for Munitions Winston Churchill, who had been appointed to that position in July 1917. Churchill was on a five-day fact-finding tour of the front and had just undertaken a battlefield tour of recently liberated Le Cateau. In Lille, Winston met his brother Major Jack Churchill, then on Lord Birdwood's staff (whom Monty would have encountered), and who would shortly be awarded the DSO.[22]

Photographs of that day in October 1918 capture the two, Monty and Winston, then unknown to one another, side by side amidst a sea of French and British top brass. Monty is in his staff officer's uniform, red tabs, no hint of the sword he would have worn in 1914. He is serious-looking, with a wiry moustache, jaw clenched, concentrating and squinting at the passing troops with those piercing eyes and a brooding, earnest expression. Monty at this time was described as 'indefatigable, with the bearing of a fanatic, his ruthless determination occasionally relieved by a sudden access of charm and generosity. He studied industriously and was, to an almost frightening degree, self-sufficient and in complete control of himself. Or at any rate, he appeared to be.'[23] Churchill, long since out of the trenches and back at Westminster, by contrast appears more relaxed, seated, and unusually for him, carrying no military adornment whatsoever; wing-collared and bow-tied, he wears a British warm, grey Homburg and holds his trademark walking cane. Of the two, Churchill looks much the younger, though he was Monty's senior by thirteen years. Winston toured more of the front that afternoon, coming under intense shellfire.

Gorringe and Monty led their division back into the fray on 31 October. There was a cautious advance: still much skirmishing and artillery duels. News very rarely penetrated the front line in 1914–18, except via newspapers, but an army on the move is a rumour-mill all of its own. Fantastic stories spread as to the state of the German nation, the future of the

Kaiser, revolutions in Germany: all pointed to one thing – that peace negotiations were under way. Many British memoirs of November 1918 concur on one point – peace emerged quite unexpectedly, after nearly four and a half years of constant war. Looking back, Charles Carrington, a Royal Warwicks Territorial, observed: 'In 1918 we had not been sure even of eventual victory till the late summer and had seen no hope of a quick ending to the war until three or four weeks before it happened. Victory was sudden and complete and the general sensation was that of awaking from a nightmare.'[24] At Pasewalk, forty miles from the Baltic, on the modern German–Polish border, a wounded soldier lay in bed, his eyes bandaged after a mustard gas attack on 14 October, near Ypres. On hearing the news from his doctor that Germany had signed an Armistice, that his beloved Reich and Kaiser were no more, he became inconsolable with grief and his sightless eyes shed tears. Healed and discharged eight days later, Corporal Adolf Hitler vowed eternal revenge for his country's defeat.

Monty reflected in his memoirs on his career as it stood in November 1918:

> It had become very clear to me that the profession of arms was a life-study, and that few officers seemed to realise this fact. It was at this stage in my life that I decided to dedicate myself to my profession, to master its details, and to put all else aside.[25]

It comes as a bit of a surprise that Monty had ever considered an alternative to a military life, and November 1918 would have been an especially odd time for a thirty-one-year-old Regular Army officer with no other work experience to seek alternative employment, when several million would soon be hanging up their uniforms and doing likewise. Acting rank and a DSO would have counted for nothing in civilian life. Army life seemed to have found him, and he had excelled at it and had already managed to put many of life's distractions aside. Reading between the lines of his later memoirs, there really were no other interests, hobbies, or outside distractions in Bernard's life at this time; he had spent the previous four years training himself to be focused and single-minded to an exaggerated degree. The conclusion one draws is that Monty had already long since made this career decision (perhaps even unconsciously) – but possibly, for the purpose of his memoirs, wanted to demonstrate an 'on the road to Damascus' moment, when he deliberately embarked on his vocation: it comes across as a more serious commitment, when the truth appears to be that he simply drifted into being a soldier and found that he was rather good at it.

The clock which started on 1 August 1914 had now stopped, after consuming 16 million lives through violence, at the rate of seven per minute. Around ten percent of the dead, who lay scattered over the world's battlegrounds and oceans, were the 'missing', definitely killed, but never seen again, or identified. For families and surviving friends – Montgomery and Rommel included – the question mark over the fate of those whom they had known, but were now 'unfound', was more painful than the bleak reality of a government-issue grave marker. The horrific proportion of 'missing' would never be repeated: modern advances in DNA testing make the concept today of the 'Unknown Soldier' redundant, and one specifically associated with twentieth-century warfare.

To reflect on the war the pair had just experienced: at platoon and company level, Monty spent about two months, until he was wounded, at the front in 1914, taking part in two actions, at Le Cateau and Méteren. After his recovery, he spent the rest of the war, some forty-six months, wearing the red tabs of an officer of the General Staff. During that time, especially when Brigade Major of 104th Brigade and GSO2 of 33rd Division, he spent much time visiting the front – every day for protracted periods – but his headquarters was of necessity away from the trenches. Rommel spent thirty-seven months at or near the front and the eleven months of 1918 very reluctantly as a staff officer. Monty's Great War was predominantly that of a staff officer, whilst Rommel's was spent primarily dodging bullets. In later life, Monty would portray himself as the trench-bound muddy-booted infanteer at odds with the clean-shaven staff; but for most of the war, he was a creature of that much-maligned red-tabbed brigade.

The staff–combat troops divide hung over Montgomery like a dark cloud throughout his post-First World War life. In a 1969 letter he observed (if not entirely accurately):

> I served on the Western Front during the 1914/18 war, as a platoon commander in 1914 and rising to GSO1 of a Division by 1918. I never once saw Haig, nor did I ever see him after the war ... I can never forgive a general who intrigues, as did Haig – against his C-in-C, and

against his political chief . . . There was a tremendous gulf between the staff and the fighting army; the former lived in a large chateau miles behind the front . . . Kiggell [Haig's Chief of Staff], who was in my regiment, had no idea of the conditions under which the soldier's lived and fought.[26]

Monty made the same allegation in a longer passage in his memoirs of 1958. The temptation is to think that he was influenced by the anti-generals sentiment that prevailed in the 1960s, or that he might have felt this throughout his life. The trouble with this passage is that we know Monty *did* meet Haig at least twice – not long after he had arrived in France with his brigade, in February 1916; writing home, he observed:

We were out route marching, and he watched us march past. His military secretary, the Duke of Teck, was with him and I talked to him for about 10 minutes. Haig looks very worn and aged: and of course he had a very harassing time at the beginning of the war in the retreat [from Mons]. That was where he made his great reputation. He has been out here all the time and been in all the big fights, and I suppose a man of his age [fifty-six] ages quicker and hasn't the same rebound as a younger man.[27]

On 13 April 1917, Haig's diary records him as visiting Major General Reginald Pinney's HQ in a quarry at Blairville where Monty was GSO2 at the time, and the Division's war diary records Monty as present.[28] Monty also claimed in Memoirs 'only twice did I see an Army Commander',[29] yet we know that he saw Herbert Plumer, his Second Army C-in-C often during the Third Ypres campaign, being (indirectly) on his staff – he wrote and spoke of Plumer in a manner that suggested frequent, personal contact. He shared a platform with Sir William Birdwood (and Winston Churchill), commanding Fifth Army on 27 October 1918.

Why might Monty be so keen to distance himself from the Great War generals? The answer surely lies in the reputation he was trying to forge for himself in 1958 as the first general who overtly cared about his men, could communicate with them and was seen by his troops, in total contrast to his predecessors. Though we know Haig cared, he was poor at expressing his concern in public. At the time of the publication of his memoirs, Monty's reputation was such that he could afford to exaggerate with little fear of criticism from the 'old guard' of the First World War – those few left alive – so exaggerate he did. Brigadier Sir John Smyth (a First War VC-winner, Second War divisional commander and MP) made the point that all senior officers he knew toured the front often, but with miles of trenches these visits were very localised and unknown to all but those in the vicinity; there was 'no opportunity at all for the general to say, à la Montgomery: *Now gather round chaps and I'll tell you how the war is going.*'[30]

Smyth also observed that with the Clausewitzian fog of war, Great War generals were where they should have been – at their headquarters. Because mobile (as opposed to static) communication devices were so primitive throughout 1914–18, by coming any further forward commanders risked being out of touch, rather than improving their battlefield awareness.[31] Other Great War generals were known for drifting around the trenches handing out chocolate bars (like Robert Fanshawe of 48th South Midland Division), cigarettes, or disguising their rank with an old raincoat and chatting anonymously: nothing Monty would do in the Second World War was new, but he was the first to publicise it, and enjoyed occasions where he could speak to huge numbers all at once.

Given that both Monty and Rommel had started the war as professional officers, the proportion of their pre-war colleagues who survived was very low – by some estimates fewer than ten percent emerged alive from the 1914 Regular forces. Whatever their views of the staff, both were extremely fortunate to find themselves (happily or otherwise) doing staff jobs during the last few months, when many of the old guard of 1914 finally succumbed. An inkling of what this meant for Monty can be had by realising that of the eighty-one Regular officers listed as serving in the first and second battalions of his Royal Warwickshire Regiment in the 1914 *Army List*, by 1918 some twenty (or twenty-five percent) had been killed and statistics suggest that at least sixty had (like Monty) been wounded at least once. Likewise, a staggering 506 old boys (equivalent to a year's intake of students) from Monty's school, St Paul's, had been killed by 1918, which must have had a profound effect on the survivors.[32] Rommel's generation suffered even more: Germany mobilised 11 million men, of whom 1.6 million had died and 3.6 million were wounded by 1918.

Both, too, were lucky to find individuals willing to foster and polish their Great War careers. Although Monty had several postings, working under a variety of senior officers, it appears to be Brigadier James Sandilands who was his key mentor and guide, and Monty was gracious to admit as much in a letter of 1945. When they first met, James Walter Sandilands was fit and constantly at the front; newly promoted himself, he was as much learning 'on the job' as was Monty. Sandilands was a protégé of Henry Wilson who was Commandant of the Staff College when Sandilands was a student (1909–10) and would have been hugely influential in steering Monty's future career. In later life he kept Monty in touch with developments in Germany, where he was military attaché, 1927–28. They seem to have clicked and liked each other; certainly Monty talked of Sandilands with a mixture of affection and respect he reserved for few others.

In Rommel's case it was clearly Major Theodor Sprösser, who taught Erwin so much, accepted his advice and of course with whom he won the *Pour le Mérite*. Perhaps these two mentors were not so dissimilar: Sandilands, aged forty-two in 1916, newly promoted from battalion command, running a brigade with a talented subordinate; and Sprösser, aged forty-five in 1915, previously commanding a battalion, running the equivalent of a brigade in the WGB, though only holding the rank of major, and leaning equally heavily on his right-hand man, Rommel.

Like Montgomery and Rommel, many of their colleagues were singled out for recognition. But they knew of many more comrades whose efforts went unrecognised or had perished. Both men had embarked on quiet careers with limited expectations and ended up in a maelstrom. They emerged from it as different people; their decision-making faculties and their mental approach to everything for the rest of their lives were irrevocably cast by the long hours of the Great War. These were not just formative years, but the deciding days in their careers, for the way they would go on to fight in the meadows of northern France, across the sands of North Africa or through the orchards of Normandy was henceforth predetermined by what they had done and witnessed. When they laid down their arms on Monday 11 November 1918 – Rommel in Colmar, Montgomery at La Tombe, on the northern outskirts of Tournai – they had already learned most of what Mars, the god of war, would teach them. It was now time to consult Minerva, the goddess of wisdom, to make sense of it.

PART TWO

INTERWAR: PREPARATIONS

The same principles of war which were employed in the past, appear again and again throughout history.

Montgomery of El Alamein

10

Coping with Peace

MONTY MAY HAVE decided to embrace fully the army as a profession, but would it truly embrace him? He had experienced a 'good war' and amassed some enviable staff experience. But the British Army in common with all the other participating nations now had to revert to a peacetime size. In Monty's case, this meant demotion to major (he was lucky it was not lower) and a staff job at the headquarters of the British Army of the Rhine in Cologne.

Despite his two-week course at Hesdin in 1917 and forty-six months as a staff officer, he had no formal staff qualifications, most conspicuously the sought-after 'psc' initials which appeared post-nominally in the *Army List* and on letterheads denoting one's attendance at Staff College in either Camberley or Quetta. He was doubtless pained to discover that his name was not on the list of students for the first post-war course at Camberley, starting in January 1919, or the second, commencing in 1920. But, with a determination which was fast becoming a hallmark, Monty contrived to get an invitation to play tennis at his commander's house, and poured out his troubles to the old soldier.

General Sir 'Wully' Robertson (the 'Wully' was a pun on his broad Lincolnshire accent) was an interesting and shrewd choice of confidant for he held the unusual distinction of being the first and only man to enlist in the British Army as a private and rise to field marshal.[1] Robertson, who had been CIGS 1916–18 (arguably the toughest years of the war), had just been made a baronet at the time of Monty's approach and would receive promotion to field marshal in March 1920. He knew all about being an outsider. He came from humble beginnings; his father was a postmaster and tailor, he was educated in the village school and his family had no military connections. Robertson held the key post of Commandant at the Staff College, Camberley, between June 1910 and October 1913, and was regarded as having been one of the three most outstanding commandants. His name had earlier been mooted as a possible contender for French's position as British C-in-C. In the event Haig received the appointment and historians speculate that Robertson's modest family background worked against him. Could he possibly do anything to help an aspiring GSO2 from within his own headquarters? The answer was 'yes'. Not long after his tennis meeting with Robertson, Monty found his name added to the list of students for the 1920 intake. The able and ambitious young officer was now making his own luck.[2]

Having brought himself to Robertson's attention, Monty found himself guiding a civilian friend of his Commander-in-Chief on a tour of the French battlefields. The frugal Monty (his chosen profession had meant that he remained impecunious) wrote to his mother on 18 August 1919: 'The trip will be Wiesbaden – Strasbourg – Nancy – Metz – St Mihiel – Rheims – Chemin des Dames – Soissons – Mons – Mauberge – Louvain – Cologne. A nice round trip and over a thousand miles by car altogether . . . He is a rich man and I get all my expenses paid.'[3] It is clear from this itinerary that the destinations were not exclusively associated with British battles: although Monty knew something of Mons (predating, as it did, Le Cateau by a few days) and the Chemin des Dames (where he had fought in May–June 1918), the other battle zones were contested by French and American troops, and would have broadened Monty's own knowledge of the war. On his return, he found himself re-promoted to lieutenant colonel, to command 17/Royal Fusiliers for three months, until a spot of leave preceding his arrival at Camberley on 22 January 1920.

The Staff College at Camberley had reopened its doors in January 1919, having terminated its last pre-apocalyptic course in August 1914 with the outbreak of war. Students on each of the three year-long post-war courses (1919–21) were specially selected on the basis of their aptitude and experience in the war, and thus spent a shorter time at Camberley (subsequently courses reverted to the pre-war model of two years). Presided over by the chief instructor Brigadier General (later Field Marshal Sir John) Dill, all the 104 students of 1920 seem to have had a 'good war': their ranks included (excluding bars) sixteen awards of a CB or CMG, seventy-eight DSOs, thirty-six MCs, and thirteen with a CBE or OBE.[4] Three were colonels and thirty-seven lieutenant colonels, but most had held higher acting rank.[5]

Amongst Monty's exact contemporaries at Camberley were Major Victor Fortune, the unlucky commander of 51st Highland Division in 1940 and – both as fellow student and later, fellow instructor and corps commander – Major Richard O'Connor, who had commanded the 2/Honourable Artillery Company and won two DSOs and an MC. College records from the time have not survived but we know that Monty's course finished in December 1920. With his wartime staff experience he sailed through the truncated one-year Camberley course to be appointed, in January 1921, as Brigade Major of the 17th (Cork) Infantry Brigade, in the midst of the Republican insurgency in southern Ireland. Brigade Major's appointments were (and remain today) the prizes awarded to the brightest emerging from Staff College, and the assumption that Monty did very well there is reinforced by the fact that he would return within five years as an instructor.

Monty's arrival in Cork coincided with the worst moments of the three-year Irish War of Independence, as Republicans strove to wrest independence from their British overlords by armed insurrection. Volunteers (mostly members of the Irish Republican Army) were attacking government property, carrying out raids for arms and money, and murdering prominent members of the British administration, policemen and soldiers. Many Royal Irish Constabulary barracks had already been burned along with income tax offices and court buildings. This had caused the predominantly Catholic police force to abandon large swathes of the countryside to the IRA, whilst the courts system collapsed. Law and order, and safe passage between towns had come to rely on regular troops like Monty's 17th Brigade and auxiliary (often ill-disciplined) paramilitary policemen.

Monty's role was a key one within the 57,000-strong British garrison which supported 17,000 Royal Irish Constabulary and auxiliaries known as the Black and Tans. The extreme violence used by the Republicans (ambushes, intimidation, doorstep murders) was initially unpopular with the Irish people, but by the time of Monty's arrival, the heavy-handed British response had made heroes out of the IRA, of whom 4,500 had been interned. Although a truce was agreed in July 1921, it was frequently abused and whilst Monty remained with the Cork Brigade, over 250

regular British troops would be killed in Ireland. Cork was thus a difficult assignment, the more so for Monty was of Irish Protestant descent,
with his family seat, Moville, near Londonderry in the north of the
island; while in November 1920 his cousin and contemporary, Hugh
Montgomery, a regular officer who had been a staff officer in the late
war, was assassinated by the IRA in Dublin just after Monty's thirty-
third birthday.

Counterinsurgency is a notoriously difficult skill to master: deploy
too harshly (even if militarily effective), and you run the risk of antag-
onising public opinion, politicians and the media; too soft an approach
will leave the rebels unsubdued and waiting for your departure. Only
two years earlier, on 13 April 1919, Brigadier General Reginald Dyer
had used effective but excessive force to quell unrest in Amritsar in the
Punjab, killing nearly 400 people and injuring at least a thousand.
Churchill called the incident 'a monstrous event' and Dyer, who appeared
to have had no sense of proportionality, was required to resign immediately and was never re-employed. In 1923 Monty wrote to Arthur
Percival about the campaign in southern Ireland. The letters show in
Monty an incisiveness and confidence beyond his years:

> My own view is that to win a war of that sort you must be ruthless;
> Oliver Cromwell or the Germans would have settled it in a very short
> time. Nowadays public opinion precludes such methods . . . if we had
> gone on we could have squashed the rebellion as a temporary measure,
> but it would have broken out again like an ulcer the moment we had
> removed the troops. The only way therefore was to give them some form
> of self-government, and let them squash the rebellion themselves; they
> are the only people who could really stamp it out . . . it seems to me
> that they have had more success than we had.[6]

Monty wrote here with much foresight and wisdom, for the issue of
public opinion would go on to vex the British authorities in Northern
Ireland in the 1970s and 1980s, while the tactic of using self-government
to repress insurgency would be employed in many post-war, post-
colonial theatres. In view of the fact that one of Monty's protégés was
Gerald Templer, the 'Tiger of Malaya', responsible for much of the post-
1945 success in containing insurgencies, there exists a direct chain of
experience between Monty in Ireland in the 1920s and Britain's post-
war record in Malaya, Kenya, Cyprus and Borneo, while the spectre of

failing to understand and correctly apply counterinsurgency doctrine hangs over British and American forces in Afghanistan as I write.

In December 1918, Erwin Rommel was posted from the staff job he hated, just as everything he had known and represented – his regiment, his king in Württemberg and his Kaiser – was being swept away by the Allied governments. This was a hugely difficult time for most Germans, on many levels, whatever their political persuasion. Some might have accepted the *Dolchstosslegende* – the Hitler-appropriated myth that the German army was 'stabbed in the back' on the home front by Jews and Communists, but the painful truth was that the Kaiser's army had been fought to a standstill by the better-trained, better-equipped and more numerous Allied armies of Britain, France and America. Standstill was the key word, for there was no surrender as such, but an armistice: in reality a ceasefire to allow time for terms to be negotiated, when German units marched home with their weapons, preceded by their bands and banners; hence the knife-in-the-back legend because the imperial army was still intact when the ceasefire occurred.

Being away from the front, the transition from war to peace was much less dramatic for Rommel than for Monty. From his Württemberg Corps headquarters behind the lines in Alsace, Rommel was merely reassigned to his old regiment, 124th Württemberg at the depot in Weingarten on 21 December. His role was to oversee the gradual disbandment of the Württembergers, with whom he had started his soldiering in 1910. It is a tribute to the efficiency and loyalty of the German war machine that there was still an organisation left to do the reassigning; the old state apparatus had not quite disappeared.

Rommel used the opportunity to travel by rail across Germany to Danzig to rescue Lucie, his bride of two years, and bring her back to his mother's lodging in Weingarten for Christmas 1918. He chose to wear his uniform for the journey (inevitably sporting the *Pour le Mérite* – he was never without it); this was probably a mistake, showing how out of touch he had become with wider society. During the trip he was questioned, mildly insulted and once nearly arrested by discontented revolutionaries – mild treatment at a time when others were kidnapped, roughed up and even murdered by piratical bands of unruly soldiery, many operating in the name of 'brotherly Communism'. It was part of the sort of violent upheaval Bismarck had always feared: the unleashing of politically inspired frustration by a predominantly urban working class. And it was all a huge culture shock for Rommel, who was a product of the Kaiser's highly militaristic society, where uniforms were worshipped.

The period up until the signing of the Treaty of Versailles (28 June 1919) continued the anxiety for Rommel: his only trade was soldiering and he did not even know if his army would continue to exist. The upheaval and uncertainty remained for Germany's partners in the war; it was only on 10 September 1919 (in the Treaty of St Germain-en-Laye) that the much-reduced, Germanic-speaking regions of the old Hapsburg dual monarchy were established as the Republic of Austria and on 4 June 1920 (in the Treaty of Trianon) that a shrunken Hungary emerged as a separate state. That summer Rommel was in southern Swabia, at Friedrichshafen and then from June 1919 in Schwäbisch Gmünd (where he had been at school), commanding a company of the 25th Internal Security Regiment. The regiment, comprising soldiers and sailors, had nowhere to go and was in an ugly mood. Without rank or uniform, Rommel was forced to rely on his strength of personality to teach, drill and discipline his band. He showed himself to be a highly adept leader, and did well enough for the head of the police in Stuttgart to recruit some of his group. The few photographs of Rommel at this time, dressed in civilian clothes, show a man ill at ease, almost shifty, and uncomfortable out of his field grey.

Although Swabia was not as torn by internal unrest as Bavaria, the atmosphere was tense. Old scores were being settled and a new class was starting to emerge from the shadows. Germany desperately needed law and order. In most cases this was provided either by the state forming soldiers into ad-hoc units like Rommel's, or by right-wingers who were able to raise their own *Freikorps* (Free Corps) units. *Freikorps* were a German tradition of volunteer military bands originating in the eighteenth century. In 1919 many soldiers, disconnected from civilian life and looking for stability within a paramilitary structure, flocked to them. Other *Freikorps* were overtly political and anti-Communist. They were totally unofficial but had uniforms and helmets and utilised the arsenals of the Kaiser's army, including machine guns and tanks. These corps were particularly active in Bavaria. In view of their right-wing leanings it is no surprise that many *Freikorps* would soon evolve into the early SA (*Sturmabteilung*) and SS (*Schutzstaffel*) units of the Nazi party. Indeed many future Nazi leaders emerged from the thuggery of the *Freikorps* days, including Ritter von Epp of the *Alpenkorps*, Heinrich

Himmler and Sepp Dietrich, and the sinister Reinhardt Heydrich, who would head the SD (*Sicherheitsdienst*). Rommel remained detached from such political groups during these years: he was an army man and just wanted to stay part of the army for which he had fought so hard. He was probably also shrewd enough to realise that the various *Freikorps* would have only a limited existence – and then what?

By the Treaty of Versailles, Germany was required to reduce its vast and long-established army to a size of 100,000 (seven divisions of infantry and three of cavalry), which included an officer corps of no more than 4,000. The implications for Rommel were profound, for, *Pour le Mérite* or not, he was faced with the real possibility of having no job. To ensure that Germany did not form any significant reserves from freshly discharged troops, privates and NCOs were required to enlist for a long period – twelve years – and officers for twenty-five in the new *Reichswehr*. The minimal officer presence and the very size of the force ordained in 1919 were guarantees that Germany's army would be too small to be any kind of effective, offensive military force. Versailles also demanded that the General Staff corps (its *Kriegsakademie*), and the officer schools (*Kriegsschulen*), where Rommel had been educated, were outlawed. But the officer tasked to design the new army structure for President Ebert, the sixty-eight-year-old General Hans von Seeckt, managed to retain the essential elements of the old, though suitably camouflaged.

Short, neat and almost dainty (known as 'the Sphinx with the monocle', no doubt for the enigmatic smile he invariably wears in surviving photographs), Seeckt was an extremely able and highly cultured staff officer. He rebuilt the General Staff within the new army's *Truppenamt* (Troop Office) and created an organisation that was flexible and could be expanded in the future. Every regiment and unit had to prepare plans for a potential seven-fold future expansion and each officer and NCO was trained, as in 1914, to be able to command 'two-up'. Seeckt was clever enough to realise he could create a very powerful force with the tools Versailles had given him, for technology was moving ever forward and his relatively small structure could hold enormous potential.

Seeckt selected 4,000 out of the existing 34,000 officers – fewer than one in eight – for his new force. Rommel was not necessarily assured of a place because, although younger candidates were preferred to older men, General Staff officers received preference over all others, followed by *Freikorps* veterans and then pre-1914 regular junior officers. In view of the social unrest of 1918–20, often class-inspired, officers from the conservative middle and upper classes were favoured over those from more humble backgrounds, which ensured a disproportionate number of well-educated aristocrats in the new officer corps: nearly fifty percent of the new army's generals and colonels were noblemen, and half the cavalry's officers (in a country where less than one percent of the population claimed noble birth).[7] Rommel must have thought he would be squeezed out by the 'vons'.

Seeckt envisaged his *Truppenamt* as epitomising silent, selfless devotion to the army; its soldiers should be anonymous, achieving greatness with minimal display. Seeckt picked the brightest and best for his officer corps whilst exploiting modern military technology to the utmost. Rommel need not have worried: he had an enviable front-line record, and experience of staff work and training; he was an excellent leader and was technologically curious – just the sort of man Seeckt required. Patronage was also important, and Generalleutnant Eberhard von Hofacker probably had a role here, behind the scenes. In December 1920, Rommel was informed he could stay on in the new army as a company commander with the 13th Infantry Regiment in Stuttgart if he wanted – which he most certainly did.

In May 1922, Montgomery's success as Brigade Major of 17th Brigade led to a second brigade major's appointment, to 8th Infantry Brigade in Plymouth. The brigade formed part of 3rd Division, whom he would take to war in September 1939 and oversee during the invasion of Normandy in 1944. During his time with the brigade he was much encouraged to develop training ideas by its visionary brigadier, Tom Hollond, a classmate of Churchill's at Harrow, who had risen from major to brigadier in the 1914–18 war, and had been Chief of Staff to the Inspector General of Training. This was a lucky break, for it gave Monty an early sense of where the post-war army was heading in terms of doctrine and training, before he moved on again in 1923 to York, as GSO2 of 49th (West Riding) Division.

In York, Monty encountered Francis de Guingand, a bright subaltern of the 2/West Yorkshire Regiment, who was thirteen years his junior and lived in the same mess. The pair had met before in Cork, where de Guingand's battalion was stationed in Monty's brigade. Despite the fact that de Guingand's 'days revolved around wine, women and gambling – in all of which he excelled'[8], the two made an important impression on one another: 'Freddie' would eventually become Monty's indispensable

Chief of Staff from 1942 onwards. De Guingand recalled of these days that Monty socialised very little, played a little golf, and when not reading (invariably military-related books, rarely fiction), engaged in the odd hand of bridge. Monty also spent his evenings 'coaching young officers in the subjects which later they would have to take in the Staff College exam'.[9]

It was during this time also that Montgomery got to know a lieutenant by the name of Basil Liddell Hart, who had also attended St Paul's School. The two became firm intellectual friends, exchanging thoughts and ideas on aspects of doctrine and training.[10] Liddell Hart (about to be medically downgraded and leave the army) found a sounding board for many of his innovative ideas on warfare in Bernard Montgomery. They would have other things in common, for Liddell Hart revised the War Office's *Infantry Training Manual* in 1920–21, as Bernard would do in 1929–30. They both commenced writing articles: Bernard in his regimental magazine the *Antelope*, later joining Liddell Hart in the pages of the *Army Quarterly*.

In York, Monty's task was to oversee all aspects of training the 49th (Yorkshire) Territorial Division, something for which he was well suited. It also gave him an opportunity to build on his experiences with the 47th (London) Territorial Division with whom he had served in 1918. Understanding the enthusiastic, part-time Territorial Army is a gift not always acquired by regular British commanders, past or present, but comprehending the volunteer ethos would prove important, given how many Territorial battalions and divisions Monty would command in the Second World War. He wrote along these lines to Basil Liddell Hart in July 1924:

> The Regular Army should without doubt give of its best to the Territorial Army ... I always tell our Division that there is only one Army in England, and we all belong to it, whether we are in the Regulars, Territorials, or OTC. In that Army there are two categories:–
> 1. Those who devote all their time to soldiering, i.e. the Regular Army.
> 2. Those whose main work is some other profession, and who only soldier in their spare time.
> Personally, I take my hat off to those in Category 2 every time.[11]

Although the Territorial Force, with whom Monty had served in 1918, had performed magnificently during the war, it was disbanded, like so many organisations, in 1919. It was restructured and reformed in 1921 by the Secretary of State for War (Winston Churchill, himself a one-time Territorial officer) as the Territorial Army, which it remains today.[12]

As was usual in an officer's career, Monty returned for a short while to his regiment, 1/Royal Warwicks, where he commanded 'A' Company from March 1925. Although he had twice held the rank of lieutenant colonel (in an inflated army), a company command – a major's job – was an important rite of passage. In fact Monty had actually only commanded a company for a month as a very immature acting captain in 1914; bizarrely he was only promoted to substantive major on 26 July 1925, although he had been granted the rank of brevet major back in March 1918 and had been a temporary lieutenant colonel twice!

So far, the main keys to Monty's success had been his focus and application. As he had few interests outside the army and socialised little, he was able to put in the long hours required to hone his professionalism. He continued this formula, which had worked so well in the war, into peacetime. There is scant evidence of any significant friendships forged outside the military until this stage. Women remained a mystery to him – perhaps the overbearing personality of his mother acted as a deterrent from pursuing girlfriends? – though he took solace in the company of his youngest sister Winsome and occasionally her friends. There certainly had been no sign of Monty being either interested in or willing to be distracted by affairs of the heart. However, at the age of thirty-seven, whilst on holiday with his CO in Brittany, his hormones seem to have belatedly kicked in and Bernard met and became besotted with the seventeen-year-old daughter of a senior colonial administrator. His idea of wooing the young Betty Anderson consisted of walks around the old walls of St Malo (highly interesting to any student of fortification, as I will readily attest) and drawing pictures in the sand illustrating the employment of armour in modern war. Miss Anderson turned him down, but the event seemed to awaken something in the newly promoted major.

Shortly afterwards, in the winter of 1926, Bernard met another Betty – Elizabeth Carver – whilst on a skiing holiday with the Andersons. Betty's husband, Oswald Carver, a Royal Engineers officer, former Olympic oarsman, and rowing blue from Trinity College, Cambridge, had been killed at Gallipoli. Since then the Carvers had fallen on hard times: the Depression had forced both the closure of the (substantial) family business and the family had to leave their equally substantial country house. Betty and her two sons from the marriage, John and Dick, had been obliged to downscale to a small cottage in Chiswick, which she filled with a vibrant assortment of poets and artists. She had ambitions to be an artist and was holidaying with her sons to celebrate her acceptance at the Slade School of Art in Bloomsbury when she and Monty first met, over breakfast at the Hotel Wildstrubel in Lenck. Betty's grandson, Tom Carver, captured the moment:

> On the second morning there . . . an Englishman walked up to their table and introduced himself. He had a neatly clipped moustache and was dressed for some reason in lederhosen . . . He was short, with vivid blue eyes and a beaky nose. When he spoke it was with a wiry, nervous energy as if he didn't have time to explain his thoughts in great detail. The image that came to [twelve-year-old] Richard's mind was of a Jack Russell dog . . . He said that he had been drawn to their breakfast table by the presence of the children, having none of his own.[13]

As it turned out, Betty's own family, the Hobarts, seemed remarkably similar to Monty's: Irish, Protestant, strong mother, service ethos, and she had four brothers, one an Indian civil servant, the second in the Royal Navy and two in the Regular Army (Percy, who attended Camberley in 1919 and Stanley, who was Bernard's contemporary at Staff College in 1920). Perhaps Betty was as keen to remarry as Bernard was to procure a suitable bride with whose family he felt he had so much in common; for, three months after they met, they were engaged – Monty having proposed, in uniform, at the unusual venue of the fives courts at Charterhouse (where the boys were at school). It seems to have been truly a love match, for many friends and relatives commented at the time, and afterwards, that they were chalk and cheese: Betty was artistic, well read and cultured (her Bohemian friends were horrified by the reactionary little colonel); whilst Monty was all army. She seems to have loved him for his energy and the passion he had for his vocation; and he loved the way she accepted him for who he was, quirks and all.

Despite the fact that Monty had earlier confided to de Guingand his belief that 'matrimony and the military life did not go well together . . . a young officer to be successful must spend so much time in getting to know his soldiers and studying his profession that a wife would receive inadequate attention,'[14] he and Betty were married quietly in Chiswick by Bernard's father, Bishop Henry Montgomery, on 27 July 1927. A son, David, followed in August the following year. Monty embraced the new roles of husband, parent and step-parent with characteristic enthusiasm, taking the children horse-riding and swimming, and attempting to plan every aspect of family life meticulously. Betty, too, embraced her new role as an army wife, forgoing her artistic ambitions and settling into army quarters in Camberley, where Monty took over the management of the household chores, dividing Betty's 'disorderly' life 'into carefully allotted spans, posting "Orders of the Day" on the dining-room door with instructions such as: "lunch at 1300 hours" . . . Betty often ignored them but she nonetheless found the routine reassuring.'[15] Monty's brother Brian similarly recalled house parties on the family estate at Moville, where Bernard would attempt to impose order by composing his daily programmes in the manner of battalion orders. Eccentric or genius?

Late January 1926 witnessed a key moment in Monty's transition from thrusting staff officer to battlefield commander as he and fellow Lieutenant Colonels Paget and Pownall[16] (both of whom had been at Staff College with Monty) arrived at Camberley as directing staff under the Director of Studies Alan Brooke and Commandant Major General Edmund Ironside (both later field marshals and CIGS). Thus began the blossoming of a relationship, based more on professional respect than social intercourse, which would result in Brooke, three years older, becoming Monty's patron – and protector – in the war years. This was one of Camberley's golden eras, when a unique and important mix of staff and students met, interacted and educated each other.

For the three years at Camberley Monty was in his element: all the traits inherited from his father, Bishop Henry, and maternal grandfather, Dean Farrar, in terms of an ability to write, lecture, explain and enthuse, came to the fore. Students were divided into syndicates, each ten-strong, supervised by a directing staff lieutenant colonel, who came from any arm or service. Each section of the two-year staff course was introduced by a series of indoor lectures, followed by syndicate discussions; concepts would be then demonstrated and enacted on the ground, through Tactical Exercises Without Troops (TEWTs) and manoeuvres. Each syndicate member was given an appointment, usually as a commander, or brigade or divisional staff; stress was laid on the writing of orders, appreciations and verbal presentations. In the first year students would learn aspects of battle (defence, attack, advance, pursuit, withdrawal) at the divisional level and in the second year theories of war and strategic issues, predominantly the defence of the Empire.[17]

Under Monty's direction his students at Staff College also explored the First World War battlefields and visited the military college at St Cyr, recording: the 'general impression gained was that France was disappointed with her late allies ... the man in the street undoubtedly considers that although both countries fought with him, they both made peace against him ... that resentment against the USA was deeper than that against Great Britain.'[18] Captain F. S. Tuker of the 2/Gurkha Rifles (who would command 4th Indian Division in Tunisia and Italy and demand the aerial bombing of the abbey at Monte Cassino) and his syndicate looked at Mons 1914 and then assessed how the battle would be fought with modern weapons in 1940 (of all years). They correctly decided to 'ignore' the limitations imposed by the Versailles Treaty, but did not foresee a great increase in the number of RAF aircraft; 'all artillery will be mechanicalised' they stated, but also 'the need is now apparent for an increased number of [horsed] cavalry [to act] as Divisional Cavalry ... to seize, resist, demoralise and destroy'. They foresaw a BEF of two corps, each of two divisions (which was the case in 1939).[19] Another of Monty's syndicates were more sceptical about the future:

> financial considerations will not permit of vast schemes of mechanicalisation ... Though the next war will undoubtedly bring about radical changes, there can be little doubt that at its commencement manpower and horsepower will not have been entirely replaced by mechanical power ... it is considered that infantry transport should remain horsed until the infantryman himself is carried in a cross country vehicle.[20]

Although the year this was written, 1926, was the year of the General Strike, the reference to 'financial considerations' is noteworthy because reductions in defence expenditure became acute only in 1932. Not even in the early stages of Normandy 1944 were British infantry housed in cross-country vehicles, though the Germans would make widespread use of half-tracks. Perhaps these scatter-gun guesses about the future can be explained by the fact that looking ahead a decade or so was not a common practice in the inter-war period; however, it certainly disproves the common belief that the Staff College (and by extension, its directing staff, including Monty) were stuck 'fighting the last war'.

Monty was joined at Camberley in 1927 by Lieutenant Colonel Richard O'Connor, when the Senior Division was sixty-two strong and included a VC, DSO, MC (Hudson, Sherwood Foresters), another twelve DSOs, forty MCs, seven with other decorations and sixteen with none. Students included Harold Alexander[21] and Douglas Wimberley[22]; the latter had already come to Monty's attention when Assistant Adjutant of the Cameron Highlanders in 17th (Cork) Infantry Brigade in 1921. Alexander's presence on the course, aged thirty-four and a full colonel, was most unusual, but he recognised he needed Camberley under his belt to progress so he agreed to be temporarily demoted to major for the two-year course. Monty and his fellow directing staff encouraged their students to explore a First World War battlefield and 'anticipate fifteen years hence, how the battle might be re-fought under the conditions likely to prevail in 1942'. This was a strikingly proactive use of history, and their students' reactions to the challenge illustrates how forward-thinking some junior officers were in 1927. When asked to cast their minds forward, Harold Alexander's syndicate[23] were broadly correct in their predictions, acknowledging the supremacy of coalition warfare in days when the Empire still – on the face of it – seemed more important:

> If . . . confined to the exploits of the students' own nation, military study becomes narrow and fails to achieve that breadth of view so necessary in those who would direct war successfully. This is particularly applicable to the British Army, which is seldom likely to be called upon to fight a European war, except in conjunction with foreign allies.[24]

But they badly misjudged the future of the tank:

> Tanks not being a commercial proposition, it is unlikely that we shall see a very large increase in their numbers, even in fifteen years time. Such increase as there is will be the product of money set aside for this form of military specialisation and will therefore be limited. On the other hand, the production of armoured cars is so much akin to that of touring and commercial vehicles that . . . a considerable increase in these weapons may be looked for in the near future.[25]

However they were correct in their thinking about anti-tank weaponry and machine guns:

> The presence of tanks, even in limited numbers, in the hands of the enemy, constitute so great a threat that every effort will be made to

procure an antidote. The next few years will see the introduction of improved *anti-tank artillery and mines . . . Automatic rifles and machine guns* proved the most deadly weapons to the infantry in the open during the Great War. They are the essence of fire-power and more economical to maintain than rifles; machine-gun units will therefore increase considerably, and a number of them will be mechanised [the Machine Gun Corps (MGC), raised in October 1915, had in 1922 been disbanded as a cost-cutting measure].[26]

The syndicates' conclusions give us the best assessment we have of the impact and accuracy of Monty's teaching whilst he was at Camberley, and provide valuable insights into British Army thinking between the wars. They also give us a window into the mind of Monty's fellow Army Group commander in Italy, 1943–45, Harold Alexander – the man with whom Eisenhower would have replaced Monty, had the choice been his. The Alexander syndicate went on to misjudge (as did many) chemical weapons, but were right about general mechanisation:

Chemical Warfare . . . will show a marked increase. Humanitarian objections will not be strong enough to prevail . . . *MOTOR TRACTION* . . . Improvements in cross-country vehicles, particularly six-wheelers, and the mechanisation of an increased proportion of field artillery will add greatly to the mobility and radius of action of formations [nearly all field artillery in 1927 was still horse-drawn] . . . Improved organisation, such as the introduction of extra machine-gun units and the provision of anti-tank and close support weapons as part of the infantry will increase mobility and firepower. [This prediction was accurate for 1940.] While the development of wireless, the increase in motor despatch riders and improvement in methods of signal routine will speed up the circulation of information and orders in the modern battle.[27]

They concluded, 'in modern war, owing to greater mobility and fire power and more rapid means of communication, the vulnerability of flanks is infinitely greater'.[28] Hindsight reveals, of course, that this flank vulnerability was exactly the German problem in their blitzkrieg through France thirteen years after Alex and his syndicate penned their thoughts.

Another 1927 syndicate under Monty studied the retreat from Mons and concluded with remarkable foresight:

As the size of . . . [Great Britain's] contingent is likely to be no greater at the beginning of the next European War than of the last, she may be early surprised by the attack of greatly superior enemy forces, and be forced hurriedly to convert an advance into a retreat; especially in view of the fact that she will probably be fighting without unity of command, and so without guarantee of coordinated action, by the side of Allies who may act independently . . . The employment of cavalry with mechanised forces also requires consideration, as the future of this arm will depend on its ability to maintain an equal relative speed to the rest of the army.[29]

This syndicate were scoring first-round hits with every observation. They went on to observe:

In Belgium, the people believe that there will be another big European War within the next ten years, and that their country will again be the scene of operations. They were evidently afraid of an invasion . . . by Germany and did not seem to be very trustful of support from England . . . They did not have a high opinion of the efficiency of their own army. As one graphically put it, they thought their army would retire to Spain if Germany invaded Belgium . . . The future of army mechanisation appears to lie with Britain and Germany, French and Belgian machinery is likely to prove unreliable.[30]

The sapper major, Brian Robertson DSO, MC[31] (son of the 'Wully' Robertson who had arranged Monty's arrival at Camberley in 1919) was at Staff College between 1926–27 and would later serve under Alex in Italy and Monty in Germany. In May 1926, his biographer states that the Camberley students were encouraged to volunteer to help keep essential services running during the General Strike, and went up to London dressed in their oldest clothes (to infiltrate the picket lines) and work as bus, tram or tube drivers and guards. What their directing staff, including Monty, did in the strike remains – alas – unrecorded. The following year, as Robertson's biographer relates,

Probably the most enjoyable part of the course took place during Easter 1927. Small groups of students were given sufficient money to travel to France or Belgium and study a First World War battle of their own choice. An interesting light was shed on the more congenial aspects of this tour in an article in *La Vie Parisienne: Chaque soir vêtus d'impeccable smokings ils dînent ensemble et discutent tactique et stratégie ... Les problèmes de stratégie sont multiplis, et les officiers anglais en soutenant leurs forces d'un peu de champagne et de brandy.*[32]

Alas, we remain unaware what the abstemious Monty made of his students' evening antics, as recorded in the Paris gossip columns. Oliver Leese of 1928 Senior Division recalled of his Camberley days:

> You were presented with almost every known military problem in the many syndicates and schemes with which you were confronted . . . you came away with complete confidence that no matter what problem might face you, either in peace or war, you would always be able to find a solution. You learned, too, at Staff College how to work. You were only given a limited time to get your exercise finished, and you knew that you had got to get it done in that time; you therefore learnt very quickly how to organise your work and how best to get it done in time, so that you were ready for whatever problem which might next be shot at you.[33]

These were the very qualities that would bring the brusque Oliver Leese to Monty's attention and cause Monty to select him first for XXX Corps in September 1942, then as his successor at Eighth Army in December 1943.

The directing staff and students' identities point up another important aspect of the Staff College, namely that the instructors would benefit hugely from getting to know a generation of staff officers and were well placed to pick trusty subordinates when the war came. Students in 1927 who would re-enter Monty's life at some stage included E. H. 'Bubbles' Barker,[34] R. C. Bridgeman,[35] E. E. 'Chink' Dorman-Smith,[36] A. F. ('John') Harding,[37] J. L. I. Hawkesworth,[38] O. W. H. Leese,[39] and W. R. C. 'Ronnie' Penney.[40] Dorman-Smith would disagree so violently with Montgomery's teaching that 'Chink' publicly burnt Monty's lecture notes on graduating – much to his detriment in North Africa in August 1942. When Monty replaced the sacked Claude Auchinleck, 'Chink' as the Auck's Chief of Staff, was dismissed immediately by Monty. The 1928 Senior Division included Captain P. G. S. Gregson-Ellis,[41] Wing Commander A. T. Harris,[42] Major R. L. McCreery,[43] and Captains G. H. A. MacMillan[44] and G. W. R. Templer.[45] The course also benefitted from the 'joint' approach to war (or 'combined' as mixed-service operations were then called), as witnessed by Arthur Harris's presence on the course. According to his biographer, 'Bomber' (as he later became) Harris was struck by Monty, and regarded him as:

> an excellent instructor and a man with advanced ideas on the likely pattern of future warfare . . . Ten years later [in 1939] . . . Harris drew on this knowledge of Montgomery and described him to his ADC as a 'very good soldier who will make a damned good general . . . he is the first soldier I have come across who has a proper grasp of the vital role of a tactical air force in land battles'.[46]

At the end of his last year, Harris was apparently asked to stay on and join the staff, but declined to do so.[47]

11

Preparing for War

MONTY'S TERM AT Camberley ended in January 1929, just after the tenth anniversary of the armistice. Around this time there was a literary outpouring of wartime nostalgia as the war's physical and mental scars began to heal. The most celebrated, Erich Maria Remarque's *Im Westen Nichts Neues*, was first serialised in the *Vossische Zeitung*[1] from 10 November 1928 and subsequently published as a book in January 1929, when it caused a sensation, selling 200,000 copies in three weeks. By March it had been translated into English as *All Quiet on the Western Front* and had sold 260,000 copies in six months. By December 1929, Remarque had sold a million copies in German and a further million in translation. Universal Pictures immediately bought an option and it became Lewis Milestone's hit feature film of 1930.[2]

The time lapse of a decade from the last of the fighting prompted many others to rush into print, such as Robert Graves, Siegfried Sassoon, Edmund Blunden, Ernest Hemingway and Ernst Jünger. *All Quiet's* arrival coincided to the month with *Journey's End* – still considered to be the best First World War play – by a thirty-two-year-old former captain in the 9/East Surreys who was wounded at Passchendaele – R. C. Sherriff.[3] The single, collective experience of the war on the Western Front seemed to have been remarkably similar, whichever side of no-man's land the authors found themselves. None of the other hard-fought campaigns (in the East, Italy, Salonika, Mesopotamia, Palestine, the Balkans, Gallipoli, Africa or even the naval war) caught the public imagination in the same way. Never before nor since have so many works of literature and memoirs from a single conflict appeared over such a short (two-year) time span.

Just as this cathartic, literary return to the 1914–18 battlefields took place, many veterans returned to the battlefields in person. By a quirk of coincidence, Monty and Rommel both revisited their First World War battlefields in the same year: 1927, Monty by bicycle with officers in his regiment, shortly before his marriage to Betty; Rommel with Lucie on the back of his motorbike, around the battlefields of France and Italy.

Erwin's reason for returning was deliberate battlefield tourism, but also – no doubt – to surrender a little of his warrior's reserve to his life partner. He took lots of photographs and made many notes which would find their way into his tactical lectures, and ultimately *Infantry Attacks*. They paused at Longarone, where he had won his *Pour le Mérite*, and Lucie discovered some headstones of her Italian ancestors, though they were moved on by an irate local official with a long memory – Rommel's deeds being less than ten years old; the German–Italian alliance was then a long way off.

Montgomery, on the other hand, used his Staff College notes to frame a two-week battlefield visit he made in April 1927, when he undertook 'a very Spartan bicycling tour of the First World War battlefields, taking with him his brother Brian and three other subalterns from 1st Battalion of the Royal Warwickshire Regiment'.[4] The trip started at Ostend, then took in Ypres, Tournai, Mons, Le Cateau, Lille, Armentières, returning via Ypres and Ostend. The tour was no mere 'swan' – the newly engaged lieutenant colonel had secured funds from the Eastern Command Training Grant and 'at every battlefield he would stop, give a lecture on what had taken place there, and ask each subaltern how, with modern firepower, the battle could be re-fought'.[5]

The return must have triggered some kind of cathartic experience, deep down, in both men. It is impossible to imagine invoking such memories without provoking a reaction, though Lucie later confided to her husband's first biographer Desmond Young that, apart from the scars of his wounds, the First World War appeared to have left no trace on Erwin: the transition from killing machine to peacetime warrior appeared to be seamless. For Betty, it was more difficult to tell, because she had not known Bernard before the war; for her, Monty had always been self-disciplined, intense and passionate. His mental focus was a product of exhaustive hours as a staff officer in the war; but he was also socially awkward, at times outspoken, and even rebellious – traits that went back further.

It may be significant that in the year following both warriors' returns to their battlefields – and confronting the ghosts of the past – each welcomed the arrival of a son: David, born on 18 August, and Manfred, born on Christmas Eve, 1928.

Now a brevet lieutenant colonel, Monty returned to 1/Royal Warwicks but was asked in October to revise the *Infantry Training Manual* for the War Office, as Basil Liddell Hart had done before him. This basic infantry doctrine was (and still is) revised every five years or so, to take account of new designs, developments and technology. Lord Gort had revised the intervening volume and in 1929 Monty determined 'to make the book a comprehensive treatise on war for the infantry officer'. The 'book when published,' Monty later commented, 'was considered excellent, especially by its author.'[6]

Battalion command followed, with 1/Royal Warwicks, who set sail for Palestine in January 1931. Normally the command of a battalion, especially in one's own regiment, is a key moment in the career of a regular infantry officer; it is the crowning moment when long service to a particular cap badge is recognised and the support the regiment has given an officer through his career is repaid. But in his memoirs Monty gave less than a page to this essential posting. In fact, his tenure in Jerusalem does not seem to have been the success widely expected: his method of promoting NCOs – on merit rather than long service – and his refusal to obey brigade directives aroused antagonism. Monty came under the GOC Egypt and Palestine, but since that general was based in Alexandria, Monty was essentially responsible for security within the whole state. Being the senior battalion commander based in Jerusalem, he also 'controlled' a second British battalion in Haifa. At the end of 1931, Monty's battalion was transferred to Alexandria and Montgomery lost his 'dictator of Palestine' status, though his GOC was kind enough to observe in Monty's Confidential Report:

> He is clever, energetic, ambitious, and a very gifted instructor. He has character, knowledge, and a quick grasp of military problems. But if he is to do himself justice in the higher positions to which his gifts should entitle him, he must cultivate tact, tolerance and discretion. This is a friendly hint, as I have a high opinion of his ability.[7]

*

January 1921 saw Rommel achieve his first *Reichswehr* posting, to 13th Regiment as a company commander, based in Stuttgart. Hans von Seeckt had effectively recreated the Kaiser's army in miniature: each of the *Reichswehr*'s small regiments and sub-units was assigned the tradition and continuity of one of the previous imperial regiments: thus 13th Regiment was also subtitled the *Alt* (old) *Württemberg*.

Although this posting was considerably below his proven command abilities in the war, Rommel seemed content with the relative stability peacetime soldiering offered, although the promotion prospects in a small army promised to be very limited (he would not be promoted major until 1932). He now had a wife and a job he loved; and so he was grateful to have been retained in the *Reichswehr*. It has been suggested that Rommel may have resented not being selected for the Kriegsakademie course (which would not, of itself, have necessarily led to a General Staff appointment, but which was seen as a distinct career enhancer) and that he 'came to regard too many of its members as remote and over-intellectual, compared to the front-line soldier schooled by his own experience'.[8] Instead, he made the most of what the army offered: he skied, canoed and rode with Lucie around Stuttgart in his beloved Württemberg. The scientifically curious Rommel re-emerged as he took apart his motorcycle, memorised logarithm tables, began stamp collecting, played the violin (none of these were pursuits recorded as attracting Monty) and became an active member of the WGB Old Comrades Association, which held annual meetings and parades. In 1924, Rommel took over the Machine-gun Company within 13th Regiment, but in every other respect his life altered little.

This stability would prove increasingly important in the turbulent years to come, when much of the Fatherland became embroiled in political chaos. One obvious cause of this was the massive inflation experienced throughout Germany which reduced the real value of wages and eroded middle-class savings. To a certain extent, Rommel – on a government salary – was better protected than most of his fellow citizens, and it is perhaps easy to see why many flocked to join parties like the Nazis or Communists, with their siren promises of work and bread. Rural Swabia – the area around Stuttgart – was less prone to extremism than neighbouring Bavaria to the east, or Prussia to the north. Its people were naturally thrifty and closer to the land; even today many families, even those with manual labouring jobs, own and tend livestock and vegetable gardens. An expression of the extreme political turbulence at this time was Hitler's attempt to seize power from the Bavarian provincial government in Munich on 8 November 1923. Although inspired by Mussolini's coup the previous year, the National Socialists had little popular support and the coup failed after intervention from an armed police unit, which killed fourteen Nazi supporters.

Seeckt was determined to keep his army out of politics and the early *Reichswehr* was designed to serve the state rather than any individual, whether this be Kaiser or Führer. Alas, the seeds were sown, innocently perhaps, that later encouraged many officers in the Third Reich to turn a blind eye to political developments they did not like on the grounds that they didn't concern them; political isolation encouraged political naivety, with many senior officers becoming apolitical rather than unpolitical. Seeckt was also scrupulous about absolute, unwavering obedience to the Fatherland, expressed by means of a sacred oath. In the context of 1920–23 it is easy to see why. Stability was confirmed for the army, and was certainly welcomed by Rommel, when Paul von Hindenburg – one of the army's own – agreed to become President in May 1925, succeeding Ebert. Often referred to as the *Ersatzkaiser* ('substitute Emperor'), though he claimed to be a simple, loyal soldier, Hindenburg had very acute political antennae. When called before the Reichstag Commission investigating responsibility for both the outbreak of war in 1914 and the defeat of 1918, he was the first to advance the 'stab-in-the-back' theory of disloyal elements at home and unpatriotic politicians being responsible for Germany's downfall. Many found it convenient to believe this figure of authority, who was never subsequently challenged.

In late 1928, Rommel was assessed by his commander at 13th Regiment as being 'a quiet, sterling character, always tactful and modest in his manner'. The report continued: 'he has shown some very good results training and drilling his company', and concluded: 'There is more to this officer than meets the eye.'[9] This, and previous reports like it, secured for Rommel the post of instructor at the Officers' Infantry School, Dresden, where he arrived on 1 October 1929, just before his twenty-eighth birthday. (Monty had just finished a not dissimilar job, instructing at the Camberley Staff College.)

The Infantry Officers' School had originally been in Munich, until some officers joined Hitler's 1923 attempted coup. Thereafter a furious Seeckt had relocated it to Dresden to prevent a future reoccurrence of this fusion of army and state politics. Perhaps inevitably, most of

Rommel's lectures on tactics at Dresden included first-hand narratives of actions he had fought in 1914–17, together with critical assessments of his decisions and actions. These ran the risk of being self-indulgent, but one of Rommel's emerging qualities was that he, like Montgomery, had become an excellent teacher and trainer of troops. He was a confident public speaker. He always liked to have pen and paper at hand and was forever doodling – a hangover from the days when all officers were required to be proficient in field-sketching. If a picture could tell a story better than words, then he would produce one, delighting his students at Dresden with quick sketches to accompany his talks. In 1931 his commandant, Generalmajor Wilhem List, wrote that 'his tactical battle lectures, in which he describes his own war experiences . . . are always a delight to hear'. A year later the senior instructor Oberst Karl-Heinrich von Stülpnagel described him as a: 'first-rate infantry and combat instructor . . . respected by his colleagues, worshipped by his cadets'.[10] Rommel's Swabian accent also marked him out: one can imagine him deploying it, loaded with all the vernacular gained from front-line soldiering, to the delight of his young students and the annoyance of his starchy seniors.

He also made some important friends at Dresden. His first commandant was Alexander von Falkenhausen (future Military Governor of Belgium 1940–44 and one of the 1944 Stauffenberg plotters), whilst the Colonel of the Training Division was Karl-Heinrich von Stülpnagel (future Military Governor of France and a future fellow plotter in 1944). Rommel got on better with his second commandant, Wilhelm List, who had graduated from the Kriegsakademie and spent his war as a staff officer in the Bavarian Army. Although List was eleven years older, their friendship stemmed from the fact that List was a native of Ulm, a fellow Württemberger and military 'outsider'; he was not a nobleman like Falkenhausen or Stülpnagel – his father was a provincial doctor. List would become an influential and rapidly promoted officer (*Generalleutnant* in 1932 and *Generalfeldmarschall* after the French campaign in July 1940); in the inter-war years he, like Hofacker, played an important role as Rommel's patron, long before Erwin encountered Hitler.[11]

At Dresden, some of the gloss was probably removed for Rommel by the reappearance as a fellow instructor of Hauptmann Ferdinand Schörner, whom he believed had acquired *his Pour le Mérite* in 1917. Six months younger than Rommel, and Hitler's last *Generalfeldmarschall*,

Schörner's name after 1945 would be forever associated with orders to execute without trial any German soldier found behind the lines without orders, as a means of combating desertion: 'he shot privates and colonels with equal zeal for the smallest infractions'.[12] But at the Officers' Infantry School in Dresden, Schörner was popular with the commandant, and was known as a joker. His favourite trick at formal functions – repeated at mess dinners around the world – was to plant the school's silver cutlery in the pockets of fellow diners and observe their embarrassment as it tumbled out of their pocket at an inopportune moment, it was a schoolboy prank to which Hauptmann Rommel had no reply.

The new *Reichswehr* trained its leaders (officers and NCOs) with infinitely more care than the Germany army had hitherto – over longer periods and placing previously unheard-of emphasis on sports and fitness. To Rommel, a wiry, natural sportsman, this was a welcome development. He had always excelled in country pursuits and enjoyed leading his young officers on hikes, rambles and exercises. A German tradition enjoying a resurgence, and one that Rommel embraced wholeheartedly, was that of the '*Wandervogel*' (meaning 'migratory bird'), which involved getting back to nature and leaving the restrictions of society behind. The movement started in the 1890s – a sort of cousin to the scouting movement – and gained enormous support in the aftermath of the First World War. It encouraged initiative, independence, resilience and a strong, young leadership culture. Although conceived of as non-political, many of its ideals would be hijacked by National Socialism and subverted into the Hitler Youth movement.

Rommel saw a younger version of himself and all that he had achieved to date in the next generation of *Reichswehr* students and he was determined to pass on all he had learned to his young followers. It helped that no matter how senior in rank he became he always behaved and appeared young at heart. At the end of the war he had shaved off the moustache which had added years and gravitas to his face; thereafter he appeared with his upper lip clean-shaven at a time when many colleagues and leading figures, including Hitler, still sported a bizarre array of facial hair. It was a small act of independence and rebellion that appealed to youth.

If the previous ten years had seen Rommel's career, though safe and stable, marking time, his four years at Dresden dramatically altered his future: he was 'a towering personality', reported the senior instructor. It takes a confident lecturer to fill the challenging slots of a Monday morning (or immediately after lunch), but Rommel was that man. 'I can guarantee they won't fall asleep on me,' he is meant to have claimed.[13] It was Rommel's time at the Officers' Infantry School that propelled him into a secondary career, as an author. He collected together many of his tactical lecture notes and published them as *Gefechts-Aufgaben für Zug und Kompanie* ('Combat Tasks for Platoon and Company') in 1934, by which time he was a battalion commander. It is not known how well this volume sold (the title suggests it was aimed at a very limited market), but he followed it with *Infantry Attacks*, which, in catching his Führer's eye, would be life-changing.

One of Rommel's officer cadets at Dresden would one day become a distinguished subordinate of his: born in Flensburg, the son of a naval officer, but of ancient Prussian army ancestry, Hans-Ulrich von Luck was twenty when he arrived at the Dresden *Kriegsschule* in October 1931. Luck recalled:

> Here, in the pearl of Saxony, I met for the first time Erwin Rommel. He was a captain, our infantry instructor and at the same time our most popular training officer . . . He was forty-two years old when I met him, tall, strong, tough, wearing a severe uniform with a high collar, but a man with a warm and sympathetic smile. He told us war stories – we hung on every word – and his book *Infantry Tactics* was our bible.[14]

I knew Hans von Luck in his later years and accompanied him on several occasions trampling through the hedgerows of Normandy. A spry, very fit, larger-than-life character with an easy command of many languages, he could drink anyone under the table and brought flair and panache to all he did. Like the Ancient Mariner, he had many splendid tales to tell of his war but could get carried away: the above passage is a good example; and a cautionary note in dealing with memoirs. Whereas Hans was tall (over six feet), Rommel was not tall, only about five feet six inches; Erwin was forty when they met, not forty-two and his book, *Infantry Attacks* (not *Tactics*) was not published until 1937 – long after Hans left Dresden.

In Alexandria, where Lieutenant Colonel Montgomery had arrived with his battalion in late 1931, Freddie de Guingand reappeared when the West Yorkshire battalion of which he was adjutant was also posted to the Canal Zone alongside Monty's. Twice he assisted Monty in an important exercise when the latter was role-playing the brigadier's job; acting as Montgomery's brigade major, Freddie first commandeered a rare RAF reconnaissance plane to provide Monty with much-needed intelligence; then he used a camel troop to guide the brigade through the desert.[15] In Alexandria, Monty was joined by Betty and David (Bernard had reverted to a bachelor existence in Jerusalem), and whilst in Egypt they were visited by old friends, including the novelist and MP Alan Herbert and his wife.

Still as commanding officer of the 1/Royal Warwicks, Monty embarked for Poona on Christmas Eve 1933, having managed to antagonise his brigadier, the neighbouring infantry battalion and the local cavalry regiment. No one doubted Monty's ability, but his brigadier, Frederick Pile, a Dubliner and a baronet, observed that whilst he 'should attain high rank in the army. He can only fail to do so if a certain high-handedness which occasionally overtakes him becomes too pronounced.'[16] Monty would continue to cause trouble in Poona, where the general view was that his habitual rudeness made him needlessly unpopular. Betty had dutifully sailed out to Egypt with young David, whilst her own sons, Richard and John, remained at home, boarding at Charterhouse. This arrangement continued with the move to India.

Bernard cannot have relished returning to India, even as a commanding officer; it was a place that held unhappy memories from his first military years, 1908–13. That experience had left him prejudiced against the languid ways of India – pink gins, cane chairs, revolving fans and a heat which sapped the energy of even the most robust and intellectually curious. It need not have been so, and may have been one of Monty's personal failings, for Wavell, Slim, Claude Auchinleck – even Winston Churchill – had all flourished in the Indian military environment. It also left him forever prejudiced against Indian Army officers, especially Auchinleck. At one stage, according to Brian Montgomery, Monty's entire battalion was turned out on full parade in front of their brigadier, an old Guardsman who had no interest in training, but an obsession with drill. When Monty's superior told him that he was not standing in the dead centre of his command, but was six paces too far to the right, Monty's reaction was simple and characteristic: instead of moving himself, he ordered 'Royal Warwickshire. Six Paces. Right Close – March'!

In the spring of 1934, Bernard and Betty left India for a six-week cruise to the Far East, having first sent young David back to Blighty with a nanny. Amongst the other passengers on board was not only his GOC (the Guardsman), but Generaloberst Hans von Seeckt – the Sphinx with the monocle – sailing out to act as an adviser to Chiang Kai-shek. Monty apparently had many lengthy discussions with Seeckt via an interpreter; he was probably the first German general Monty had ever encountered (the next would be the acting head of the Afrika Korps, Ritter von Thoma, in 1942) and one wonders who dominated these proceedings. With the mail waiting for Monty at Hong Kong was an invitation from the Indian Army HQ at Simla for Monty to take up the position of Chief Instructor at the Indian Staff College, in the cool foothills of Quetta (still the Staff College for the Army of Pakistan today). Although Bernard was desperate to leave Poona and his unsympathetic GOC – he feared another three years in India might be detrimental to his career – he took up his duties on 29 June with promotion to full colonel. Also stationed there, as Adjutant of 2/West Yorkshires, was Captain Freddie de Guingand, who had just sat – and failed – his entrance exam for a Staff College place at Camberley. Confiding his misery to Monty, the Chief Instructor at Quetta wrote immediately to the Commandant at Camberley to warmly recommend his protégé. De Guingand was accepted for the 1935 intake and would more than repay Monty's kindness in the years to come.

In contrast to his stint as CO of 1/Royal Warwicks, Monty's time at Quetta was judged by all concerned to be an outstanding success, on account of his flair for instruction and the innovative attempts he made to liven up training and the exercises – though few knew what to make of his introduction to one lecture 'When I am Commander-in-Chief of the Allied Armies in the next war . . .'.[17] Monty and Betty had a blissful stay at Quetta and their quarters there are still known in Pakistani army circles today as the Monty House. He returned with his family to England in May 1937, having been offered command of 9th Infantry Brigade, of 3rd Division, based at Portsmouth.

Much had changed during Monty's six-year absence from Europe. After the First War, Lloyd George's cabinet had imposed stringent limits on defence expenditure, and stipulated that future personnel and equipment budgets be based on the assumption that there would be no major European war within ten years and that no expeditionary forces would be needed. In the aftermath of a terrible war, with the German army defeated and dismembered, and the Imperial German Navy scuppered at Scapa Flow, this was not an unreasonable supposition to make. The 'ten-year rule' remained until 1932, and was linked with the policy of successive governments to resist any increased spending on defence through fear of financial disaster, coupled with the personal moral repugnance of leaders like Baldwin and Chamberlain of preparing for a war if they could possibly avert it. It was those politicians who had fought in the Great War and seen its horrors, like Churchill, Eden, Macmillan and Duff Cooper, who took the view that Britain needed to prepare to fight another war. Arguably if Baldwin and Chamberlain, with no military service, had experienced the Western Front, then their adherence to appeasement – and financial prudence – might have been less likely.

Between 1933 and 1935 there was government reluctance to face up to the realities of defence. Chamberlain in particular trusted that the 'veneer' of a strong air force would act as a robust enough deterrent to Germany (the chief, perceived aggressor), without the need to deploy land forces to the continent. It was only external pressures, brought about by the withdrawal of Germany and Japan from the League of Nations in 1933, the Italian invasion of Abyssinia in October 1935, and the possibility of civil war in Spain, that were instrumental in persuading the cabinet that defence expenditure could no longer be avoided. There was, however, a mood of pacifism in Britain at this time that had been emphasised by the decision of the Oxford Union in 1933 to pass a motion following a debate, declaring that: 'This house will under no circumstances fight for its King and Country'; whilst only passing attention might have been given to the extreme views of some university undergraduates, Oxford and Cambridge were widely accepted as being an international 'finishing school' for future leaders in politics, government, diplomacy, law, industry and the arts.

In May 1937 the fortunes of the British Army began to change with the appointment as War Minister of Leslie Hore-Belisha, who had served as a major in the First World War. Unconventional, and alive to the potential of publicity, over the next two years he would raise the plight of the army before the public, coupled with the message that support for the military could redress the military balance in Europe. (Alas, he also aroused considerable personal animosity from the establishment, not only for his innovations, but also on account of his Jewish ancestry.) He took advice from Montgomery's friend, Basil Liddell Hart, military historian and defence correspondent with *The Times* (who also lobbied hard to expand the armoured corps, with J. F. C. Fuller) and proposed some original ideas, which antagonised the War Office not least because Liddell Hart (a mere former captain) was also advising Hore-Belisha on the highly contentious issue of the promotion and appointment of senior army officers. The look of the British Army of 1939–45 would owe much to Hore-Belisha: he introduced, among other things, battledress, 1937-pattern webbing, the Bren machine gun, 40mm Bofors anti-aircraft cannon (and also, as Transport Minister, the Belisha Beacon for pedestrian crossings). In 1939, quite by chance, Hore-Belisha chose as his military assistant the young, thrusting Francis de Guingand.

One of the final pieces of Monty's persona enigma falls into place with the tragic events of the summer of 1937. On 21 August, Montgomery was in Portsmouth busily training his brigade for a future war as Betty holidayed with ten-year-old David at nearby Burnham-on-Sea. While on the beach she was bitten by an insect, and afterwards contracted septicaemia. For two painful months she battled, during which time she had both her legs amputated in an attempt to halt the blood-poisoning. But she succumbed on 19 October, dying in Bernard's arms. Later that same day, Monty wrote to his stepson, Richard:

> Life is very black at the moment. I do not know what I shall do without her. It will be very hard for you to bear . . . Mummy looks very peaceful now. There is a look of complete calm and rest on her face.[18]

Though they had married late, he and Betty had been blissfully happy for just over ten years. He was devastated and refused to let any other family members attend the funeral, recording very privately:

> I sat in the room at the hospital until they came to screw the lid on the coffin. I kissed her dear face for the last time just before the lid was put on . . . I tried hard to bear up at the service and at the graveside. But I could not bear it and I am afraid I broke down utterly. I feel desperately lonely and sad. I suppose in time I shall get over it, but at present it seems that I never shall.[19]

He hid away for four days, then concluded that work was to be his best (and perhaps only) therapy; he leant heavily on his Brigade Major Frank 'Simbo' Simpson for emotional support (Simpson would play a key role several times in his future career), and then immersed himself completely in his vocation. Alun Chalfont astutely asserts that Monty, for the first time in his life, was literally beaten into submission, which rings true. Everything he had so far attempted had succeeded, and now, with Betty's death, he was powerless to influence events. If this drama threw Monty into work on an even more impressive scale than in the earlier war, it caused a breach in his relationship with his son, David. Monty had sent him back to boarding school whilst his mother was still alive, but obviously (to him) dying. Henceforth, there would be respect, but distance, and David was brought up by a series of distant relatives, in between long bouts of boarding at Winchester. Monty perhaps felt he could not cope with the pain and loss of anyone else dear to him, so he kept the rest of the world at bay. This self-fashioned armour plate would insulate him from hurt, but also from empathy, for the rest of his days.

Bernard's elder stepson, John, mused in later life that had Monty never met his mother then his eccentric behaviour and aloofness might have developed 'to such an extent as to have rendered him unsuitable for high command'. There is evidence that Betty brought him added confidence in himself, where previously he might have succeeded through sheer bravado; she was a steadying, but also maturing influence; she introduced him to a world beyond the army and to friends, including A. P. Herbert and the painter Augustus John. And had loved him unconditionally for who he was.

After fourteen years a captain, Rommel was promoted to major in April 1932, remaining at the infantry school in Dresden until September of the following year. 1933 was an ominous year: that January Hitler had been sworn in as Chancellor of Germany. In February the Reichstag had been set on fire, and the freedom of the press had been limited, whilst the Enabling Act – allowing Hitler to issue decrees with the force of law through his cabinet, thus circumventing the elected deputies in the Reichstag – had been signed in March. While an increasingly senile Hindenburg lingered on (he would die from lung cancer on 2 August 1934), Hitler converted Germany from a weak, strife-torn democracy into a strong-willed dictatorship. When on Hindenburg's death Hitler merged the positions of head of state and head of government, declaring himself Führer and *Reichskanzler*, ninety percent of the electorate (the army, as Seeckt intended, did not vote) endorsed the move in a plebiscite. Rommel and his colleagues remained detached from politics, regarding (if they had a view at all) the Nazis either as lower-class thugs or crude operators whom they could manipulate as they wished. The *Reichswehr* (comprising the *Reichsheer*, the army, and the *Reichsmarine*, the navy), however, welcomed many aspects of Hitler's policies – stability at home, a strong military and robust reputation overseas – and felt they could control 'the Austrian Corporal' (as Hindenburg had once dubbed Hitler in private).

On 1 October 1933, Rommel was promoted to *Oberstleutnant* (lieutenant colonel) and posted as commanding officer of III Battalion, 17th Infantry Regiment, based at Goslar, an old garrison town deep in the Harz Mountains of central Germany with scenery not unlike Swabia. Heinz Guderian had been commissioned into the regiment in 1908, commanded one of its companies in 1920 and is buried in the town, so the area today remains steeped in military tradition. The local barracks proudly records the coincidence that two of Germany's most important military sons commanded troops here. Rommel's command was the distinguished *Goslar-Jäger-Battalion* (equating to Light Infantry), a cut above the average foot soldiers, and as close to Erwin's experience with Major Theodor Sprösser and the WGB as he was likely to find. In this pre-conscription era, the ranks comprised the countrymen of Lower Saxony – hardy types who were natural sportsmen and good skiers. The *Goslar-Jäger-Battalion* was descended from a Hanoverian regiment which had fought under Wellington as part of the King's German Legion in the Peninsular campaign and at Waterloo. *Jäger* battalions were hunters and Rommel insisted that all his officers must be able to stalk and shoot. Then (as now) this was a serious, very German affair of practical and written exams, certificates and permits, allowing the individual to hunt with a game rifle. Rommel filled his own house in Goslar with hunting trophies, taken in the surrounding woods.

His regimental commander, Oberst Kurt von der Chevallerie (a future First Army commander in France 1944), reported that Erwin was head and shoulders above his fellow commanders, and that the *Goslar-Jäger-Battalion* 'is the Rommel Battalion' – a well-meant, but double-edged compliment in an era when Seeckt had taught his officers to be modest and invisible. Indeed one battalion anecdote describes how in Rommel's first few days his officers invited him to ski with them down a local mountain, a test to see whether the almost-forty-two-year-old CO could match their athleticism. There being no ski-lift, they hiked up and Rommel led the way down at a breathtaking speed that compelled them to agree that he was indeed an impressive sportsman. Rommel then invited them to a second, and a third descent. At this point, the remaining officers dropped out, except Rommel, who set off for a fourth.

Rommel was in Goslar during the dramatic events of 30 June 1934, when Hitler ordered the massacre of his senior *Sturmabteilung* (SA, or brownshirt) leadership, in what became known as the Night of the Long Knives. Under Ernst Röhm, the SA movement had emerged (often from *Freikorps* detachments) in most towns as Hitler's thugs, and by 1934 was a highly organised, uniformed group, numbering some half a million. But Röhm's language within the brownshirt movement – full of Communist-inspired terminology talk of revolution, class war and the overthrow of the status quo – alarmed middle-class Germans and the army. Indeed Röhm saw himself as an alternative to the regular army and may have been about to launch some sort of coup; but Hitler struck first, using Himmler's SS to round up and execute Röhm and his colleagues with weapons supplied by the army. There was outrage abroad at Hitler's ruthlessness, but rather less than might have been expected at home. At the Nuremberg rally that September, Leni Riefenstahl's celluloid showed the presence of senior army officials at a Nazi party event for the first time, confirming the support the *Reichswehr* had started to give the party – a devil's bargain in exchange for the SA threat having been removed. The brownshirts thereafter remained in being, but carefully controlled.

The day President Hindenburg died, 2 August 1934, the entire German armed forces, of every rank, was required to swear a new oath of allegiance to their Führer:

> I swear by God this holy oath that I will render to Adolf Hitler, Führer of the German Reich and People, Supreme Commander of the Armed Forces, unconditional obedience, and that I am ready, as a brave soldier, to risk my life at any time for this oath.

The essential point here is that the old oath was to the state, whereas this was to an individual; the office did not feature. No one disagreed, there was no debate; the wording had obviously been prepared with great care some time before. Hitler, the ex-corporal who had sworn a similar oath to his Kaiser, understood better than most how binding this oath was: it was sacred and unbreakable (a point that German generals would make at Nuremberg in 1946–47 when defending the most un-soldierly of activities).

Politics began to claim Rommel's attention from 30 September 1934, when the Führer visited the eleventh-century *Kaiserpfalz* (Emperor's Palace) in Goslar, shortly after that year's Nuremberg rally. Rommel had been asked to provide the guard of honour. Even this event had an odd

Rommel twist to it. He was stung when told that a single file of SS men would stand in front of his troops as security. Never short of moral as well as physical courage, Rommel threatened to cancel his battalion's parade if this were the case. In the event, he was invited to discuss the arrangements directly with *SS-Reichsführer* Himmler, who in 1934 was already in possession of a fearsome reputation (as a result of the Night of the Long Knives), and the *Reichsminister für Volksaufklärung und Propaganda* (Public Enlightenment and Propaganda), Josef Goebbels. Rommel got his way and the parade went ahead as originally planned. It was Rommel's first contact with the gang – Himmler, Goebbels and Hitler. Whether he liked it or not, his beloved army and the party were growing closer together.

On 16 March 1935, in outright defiance of the Versailles Treaty, Hitler announced the return of conscription – news that was welcomed throughout Germany. The expected international condemnation was muted, however: countries watched Germany enviously as the Fatherland emerged from the extreme political and financial crises of just a decade earlier. To Rommel and the army as a whole, these promised to be busy and interesting times: conscription meant a larger army, which implied much better promotion prospects. The net effect would be to expand by six-fold the 4,000-strong officer corps permitted under Versailles into an institution that was 24,000-strong in 1939. Rommel would soon see the impact, for in October 1935 he was posted to the Potsdam *Kriegsschule*, where he would stay for three years as a senior instructor.

This institution had nothing to do with the old *Kriegsakademie* (as some historians have erroneously suggested), which had trained military staff officers for the higher direction of war and was banned by Versailles. On expanding the army, four *Kriegsschulen* were created at Potsdam, Dresden, Munich and Hannover; a fifth in Austria at Wiener-Neustadt was added in 1938. At each, officer candidates undertook an extensive basic course, after which they were posted to their chosen service school – infantry, artillery, engineers, or armour, for example. The Potsdam school contained about a thousand officer cadets; all were treated as infantry and divided into two detachments, each of 500. Rommel's role was to teach the young cadets how to lead an infantry platoon, company and battalion in combat. The cadets role-played most of the jobs they would encounter, including that of battalion commander.

This was Rommel's first posting to the heart of Prussian militarism, and he found he was still a countryman at heart: he and Lucie met and mingled with students and fellow instructors, but neither felt they were suited to the fast-lane life of a capital city. As far as the army was concerned, Hitler enabled it to reclaim the status in German society it had surrendered at Versailles – perhaps not so surprising in an era when the military was so disproportionately dominated by the aristocracy. Great stress was laid on the social aspects of being a military officer in the Third Reich. Potsdam's smart functions would probably have brought back fond memories for Erwin and Lucie of their early days together at the *Kriegsschule* in Danzig, where they had first met. Arriving on the course at Potsdam in mid-October 1937, twenty-year-old *Fahnenjunker* (cadet) Siegfried Knappe recalled that

> Social affairs were arranged so we could practise our social graces and be observed and evaluated ... There were perhaps four dances during the nine months I was at Potsdam. At one of [these] all the thirty-two men in my platoon were introduced to Rommel's wife. We had to kiss her hand properly and exchange a few words.[20]

On the first day, 'Rommel was introduced and gave a little speech about the importance of a strong officer corps and what an honour it was for us to be selected to become officers in the German Army.' Knappe remembered Rommel's instruction:

> Every week we would have a test that was graded ... [and] got home-work every second or third weekend. In an attempt to put us under stress similar to a combat situation, they gave us very little time to do the assignment. They would give us a situation in which we were a battalion commander. Our battalion was given a certain goal for the day and we were marching to reach that goal. Suddenly we would receive a message that the enemy had been spotted. Then we might get a contra-dictory message. Then we would encounter something else that would alter the situation ... From all the information given to us, we had to make our decision. Three of four possibilities might be equally correct. We had to judge the situation and make a decision on the basis of what we knew.[21]

We have seen in the First World War how Rommel habitually summed up situations and made decisions at lightning speed. He understood that there was a cost to every minute spent mulling over decisions and that his decision-making loop had to be faster than that of his opponents in order to set the pace and thus the course of combat. He taught – as do military schools around the world today – that if your tempo exceeds that of the enemy, then you have to accept a higher degree of uncertainty. If you strive for the perfect truth, the best intelligence picture, you'll never achieve it. So Rommel and his *Kriegsschule* colleagues embraced and taught the concept of making quick decisions in the face of incomplete or uncertain information. The Rommel approach was that small, frequent, rapid decisions often avoid a major decision at the eleventh hour. Perhaps this instinctive approach to combat inspired – or was shared by – George S. Patton, who memorably observed: 'a good plan executed violently now is better than a perfect plan executed next week'.[22]

It was whilst at Potsdam that Rommel began adapting his past lecture notes, based on his war diaries, rewriting and updating them in the present tense for what would become the military textbook, *Infantry Attacks*.[23] With conscription and Germany's rapid military expansion, Rommel's experiences of the First War now had a clear application and purpose: to prepare for the next. As his book portentously concluded:

In the east, west, and south are to be found the last resting places of those German soldiers who, for home and country, followed the path of duty to the bitter end. They are a constant reminder to us who remain behind and to our future generations that we must not fail them when it becomes a question of making sacrifices for Germany.[24]

Note that Rommel employed the word 'when', rather than 'if' in his final sentence, as though the issue of another war – for him – was already decided.

Infantry Attacks quickly became required reading at Potsdam and the other officer schools, and half a million copies were sold within Rommel's lifetime. It was often a handy gift for teenage Nazis and an extremely popular read amongst the Hitler Youth and Reich Labour Service. Not only was Rommel reaching a wide readership; during his three years at Potsdam, some 3,000 officer cadets passed through his hands, allowing him a very high profile amongst the officer corps of new army.

The cult of youth in Germany reached its peak in 1936 with the summer Olympic Games, held in Berlin. They were opened by the twenty-six-year-old German athlete Siegfried Eifrig, bearing the Olympic torch. Tall, blue-eyed and blond, Eifrig – the personification of the Aryan race – would eventually fight under Rommel in Africa. At the eighth Nuremberg rally, held over 8–14 September 1936 – hard on the heels of the Olympics, an extremely high-profile event – Rommel, now in his second year at Potsdam, was put in charge of Hitler's military escort. This was the first occasion Rommel came to Hitler's specific attention, for the Führer would have had no reason to recall him at Goslar. The rally was significant for one whole day was taken was taken up as the *Wehrmacht*'s showcase: after a Luftwaffe flypast, mock battles involving 18,000 soldiers took place, concluding with an address by armed forces chief Werner von Blomberg, who had five months earlier been appointed Hitler's first *Generalfeldmarschall*. We do not know why Rommel was given such a prominent role in 1936 (his influential book had yet to appear), but it would prove a lucky break. On one particular day, Hitler asked that no more than six cars follow his (the Führer's automobile outings frequently degenerated into a scramble of open-topped tourers and staff cars competing for road space behind Hitler). Rommel, with great presence of mind, simply let the first six cars pass and then boldly stepped out into the road, blocking the path of the rest, to career-threatening howls of outrage from party officials; while Hitler was delighted with this display of pluck from the diminutive *Oberstleutnant*.

Although Montgomery's 9th Infantry Brigade was running very well indeed, outside events had begun to shape Monty's fortunes. On 7 March 1936, Hitler marched German troops back into the demilitarised Rhineland without the expected protest from the international community. On 12 March 1938, he initiated the *Anschluss* (union) with Austria. With war looming, the War Office launched a very public amphibious exercise – partly as a test of their own readiness, and partly a demonstration of resolve to the nation, and anyone from the international community who happened to take notice. Watched by the press, and many VIPs including Lord Gort (the new CIGS) and General Wavell (C-in-C Southern Command), Monty's three battalions from 9th Infantry Brigade were landed at Slapton Sands in south Devon, supported by HMS *Revenge*, the cruisers *Southampton* and *Sheffield*, five destroyers, two troopships and a dozen Swordfish aircraft. The amphibious aspect of the proceedings involved the Royal Navy's only landing craft; most

troops rowed themselves ashore – as they had done at Gallipoli.*
Montgomery's next experience of littoral warfare would be Dunkirk.
Even though the exercise underlined Britain's pitiful state of unpre-
paredness, Monty was in buoyant mood.

He had good reason: he had received early promotion to major general
(very much pushed by Wavell, who had written, 'Brigadier Montgomery
is one of the cleverest brains we have in the highest ranks, an excellent
trainer of troops and an enthusiast in all he does') and was sent out to
Palestine to lead 8th Division, one of two formations formed to counter
the Arab insurgency he had seen brewing when there in 1931. The other,
7th Division, was led by Dick O'Connor. Both men arrived in October
1938 and, with war seen as imminent, relished the opportunity to handle
a division before the 'balloon went up'. O'Connor commanded in the
south and became Military Governor of Jerusalem, whilst Monty took
the north, based in Haifa. The senior airman in Palestine was Air
Commodore Arthur Harris.

As might now be expected, the arrival of the slight figure with the
energy of a coiled spring immediately caused consternation in Palestine.
William Battershill, chief secretary in Jerusalem, confiding in his diary
of Monty:

> He called on me last week and put me through a cross-examination
> which would have come rather amiss from the GOC himself . . . He is
> sharp featured and evidently has a brain. I fear this man will try to be
> a new broom and will successfully put the backs up of most people.[25]

Soon Monty and the local district commissioner were no longer on
speaking terms, a situation exacerbated by Montgomery's imposition
of an immediate night-time curfew, which was effective but halted
round-the-clock work on a new oil refinery at Haifa. (It took Whitehall's
intervention to override Monty's blanket curfew and permit the refinery
workers through to Haifa.) He had more success in rural areas and by
the summer of 1939 the rebellion had been largely suppressed. But
Monty's achievement was largely a military one, attained at the expense
of civil–military cooperation in Palestine. He commented later that he
found this type of soldiering distasteful, and even claimed that such

---

*Slapton Sands would also witness a disastrous pre-D-Day exercise in April 1944.
See chapter 16.

counter-insurgency work was more of a professional challenge than high-intensity world war. Writing of counter-insurgency in Ireland, he observed:

In many ways this war was far worse than the Great War which had ended in 1918. It developed into a murder campaign . . . such a war is thoroughly bad for officers and men; it tends to lower their standards of decency and chivalry, and I was glad when it was over.[26]

Monty's 8th Division was a colonial expedient numbering only some 10,000 men, but as soon as he arrived in Palestine, Monty was informed he had been nominated to take command of 3rd Infantry Division in December 1939. This was a real prize, as it was one of the regular divisions earmarked for an expeditionary force to France in the event of war. However, the strain and the climate were taking their toll, and Monty was taken seriously ill. He ignored his symptoms at first but grew weaker and weaker. Tuberculosis was suspected; Bernard's old lung wound from Méteren 1914 did not help matters. He demanded to be sent back to England, insisting a sea voyage would do him good. A cabin was reserved for him and by good fortune his sister, Winsome, to whom he was closest of all his siblings, was on the same liner, the SS *Ranchi*. Her husband, Lieutenant Colonel John 'Wangy' Holderness, was commanding the resident guard battalion, 1st Royal Sussex, at Ismailia in the Suez Canal Zone, and she was travelling home on periodic leave. Winsome nursed Bernard throughout the three-week voyage home, via Malta, Casablanca and Gibraltar.[27] The illness remains a mystery for Monty walked down the gangplank at Tilbury on 14 July a relatively fit man. His sister put his rapid recovery down to willpower. His biographer, Nigel Hamilton, observes that 'it would have been most uncharacteristic of Bernard to have exaggerated his condition'. However, he was now where he felt he needed to be, back in England and able to take over 3rd Division, which he did, after a lot of indiscreet lobbying, on 28 August 1939.

Despite the fact that Montgomery was reasonably well prepared mentally for the next war (though his 3rd Division most certainly was not), there was a vital element of a senior officers' education which had passed Monty by. In January 1927, the new Imperial Defence College had opened its doors at 9 Buckingham Gate, Belgravia, to educate selected students in the higher levels of war. There were usually only eight army students (out of twenty, later increased to thirty-six) on each year-long course; the rest came from the Royal Navy, RAF, the Empire, and (eventually) from the USA, the Territorial Army, industry and government organisations. Wing Commander Sholto Douglas[28] had been on the first course as a student; Auchinleck attended in 1927, Tedder in 1928, the future Admiral Andrew Cunningham arrived in 1929, Alexander in 1930, Leigh-Mallory in 1934, with the Canadian soldier Harry Crerar (who would command First Canadian Army in Normandy), O'Connor in 1935, and Slim with the New Zealander, Keith Park, in 1937. Captain Bertram Ramsay had been on the directing staff there in 1931–33, with Colonel Alan Brooke and the returning Group Captain Sholto Douglas, underlining the real value of the tri-service college as an arena for 'networking'. Ramsay's predecessor was Captain James Somerville,[29] who would later sink the *Bismarck* and achieve five-star rank.

The curious omission from this higher college of war (which may explain much) was Monty. His three British colleagues at Supreme Headquarters Allied Expeditionary Force (SHAEF) in 1943–45 (Ramsay, Leigh-Mallory and Tedder) all attended the Imperial Defence College at some stage as students or directing staff, whilst Eisenhower had attended the US equivalent of 'war college' in 1927–28. Commander Andrew Cunningham, who thought his year at the Defence College was 'one of the most interesting and valuable I have ever spent', recalled in his memoirs:

Besides the most fascinating lectures and addresses the course took us on many interesting expeditions. We visited tank schools, gas schools, naval establishments at Portsmouth and elsewhere, RAF stations, experimental establishments and many others. One most informative trip was to the 1914–18 battlefields under the guidance of General Sir A. Montgomery-Massingberd. His astounding memory, the way without notes or a map he could point out prominent features of the battle areas and could name all the commanders engaged, were matters of wonder to us all. At Le Cateau the General even knew the names of some of the platoon commanders.[30]

No doubt Monty could have helped with Le Cateau had he been present, but the decision not to send him to the Imperial Defence College (which may have been, of itself, a War Office comment on his suitability for high command) would prove a grave error in preparing him for coalition warfare at the highest levels. Monty's character had always swayed between eccentricity and genius and it was almost inevitable that he would continue to rub people up the wrong way. A year at the Defence College might have better equipped Bernard with the skills of diplomacy he would shortly need.[31]

In February 1937, while still at Potsdam, and as a direct result of the success of *Infantry Attacks*, Rommel was given a special assignment. He was to be the War Ministry's liaison officer at Hitler Youth headquarters in Berlin to advise on their paramilitary training, which included weapons handling, forced marches and basic field tactics. In some ways Rommel was the natural choice for the task. He was a gifted instructor; he loved working with young people; he was a high-profile author and a hero of the First World War with an exciting story to tell. What could possibly go wrong?

Neither the party nor the army foresaw that Rommel and the movement's twenty-nine-year-old leader, Baldur von Schirach, would fall out. Their sixteen-year age gap certainly contributed, as did background. The rather exotic young aristocrat had a theatre director father and an American mother; he was bilingual in English and German, was well travelled and sophisticated – a world completely at odds with Rommel's. Schirach was married to the teenage daughter of Hitler's court photographer, Heinrich Hoffmann, giving him a place in the Nazi 'inner circle'. At times Schirach was made to feel like one of Rommel's cadet officers, rather than the Reich's leader of 5.4 million uniformed young men. Rommel's failure seems to have been due partly to his personality; either a lack of empathy or inability to compromise contributed towards the impasse with Schirach; the experience may have left Erwin wary of pursuing too close a relationship with the upper echelons of the regime.

Though the assignment ended in failure, Rommel remained on the staff at Potsdam throughout and it did not seem to prejudice his promotion to *Oberst* in October 1937 and subsequent move to the Maria Theresa Kriegsschule in Wiener Neustadt, near Vienna, as its commandant in November 1938. Some of the party rhetoric was beginning to permeate, and in 1937 and 1938 Rommel attended two nine-day Nazi indoctrination courses for senior officers. He had even started signing cards and letters '*Heil Hitler!*'[32] And he approved of the

Anschluss in 12 March 1938, as did most Germans – after all, he had served with many good Austrian mountain troops in Romania and Italy in 1916–17 and had a high opinion of their fighting prowess.

Shortly after the debacle with Schirach, Hitler proposed a scheme involving Rommel, and he was hauled back from Potsdam to do yet another job for the party. He was to command the *Führerbegleitbattalion* – the special detachment of sixteen officers and 274 men responsible for Hitler's personal safety – for the occupation of the Sudetenland in October 1938. It was not much of a command – Rommel had already commanded thousands in the First World War – but it brought him access and proximity to the Führer, a career-making prize that many desired. Rommel performed his duties in the Sudetenland so well that between 1938–39 he was twice more summoned from Austria to command the battalion. A bond was clearly developing, with the Führer depending on Rommel's confident, soldierly advice. On 15 March 1939 he drove with Hitler to Hradcany Castle for the occupation of Prague in an open-top car, without the planned SS escort. Later that month, he accompanied his Führer in the occupation of the Baltic port of Memel from Lithuania, Hitler's last bloodless conquest.

The German occupation of Prague on 15 March 1939 proved the last nail in Chamberlain's appeasement strategy and on 30 March he announced a doubling in size of the Territorial Army to 340,000, probably in the hope of avoiding a resort to peacetime conscription. The news came as a bolt from the blue to the War Office who thought conscription was being considered. Each Territorial unit was halved to create a duplicate, thus creating twelve new infantry divisions, which stretched their resources to the limit. Battalions lost their best officers and NCOs, who were promoted to run the new additions.

Eventually, under huge pressure from the French, and the announcement on 22 March 1939 that the Dutch had extended their period of conscription from eleven to twenty-four months, Chamberlain reluctantly followed suit. It was the only answer, and barely three weeks after the enlargement of the Territorial Army was announced, on 26 April, conscription was introduced under the guise of 'national service', the very word 'conscription' being an anathema to Chamberlain. British conscripts would receive six months' full-time training, followed by three and a half years in a Territorial unit.

By August 1939 the British military was at last regaining its self-confidence: there was a surfeit of numbers, a job to do, and the knowledge that if not immediately forthcoming, full supplies of equipment were on their way. Nevertheless, in the minds of many, there seemed little sense of urgency. In June 1939 there was a predicted requirement of 1,770 twenty-five-pounder field guns to replace the artillery's eighteen-pounders, yet only 140 were delivered by September.[33] The War Office had several major challenges, none of which had been foreseen, each of which required much careful planning; these included plans to move a British Expeditionary Force to France in the event of war. Other important decisions needed to be made: a comb-out of the medically unfit, the under-age, and those in reserved occupations was needed, but government deferred the matter, losing the opportunity to address the problem during peacetime. Mechanisation of the army stumbled because few civilians could drive.

Chamberlain's government had had as long a warning scenario as it is possible to envisage, yet the Prime Minister refused to face reality until the last moment. Then he concluded that the very announcements of doubling the Territorial Army and of conscription would be enough to avert war. Further decisions affecting the army that should have been taken were not, and the twin acts of doubling the TA and conscription – with all the accompanying administrative nightmares – arguably made the British Army less ready than it would otherwise have been. Hitler understood this well: on 22 August 1939, in an address to his commanders, he stated: 'There is no actual rearmament in England, just propaganda.'[34] The French summed up their colleagues thus: 'the British Army consists of generals and lorries, and always attacks at dawn'. There had been cuts to the British Army's budget in every year between 1922–36, and at the outbreak of war eleven Regular and reserve cavalry regiments still soldiered on horseback. The army was undermanned and equipped largely for an imperial policing role. Provision of regular troops for India had been the main priority – hence Monty's postings in the 1930s – as troops garrisoned there were fully funded by the Indian government. German intelligence did not have to work hard to report that obsolete First World War equipment was not modernised or replaced, that soldiers' pay and allowances, likewise, were frozen between 1920 and 1937, and that plans for mechanisation were deferred because of lack of funds. For this lamentable situation, Neville Chamberlain must personally shoulder much of the blame; prior to becoming Prime Minister in May 1937, he had served as Chancellor throughout 1931–37, withholding much-needed cash from the forces' budgets. In 1939 one artillery officer arriving in northern France remembered:

> On the edge of the village stood a small farmhouse in a very ruinous condition. It was utterly deserted, and obviously hadn't been tenanted for years. I opened the door, which led straight into a room. Chalked in big letters on the opposite was the command: 'WIPE YOUR FEET, PLEASE. 1918.' It was so peremptory that I almost obeyed, forgetting that it was not 1918 but 1939, and that only ghosts of a past officers' mess in a past war would grouse if I didn't ... it stood there exactly as it had been left after the Armistice. I could almost imagine that, with terrible clairvoyance, its occupants had foreseen that it might be wanted again. In a corner were a rusty bayonet and a Mills bomb. The stained deal table and three wooden chairs were thick with the dust of twenty-one years of peace ... There was something pathetic about this little heap of forlorn-looking debris, a legacy from one British army to another.[35]

Both Montgomery and Rommel realised that the coming war would require new tactics and thinking; however, how they would fight and were directed to fight over the next few years would be influenced by the bloodletting of 1914–18. Whilst some British politicians were cowed by memories of the Somme and Passchendaele, by contrast Staff College students, shaped in their thinking by outstanding instructors like Monty, were profoundly forward-looking and emphatically not stuck fighting the last war. Monty spent the years in the run-up to the Second World War recording his own doctrine in journals like the *Army Quarterly* (where he wrote on 'Major Tactics in the Encounter Battle' in July 1938) having preceded with a similar piece in the *Royal Engineers Journal* in September 1937. Here Monty had written prophetically and in his own unique phraseology:

> About one thing there can be no doubt – we have got to develop new methods, and learn a new technique. And in thinking out the problem we shall do no good if we are going to be influenced too much by the past when conditions were different. There is no need to continue doing a thing merely because it has been done in the Army for the last thirty or forty years – if this is the only reason for doing it, then it is high time we changed and did something else ... We are a very conserva-

Montgomery was standing near where this photograph was taken at Méteren when he was severely wounded in the chest by a sniper's bullet, 13 October 1914.

Troops from Rommel's 124th Würtemberg Regiment outside their wooden shelter in the Vosges mountains, early 1915.

Brigadier J. W. Sandilands (*left*) would have a huge influence on his young brigade major, Captain Montgomery (*right*) in 1916. Monty wrote later that Sandilands was the best senior officer he ever served under.

Major Theodore Sprosser (*left*), Rommel's mentor and superior in the Alpenkorps, was the soldier most responsible for shaping the way the future Desert Fox fought and thought.

Then unknown to each other, Lieutenant Colonel Montgomery and Winston Churchill, Minister of Munitions, watch the 47th London Division march through Lille, 28 October 1918.

Rommel's first encounter with Hitler was in 1934, when his battalion paraded at the Goslar Kaisenpfalz. The photograph reminds us how small Rommel was (about five feet six inches). The location (*right*) remains exactly the same today.

Oberst Erwin Rommel

Infanterie greift an

Erlebnis und Erfahrung

3. Auflage

367 Seiten, über 70 Abbildungen — Kartoniert 4.80 RM — Ganzleinen 5.80 RM

Ludwig Voggenreiter Verlag Potsdam

Infanterie Greift an ('Infantry Attacks') brought Rommel to Hitler's attention and made his career in a way that ordinary military service would not have done. This advertising leaflet portrays a confident Colonel Rommel at the time of publication in January 1937.

After the war, Lucie Rommel would help to rehabilitate the name of her husband and become one of the central figures in the Afrika Korps Old Comrades Association. She is pictured here with Rommel in 1941.

Albert Kesselring – saluting Hitler in 1939 – made a remarkable transition from
Air Fleet Commander in 1940 to Supreme Commander in the Mediterranean, where he
clashed frequently with Rommel (to his right), when commanding Hitler's bodyguard.

Rommel's 7th Panzer Division was equipped with thirty-four Panzer Mk 1s in May 1940. The soldier demonstrates its puny size and underlines the fact that tank warfare then was less about numbers and more about psychology, surprise and daring – qualities Rommel possessed in abundance.

Symbols of defeat: British staff cars abandoned on the beach at Dunkirk, May – June 1940. The vehicle on the right bears the triangular badge of Montgomery's 3rd Division, and may have been used by Monty himself during his final hours, when promoted to command II Corps.

This Willrich portrait, made during the 1940 campaign in France, shows Rommel's new Knight's Cross, awarded on 16 May of that year. A poster-sized copy of this evocative image hung in Montgomery's campaign caravan throughout 1942–45.

Monty's Staff College contemporary Major General Victor Fortune (*second left*) has just surrendered his division to Rommel, but it could just as easily have been Monty's own troops caught against the coast at St Valery-en-Caux by the Ghost Division. This section of the little fishing port has not changed since 12 June 1940.

'The sun never goes down on us.' This Afrika Korps sundial set in a German war cemetery underlines the ruthless determination Rommel brought to the desert campaign.

If the Afrika Korps became one of the best-known 'brand names' of the Second World War, then its badge – the swastika and palm tree – was one of the most successful 'trade marks' of the war.

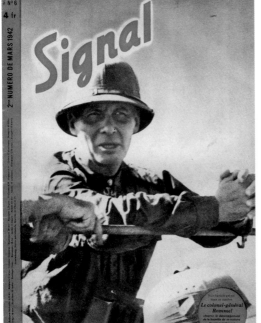

Rommel's exploits were promoted via German weekly cinema newsreels, newspapers, radio and mass-circulation propaganda magazines, like *Signal*, whose cover he adorned several times. This edition dates from March 1942. He was the only military commander to be featured in this way, which raised his profile but excited envy amongst his fellow generals.

tive army . . . It is felt that if we could get rid of the terminology which has been in use in our Army for generations, the Army as a whole would find less difficulty in adapting itself to the new conditions that now confront it.[36]

Conversely, the younger generation, too young to remember the First World War, were brought up on its horror stories. These were men like Sydney Jary, soon to be a young lieutenant in Monty's 21st Army Group, who would later recall of Normandy:

It seemed to me that the German infantry still followed the same defensive tactics which they had used so effectively in 1916 . . . I remember with horror being 'locked' into the timetables of meticulously planned large battles . . . Undoubtedly, far shadows from the Somme clouded my emotions, but instinct told me that this kind of show would be unlikely to succeed . . . The Somme had also cast its shadows on our artillery and armoured commanders. Both genuinely believed that in their hands they had the panacea which would protect us, the infantry, from the terrible slaughter of 1916 . . . Now in my sixties, I do not underestimate the influence that the Somme had on the British military psyche.[37]

Much later, basking in the aftermath of his unexpected victory over France in June 1940, Hitler celebrated by visiting the First World War battlefields that had so altered his life; he undertook two mini-battlefield tours of Flanders and northern France, where he had fought. Initially, he visited the Menin Gate at Ypres, Langemarck, Lens, Vimy, Arras and Cambrai. Subsequently on 25–26 June, in the company of two First World War comrades, he visited former billets at Fromelles and Lille, before inspecting Dunkirk, recently evacuated by another British Expeditionary Force.[38]

Ironically, it was during an earlier battlefield tour that Winston Churchill came closest to meeting Hitler. In late August–September 1932, Churchill was visiting the battlefields fought over by his ancestor, John Churchill, Duke of Marlborough, whilst researching his acclaimed biography. He visited battlefields from Flanders to Bavaria but before going to Blenheim spent a week at the Hotel Regina in Munich, a favourite daily haunt of Hitler's, where an intermediary tried to arrange for the two to meet. Alas for the future of Europe, the meeting of these two colossi never happened, but military history might have provided interesting 'neutral' ground from which these two could have launched into wide-ranging debate, for ex-Corporal Hitler and former Lieutenant Colonel Churchill served for a while in the same small area of Ploegsteert and Messines, in the south-west corner of the Ypres salient, though at different times. The conversation might have followed similar lines to one of two years later in June 1934, when the Grenadier Colonel Andrew 'Bulgy' Thorne, British military attaché in Berlin, was summoned to meet the Führer, who realised they had served in the same sector of the front at Ypres in 1914. Thorne observed:

> He discussed particularly the fighting at Gheluvelt in 1914 where he had his baptism of fire ... For a long time we stood over a map exchanging reminiscences over the area where the fighting had taken place, the road along which the Germans had attacked, the various fields and hills. Hitler was delighted and stressed that it was not pale-faced staff officers but the ordinary front-line soldiers who could understand war. I tactfully refrained from telling him that I had actually been a staff officer at that time.[39]

Biographers of Prime Minister Anthony Eden observe that on 21 February of the same year an Eden–Hitler discussion in Berlin had already taken place, when the pair discovered they had faced one another in exactly the same sector in 1914 at Ypres, and were within hand-grenade-throwing distance on the Somme. They refought the latter battle at length on the back of a menu card, leaving Eden (who himself had won an MC and was initially taken in by the Führer's 'charm') impressed with Hitler's knowledge of that struggle. He travelled to Berlin by train and noted cursorily in his diary:

> The journey was interesting, more especially the first part which took us through the battlefields. I sat next [to the] Chancellor. We talked freely enough with the help of an interpreter and my limited German. Hitler thawed materially, especially when we discussed the war which he likes to recall, like most Germans. We discussed the various sectors where we had each been. I took the chance to rub in that ex-soldiers should be the last ever to wish for another war. He assented heartily.[40]

PART THREE

THE MAKING OF MODERN MAJOR GENERALS

This was a battle, but its significance as an event in human experience transcends the military. The tactical, technical and logistic problems it raised are often of high and absorbing interest. More and more, however, those of us who fought through this battle have become aware that what remains with us can best be described as a spiritual experience.

General Sir John Hackett

12

Phoney War

OBERST ROMMEL WAS in his element as Commandant of the Maria Theresia Kriegsschule in Wiener Neustadt, but on 23 August 1939 he was recalled from his schoolmasterly job as Germany prepared for another war.[1] Rommel was once again put in charge of the special army unit responsible for Hitler's security – the *Führerbegleitbattalion*. His principal task was protecting Hitler's mobile war headquarters, the *Führersonderzug*, and he received promotion to *Generalmajor* for this illustrious task.

The *Führersonderzug* was a specially fitted train (whose code name, bizarrely, was Amerika) comprising around twenty carriages, which included Hitler's own carriage, the *Führerwagen*, and the *Befehlswagen*, command car, which contained a conference room and communications centre.[2] While Rommel's mission was to protect the entire headquarters (*Führerhauptquartier*), Himmler's black-uniformed SS were still responsible for Hitler's immediate personal security; they had expanded from a bodyguard unit of eight men in early 1923 to a Berlin ceremonial guard of a hundred in 1933. By 1939 the SS-*Leibstandarte* (Lifeguard) numbered nearly 5,000 and was the size of a full infantry regiment with three infantry battalions, an artillery battalion and anti-tank, reconnaissance and engineer subunits. Some still lurked in Hitler's vicinity, but the majority went to war alongside the army. The Leibstandarte was eventually expanded to a division of over 20,000 and by 1945 Himmler's empire had absorbed so many other organisations and rebranded them as SS (police, security services and foreign volunteers), that it totalled nearly a million.

This new task was an extension of the bodyguard role Rommel had already assumed, but carried the enormous advantage of allowing him privileged access to Hitler. At times Rommel was summoned to sit next to his Führer at lunch, and on 19 October he wrote home to Lucie of the honour, how he 'was allowed to chat for almost two hours with him about military problems yesterday. He was extraordinarily friendly towards me.'[3] Rommel had become a 'face' to Hitler; in the extremely competitive world that was the Third Reich – essentially a vast ocean of glittering uniforms – Rommel stood out. The cheek-by-jowl proximity brought him fascinating insights into the practical application of modern warfare; he gained an understanding of the way tanks could be deployed and of the necessity of close ground-to-air cooperation. It also exposed him to other prominent Nazis: at one stage he had a disagreement with the ever-possessive Martin Bormann, who, when Rommel blocked his way, swore and screamed at him. It was a foretaste of the problems to come: such close access to Hitler would make others insanely jealous and bring as many enemies as advantages.

Apart from Hitler's own HQ, the German armed forces had several rival and contradictory command chains. Although many of its staff officers remained in central Berlin at Bendlerstrasse 14, Army High Command (*Oberkommando des Heeres* – OKH) moved on 28 August 1939 to a sprawling complex of deep underground bunkers and armoured telephone cables at Zossen, twenty-five miles south of Berlin. Policy was administered in Berlin, but the detailed orders originated from Zossen.[4] OKH eventually came to administer operations on the Eastern Front. The superior – and rival – tier in Hitler's hierarchy, the Armed Forces High Command (*Oberkommando der Wehrmacht* – OKW), also based in the Bendlerstrasse, came to oversee all military operations elsewhere, and likewise moved to a parallel compound at Zossen on 29 March 1943. OKL, representing the Luftwaffe, and OKK, the *Kriegsmarine*, operated completely independent headquarters, indicating a lack of inter-service cooperation and planning capability.[5]

Rommel thus had only a week's warning before the launch of *Fall Weiss* (Plan White) – the invasion of Poland – on Friday 1 September 1939. Hitler's *Führersonderzug* did not deploy every day; only when Hitler left state business in Berlin to review progress in Poland did the train, under Rommel's watchful eye, trundle over the border to war. At other times, Rommel escorted Hitler's motorcade into Poland. The young Prussian aristocrat Alexander Stahlberg, then serving in 2nd Infantry Division, remembered being ordered to pull over:

> An armoured reconnaissance car came up behind us, travelling at high speed . . . From the turret protruded the figure of a general, unrecognisable in his dust-goggles, but round his neck we spotted the highest Prussian military order, the *Pour le Mérite* . . . a large, shapeless open Mercedes appeared, travelling at the same high speed – a model we had not seen before, a three-axle cross-country vehicle. In the front passenger seat was Hitler, in a grey uniform . . . A few more vehicles followed behind, at break-neck speed.[6]

On 5 October 1939, Hitler reviewed his victorious troops in a two-hour victory parade. Rommel must have itched to be with the marching troops instead of playing bodyguard, standing on Hitler's left, below the viewing stand – where the newsreels and photographers caught him – adrift in a sea of other dignitaries. Combat troops have always privately derided the staff; Rommel would have felt this, and the Warsaw parade might have been the moment he realised that his place was at the front, leading men towards the gunsmoke. He had not commanded a battalion since October 1935; he had been a *Kriegsschule* instructor since. Comfortable headquarters life, as in 1918, was proving frustrating; he was unhappy loitering in Hitler's stuffy headquarters, with security men breathing down his neck. It all confirmed that he was not by temperament a desk warrior.

At some stage the subject of Rommel's next command emerged whilst overseeing the *Führerbegleitbattalion*. This was unlikely to have been before the end of the Polish campaign in early October; Rommel would have had no idea of Hitler's strategic plans or timescale. When, from private conversations with Hitler, Rommel gathered that a military confrontation with the old enemy, France, was inevitable, he made up his mind to ask for a transfer to the sharp end. Given Rommel's background to date, the next logical stage in his military career would have been command of a three-battalion regiment, and then a division. However Rommel was impatient by nature: with Hitler's backing he might have hoped to be in line for a divisional command straight away. As an infantryman, it would have made sense to give him an infantry division or an elite mountain division.

Rommel's patron and fellow Württemberger, Wilhelm List, was at this time in a position of great influence, commanding the Fourteenth Army which had led the invasion of southern Poland, and would no doubt have welcomed a friend in Hitler's court. List had been near to Rommel in Austria, commanding the post-*Anschluss* Army Group V, based in Vienna. Rommel, too, would have encountered General Heinz Guderian, father of Germany's tank arm and then commanding XIX Panzer Corps in Poland with great success. Rommel and Guderian had common ground in that Guderian had commanded a company in III Battalion of 17th Infantry Regiment in Goslar between 1920–22, the same battalion Rommel commanded between 1933–35. Rommel had gleaned enough of recent warfare, from *Wehrmacht* deployments in Spain during the civil war and Guderian's performance in Poland, to realise that the up-and-coming men commanded panzer divisions. Rommel was thus well placed, with the backing of Hitler and List, possibly Guderian also, to request an armoured command. Whether Hitler offered (possible), or Rommel asked outright (more likely) remains a matter for speculation. The conservative army personnel office might have demurred and Hitler at some stage might have insisted that Rommel should have his wish, but the upshot was that on 6 February 1940 he received the news that a newly formed division, 7th Panzer, was his.

Three months later, at 4.30 on the morning of Friday 10 May, Rommel's panzers clattered over the Belgian border as part of *Fall Gelb*, the invasion of France. The prospects for outshining his colleagues, if such was his intention, were not good: he had no experience of armoured warfare, let alone command of a single tank in combat; he had last heard the sound of shots fired in anger twenty-two years previously, in 1917, and he was leading an untried formation more used to horses than panzers.

His attitude of 'insufferable arrogance' did not help matters. When he first arrived at the divisional HQ at Bad-Neuenahr, astride the Rhine, he made a terrible first impression. Fresh from Hitler's headquarters, Rommel greeted his new colleagues with the raised arm of the Hitler salute, instead of the preferred army touch to the cap. The situation was exacerbated by the popularity of the outgoing commander, Generalleutnant Georg Stumme, who had successfully led the division's predecessor, 2nd Light Division, in the Polish campaign and was now being promoted up the chain of command to XXXX Corps. Hence the vacancy for *Generalmajor* Rommel, though there is no evidence that

Stumme was promoted to make way for Rommel. Days after taking over his command Rommel disappeared back to Berlin for a farewell interview with Hitler. Meanwhile a cabal of the division's officers met to discuss what to do about their new commander; they knew he had absolutely no armoured experience, and despite his earlier war record, considered him merely a toady of Hitler's. Rommel returned from Berlin with a gift for each: an inscribed copy of *Infantry Attacks*, which only annoyed them the more.

Rommel's first challenge was to train up an unskilled and untried tank formation, instil in them an esprit de corps and inspire loyalty and trust in him, their commander. The 7th Panzer Division had been formed in October 1939 from one of four cavalry divisions which had lost its horses immediately after the Polish campaign. Morale had dipped and the division was relatively poorly equipped. With the success of armour in the invasion of Poland, the tank arm of the *Wehrmacht* had expanded hastily. As divisions were converted to an armoured role, with more panzer units, the available vehicles had to be spread around, like good marmalade on toast, more thinly.

The Division was a self-contained fighting formation which included a reconnaissance battalion of fifty armoured cars, a motorcycle battalion, three battalions of 105mm artillery pieces, an anti-tank regiment with thirty-six guns, battalions of light and heavy anti-aircraft guns, a signals and an engineer battalion. The division fielded eight command tanks, fitted with extra radios, from one of which (Number 01) Rommel led and commanded.[7] However, 7th Panzer was more poorly equipped than most, for it possessed no Panzer Mk IIIs, the premier German tank of 1940, but had instead thirty-four two-man Panzer Is (equipped only with a pair of machine guns), sixty-eight three-man Panzer IIs (which carried an obsolete 20mm main gun) and twenty-four five-man Panzer IVs (a much bigger tank of 18–20 tons, whose short-barrelled 75mm main armament was of too low a muzzle velocity to penetrate the best British and French armour plate). The balance of the division's tanks (including Rommel's own command vehicle) was made up of ninety-four, four-man Skoda-built Panzer 38(t) – the 't' standing for *tschechisch* (Czech) which gave away the vehicle's origins. These were part of the booty Hitler acquired in Czechoslovakia. It was, however, a popular and reliable tank, recognisable by its distinctive, riveted armour plate, though its maintenance manuals had yet to be translated from Czech.[8] In future years, German equipment would include the suffix 'f', *französisch*, and 'r', *russische*, illustrating how the Wehrmacht not only defeated opponents, but absorbed their equipment also.[9]

The majority of Rommel's tanks were not the giant metal monsters we might associate with the armoured vehicles of the later war years (like the 44-ton Panther or 68-ton King Tiger, all crewed by five), but tiny iron boxes that were relatively easy to disable. Standing beside the Panzer I and Panzer II at Britain's Bovington Tank Museum, Dorset, it is immediately apparent that neither is much taller than a man: the 6-ton Mk I is only 1.7m (5ft 7ins) tall, whilst the 9-ton Mk II measures 2m (6ft 6ins) high. When one considers just how puny some of Rommel's tanks actually were, it is obvious how important the psychological dimension was to his success in 1940.

Tank. The very word was designed to mystify and conceal. The original term landship concealed nothing and rather gave the game away, so they had been code-named by their inventors at the British War Office 'mobile water tanks for Mesopotamia'; the name designed to mislead, stuck. Shortly after their first use in 1916, the public were positively bombarded with images, patriotic songs and a new cinematographical film reinforcing the concept of the tank as a means of defining victory, especially as the Germans, until 1918, had none. By the war's end, the French automobile manufacturer Renault had manufactured a little two-man tank with a revolving turret and the British had produced other tanks like the 18-ton, eight-mile-an-hour Whippet, but the image of a tank retained by most soldiers and civilians, and incorporated in armoured corps badges around the world, was the Big Lozenge.

By 1939, whatever the military had been developing in relative secrecy, the world's citizenry understood a tank to be a big, slow, cumbersome lozenge. New, turreted models might have been sighted in cinema newsreels, but it seems that the world was almost in as much ignorance about tanks in 1939 as they had been before 15 September 1916, when tanks first emerged from the morning mists of the Somme battlefields around the villages of Flers and Courcelette. Despite Poland – where in September 1939 tanks had been employed extensively and with much success – when the panzer divisions cascaded out of the Ardennes in May 1940 they were preceded by ignorance and mystery. This applied to soldiers as much as civilians in armies where conscript troops had been civilians a matter of days earlier. And what you didn't understand, you feared.

In 1939–40 even Hitler was awestruck by the power of tanks and the relatively light casualties incurred in armoured warfare, compared to his personal experience of 1914–18. On 5 September 1939 he (with Rommel in tow) paid a surprise visit to Guderian's XIX Corps, in the vanguard of the advance into Poland. At the sight of a smashed Polish artillery regiment, Hitler asked Guderian if the destruction was the work of Luftwaffe dive-bombers.

> When I replied, 'No, our panzers!' he was plainly astonished . . . I was able to show him that the smallness of our casualties in this battle against a tough and courageous enemy was primarily due to the effectiveness of our tanks. Tanks are a life-saving weapon. The men's belief in the superiority of their armoured equipment had been greatly strengthened by their successes . . .[10]

The power of Rommel's tanks lay not only in the shock of their appearance, especially in sectors where there were no opposing tanks, but also in their speed. All of Rommel's tanks had two key characteristics that their French and British armoured counterparts lacked: a high road speed (typically 35mph) and long range, usually being capable of over a hundred miles before refuelling. Significantly, each German vehicle was equipped with a wireless radio set, an innovation which had yet to spread to all Allied tank formations, some of whom still attempted to communicate by flag. The radios were the gift of Guderian, considered the 'father' of the panzer arm, who had started his own service in the signals corps. These gave a commander the ability to control and direct his tanks in combat, as well as summon air or ground support.

Recent research has shown that Rommel was also extraordinarily lucky in the set of commanders with whom he invaded France. Amidst his cast of senior officers were twelve future generals and several Knight's Cross winners, whilst others, who were clearly destined for promotion and the laurels of war, died in combat.[11] Amongst them was Oberst Georg von Bismarck, a descendant of the Iron Chancellor, who commanded Rommel's motorised infantry, the 7th Rifle Regiment*. He would go on to win the Knight's Cross in command of 21st Panzer Division in Africa under Rommel and seemed destined for higher things, but died on 31 August 1942 during the battle of Alam Halfa. Major Eduard Crasemann commanded Rommel's heavy artillery in France and also followed Rommel to North Africa, where

---

*Rommel had a second motorised infantry brigade, 6th Rifle Regiment, commanded by Oberst Erich von Unger, later killed in Russia in August 1941. It was the balance of motorised infantry and armour that made a Panzer Division such a potent force.

he led 15th Panzer Division, before commanding a corps in Germany in 1945, where he was captured. Another future Africa veteran, promoted by Rommel to command the 37th Reconnaissance Battalion during the French campaign, was Hans von Luck, one of Rommel's former students, who would rise to colonel and go on to win a Knight's Cross. (Luck would resurface under Rommel's command in Normandy four years later.)

Rommel's success in 1940 begs the question: how did he, with no previous experience of tanks, manage to outshine his fellow divisional commanders in France? The answer lies surely in forty-four-year-old Oberstleutnant Karl Rothenburg, his key subordinate and fellow *Pour le Mérite* holder, who commanded the tanks of 25th Panzer Regiment. Whilst Rommel provided the drive and direction, Rothenburg supplied the expertise in handling the armour. Using his experience of fighting in Poland, leading 6th Panzer Regiment (in Guderian's XIX Panzer Corps), Rothenburg juggled his armoured vehicles about to achieve two light tank companies and one medium company per battalion. Rothenburg would go on to win the Knight's Cross and, had he not been killed on the Eastern Front in June 1941, would probably have risen to the top as a great panzer general. One of Rothenburg's battalion commanders, Oberstleutnant Rudolf Sieckenius, eventually rose to lead 16th Panzer Division, fighting in the east and at Salerno. A company commander under Rothenburg, Adalbert Schulz, was himself commanding the 7th Panzer Division in January 1944. By this time, he had been promoted to *Generalmajor* and awarded the very rare Knight's Cross with Oakleaves, Swords and Diamonds. Major Joachim Ziegler, a General Staff graduate, was Rommel's intelligence officer in France and eventually transferred to the Waffen-SS, winning promotion to general and a Knight's Cross with Oakleaves, whilst Frido von Senger of 2nd Cavalry Brigade, attached to Rommel's command in 1940, would make a name for himself as commander of XIV Panzer Corps in 1944 and defender of Monte Cassino. A Rhodes Scholar at Oxford, where he read history, Senger would go on to win a Knight's Cross with Oakleaves.

Rommel also had the unusual figure of Karl Hanke assigned to him. Hanke was a young Nazi who had joined the party in 1928. Keen (in photographs he exudes an air of bustling efficiency) and with good looks, charm and a good understanding of public relations, he had become number two in the Propaganda Ministry by 1938. He was a friend of Albert Speer and an *Oberführer* in the SS. Hanke had joined the army reserves and served with 3rd Panzer Division in Poland, where he was wounded. He was then assigned to Rommel as *Ordonnanzoffizier* (a mix of ADC and military secretary) in 1940.

The pair initially hit it off, though others in the division regarded Hanke as a Nazi plant. Hanke allowed Rommel easy access to the influential figure of *Reichsminister* Goebbels, whilst it was clear that Rommel was the coming man and a favourite of Hitler's. Rommel gave Hanke one of Rothenburg's tank companies to lead, which he did very well, and Hanke would save Rommel's life on 14 May at Sivry. Like a first-rate orchestra led by a world-class conductor, 7th Panzer Division's achievements in France would be the result of a talented team effort, directed by an outstanding leader.

Three days after Bernard Montgomery had taken over 3rd Infantry Division on 28 August 1939, German troops crossed into Poland, triggering full mobilisation in Britain. The contrast between the tiny BEF, starved of funds, equipment and proper training for well over a decade, yet earmarked to cross the Channel, and the German war machine, rearming and conscripted since 1935, could not have been greater. Monty would write in September 1939, 'the British Army was totally unfit to fight a first-class war on the Continent of Europe'.[12] Another future field marshal tellingly observed:

> It is impossible to understand why the British army was so much less well prepared to fight the German army on the Continent in 1939 than it had been in 1914, unless one realises that the Government policy, up until almost the last moment, had been that it should not do so.[13]

What it was equipped to do very well was the mission from which Montgomery had just returned: policing the Empire, where there was no aerial challenge to the primitive RAF, or organised ground force to take on the army at anything above a company. The British Army of mid-1918, with able leaders being commissioned from the ranks, and the development of the all-arms battle, of tanks, aircraft, artillery and infantry working in harmony, had been the most professional in the world. With the onset of peacetime soldiering, the force seemed to turn its back on the lessons just learnt, and re-adopted a cosy turn-of-the-century existence. Class and poor pay excluded some able potential officers from poor backgrounds, whilst the money frequently attracted only a dubious quality of soldier, who required close supervision. The result was that, like the French, initiative at junior level was squashed. This was also partly the effect of a purely voluntary army: both the French and Germans fielded conscript armies. Nevertheless, a 1934 War Office pamphlet, 'Notes on Certain Lessons of the Great War', had concluded:

> In looking back at the War and all its lessons, we must not overlook the most important lesson of all, viz., all wars produce new methods and fresh problems. The last war was full of surprises: the next one is likely to be no less prolific in unexpected developments. Hence, we must study the past in the light of the probabilities of the future, which is what really matters. No matter how prophetic we may be, the next war will probably take a shape far different to our peace-time conceptions. In order to cope with this upset to our preconceived ideas, our leaders must be versatile, mentally robust and full of common sense and self-reliance. To produce this sort of mentality must be the object of our training.[14]

When Britain finally adopted conscription in April 1939, even less initiative was allowed with so many poorly trained soldier-civilians about of unknown quality. British doctrine largely followed the French, though was moderated by the fact it was not foreseen that Britain would ever fight alone. Financial cutbacks throughout the 1920s and 1930s had ensured the development of little new equipment: British armour, for example, was not only inadequate, but rarely even seen by the infantry. British tanks were designated either infantry (relegating them to an infantry support role), or cruiser (emphasising speed). Cooperation with the RAF was non-existent; there was no ground-to-air radio link or inter-tank communication (other than by flag). The few radios in service were often unreliable. In any case – unlike the Germans – artillery, engineers and infantry had no cross-country vehicles with which to keep pace with armour in a fast-moving battle. When it came to war, dismounted troops advancing with tanks either fell behind, or held the tanks back, but this problem was never apparent until too late, because there were so few pre-war joint exercises.

As the British Army possessed precious little mobility, Monty's 3rd Division (nicknamed the Iron Division in 1916) was forced to requisition civilian vehicles as it trained around Crewkerne in Somerset. On 19 September 1939 it was inspected by George VI, before trickling over the Channel. In a sense it was depressingly similar to 1914 all over again. Another khaki-clad British Expeditionary Force (with the same three initials – BEF), heading over to France, wearing the same soup-bowl Tommy helmets as their predecessors, carrying identical .303-inch Lee Enfield rifles and a very similar box respirator. Monty had to rely on a huge influx of Reservists (as many as fifty percent in some infantry battalions) to make up numbers and was forced to use impressed laundry vans to move his division's stores.

If Montgomery was powerless to solve these larger issues, he could deal with lesser ones: in expending 100,000 rounds of ammunition he at least ensured that his reservists had all fired their weapons and thrown at least three hand grenades. They then crossed the Channel: vehicles to Brest (where they were looted by French dockers) and troops to Cherbourg, where no transport awaited them. It was all sadly amateur. Monty knew his superior, the II Corps commander Alan Brooke very well (they had been at Camberley together). Brooke's other division was the 4th, commanded by Major General Dudley Johnson, who had experienced an extraordinarily good First World War, winning a VC, two DSOs and an MC. Though undoubtedly very brave and 'charming, amiable and capable of evoking great loyalty', with hindsight it seems he may have been over-promoted as a divisional commander (though it is often difficult to get rid of a VC-winner if they don't wish to go). Although only two years older than Monty, he surrendered his command after Dunkirk, having found the campaign tiring and upsetting in the extreme. Brooke recalled of the pair: 'My divisional commanders amuse me, it would be hard to find two more different types, both most efficient in their own way.'[15]

Allied strategists reasoned that the *Wehrmacht* would attack through northern Belgium, giving their tanks room to manoeuvre in open country; southern Belgium and the hilly Ardennes were not thought suitable for the chosen German tactic of tank thrusts. Thus the Allies, under the supreme command of the aged French general Maurice Gamelin, adopted a defensive plan for Belgium in November 1939, Plan 'D'. It was based on the concept of rushing into northern Belgium and creating stop lines to slow down or halt the blitzkrieg, allowing time to organise a counterattack, corresponding to French military doctrine.

These stop lines ran along rivers, particularly the Escaut (Scheldt) and the Dyle (hence Plan 'D'). Gamelin commanded from his HQ, the gloomy Château de Vincennes outside Paris, and had no direct radio links with his field forces. Using an unnecessarily tangled chain of command, Gamelin, who was also responsible for troops in the Alps, Syria and

North Africa, issued orders via his deputy, General Georges, C-in-C of the North-Eastern Front. Georges had three Army Groups under him, of which General Billotte's No. 1 Group was earmarked to move into Belgium. Billotte commanded four French armies, and the BEF.

The supremely courageous, if dull-witted, fifty-three-year-old Grenadier Guardsman, John Vereker, 6th Viscount Gort, VC, DSO with two bars, MC, thrice-wounded and eight times Mentioned in Despatches, commanded the nine divisions of the British Expeditionary Force. Six divisions followed later on. His undoubted credentials fighting the Germans left the French in no doubt that Britain was finally taking the war seriously. Though under French command, he also had the right of appeal to the British government before executing any order 'which appears to you to imperil the British Field Force'. However, there had been no pre-war exercises, tactical cooperation or even plans drawn up to work between the coalition partners. Whilst the First, Seventh and Ninth French Armies of Billotte's Group were to advance into Belgium, General Charles Huntziger's Second Army was to remain in France covering the Ardennes sector as far south as the Maginot Line. French military strategy relied on a shield of Jules-Verne-like underground fortresses, interconnected reinforced concrete bunkers mounting heavy artillery, anti-tank and machine guns, protected by acres of anti-tank obstacles, ditches and barbed wire, which stretched from the Swiss frontier to the Belgian border – the Maginot Line, named after the French defence minister who initiated the project and who as an infantry sergeant had seen how the strong forts at Verdun had held out against German attacks throughout 1916.

Unfortunately for the Alliance partners, the Maginot Line did not stretch any further north than the southern Ardennes: at the time of building, Belgium was an ally, and a Maginot Line along the Franco-Belgian border was considered diplomatically impossible. Shortly after the line's completion in 1936, Belgium had opted for neutrality, and it was tacitly acknowledged that the moment she was invaded by Germany, (this was a 'when', not an 'if'), France and Britain would roll into Belgium to halt the next generation of field-grey invaders. In the meantime the Belgians pursued a policy of aggressive neutrality towards their future war partners, refusing permission for reconnaissances and shunning military contact, so as not to offend Berlin.

Monty's 3rd Division was immediately deployed to defend the area east of Lille, which is featureless and where it would have been impossible to halt, even temporarily, a determined opponent. Monty had under his command three infantry brigades, one of which, the 9th Brigade (2/Lincolns; 1/King's Own Scottish Borderers; 2/Royal Ulster Rifles) he had commanded for eighteen months in 1937–38, whilst he had been Brigade Major of 8th Brigade. (1/Suffolks; 2/East Yorkshires; 4/Royal Berkshires) in 1922–23. The third was 7th Guards Brigade (1/Grenadiers; 2/Grenadiers and 1/Coldstream Guards). Only the 2/Lincolns of 9th Brigade had trained with tanks before, on a single armour–infantry exercise at Bovington in early 1939.

Despite Brooke and Monty's wishes to the contrary, 3rd Division was tasked to dig frontier defences, the 'Gort Line', throughout October and November in the wet, clinging Flanders mud. The adoption of Plan 'D' in November required on invasion that Monty and his division dash seventy-five miles east then occupy a defensive position along the River Dyle covering Brussels; so trench digging was abandoned, sensible numbers of military vehicles began to trickle through to the brigades and the move was practised many times, but only in France. Undoubtedly Monty did more than his fellow divisional commanders to correctly anticipate the nature of future war, but he was part of a larger machine and could only initiate his ideas within his division. In this sense each division developed its own procedures and no one at BEF headquarters (in Arras) was overseeing a special theatre-wide doctrine.

Monty first made himself known to as many men in his division as possible, touring around in his staff car and talking to them, trying to gauge the calibre of men he was leading. He soon found that very few sported ribbons from the First World War; this was both good and bad: good because it meant his division was relatively young; bad because very few had seen action before. He believed that sniping and patrolling were essential battle skills lost since 1918, and insisted that each battalion form a specialist fighting patrol of one officer and twelve men, and a sniper section.

He made a point of knowing everything about his command and it became a joke amongst visitors to try to find a question he could not answer.[16] Thus, he began to develop his hallmark techniques of leadership – almost showmanship – to infuse in his men the enthusiasm and confidence he genuinely felt himself. He was ruthless, too, about purging those he didn't like, or felt were not up to the challenge – a task made easier when his old colleague 'Simbo' Simpson (on whom he had leant heavily after Betty's death in 1937) arrived at GHQ in Arras as Military

Secretary, in charge of officers' postings. Monty began to use him to move on older captains, majors and lieutenant colonels he regarded as 'useless' (very much a Monty word), largely due to their age or poor fitness. 'Simbo' was able to place them in other, less-taxing roles, often where their talents were better employed, sometimes so they did not realise they had been sacked.

Monty's insistence on fitness was already legendary within the division, and stemmed from his experiences after Le Cateau and his robust physical condition on being wounded in October 1914, which significantly aided his recovery. Such fitness was uncommon in the French and British armies, where officers still made a point of lunching and dining well. This was Monty leading by example, though achieved at personal cost, for his left knee and lung wounds never quite healed, the latter leaving him short of breath. Later, when back in southern England, Brigadier Brian Horrocks recalled one of Monty's rants on fitness:

> 'Too many officers spend too much time in their offices and are becoming fat and almost permanently chair-borne. No good for war,' Monty said. 'Every officer in this command must carry out two cross-country runs weekly, irrespective of age or rank.' His senior medical officer protested against this no-exception rule and mentioned a senior administrative staff officer. 'Colonel X must not run, sir. If he runs he will probably die.' Monty replied, 'Let him die. Much better to die now rather than in the midst of a battle when it might be awkward to find a replacement.' Colonel X did run and Colonel X didn't die.[17]

In France, Monty also tried to read ahead and anticipate what the coming war would mean for him and his division. He insisted that 3rd Division trained for withdrawal, day and night, in contact with the enemy. He introduced scenarios where the withdrawal would be dogged by enemy aircraft, mobile ground forces; where roads were blocked with refugees (as he had witnessed after Le Cateau) and his infantry would have to march, dig, fight, and march again. All of these training initiatives were Monty's own: at no stage was any attempt made to make any of this practical preparation for war mandatory across the BEF.

In November, Monty found his career on the line in unusual circumstances. Alerted to the growing incidence of venereal disease amongst his troops, he issued an Order of the Day which required battalion COs to educate their men in matters of sexual hygiene, promote the safer brothels that had sprung up in towns (patrolled by British military policemen), and encourage the sale of condoms in the NAAFI, or teach soldiers how to ask for one in French.

> My view is that if a man wants to have a woman, let him do so by all means; but he must use his common sense and take the necessary precautions against infection – otherwise he becomes a casualty by his own neglect, and this is helping the enemy ... There are in Lille a number of brothels, which are properly inspected and where the risk of infection is practically nil. These are known to the military police, and any soldier who is in need of horizontal refreshment would be well advised to ask a policeman for a suitable address.[18]

A copy of this perfectly sensible circular ended up on the desk of the BEF's senior chaplain, who demanded action from Gort. The problem seems not to have been the subject matter, but the vernacular language used by Monty. The C-in-C of the BEF, Lord Gort – a shy, conservative Guardsman, who, though also of Anglo-Irish extraction, seems to have stood for everything that Monty found old-fashioned and counter-progressive – does not seem to have been a fan of Monty's and required him to withdraw the document, which (in Monty's eyes) would have so undermined his position as to require him to be sent home. Brooke, who had fought hard to get Monty into 3rd Division, was reluctant to lose him, and interceded, noting in his diary on the 23rd,

> Started the day by having to 'tell off' Monty for having issued a circular to his troops ... I therefore pointed out to Monty that his position as the commander of a division had been seriously affected by this blunder and could certainly not withstand any further errors of this kind. I also informed him that I had a very high opinion of his military abilities and an equally low one of his literary ones! He took it wonderfully well, and I think it ought to have done him some good. It is a great pity that he spoils his very high military ability by a mad desire to talk or write nonsense.[19]

Of the occasion Monty later wrote:

> I thought in my innocence that some urgent tactical problem was to be discussed. I was mistaken! He [Brooke] arrived, and I could tell by the look on his face and his abrupt manner that I was 'for it' – and I was

right. He let drive for about ten minutes, and I listened. When he had run out of words he became calmer, and even smiled. I then plucked up courage and said that I reckoned my circular letter was a rather good one, and extremely clear. That finished it; he began again and I received a further blasting![20]

The winter of 1939–40 was the coldest of the century, and the English Channel froze at Boulogne. British troops struggled to keep warm in poor billets and in trucks with open cabs, whilst the little steel-tracked Bren carriers frequently skated on the icy roads. The weather seemed to sap the Allies' will, too. Although German blitzkrieg tactics had already been demonstrated in Poland, the Allies appeared reluctant to develop new ideas to counter them, or revise their doctrine. The Siberian climate (though nothing compared with the following winter the Wehrmacht were to face in Russia) severely curtailed the training of the BEF and French armies, who spent much of their time digging defensive lines. Consequently the initiative was surrendered to the traditional enemy, lurking behind their Siegfried Line, whilst throughout France troops who desperately needed additional training focused most of their energies and resources on improving and building fortifications, rather than developing their fighting skills. The fascination with fortifications spread to the British, and in January 1940 brigades of the BEF were sent in rotation to positions forward of the Maginot Line, to give troops a taste of contact with their opponents. Each battalion of 8th Brigade spent a week up in the 'Ligne de Contact', sending out fighting patrols.

During the same month, mobilised divisions of the Territorial Army started arriving in France and a 'stiffening' process began whereby Regular brigades were given a Territorial battalion and vice versa. Monty's 3rd Division received 4/Royal Berkshires and 76th (Highland) Field Regiment, RA. Whereas many commanders had been resistant to having Territorials foisted on them, Monty, with prior experience of Britain's part-time army (gained with 47th Division in 1918 and at York with 49th Division during 1923–25), had no such objections.

To keep the troops interested during the lull before the coming storm, some units were sent off to tour the nearby Great War battlefields, where the older soldiers amongst them had fought. A young platoon commander in a Territorial battalion of Monty's Royal Warwickshire Regiment wrote of this to his mother in March 1940:

Went off in trucks for a long drive round the old battlefields conducted by an officer, Major Harding, who went all through the last war with this battalion. He's a good bloke, one of the best in the battalion and a very good talker. He remembered all sorts of little details and made it intensely interesting. In a village where he had been for eight months, he showed us the house where his dugout had been, the communication trenches and the old front line, and related a variety of stories about people and places. To have history told like that, by an eyewitness on the spot, of a regiment one is now in, and simultaneously to be able to pick up souvenirs, made it an unforgettable day. It's extraordinary how much one finds lying about still, even in cultivated fields: rusty clasp knives, bits of bayonet, old bullets and bits of shell, even helmets . . . all the chaos of those shows which I am absolutely convinced will never happen again.[21]

The letter's author, Lieutenant Gordon Potts, did not return home and lies buried in a war cemetery next to his comrades from the former battlefield he was exploring.

Still the BEF was barely fit for purpose, and had sadly spent too much of the winter digging defences rather than rehearsing tactics, though 3rd Division perhaps rather less than the rest. The portentous diary entry for 28 November 1939 of Monty's corps commander, Brooke, provides a flavour of the whole Expeditionary Force at this time and well into the New Year:

On arrival in this country and for the first two months the Corps was quite unfit for war, practically in every respect. Even now our anti-tank gunners are untrained and a large proportion of our artillery have never fired either their equipment or type of smoke shell they are armed with. To send untrained troops into modern war is courting disaster such as befell the Poles.[22]

13

Blitzkrieg

THE ORIGINAL GERMAN strategy for the invasion of the west in 1940, named *Aufmarschanweisung Gelb* ('Deployment Directive Yellow'), envisaged a Schlieffen-like thrust into northern Belgium and Holland, rather as the Germans had done in 1914. This is what the Allies, the British and French, expected, and the capture of German war plans from a crashed plane in January 1940 reinforced this perception. Whilst Directive Yellow originated from the German Army's high command, a brilliant tactician, fifty-three-year-old *Generalleutnant* Erich von Manstein, proposed an alternative. A northern army would still thrust into Belgium as it was expected to do, thus distracting the Allies' attention, whilst a southern army would attack through the poorly defended Ardennes, then turn north and cut off the French and British armies. The Germans expected the cream of the Allied armies – France's best armoured divisions and the whole of the BEF – to be sent into Belgium, where they would be encircled and destroyed by the Ardennes thrust materialising behind them. Basil Liddell Hart labelled the German advance into Belgium 'the matador's cloak', at once enticing the Allies into Belgium, and distracting them from the real attack further south. The Germans called Manstein's manoeuvre *Sichelschnitt* ('cut of the sickle, or scythe').

Apart from excellent pâté, the Ardennes region of Luxembourg, southern Belgium and north-east France is renowned for its hills, valleys and above all, forests. Rommel had served here with the 124th

Württemberg Regiment in 1914. Since only few minor roads wind their way through the trees the French considered the region to be impractical as a fast attack route for motorised columns. French war games of the 1930s calculated that an attacking force might take nine or ten days to work its way from the German border through the Ardennes to the River Meuse (the River Maas to the Germans), a distance of about seventy miles. German war games during the autumn and winter of 1939–40 suggested that it might take as little as sixty hours to penetrate the forests and reach the Meuse, the gateway to France. In the event, the panzers managed it in fifty-seven.

German tactics of 1940 had evolved from those of the *Sturmtruppen* of 1917–18, which Rommel knew well, whereby small bands of what were effectively special forces probed the front, looking for a weak spot, then attacked. In 1940, the weak spots had been already identified and troops trained beforehand. In contrast to the Allies, German doctrine, as taught by Rommel at the officer schools, also embraced decentralisation and personal initiative: identifying a commander's intent, and acting to achieve that intent even if beyond the strict remit of orders issued. Momentum was another key principle to understanding German doctrine; it won them the day in France and applied to every arm: assault pioneers, artillery, logistics, as well as armour or infantry. The use of reserves to reinforce success and maintain the momentum of the advance was demonstrated often in the 1940 campaign, and again in 1941 in Russia. The concept of the all-arms battle was fully imbued into the Wehrmacht by 1940, field commanders being able to call on Luftwaffe dive-bomber support in neutralising centres of opposition, whilst on the ground engineers, artillery, infantry and tanks worked together. Full use was made of radios between the various arms, ground-to-air, and tank-to-tank, making an enormous contribution to control of the battle.

The year the Maginot Line was completed, 1936, Heinz Guderian had published a textbook on armoured warfare, advocating fully mechanised divisions capable of punching a hole in a front line. He advocated large formations of tanks, fighting en masse, that would break through their opponents' defences, followed by motorised infantry who would mop up the enemy defenders. Up with the tanks would be engineer units, dealing with obstacles, whilst mobile artillery and anti-tank units would support the advance, and protect the flanks. Guderian saw a tactical air force interdicting the deployment of the enemy's reserves, and providing close support to the armoured thrusts. His book, *Achtung-Panzer!*, is now famous, but at the time of its publication it was ignored by the British and French, though after the war Guderian himself generously (perhaps over-generously) acknowledged the influence of the British advocates of armoured warfare, Captain Basil Liddell Hart and Major General J. F. C. Fuller. Nevertheless, it is also true to say that blitzkrieg, whilst based on the ideas of General Heinz Guderian, the father of the panzer arm, was not an established doctrine but rather one that evolved and was fine-tuned during combat operations.[1]

The final plan called for *Generaloberst* Fedor von Bock's Army Group B to advance into Belgium and Holland on 10 May 1940 with twenty-six infantry and three panzer divisions. The main German thrust through the Ardennes would be undertaken by *Generaloberst* Gerd von Rundstedt's Army Group A, with forty-five divisions under his command. His cutting edge would comprise three panzer formations: 5th Panzer and Rommel's 7th Panzer Division forming Hermann Hoth's XV Corps; Georg-Hans Reinhardt's XLI Corps (of 6th and 8th Panzer), and Guderian's XIX Panzer Corps (consisting of the 1st, 2nd and 10th Panzer Divisions). Rundstedt's Chief of Staff was none other than Manstein, architect of the *Sichelschnitt*. Clearly, Guderian possessed the largest corps, including over 800 tanks of the *Wehrmacht*'s senior panzer divisions. Both XLI and XIX Corps were grouped together into a Panzer Group under the conservative General Ewald von Kleist, the appointment being a precautionary measure by Army High Command, still unsure, suspicious, and probably jealous of the new arm and its potential. Kleist in turn was under the newly created Twelfth Army, led by Rommel's old patron, Wilhelm List. It would not be numbers that defeated the Allies in 1940, for the Allies possessed greater physical power than the Germans – in tanks, men and aircraft. The French, British, Belgian and Dutch armies together fielded 151 divisions, to Germany's 135; had 14,000 artillery pieces to the German 7,378; possessed a total of 4,204 tanks of all kinds, compared with the Wehrmacht's 2,439 and had 4,469 aircraft available to the Luftwaffe's 3,578.[2] More than anything else, the French collapse would underline the importance of the inner will to fight, succeed, or triumph. Even Napoleon recognised that the inner will was more important that raw power, when he stated 'The moral is to the physical, as three is to one.'

On receiving the code word Danzig, the campaign started in the early hours of Friday 10 May with extensive raids of Allied airfields, major arterial roads and specific railways, the aim being to paralyse military movement and spread confusion. The activity of the Luftwaffe, one of the hallmarks of the campaign recalled by all its survivors, was evident from the first moment.

The majority of the panzer divisions thrust into the Ardennes at 4.35 a.m. on the morning of 10 May and fought their way past Belgian *Chasseurs-Ardennais* and French light cavalry units in well-rehearsed moves. Their initial advances on 10–11 May were assisted by special forces, previously infiltrated in civilian dress, who secured key road junctions and neutralised the few strongpoints scattered through the hills. Glued to the roads as they were on account of the terrain, the long columns of German vehicles were not vulnerable to hostile aircraft or fuel shortages, as they would be when battling over the same ground in the December 1944 Battle of the Bulge, for the Luftwaffe ruled the skies. Despite French time-and-space studies anticipating a German arrival on the Meuse on or about 20 May, 1st Panzer Division reached the Meuse by 2 p.m. on 12 May, their first significant obstacle, which was well defended by French bunkers.

Crucially, although the French detected panzers in the Ardennes on 10–11 May, the German attacks of Army Group B to the north, including the spectacular assault by glider-borne troops on the Belgian fort of Eben Emael (a very modern touch, with a *coup de main* strike launched by the 1940 equivalent of an SAS squad, and the first such assault in military history) and the presence of some tanks and large numbers of paratroops in Holland, persuaded the Allies that the Ardennes was not the focus of a major attack. Even reports from aerial reconnaissance over the Ardennes of long motorised columns on 12 May did not disrupt this view. At this stage, the Germans, aware of Allied expectations, were still showing the Allies what they expected to see: much activity to the north, and far less in the centre – a classic intelligence ploy. There was little artillery shelling accompanying Army Group A's advance, trad-itional before launching an attack, and Luftwaffe activity in the Ardennes was kept to a purposeful minimum.

By the evening of 12 May Rommel was at Dinant and Guderian at Sedan. Whilst Guderian's three divisions that made up XIX Panzer Corps were obliged to mount a set-piece assault river crossing, Rommel's recon-naissance troops, the 7th Motorcycle Battalion, had found an unguarded weir at Houx, further north, which led to a narrow island in midstream. In the dark, his men stepped gingerly across a weather-worn stone weir and occupied the island; they then crept over a pair of unguarded lock gates and established a small bridgehead on the far bank. Incredibly, the Germans had discovered and exploited a gap between two defending French units. (The island is best viewed today from the ruined mediaeval castle at Poilvache, from where Rommel briefly assessed his men's triumph the following morning, before hurrying on.) Already Rommel was setting an impressive pace, ahead of his colleagues. Pontoon bridges still needed to be constructed and 13 May saw Rommel at the bridging sites, first taking his place in one of the rubber assault boats, then inspiring his engineers, at times standing up to his waist in water, shifting timber. It was at another crossing site, Leffé, according to his memoirs, that Rommel crossed in one of the first assault boats and commanded an infantry company in combat for over an hour, before returning.

Once through the Ardennes and over the Meuse in a series of well-rehearsed bridging operations, exhaustively practised on the Mosel the previous autumn, the panzer divisions opened out like a fan across northern France, and began to cover astounding distances, there being little or nothing to stop them. Like a latter-day Wellington, Rommel was always at the front, unerringly just where he was needed, but within range of enemy fire. This was demonstrated on 14 May when, under fire at Sivry, Rommel's command tank drove into an unseen ditch and he was wounded in the face by a shell splinter and had to wait whilst Karl Hanke's tank chased off his attackers.

On 16 May, Rommel's colleagues in 1st Panzer Division covered forty miles. That night Rommel's men captured a complete French armoured battalion of forty-eight tanks in Avesnes, parked by the roadside. But behind the panzers there was nothing, for most German infantry divisions relied on horses for mobility and were struggling to keep up: soon they would be days behind the armour. If we use historian Karl Heinz Frieser's metaphor of the *Wehrmacht* as a lance, there was a danger that the 'steel tip' – ten panzer divisions and six motorised infantry divisions – would become separated from the 'wooden shaft' – the remaining 127 horse-bound infantry formations (which contained both the infantry support and vital logistics).[3] At Le Cateau on 17 May, for example, Rommel had to halt his own panzer battalions whilst he personally went back in an armoured car to shepherd the rest of his division forward. With completely vulnerable flanks, a hint of the missed opportunities open to the Allies came on the 17th, when the French 4th Armoured Division, led by Brigadier General Charles de Gaulle, attacked Guderian's flank at Montcornet. De Gaulle's two battalions of light Renault R-35

tanks, one of heavy Char-Bs, and a single battalion of infantry in requisitioned buses caused havoc amongst the German support troops.
However the R-35s were massacred, and although the Char-Bs were
impervious to German anti-tank shells, they turned back, being without
infantry support. Guderian neglected to inform his higher command of
this attack to his flank, doubtless fearing a repetition of the halt order
he had just received and ignored.

It was at Landrecies, scene of a dramatic British cavalry charge in
1914, that in the small hours of 17 May hundreds of French troops,
many vehicle-borne, surrendered. Rommel recalled:

> particularly irate over this sudden disturbance was a French lieutenant
> colonel whom we overtook with his car jammed in the press of vehicles.
> I asked him for his rank and appointment. His eyes glowed hate
> and impotent fury and he gave the impression of being a thoroughly
> fanatical type. There being every likelihood, with so much traffic on the
> road, that our column would get split up from time to time, I decided
> on second thoughts to take him along with us. He was already fifty yards
> away to the east when he was fetched back to Colonel Rothenburg, who
> signed to him to get in his tank. But he curtly refused to come with us,
> so after summoning him three times to get in, there was nothing for it
> but to shoot him.[4]

This often overlooked and shocking passage puts Rommel, as the senior
officer present, in the frame in May 1940 as a war criminal, condemned
by his own words. He did not pull the trigger, but nowhere in his memoirs
(written for a German readership) was he repentant for this specific act
either. Indeed, this incident must challenge his later reputation from
the Desert War as one of a clean fighter.

Racing for the coast, Rommel's 7th Panzer Division was joined by
Guderian's XIX Panzer Corps, together with XLI and XV Panzer Corps.
Nine divisions of panzers now forged a corridor heading north-west to the
Channel coast, with almost unstoppable momentum. Their axis of advance
was deliberate, though not innovative, as Sir John Keegan points out: the
'unconfined romp across open country . . . followed very closely the line
of *Route Nationale* 43, which for much of its length is the Roman road laid
out soon after Caesar's conquest of Gaul in the First Century AD'.[5]

Tearing through the First World War battlefields, Rommel's panzers
captured Le Cateau on 18 May, scene of the 1914 battle and Monty's

first introduction to war, then encountered French tanks outside Cambrai. According to aide-de-camp Captain Hermann Aldinger, who had served with Rommel in the WGB 1916–17: 'If Rommel was on your flank you knew you had nothing to worry about on one side at any rate.'[6] Demonstrating some of the guile that was to find him fame in North Africa, Rommel ordered a column of soft-skinned vehicles to advance on the town from a flank, in wide formation, firing as they went. The dust storm they created fooled the defenders into thinking they were being assaulted by a large panzer formation, rather than armoured cars and trucks, and they surrendered. By the evening of the 20th, 2nd Panzer Division glimpsed the River Somme (a name of ill omen for the British, but not so for their opponents), and followed it to the coast, at Noyelles. The rewarding sight of the English Channel for the black-suited, grimy and exhausted panzermen was proof that they had cut the Allied armies in two.

Two days earlier Hanke surprised Rommel in his headquarters, to award him the Knight's Cross on behalf of their Führer. At the same time, Hanke made Rommel aware that he was the first divisional commander in France to be so honoured. Rommel's adjutant, Hauptmann Hans-Joachim (Jochen) Schraepler, wrote to Lucie on Rommel's behalf with news of the decoration:

> My Dear Frau Rommel,
>
> May I be permitted to inform you that the Führer has instructed Lieut. Hanke to decorate your husband on his behalf with the Knight's Cross. Every man of the division – myself particularly, who has the privilege of accompanying the General – knows that nobody has deserved it more than your husband. He has led the division to successes which must, I imagine, be unique. The General is now up with the tanks again. If he knew that I were writing to you, *gnädigste Frau*, he would immediately instruct me to send you his most heartfelt greetings and the news that he is well.[7]

The tall, distinguished-looking and ever-so-correct Hauptmann Hans-Joachim Schraepler, whose diaries have recently been rediscovered and published, would follow Rommel to North Africa. He lies buried there still, having served his general faithfully until his own death in December 1941.

Hanke was himself recommended later that month for a Knight's Cross, but never received it. He had (maybe when drunk) boasted to his colleagues that as Goebbels's deputy he technically outranked Rommel and could have him replaced if he wanted to. He had overplayed his hand. Rommel heard of his boasting and in a gesture that smacks of Montgomery's later behaviour, sent Schraepler to Hitler's HQ to overturn the award, and sacked Hanke immediately. The division was split between loyalty to Rommel, and admiration for Hanke as a good leader of men in battle, some believing he should have retained the award. Hanke later became a *Gauleiter*, defending Breslau with fanatical zeal, and was named the last *Reichsführer-SS*, replacing Himmler in April 1945. One is struck by a touch of unnecessary meanness here: Rommel had lobbied hard in 1917 for his own *Pour le Mérite* but in 1940 strove equally hard to prevent Hanke's award, which would, in any case, have reflected well on his division. Josef Goebbels may not have been so concerned, for Hanke was romantically involved with Magda Goebbels whilst Josef embarked on a series of extramarital affairs of his own.

Friday 10 May saw Monty's 3rd Division begin to trickle across the frontier into neutral Belgium from 6.45 a.m. Amongst the units allotted to Monty was 2/Middlesex – a machine-gun battalion commanded by Lieutenant Colonel Brian Horrocks. This began their long wartime association. Horrocks recalled:

> I saw him every day, sometimes several times a day, and he was always the same; confident, almost cocky you might say, cheerful and apparently quite fresh. He was convinced that he was the best divisional commander in the British Army and that we were the best division. By the time we had reached Dunkirk I had come to the same conclusion![8]

With the armoured cars of the 15/19th Kings Royal Hussars (Monty's divisional reconnaissance regiment) probing forward, 3rd Division raced to take up their positions along the Dyle, only to find a jumpy Belgian division in place, who shot and wounded one of his men. At 4 p.m. on the 14th, the Dyle bridges were blown as German troops hove into view and the division came under shell- and machine-gun fire. Third Division remained until ordered to withdraw on the 16th and fought their way back to a second river line, the Escaut, on the 19th. The infantry of Monty's 9th Brigade were carried in an eccentric collection of requisitioned Belgian vehicles ranging from bicycles to large, brightly painted circus lorries, the withdrawal being covered by Horrocks's machine guns.

Monty's routine of command evolved clearly during this time: he would spend all day out with his troops, spreading calm and common sense, returning each afternoon to wherever his headquarters was located; likewise, his habit of going to bed immediately after dinner, at around 9 p.m. Monty even looked different: though the trademark beret was still a couple of years away, he was the only senior officer to wear the new, scratchy but practical battledress worn by all troops, whilst his senior officer colleagues preferred cavalry twill riding breeches and tunics. One photograph of him visiting his 7th Guards Brigade in November 1939 emphasises the point and must have sent shock waves through the old guard: next to the ramrod-straight, tall guardsmen, behatted, booted, polished and jodhpured, the diminutive Monty almost slouches, hands in pockets – as always, going against the tide.

3rd Division was allotted the defence of a 12,000-metre stretch of the Escaut between Pecq (about six miles north of Tournai, where Monty had finished the Great War) and Avelgem in Belgium. They had time to dig in on 19–20 May, with all three brigades in line and one battalion in reserve, before the Germans closed up and started shelling, prior to a general assault. Throughout 21 May the brigades broke up German attempts to cross in rubber boats. Eventually, because the British front had been penetrated elsewhere, orders were relayed in the afternoon of 22 May for a further withdrawal ten miles back to the original Gort Line, reached at first light the following morning. Now the division was back where it had started nearly two weeks earlier. For the next five days they dug and manned trenches. During this lull, when the Germans shelled but did not attack 3rd Division, Monty presented medal ribbons awarded to his men for bravery or leadership in the campaign so far: two DSOs, three MCs, a DCM and six MMs, evidence that the notion of the immediate award, adopted in 1914–18, had not been lost and was able to function in relatively adverse circumstances.

On 18 May, Gort ordered Major General Roderic Petre, commanding the British 12th Infantry Division, to hold Arras, now directly in the path of the panzers, using all available troops in the vicinity – which he called 'Petreforce'. At this stage there was an Anglo-French aspiration to seal off the panzer corridor (with Rommel's division at its centre) that lay to the south-west of Arras, with a two-division French attack from the south and a simultaneous British two-division attack from the north. Monday 20 May saw Rommel's 7th Panzer approaching the south-west outskirts of Arras, heading for the sea. On his left (western) flank was 8th Panzer, with the SS-*Totenkopf* Motorised Infantry Regiment behind it.

The day was confusing and critical for the Allied high commands, as the French C-in-C, Maurice Gamelin, had just been replaced by Maxime Weygand (formerly in charge of French troops in Syria). Desperate measures: a seventy-three-year-old replacing a sixty-eight-year-old. The same day, 8th Panzer Division had just lurched over the nearby First World War battlefields of the Somme, attacking two British Territorial divisions, the 12th and 23rd, south of Arras. The British were much understrength, and so poorly equipped that there were not enough rifles and pistols to issue a personal weapon to everyone; no one intended them to be thrown into battle as fighting formations and they had been sent over as lines-of-communication troops to help guard rear areas. Consequently they had no transport, signals or artillery – an unthinkable occurrence in modern, professional armies.

With the French disorganised and a new commander taking over, Gort was still determined to initiate the British arm of the attack, nonetheless. To undertake his assault, Gort created 'Frankforce', under Major General Harold Franklyn, of 5th and 50th (Northumbrian) Divisions, and 1st Army Tank Brigade. After units had been despatched to defend the line of the River Scarpe, east of Arras (these included 13th Brigade of Brigadier Miles Dempsey – Monty's Second Army commander of 1944), the only troops available to take the actual offensive were two Durham Light Infantry (DLI) battalions of 50th Division, and 4th and 7th Tank Battalions of 1st Tank Brigade. The DLI had only arrived during the early hours of 21 May, having had very little sleep for several days; although motorised, their transport had been left in Belgium, and they had been marching on foot ever since. Due to the fast-nearing panzers, Franklyn ordered an attack for the 21st, using the Durhams, the RTR tanks and the French 1st DLM (*Division Leger Mechanisé*, light mechanised division), equipped with Somua S-35 tanks, covering their right flank, whilst the River Scarpe masked the left.

The general British plan was to attack north to south, mopping up any German troops in the vicinity of Arras. No accurate intelligence about the advancing German units (actually Rommel's 7th Panzer Division) was passed on to the attacking formations. The two tank battalions had covered over 120 miles in the preceding five days on their tracks. Consequently, they had both lost about twenty-five percent of their strength through mechanical breakdowns, rather than combat. Fourth

Royal Tank Regiment (RTR) was equipped with the 11-ton Infantry Tank (known as the Matilda) Mark I, sporting a single machine gun, and a two-man crew. Seventh RTR, by contrast, had the 26-ton Infantry Tank Mark II, which was then the most heavily armoured tank in operation anywhere, armed with a two-pounder gun and manned by a crew of four. After their various moves, the two battalions had fifty-eight Mark Is, but only sixteen Matildas IIs serviceable.

The assault was thrown together with great haste: there was no time for a proper ground reconnaissance, and very few adequate maps were available. Tank wireless sets had been on radio silence for several days to conceal their movements, and had drifted 'off net' since: consequently when the battle started few could talk to each other. There was no means of communication with the infantry, nor was there any air support. No tank officer was present at the final infantry orders group (by the Canadian War Memorial on the summit of Vimy Ridge) that morning, and – astonishingly – no one had any idea of the extent or location of the opposition, being assured by GHQ they would encounter only weak German detachments. Additionally, though the infantry battalion commanders were placed in command of everything, this was not made clear to the two tank battalions, further paving the way for confusion.

Two mixed armour–infantry battlegroups were formed; the inner column had the shorter distance to cover, and left Vimy at 11 a.m., but after only a short way it was clear that the infantry could not keep up, and Lieutenant Colonel Miller of 6/DLI (who had fought in the same area in 1917 with the same battalion) let the tanks go on ahead. Their route took them straight into the right flank of Oberst Erich von Unger's 6th Rifle Regiment in Dainville, before even reaching the start line. Second Lieutenant Peter Vaux of 4/RTR described the clash:

> To our great surprise, we found we had come straight into the flank of a German mechanised column moving across our front. They were just as surprised as we were, and for a quarter of an hour there was a glorious free-for-all.[9]

Further on, the armour was momentarily held up at a level crossing, because the barrier was down, and the column halted out of force of habit. Eventually, an officer crashed through the gates, and the column proceeded. Between 3 and 5 p.m., the British tanks advanced through the close country south-west of Arras, overrunning Rommel's defensive screen, because his 37mm anti-tank guns were unable to penetrate their armour. In some cases, a dozen or more shells simply bounced off. *Oberstleutnant* Johann Mickl's anti-tank gunners of the 42nd *Panzerabwehr Abteiling* (anti-tank battalion) fled or were shot down, while Rommel's infantry took cover in the neighbouring houses. It was an inspiring period for the British tank crews, who briefly held the upper hand as they advanced, knocking out trucks and anti-tank guns, whilst infantry cleared the surrounding villages taking many prisoners.

The luck would not last; as the tanks reached Telegraph Hill, south-east of Arras, Rommel's 105mm field guns were waiting, and firing over open sights, twenty were despatched, the CO of 4/RTR, Lieutenant Colonel James Fitzmaurice, being killed. The British guns should then have engaged Rommel's in counter-battery fire but this required observation officers forward, and they did not yet have any armoured vehicles of their own from which to direct fire.

With 7th Panzer Division now fully alert, and no reserves to consolidate the British gains, the inner column withdrew. The situation was so confused that as they moved, the adjutant of 4/RTR flagged down a passing tank to speak to its commander, only to find it belonged to Rothenburg's 25th Panzer Regiment, which had been ordered back by Rommel to stabilise the rear; the emerging crewman and British officer starred at one another in horror, until the latter escaped in the dark.

Also leaving Vimy at 11 a.m., the outer column were to wheel west of Arras, but before even reaching their start line, Rommel's men were discovered in the neighbouring villages, which had to be cleared by tanks and the DLI. At this stage the tanks took the wrong turning, veering towards the inner wheel, and 7/RTR soon found themselves entangled with their colleagues of 4/RTR, on their correct route, but busy with Rommel's surprised infantrymen. Here both units saw the daunting picture of Rothenburg's entire regiment (over a hundred tanks), throwing up dust along the skyline. Had 7/RTR followed their correct route, they would have undoubtedly encountered the panzers with disastrous consequences. As it was, communications within 7/RTR soon broke down, the CO and his adjutant were killed, and French tanks from 1st DLM guarding the right flank were attacked in error.

Some tanks approached the village of Wailly and found it packed with startled panzer troops; at this juncture, it was Rommel's personal arrival that stabilised the situation. He assessed the situation instantly and speedily ordered a defensive screen around Wailly, using 88mm anti-aircraft guns in an anti-tank role for the first time, there being no other weapons available to stop the thickly armoured, lumbering Matildas. Rommel drove up to the high ground north of Wailly, where some of his artillery was deployed along a wood line.

> It was an extremely tight spot.... With Most's [his ADC] help, I brought every available gun into action at top speed against the tanks. Every gun, both anti-tank and anti-aircraft, was ordered to open rapid fire immediately and I personally gave each gun its target. We ran from gun to gun ... Soon we succeeded in putting the leading enemy tanks out of action. About 150 yards west of our small wood, a British captain climbed out of a heavy tank and walked unsteadily towards us with his hands up. We had killed his driver ... Although we were under heavy fire from the tanks during this action, the gun crews worked magnificently. The worst seemed to be over and the attack beaten off, when suddenly Most sank to the ground ... close beside me. He was mortally wounded.[10]

Now surrounded, the British were wondering how to escape, when six French tanks burst through Rommel's cordon, and at 3.30 a.m., the remnants of both columns managed to make it back to Vimy, unsure of the damage they had inflicted.

Ignorance, due to a lack of fundamental battlefield intelligence, had marred the action for both sides. The British had no idea of the size, identity or weaponry of the formations they were attacking, whilst Rommel in his war diary suggested that he had been attacked by *five* British divisions, rather than the actual 100 tanks and 1,000 infantrymen involved. In fact, Rommel lost between thirty and forty tanks, nearly 400 casualties, and a similar number taken prisoner. The Royal Tank Regiment lost considerably more, over half their strength. The counterattack had an effect on German strategy out of all proportion to its operational value. This was virtually the only offensive operation undertaken by the BEF in the 1940 campaign, yet the surprise of the Arras counterstroke, and its concentrated use of armour, threw Rommel off balance for the first (and only) time in May 1940. De Gaulle's flank attack on 17 May had shown what was possible, and alarmed the *Wehrmacht*'s high command. After all, the panzers were taking a huge risk, advancing at their heady pace, with little infantry in support and virtually no flank protection, their horse-drawn infantry divisions toiling far behind.

Three days later (24 May), a clearly nervous Hitler halted his panzers, which were only allowed to roll again on 26 May. Although the two-day halt was a much-needed respite for the panzer troops, who had been in continuous action since 10 May, the halt also slowed down the eventual advance on Dunkirk. For Rommel, this was the first time in either world war that he had encountered British soldiers, and caused the toughest day's fighting of the entire campaign. Perhaps this was to instil in him a respect for the Tommy that would be confirmed in North Africa. Rommel's hectic pace continued until 7th Panzer Division was ordered to rest on 29 May.

Although Arras has grown considerably since 1940, some evidence of the counterattack survives: walls lining the route of the counterattack still bear the impressive scars of artillery or bomb bursts from 21 May 1940. Looking north from Wailly, the wood line where Rommel directed his guns against the nearby Matildas, and where Lieutenant Most died, is immediately obvious. Marked on maps as La Ferme du Belloy, just a few bushy-topped trees remain. The last time I visited the spot, the plough had turned up a 20mm cartridge case from one of the anti-aircraft guns whose fire Rommel so accurately directed.[11]

Lord Gort was still under orders to attack southwards, in support of the French, on 25 May, but it became increasingly clear throughout the day that to continue doing so would isolate his troops still centred on Arras. At 6.00 p.m. that evening Gort made a courageous decision – on his own authority – ordering 5th and 50th Divisions to move from the southern end of what had become a British pocket, to reinforce Brooke's II Corps in the north. They would have the job of holding the northern flank of a corridor to the sea if (as looked likely) the Belgian army surrendered. Gort also knew that his reinforced garrison in Calais, further west along the coast, had been surrounded and would be soon obliged to surrender.[12] By 26 May the British and French had both decided to form a perimeter around Dunkirk, but for different reasons: while the BEF now intended an evacuation, the French still aspired to continue the fight.

Whilst Montgomery's 3rd Division was deployed along the Escaut Canal, Gort issued orders for the withdrawal to Dunkirk, which necessitated 3rd Division taking up new positions astride the Yser Canal, north-west of Ypres. This was a dauntingly challenging operation because it required the division to execute a night march in poor weather across the front of an enemy attack, over twenty-five miles, following a maze of minor roads, all congested with refugees and French troops. Furthermore the division possessed barely enough transport to move all in one lift. Establishing an armoured car screen first, the division was guided, under shellfire, by military police at every junction, and by first light on 28 May, 8th and 9th Brigades were dug in and ready.

'It was with a feeling of intense relief that I found Monty in position ... I found he had, as usual, accomplished almost the impossible,' wrote Brooke afterwards. This is where Monty's Phoney War training of his division was seen to pay off. It is impossible to envisage any formation undertaking such a complex manoeuvre without thorough rehearsal beforehand. Indeed, Monty's division was the only one that had managed any meaningful training. That evening, Monty lost his GSO1, Colonel 'Marino' Brown, RM (very highly rated by Brooke), who was accidentally shot dead whilst making his way through a French roadblock. Monty would have felt this loss keenly, having served as a divisional GSO1 in 1918. Thereafter, his GSO2, Charles Bullen-Smith (a future divisional commander under Monty), took over the post. His ADC, Captain Charles Sweeny, was wounded in the head on 30 May.

Over two nights, 28–29 and 29–30 May, 3rd Division withdrew back to the Dunkirk perimeter, covering Furnes in the eastern sector of the beachhead. By this time, the infantry battalions were down to about half strength, mostly through battle casualties, with a disproportionate number of senior officers down. On 29 May, 7th Guards Brigade came across 'a dejected figure, recognised as Montgomery', standing in the Market Place of Furnes. He was saddened at the loss of his driver earlier in the day. As the Guards marched past, he was given a terrific 'eyes left'; he suddenly straightened up and gave a great salute in return.[13]

All units were now obliged to abandon their vehicles and immobilise them along the line of the Furnes–Bergues Canal or on the beaches, which would create very dramatic images for German journalists and cameramen after Dunkirk's capture. If ever the *Wehrmacht* had in their mind's eye an image of victory in France, it must surely have been the countless thousands of abandoned military vehicles they discovered at Dunkirk.

By the time 3rd Division arrived in the perimeter, around 100,000 troops had already been evacuated. During 30 May all divisional troops were occupied in repulsing attempts by the German 56th and 216th

Divisions to cross the Bergues–Furnes Canal and reach the evacuation beaches. Meanwhile Gort had instructed the BEF to thin out senior officers in the beachhead to ensure a nucleus from which to rebuild the army in the future. As part of this plan, Brooke was told to leave for Britain. He handed command of II Corps (then with several other divisions under command) to his own nominee. Monty, his junior but most able divisional commander, was flattered and surprised to be chosen as his successor. Montgomery inherited Brooke's Chief of Staff, Neil Ritchie (a future Eighth Army commander). Brigadier Kenneth Anderson, the future First Army commander in Operation Torch, took over 3rd Division, and Horrocks of 2/Middlesex was sent along the beaches to find Anderson and take over his brigade (14th Brigade of 4th Division). Monty recalled Brooke's departure, where he also learned of his promotion:

> He arrived at my headquarters to say goodbye and I saw at once that he was struggling to hold himself in check; so I took him a little way into the sand hills and then he broke down and wept – not because of the situation of the BEF, which indeed was enough to make anyone burst into tears, but because he had to leave us all to a fate which looked pretty bad. He, a soldier, had been ordered to abandon his men at a critical moment – that is what disturbed him.[14]

This passage might have smacked of Monty hamming it up, post-war, had not Horrocks witnessed the same poignant scene:

> As I approached [3rd Division's HQ], I saw two figures standing in the sand dunes. I recognised our corps commander, General Brooke, and my divisional commander, General Montgomery. The former was under a considerable emotional strain. His shoulders were bowed and it looked as though he were weeping. Monty was patting him on the back. Then they shook hands and General Brooke walked slowly to his car.[15]

The perimeter held through 31 May, despite furious assaults all day, whilst plans were drawn up for the division's evacuation that evening. The majority were rescued from one of those amazing vehicle piers built out from the town of Bray Dunes that even now, in a flash, encapsulate the initiative and tragedy of the evacuation, or via the East Mole in Dunkirk Harbour, from where Monty himself would leave aboard the destroyer HMS *Codrington* at 3.30 a.m. on 1 June*. In the final hours, Gort (being recalled himself) was ordered to nominate a corps commander to oversee the final evacuation. As I Corps was to be the last to leave, Gort nominated its GOC, General Michael Barker. Monty, with all the self-confidence of a twenty-four-hours-old corps commander, assured Gort this was a mistake and suggested his fellow guardsman, Harold Alexander of 1st Infantry Division – advice Gort accepted. In attendance on Gort was his deputy CGS, Oliver Leese (one of Monty's Staff College students at Camberley), whose performance at Dunkirk left a lasting impression, and would lead to his being summoned by Monty to Egypt to lead XXX Corps in November 1942.

Whilst all of the 63,879 vehicles, 20,548 motorcycles and 2,474 artillery pieces that accompanied the BEF to France were left there or abandoned and destroyed in the Dunkirk perimeter, it was saving men and manpower that was most important.[16] Machines could be replaced; men could not. Thus Alexander it was who helped preserve what was left of the army and who brought Operation Dynamo to a close in his own calm and skilful way, staying on the beach until 3 June, to ensure all British troops had been evacuated.

Like Rommel, Monty had some exceptionally talented officers under his command in 1940: Roy Urquhart (the future 1st Airborne GOC) was his DAAG (Divisional Assistant Adjutant General), Charles Bullen-Smith (later 51st Highland Division commander) his GSO2; the Divisional Intelligence Officer was Christopher 'Kit' Dawnay, seconded from 7th Guards Brigade, who would serve with Monty throughout the war, latterly as his Military Assistant; and, of course, Brian Horrocks was one of his battalion commanders. One of the subalterns in 2/Royal Ulster Rifles of 9th Brigade was Lieutenant 'Bala' Bredin (a future major general), who had already won an MC and bar in Palestine and would win three DSOs in the Second World War. Scrambling exhausted aboard a cross-Channel ferry lying off one of the Dunkirk beaches, a passing steward asked him if he would like a drink. Bala responded he would very much like a beer. 'I'm sorry sir, I cannot serve you alcohol until we've passed the three-mile limit,' the steward replied. 'With that demonstration of British grit in the face of adversity,' said Bredin, 'I knew we would win the war.'[17]

*

* HMS Codrington would be sunk by the Luftwaffe within the month, on 27 July 1940.

Powerless to prevent the escape from Dunkirk, Rommel's 7th Panzer Division was allowed a six-day break, recognition that his men and machines badly needed rest. On the day that Operation Dynamo, the Dunkirk evacuation, was completed – 3 June – Rommel was summoned to meet Hitler, who was then paying a lightning visit to the front, and to his old battlefields of the First World War. The encounter saw the Hitler–Rommel relationship at its height: Rommel was the only divisional commander summoned to his Führer in mid-campaign, but at the meeting Hitler also exclaimed very publically: 'We were all very worried about you.' Nothing could better illustrate the mutual regard the pair felt for each other.

Although most of the BEF had escaped from Dunkirk there was an opportunity to prevent the escape of 51st Highland Division, heading for evacuation from Le Havre, if Rommel could get there in time. He led 7th Panzer across the Somme on 5 June and raced for the port. Rommel and his command group reached the coast for the first time at Les Petites Dalles in the early afternoon of 10 June.

> With my signals section I drove on in advance of the [25th Panzer] Regiment ... down to the water. The sight of the sea thrilled and stirred every man of us; also the thought that we had reached the coast of France. We climbed out of our vehicles and walked down the shingle beach until the water lapped over our boots. Several despatch riders in long waterproof coats walked straight out until the water was over their knees, and I had to call them back. Close behind, Rothenburg came up in his command tank, crashed through the beach wall and drove down to the water.[18]

Rommel had been tearing through France for exactly a month, but the sight of the English Channel made him halt, if only temporarily, for what became an almost spiritual moment. Rommel would inspect Les Petites Dalles under somewhat different circumstances on 14 April 1944. Today it has changed little from those minutes in 1940: the sea wall has been repaired, but it is evident exactly where Colonel Rothenburg burst through and drove his tank onto the shingle.

Rommel doubled back to interrupt evacuations from Fécamp, then cut along the coast to seize the western cliffs overlooking St Valéry-en-Caux, a small fishing port between Fécamp and Dieppe, to prevent the Royal Navy from approaching. Rommel had to fight for his prize, but there, on 12 June, Major General Victor Fortune, a contemporary of Monty's at Camberley in 1920, surrounded and isolated, was forced to capitulate with

8,000 of his Highlanders and 4,000 Frenchmen of General Ihler's IX Corps. Amidst the debris of surrender, 7th Panzer's band paused to give an impromptu concert on the seafront at Fécamp that evening, whilst the tank crews frolicked in the summer surf. After the war, a memorial obelisk to the Highland Division, hewn from Scottish granite, would be erected on the eastern cliffs overlooking St Valéry, and guards the little port still. One brigade of Highlanders, however, managed to elude Rommel and escape from Le Havre, and from it a new 51st Division was forged, which would fight under Montgomery in North Africa, Sicily and Normandy.

Meanwhile Rommel had been ordered on to Cherbourg, through the Normandy towns and countryside he would get to know so well four years later. Cherbourg fell on 19 June, marking the end of Rommel's remarkable campaign. The first German troops, meanwhile, had entered Paris (declared an 'open city') on 14 June, and two days later, the 30th Infantry Division marched down the Champs-Elysées in front of their general, Kurt von Briesen, an event covered by Goebbels's newsreel cameramen and swiftly transmitted around the world.

Marshal Philippe Pétain, hero of Verdun, had succeeded Paul Reynaud as Prime Minister and sought an armistice with Germany, principally to end the plight of the several million refugees still tramping the roads of northern France. With destiny breathing down his neck, Hitler selected the Compiègne Forest as the site for the 22 June negotiations, using the same railway carriage in which the 1918 Armistice had been signed. Hitler visited Paris the next morning (for the only time in his life), and on 10 August Parisian streets echoed to the jackboots and brass bands of a formal victory parade, at which a helmeted Rommel was pictured, with other commanders, celebrating this unprecedently swift victory.* The march followed the exact route taken by the Allies in their victory parade of 14 July 1919, which, in turn, was a riposte to the hated Prussian victory parade in Paris of 17 February 1871.

Journalists and soldiers grew excited at the extent of 7th Panzer's eventual haul, stated to be ten assorted generals and admirals, 97,486 prisoners, 458 tanks and armoured vehicles, 341 anti-tank and field guns, 4–5,000 trucks; whilst they had captured or destroyed 79 aircraft. The exchange rate was not cheap: Rommel's 7th Panzers lost more casualties than any other German division, and these amounted to 2,594 men,

---

* In Paris, Hitler was welcomed by Oberst Hans Speidel, then on the Military Governor's staff, and in 1944, Rommel's chief of staff.

of whom 682 were killed, 1,646 wounded and 266 posted as missing – about twenty percent of his strength, plus thirty-nine tanks.[19] The figures hide the fact that almost one third of all Rommel's casualties were leaders – officers and sergeants – difficult men to train and replace quickly.

How to interpret Rommel's 1940 campaign? On the one hand we are left with the impression of a supremely talented, effective and successful commander, with a gift for war; on the other, a maverick, cast in the mould of George Armstrong Custer, who rode to glory on a mixture of luck and guts, exhibiting a dangerous lack of concern for his men, and propelled to victory by his talented subordinates. Army high command in Berlin would have observed his lack of caution also. Fourth Army's Günther von Kluge (with whom Rommel would clash in Normandy) went so far as to criticise Rommel for claiming as his own the achievements of neighbouring units and downplaying the support of the Luftwaffe. Rommel certainly requisitioned 5th Panzer Division's bridging equipment to cross the Meuse on 14 May when his own was unavailable, which delayed that flanking formation for crucial hours. Whilst this bred mistrust amongst commanders, exactly the same criticism can be levelled at George S. Patton four years later, whose US Third Army units were renowned for demanding fuel at gunpoint from neighbouring, friendly units.

Apart from the immediate fame the 1940 victory brought Rommel (because of the presence of Karl Hanke and the propaganda men attached to his division), it earned 7th Panzer Division a nickname, *Die Gespenster-Division*, the 'Ghost' or 'Phantom' Division, reflecting the formation's seeming ability to materialise at will behind enemy lines. The name was the creation of German war correspondents, who were always made to feel very welcome at Rommel's headquarters. Recognition of this sobriquet was lacking on the Allied side: the campaign was over too quickly for the Allies to be able to attribute their defeat to individuals or specific formations. Indeed it is likely that, just as the Ghost Division moniker was a German propaganda invention, the same correspondents, who followed Rommel to North Africa, were the originators of his personal nickname, *Der Wüstenfuchs*, the Desert Fox.

Both Hitler and the Allies drew the wrong conclusions from the fall of France. Hitler had backed Manstein's plan personally, in the face of opposition from his own generals: he would henceforth lay claim to being the architect of the incredible victory in June 1940. Hitler gradually came to regard himself as a latter-day Napoleon and began to interfere increasingly with the day-to-day management of his armies, even down to tactical level. It also encouraged Hitler to misinterpret the power and effectiveness of his war machine, particularly the tank arm, and was thus encouraged to wage war on Russia with disastrous results.[20] Meanwhile, on 19 July, Hitler rewarded his top commanders, promoting twelve of them to the exalted rank of *Generalfeldmarschall* – amongst them was Rommel's old patron, Wilhelm List. The Allies, for their part, attributed to the *Wehrmacht* an invincibility it did not necessarily deserve, a myth that it took until El Alamein, nearly thirty months later, to dispel.

After Dunkirk, Monty's 3rd Division, having fought well, and survived relatively intact, was the only one to be immediately rearmed and re-equipped. Now they dug in along the Sussex coast to await the Germans, including Rommel's 7th Panzer Division. In a minute that reached Churchill's desk, Monty lobbied for permission to requisition buses and lorries to give his troops a mobile counterattack capability. The Prime Minister (who quite by chance had assumed office on the same day the Germans launched their attacks into France and Belgium) visited Montgomery on 2 July, the first time the two men had actually met – though of course they had unwittingly stood next to one another at the Lille victory parade in October 1918. Afterwards they dined together at the Royal Albion Hotel in Brighton, as Monty recalled:

> He asked me what I would drink at dinner and I replied – water. This astonished him. I added that I neither drank nor smoked and was 100 percent fit; he replied in a flash that he both drank and smoked and was 200 percent fit.[21]

Monty's continuing preparations for invasion included visiting the enterprising Auxiliary Units in his sector. These were guerrilla-type units of volunteer civilians trained by Colonel Colin Gubbins and Captains Peter Fleming (brother of Ian) and Mike Calvert, who would go into underground, pre-prepared operational bases in the event of an invasion, and disrupt the occupiers with a mixture of sabotage and assassination activities. Captain Norman Field, of 2/Royal Fusiliers, who had escaped from Dunkirk, conducted Monty around.

> I saw the small figure of Monty pacing up and down in front of a large staff car; when I apologised for being late, Monty said, 'You're not late.

I'm always two minutes early. Get in the car!' I took him to a little wood, where one of my patrols was hidden and pointed out what looked like a mouse-hole and dropped a marble down it. The marble ran down a length of concealed pipe and fell into a biscuit tin. This was the signal for the patrol leader to open the camouflaged entrance. Monty went in and was so impressed by what he saw that he promised to bring Churchill to look at the hideout and meet the patrol.[22]

Illustrative of how attentive Monty could be to the careers of bright young officers, especially Regulars, he took Field under his wing, and after his service with the Auxiliary Units, Monty insisted he join his own staff (before sending him off to Staff College in 1942). In time he would foster the careers of a select band of several young officers. This was also Monty's first exposure to special forces. Whilst relatively common today, and regarded as a significant force multiplier by all armies, in 1940 they were a novelty. It is surprising in retrospect that neither he nor Rommel saw and exploited the full potential of special forces in the desert, or elsewhere. True, the SAS (Special Air Service) and LRDG (Long Range Desert Group), amongst other bands, would emerge in the Western Desert campaign, but they had constantly to beg for resources and justify their existence. After the war, it was Miles Dempsey, not Montgomery, who rescued the SAS and became its first Colonel Commandant. Rommel, likewise, failed to imitate the SAS and LRDG. If he had done so it would have reaped rich rewards. He had many Arab sympathisers in Egypt, and apart from utilising the services of the shady Count László de Almásy (the real 'English Patient') seems to have made little effort to use unconventional troops.

Soon afterwards, Monty was promoted to command V Corps in Dorset and Hampshire. In April 1941 he took over XII Corps in Kent and that December South-East Command (which he rechristened 'South-East Army' on his letterheads). Brian Horrocks remembered Monty's tremendous energy in this role:

Monty used to pay constant visits. 'Who lives in that house?' he would say pointing to some building which partly masked the fire from one of our machine-gun positions. 'Have them out, Horrocks. Blow up the house. Defence must come first' . . . I was unprepared for his aston-ishing activity as the GOC-in-C South Eastern Command. It was as though atomic bombs were exploding all over this rural corner of Britain. Before his arrival a distinctly peace-time atmosphere had prevailed.[23]

This was also the origin of the 'no coughing rule' imposed with increasing emphasis on almost every audience Montgomery ever addressed (as a sort of trademark), well into his later life:

> Suddenly the audience [of military officers] would be called to attention, as the well-known figure of the army commander wearing battledress advanced to the centre of the stage. 'Sit down, gentlemen,' he would say in a sharp, nasal voice. 'Thirty seconds for coughing – then no more coughing at all.'[24]

So the Monty legend, and the Monty stories, had begun. In May 1942, during one of these military conferences, a visiting American major general made the mistake of lighting a cigarette in Monty's presence; he was told forcefully to put it out immediately. This was the first recorded exchange between Monty and his future boss, Eisenhower.[25]

During his tenure with South-Eastern Command preparations were made for a coastal raid on Dieppe, Operation Rutter. Monty clearly had a hand in some of the plans, or proposals, but the project was initially cancelled, then hastily resurrected and mounted on 19 August 1942 with 2nd Canadian Division, as Operation Jubilee. John Hughes-Wilson argues persuasively that by then Mountbatten had hijacked the whole enterprise and, claiming authority he did not have, launched the raid (when Churchill was conveniently in Cairo reshuffling military commanders) as an advertisement of what his Combined Operations organisation could achieve. That its abject failure did not rebound in any way on Monty was fortuitous, and largely because he had been summoned out of the country by Churchill to take over the Eighth Army at exactly the same time.[26]

The Rommel legend had also clearly begun. However, we would view him today in a very different light had Operation *Seelöwe* (Sealion), the German plan to invade Britain, been launched as intended in September 1940. According to Hitler's War Directive, dated 16 July 1940, *Generalmajor* Rommel and his division were scheduled to land as part of General Adolf Strauß's Ninth Army, in its second wave under *Generaloberst* Hermann Hoth's XV Panzer Corps, between Bexhill and Eastbourne. Churchill would not have been carried away with noble words, as he would be in 1942, for the captor of the Cotswolds, Camberley or Coventry.

14

Duel in the Desert I

THE NORTH AFRICAN campaign, with its constant changes of fortune, remains indelibly associated with Monty and Rommel, and was unlike any other during the Second World War. It was, moreover, an unlikely theatre, where neither key protagonist, the British or the Germans, would have envisaged themselves fighting major battles when they first went to war in September 1939.

Long before oil dominated geopolitical thinking about the Middle East, British involvement in the area was motivated by a requirement to control the shortest route to India, via the Suez Canal. Cairo was the regional centre of British military activity and thinking in both world wars; in the First World War, much of the campaign against Turkey was run from there. It was only in 1939 that the separate commands of the Sudan, Egypt and Palestine were united under a single C-in-C Middle East, in the person of General Sir Archibald Wavell, an exceptionally clever, staff-trained officer, who oversaw a collection of British and colonial forces scattered across the Eastern Mediterranean, from Palestine, Transjordan and Iraq to Egypt, Sudan, Aden and Cyprus. This was certainly not one homogeneous army, but a hotchpotch of scattered garrisons and locally raised levies, of variable quality, supported by obsolete aircraft with a few light tanks and armoured cars.

Archibald Wavell, four years Monty's senior, was the brightest man ever promoted to the five-star rank of field marshal. Son of a major general and schooled at Winchester, he was commissioned into the Black

Watch in 1901 and as a twenty-six-year-old lieutenant attended the Staff College in 1909. He learned Russian, spending a year with that army in 1911, lost an eye and won an MC at Ypres and was a temporary brigadier by 1919. Often caricatured as a member of the old guard, with his monocle, he was anything but, and a prolific author, in an era when officers read rather than wrote. He pushed Monty's career several times: when GOC at Southern Command, penning the comment about Monty possessing 'one of the cleverest brains we have in the higher ranks', and consulting Monty on training issues before the war. Monty's papers have preserved correspondence between the two about the advisability of wearing kilts in a chemical warfare (gas) environment.[1]

By July 1939, Wavell had been appointed GOC Middle East Command as a full general and saw his main challenges as a possible uprising against British rule in Palestine and to a lesser extent, Egypt or Iraq. Germany had no foothold in the area and Italy was not regarded as a threat: Mussolini was even considered a dove, rather than a hawk, who might intercede with Germany for peace.

The Italian declaration of war in June 1940 altered the strategic balance dramatically against Wavell, for, all of a sudden, he was faced by the prospect of half a million Italian and native levies split between the Italian colonies of Libya and Abyssinia, far outnumbering the British on either front and able to launch a two-pronged attack that would soon stretch Wavell's tiny force beyond breaking point. Although the Italians fielded even more outmoded armour and aircraft than the British, they had far more of them, and also knew that British defensive plans rested on cooperation with the French, who had just been defeated. As the Vichy French refused to surrender their fleet or fight alongside the British, Churchill, whose grasp of geostrategy was undisputed, feared the Mediterranean naval balance would be upset by the addition of the war fleets of France and Italy to Germany's modest *Kriegsmarine*. He therefore ordered the French fleet to be seized on 3 July 1940 in Operation Catapult. Those in British waters and at Alexandria agreed to surrender but much of the remainder, at Mers-el-Kébir (Oran) and Dakar did not and were attacked and sunk or damaged at anchor. If Frenchmen had any wavering doubts about which side to join, the blood of French sailors being spilt (1,297 were killed) by their traditional foes, the British, offered Germany wonderful propaganda opportunities which *Reichsminister* Goebbels was not slow to exploit.

This was followed on the night of 11–12 November 1940 by Operation

Judgement, when the Royal Navy won back control of the Eastern Mediterranean by scoring a notable success against the Italian fleet at anchor in the southern Italian naval base of Taranto, instantly neutralising much of Italy's latent maritime threat. However, British convoys were still obliged to sail the extra distance around South Africa, depleting valuable fuel stocks, adding time and increasing danger in U-boat-infested waters. (Taranto was an operation much studied over the next year by the Japanese high command, who would use it as a template for what would emerge as the strike against Pearl Harbor a year later, on 7 December 1941.)

In the early war years, there was also a real danger that Franco would repay the debt he owed Germany and Italy for their support of his Nationalists in the Spanish Civil War and join the Axis powers, thereby further decreasing Britain's ability to commit to anything in North Africa, for Gibraltar would surely have fallen.[2] Franco realised, however, that his poorly industrialised nation would have had no stomach for voluntarily entering a world war, having just emerged from long civil strife. Hitler never understood Spain's dilemma – this is where his lack of awareness of the world outside Germany let him down: he travelled so little, spoke no foreign languages, and thus had only a naïve and rudimentary grasp of geostrategy, compared with Churchill who had ranged the globe far and wide, or Roosevelt and Stalin, compelled to have a world view by the size of their nations. However, Hitler's planners at OKH had realised that if the Mediterranean could be shut by mining the Suez Canal and seizing Gibraltar, Britain's interests would be irrevocably harmed.

North Africa was Italy's main theatre of war and represented Mussolini's pride and bombast, though it was far from Hitler's mind. At every stage of the coming campaign, the Italian army would outnumber the German; Rommel would be, technically, a subordinate commander. For every three Italians eventually taken prisoner, one was killed or wounded, a higher rate than for the Germans.[3] Yet, somehow history has consistently airbrushed the Italians out of the picture. Ignoring the Italian presence in North Africa is like studying Waterloo without the Prussians: every aspect of a coalition has to be acknowledged.

As early as the summer of 1940, OKH had been exploring the idea of sending an armoured regiment of two tank battalions plus supporting troops to North Africa, as a way of attacking Britain, fighting alongside an Italian force. For Hitler it would be a demonstration of Axis unity, but OKW insisted on multinational planning. Benito's generals had been appalled at the prospect of war in 1940, knowing how poorly prepared the Italian army was for modern war; they at least wanted a delay until Germany was physically landing troops in England, which would distract any British effort from North Africa. Mussolini ignored all their objections and on 6 September 1940, for reasons of pride, rejected any joint planning and 'turned down the German offer ... to send Panzer units to North Africa, for fear that his ally would insist on a degree of command and control'.[4] He instructed his generals to begin their own offensive shortly afterwards. The Battle of Britain was at its height, and *Il Duce* fully expected Britain to sue for peace shortly; therefore his attack on British interests in North Africa was designed to enhance his bargaining power at the conference table in any negotiations with Britain, which he thought were imminent. He had already adopted a similar tactic in his declaration of war on France, as late as 10 June, when most of the fighting was already over.

The Italians were, on the whole, poorly equipped, less motorised, had inferior tanks, and their officers much less well trained; but they were far more resilient than popular myth. If not as aggressive as the Germans in attack, they were frequently as good in defence. Perhaps the Italian army's poor showing in December 1940 started the idea (though maybe it reached back to Caporetto in 1917) that the Italians were poor soldiers, but the myth was really fuelled by Rommel, who blamed his allies for every setback; they were a convenient scapegoat – and Ultra intercepts picked this up and passed it on, spreading Rommel's half-truths at a generous strength of magnification. Whilst praising the Italian fighting soldier ('willing, unselfish and a good comrade'), Rommel later claimed:

> The cause of the Italian defeat had its roots in the whole Italian military and state system, in their poor armament and in the general lack of interest in the war shown by many of the leading Italians, both officers and statesmen. This Italian failure frequently prevented the realisation of my plans.[5]

MacGregor Knox argues that Italy had industrialised very late, lacked raw materials and industrial capacity and had one of the lowest rates of vehicle ownership in Europe by 1939, and was thus very short of key technical skills (except when it came to aviation matters) – so the armies it sent to war in 1940 were closer to 1918 in appearance and tactics, but usually performed better than has been reported.[6] The official German line, though, was of unqualified praise for their Fascist allies, and certainly Rommel proved as popular with Italian troops as he was with his own. Paolo Colacicchi, an officer in the Italian Tenth Army, recalled Rommel's impression on his compatriots:

> Rommel himself became a sort of myth to the Italian soldiers just as much to the German soldiers. In fact one regiment, the *Bersaglieri* – they are the ones with a lot of feathers in their hats – fighting out of Tobruk baptised Rommel '*Rommelito*', which in Italian means little Rommel and also refers to Romulus. This was a Roman regiment and they liked him, and on one occasion he even put some of their feathers in his own colonial helmet and wore it because he was pleased with them.[7]

Rommel's ADC, Jochen Schraepler, echoed this in a letter home of 3 May 1941:

> If the Italian officers were more courageous, their troops would be pretty good. They are always delighted when they see German officers, welcoming us with '*camerada*'. Their soldiers, extremely helpful, love to be commanded. If there is no leadership and no model, the result cannot be good. For us, this is not enough. We require proper training, education and a sense of duty.[8]

*

With the Italian invasion of 13 September 1940, from Libya eastwards into British-protected Egypt, the Desert War began. The Italian force trundled over the border, then – contrary to any logical military doctrine – halted at Sidi Barani after three days, establishing a series of colonial-type fortified camps – strong, but too far apart to provide mutual assistance. Such timidity may have reflected the realities of Italian mobility and logistics; it certainly underlined Mussolini's lack of strategic planning. Other than the sudden idea of *Il Duce* entering Cairo on a white horse, the Italian government had given no thought as to how and when to invade Egypt, or where to halt.

On 9 December 1940, Major General Richard O'Connor's tiny Western Desert Force of two divisions from Mersa Matruh counterattacked Mussolini's desert army in Operation Compass. Within two months, greatly to his and the world's surprise, the Western Desert Force had managed to push the Italians out of Egypt, and across Cyrenaica, destroying ten Italian divisions and taking 130,000 prisoners, 400 tanks, 900 guns and 1,000 aircraft (for a British loss of under 2,000 killed and wounded). Perhaps Wavell should not have ordered O'Connor to halt at El Agheila, but concerns about Abyssinia and Greece meant that he was obliged to divert resources elsewhere, and the Italians appeared subdued. Wavell disagreed with the decision but followed his orders.

By then, with great reluctance, Mussolini had asked Germany for help after all. Hitler deliberated over Christmas and announced on 9 January he would despatch a small armoured force. The army selected the Prussian aristocrat General Hans Freiherr von Funck for the assignment: he carried out a reconnaissance and reported that a larger force than the previously proposed single division would be required. This required leadership of a different order with a corps staff. Hitler felt he had just the man: a young general with whom he had built a special bond. Funck, an admirable candidate in every way, would be compensated by receiving command of Rommel's panzer division instead.

Rommel, who had been kicking his heels since the fall of France, was summoned, without warning, to a personal briefing with Hitler in Berlin on 6 February 1941, and his new mission, Operation *Sonnenblume* (Sunflower), was revealed to him. Hitler decreed that Rommel would lead a new force named the *Deutsches Afrika Korps* (DAK), a reference to Rommel's own First World War service in the *Deutsches Alpenkorps* – a skilful fusion of the traditions of the old and new German armies. So secret was the assignment that Rommel had to write home and ask that his kit be forwarded, rather than risk trying to slip home incognito to collect it. He was promoted to *Generalleutnant* for the campaign, and given 5th Light and 15th Panzer Divisions to take with him to North Africa. These would fight alongside some first-class Italian formations, including the well-motivated Ariete and Littorio armoured divisions, the tough Folgore parachute division and resourceful Bersaglieri infantry battalions.

The German army's experience of tropical campaigning had been restricted to the colonial operations of 1914–18, so responsibility for designing a new uniform for the new Afrika Korps was given to the Hamburg Tropical Institute. Under pressure of time, the Institute allegedly chose to model its requirements on British uniforms worn in India; but if so, they chose poorly. Although the high, laced desert boots were popular, the cotton tunic and half-breeches resembling jodhpurs, which the Afrika Korps wore in its earliest battles, were cut too tight for comfort in the heat and had to be quickly modified and made loose-fitting by the small army of tailors that always mushroom around military camps. However, they got the headgear exactly right, the comfortable, stylish and iconic desert field cap, which was extremely well liked and worn almost unanimously in preference to the alternative tropical sun helmets. That vast quantities of these new uniforms were ready at short notice suggests that OKH in Berlin had been planning an African 'adventure' for some time (not unlike von Lettow-Vorbeck's legendary East African campaign of 1914–18); the tropical uniforms and sun helmets, or their design, were already on the shelves.

Rommel's arrival in Tripoli at noon on 12 February 1941 would alter the course of the Desert War. With him was Leutnant Hans-Otto Behrendt, an Arabic-speaking Egyptologist, and a 'minder' – Oberst Rudolf Schmundt, Hitler's chief *Wehrmacht* adjutant, whom Rommel knew from his time with the *Führerbegleitbattalion*. Though their relationship then had been strained, it would develop into one of mutual regard, even friendship. Five years younger than Rommel, Schmundt would go on to become Chief of Army Personnel in October 1942, and Rommel's key ally on Hitler's staff.

Though glowing with self-confidence, Erwin confided in a letter to Lucie on 14 February 1941: 'I hope to be able to pull it off.' False modesty, or was the magnitude of his mission beginning to sink in? The same day, his first troops sailed into Tripoli harbour. The *Deutsches Afrika Korps* was officially established on 21 February, succeeding the staff of the *Befehlahaber der deutschen Truppen in Libyen*, and a *Wehrmacht* communiqué first announced German activity on the 'Libyan Front' five days later. Later, after returning from a two-day visit to the front (450 miles east), on 5 March and in high spirits, Rommel wrote home:

A lot to do . . . Too much depends on my own person and my driving power . . . my troops are on their way. Speed is the one thing that matters here. The climate suits me down to the ground. A gala performance of *Sieg im Westen* was given here today. In welcoming the guests – there were a lot – some with wives – I said I hoped the day would come when we'd be showing *Victory in Africa*.[9]

Rommel knew all about the 120-minute *Sieg im Westen* (*Victory in the West*), just released in Berlin, for his division had re-enacted some of the May 1940 battles for it, starring against a battalion of French North African troops released from prison camp for the purpose. A *Propagandakompanie* detachment had also accompanied his 7th Panzer Division in their romp across France, so the division appeared twice, in newsreel and re-enacted footage. Part of this was Karl Hanke's doing: some of Rommel's snapshots (he was given a Leica camera by Goebbels for the purpose) of the campaign show a movie camera on a tripod, fixed to the deck of a panzer during the French campaign. Waiting for a new assignment after the defeat of France, Rommel had also busied himself writing a campaign history of 1940. He also toyed with the idea of writing a sequel to his bestselling *Infanterie Greift an*; (allegedly *Panzer Greift an*). But if it was ever commenced, it is lost to us now.

As soon as the first tanks and other vehicles of 5th Light had disembarked, Rommel, master of 'smoke and mirrors', cunning and deception in Italy in 1917 and France in 1940, ordered a parade in Tripoli. Knowing the town was awash with Allied spies, he ordered his new division to drive through and round the town several times, magnifying his strength at a stroke. Leutnant Heinz Werner Schmidt observed the subterfuge:

> Singly and at regular intervals the panzers clattered and rattled by. They made a devil of a noise on the macadamised streets. Not far past the saluting base the column turned into a side-street with mighty squeaks and creaks. I began to wonder at the extraordinary number of panzers passing, and to regret I had not counted them from the beginning. After a quarter of an hour I noticed a fault in one of the chains of a heavy Mark IV Panzer, which somehow looked familiar to me although I had not previously seen its driver. Only then did the penny drop, as the Tommies say, and I could not help grinning.[10]

Whilst his forces gathered strength, Rommel flew to Berlin to report and was there awarded Oakleaves to his Knight's Cross (only the tenth recipient of an eventual 900) for his achievements in France the year before. Martin Kitchen makes the important point that the key to anyone's fortune in the Third Reich was access to Hitler, and whilst part of Rommel's success was due to his forceful personality and sheer military achievements and ability (as demonstrated in France), the relationship with his Führer was even more important. But it was always a double-edged sword, for it incurred the envy and suspicion of others, particularly those hard-working members of the General Staff, who were, in effect, being leapfrogged by Rommel, whom they had begun to regard as a self-promoting upstart. It was therefore all-important to Rommel that he use Goebbels's propaganda staff seconded to him, to maximum effect. At the Führer's briefing on 19 March 1941, the army chief Walther von Brauchitsch took Rommel aside and told him there was no intention to strike a decisive blow in Africa in the near future and that after 15th Panzer Division was complete in theatre (by the end of May), no more reinforcements would be made available. This was a warning. Rommel saw the message for what it was: an attempt to limit what many saw as his own inherent recklessness.

One of OKH's checks against Rommel's rashness was the appointment of the cautious Oberst Klaus von dem Borne as his first Chief of Staff. Whilst Rommel would eventually replace him with his own appointee, Fritz Bayerlein, in October 1941, Jochen Schraepler insightfully observed of the two: 'the corps [DAK] chief is considered to be too hesitant. That is not correct, in my opinion. He is clever and considered and an excellent counterweight to Rommel.'[11] Almost certainly Rommel had already made up his mind to ignore Brauchitsch's warning, and maximise whatever opportunities came his way. What no one told Rommel was that the invasion of Russia had already been decided for that summer, and whatever he did would be overshadowed by this larger enterprise, and its massive call on logistics.

As soon as Rommel returned to the desert, the great Anglo-German battle for North Africa began – somewhat quietly – when his troops captured three British soldiers and destroyed two scout cars on 24 February. Perhaps *because* of Brauchitsch's warning, he immediately disregarded instructions and ordered an attack, without waiting to build up his forces, though only once Schmundt had returned to Berlin. He struck on 24 March, and over the next thirty days drove the British from El Agheila right back across Cyrenaica and into Egypt at the Halfaya Pass. This attack, which had all the Rommel hallmarks of daring, might not have succeeded had the competent Wavell kept all his forces in theatre. But under irresistible pressure from Churchill he had been obliged to split his forces and send the better half to Greece, where German invasion was feared, whilst holding the Libyan front temporarily with the remaining, weaker half.

On 3 April 1941, Rommel dashed off a quick note to his 'Dearest Lu':

> We've been attacking since 31st [March] with dazzling success. There'll be consternation amongst our masters in Tripoli and Rome, perhaps in Berlin, too. I took the risk against all orders and instructions because the opportunity seemed favourable. No doubt it will all be pronounced good later and they'll all say they'd have done exactly the same in my place. We've already reached our first objective, which we weren't supposed to get to until the end of May. The British are falling over each other to get away. Our casualties are small . . . You will understand that I can't sleep for happiness.[12]

That evening, Rommel entered Benghazi, whilst the port resounded to the detonation of British stocks of explosives, fired by the departing rear-guard. Berlin was predictably fired up too, and the chief at OKW, *Generalfeldmarschall* Wilhelm Keitel, Hitler's senior military adviser, sent Rommel a sharp message warning him not to stray from his instructions. Keitel, promoted field marshal (along with Rommel's patron Wilhelm List) the previous July, would respond to Hitler's every military whim, supported by his deputy, Alfred Jodl. Keitel and Jodl, a pair of ruthless desk warriors, made and broke countless military careers, and would number amongst the architects of Rommel's eventual downfall.[13]

Rommel probably escaped further censure because in early April Hitler had invaded the Balkans, so Berlin's eyes were elsewhere. Prompted by an anti-Nazi coup in Belgrade, the Yugoslav capital was first bombed on 6 April, before ground troops were committed two days later. Operation *Strafgericht* ('Retribution') concluded with the total collapse of Yugoslav forces on 12 April. Meanwhile German ground troops had simultaneously invaded Greece in a lightning thrust which involved many of the best armoured and air force formations practising their now highly polished blitzkrieg tactics. Operation Marita saw Athens captured on 27 April, but the British managed to extract 50,000 troops before the end.

An Allied evacuation to Crete was followed by an airborne invasion of the island on 20 May, which succeeded, albeit at a cost of almost half the paratroopers involved. In strategic terms, the invasion of Yugoslavia and Greece robbed the Germans of many vital weeks of good campaigning weather which they had been about to devote to the invasion of the Soviet Union (and would prove fatal by the end of the year, with Moscow just beyond reach; had the Germans started earlier their result in the

Russian Motherland might have been very different). Whilst Crete inspired the British to found an airborne arm of their own, the Parachute Regiment, the losses sustained by the German *Fallschirmjäger* in Operation *Merkur* ('Mercury') persuaded Hitler never to deploy these troops by air to battle again. Thereafter, some of the most efficient and hardy special forces the Germans possessed were denied the role for which they had been trained, and henceforth squandered as a land-based fire brigade, attacking and defending sectors where ordinary infantry were found wanting. They sprang from exactly the same hard-living, hard-fighting origins as Rommel and the WGB in 1915–17, and he could have used their airborne capability and prowess many times over.

Rommel, meanwhile, was lost in a desert world of his own. He had launched his force across the wastes of the Cyrenaica peninsular, which would take a week and require sheer guts and determination in coping with the 120-degree heat. He shepherded his force with his Fieseler Storch spotter plane, dropping paper notes to those of his formations he found halted: 'If you do not move off again at once, I will come down! – Rommel.' At this juncture he wrote to Lucie:

8 April [1941]
I've no idea whether the date is right. We've been attacking for days now in the endless desert and have lost all idea of space or time. As you'll have seen by the communiqués, things are going very well . . . It's going to be a 'Cannae'* modern style. I'm very well. You need never worry.[14]

Meanwhile, Rommel's adjutant, Major Hans-Joachim Schraepler, noted in a letter home that summer the problems of living under and fighting for a dictatorship which tried to police thoughts and opinions:

Six months and we are still in Africa . . . the court martial has now sentenced to death a man who, in a letter to his father, vehemently criticised the war. Rommel must confirm the sentence, though I hope he won't. Who of us, gripped by the heat of the climate, the sirocco, artillery fire such as the members of the court martial have not seen, has not expressed his feelings on this kind of war which is not a war in a palm grove, in a letter, to feel better?[15]

---

*Fought on 2 August 216 BC in Italy, Cannae is the archetypal battle of annihilation which many generals have since sought to imitate.

Cleary unsettled by the business, the following day Schraepler continued to muse:

> Walter [von Neumann-Silkow, commander of 15th Panzer Division and his cousin] confirmed the sentence, but said the soldier could present a plea for clemency. For Rommel and me, the sentence seems too harsh . . . The man had spilt his anger, but who of us has never done the same? We also have to understand the writer of the letter. He took part in an attack, exposed for hours to strong artillery fire, facing the enemy on the difficult front [at] Sollum, standing the hardest climate . . . When the soldier had half an hour to himself, he picks out a piece of paper and confides to his father, whom he trusts. That the letter cannot be positive is understandable. The soldier probably could not control his nerves . . . In his war diary he made reasonable comments. Rommel wishes to meet the judge of the court martial.[16]

In the war all armies deployed military censors who routinely opened personal letters, looking for indiscreet comments about force numbers, locations and intentions. In Germany – even at this early stage of the war, in June 1941, when success was at its height – censors were less worried about the disclosure of military secrets, than the revelation of anti-Nazi sentiment, which could cost the writer his life. Though I tried to discover the outcome of this story, it remains lost in the Afrika Korps files scattered between London, Maryland and Freiburg. Rommel was a fair commander, but hard, so the outcome could have gone either way. But it illustrates the daily stresses under which the Desert Fox was labouring.

Some stress Rommel felt was self-induced; his inclination was always to be at the front (taking his personal staff with him), leaving logistics and planning to his headquarters team. A colleague from his Dresden days later observed: 'You can understand Rommel only by taking his storming of Mount Matajur [1917] into account. Basically he always stayed that lieutenant, making snap decisions and acting on the spur of the moment.'[17] Schraepler echoed this, complaining that Rommel 'often went to the front . . . presenting himself more like the head of a commando [i.e. small combat unit] than a commanding general'.[18] His complaints continued with the observation that Rommel's command car was 'attacked by English low-flying aircraft. Rommel's driver and another were killed, a third was wounded. Maybe Rommel takes it as a further sign that he has been lucky again. His place is not at the front.'[19] Even this did not deter the Desert Fox, for later on, 'Although Rommel had promised not to go to the front line any more as he did yesterday, he did it again. And again he was fired at.'[20]

Greece had drawn away the cream of Wavell's force, and Churchill's policy began to unravel, confirming failure in Greece and threatening disaster in Libya. The key to Wavell's decision (if it was ever his to make), to switch forces within his massive theatre of operations was not just because of Churchill. He felt able to do so because of strategic intelligence known as 'Ultra'. This was the product of the Government Codes and Cypher School at Bletchley Park, where data, gained via deciphering the messages sent between German headquarters, was analysed and sent on to appropriate recipients. The massive code-breaking effort (at its peak, Bletchley ran on three 8-hour shifts of 3–4,000 people each) was possible because every German headquarters (army, navy and airforce, railways and diplomatic service) used slightly different versions of the same basic model of Enigma enciphering machine. Outside Bletchley, very few were 'Ultra-cleared', with access to its intelligence products, and even they were given to understand that the material came from top-secret agents, not an elaborate code-breaking process. Churchill visited the site on 6 September 1941 and later honoured the tireless analysts with the phrase 'the geese that laid the golden eggs and never cackled'.

Further grief was heaped upon Wavell when, on 7 April 1941, a fluke Afrika Korps reconnaissance patrol captured General Dick O'Connor, commander of Western Desert Force, along with fifty-five-year-old Lieutenant General Philip Neame, Military Governor of Cyrenaica, as they were driving to their headquarters. O'Connor's replacement was Alan Cunningham, a Dublin-born former gunner and Montgomery's direct contemporary in age. A remarkable family, Cunningham's elder brother, Andrew, was at the same time C-in-C of the Mediterranean Fleet. The loss of O'Connor and Neame, two very valuable and experienced commanders, was extremely unfortunate.[21]

The port of Benghazi fell to Rommel on 4 April 1941, but he was unable to take Tobruk, which remained isolated one hundred miles behind the front. The continued resistance of its largely Australian garrison assumed a symbolic significance for the British when the war elsewhere was going appallingly. By the end of April, the Army Chief of Staff at OKH, *Generaloberst* Franz Halder, had become enraged by Rommel's flagrant disobedience of his orders not to overextend his fledging Afrika Korps – which was effectively only a single division, 5th Light, for 15th Panzer Division was still trickling into theatre. Halder despatched to North Africa his Quartermaster General, Friedrich Paulus (Rommel's colleague, stretching right back to *Alpenkorps* days and a fellow company commander in the 13th Infantry Regiment during 1927–29) to 'bring the soldier who had gone insane back to reason'. Arriving on his inspection tour on 27 April, Paulus voiced concern about Rommel's proposed offensive against Tobruk, which had now become a fortress. Paulus reported back that Rommel's supply lines were overextended, his troops exhausted and his reserves nonexistent. With OKH's authority, he ordered Rommel to forget about Tobruk and stockpile resources for a major offensive, for it was clear to him (the epitome of the perfect staff officer), if not to Rommel, that the port of Tripoli could not handle enough supplies to feed the Afrika Korps. Henceforth, Rommel was to withdraw and operate within his resources.

Wavell was Ultra-cleared and thus knew of Rommel's difficulties and the Paulus report; but Ultra couldn't give Wavell the measure of the man.* It never crossed that honourable warrior's mind that Rommel would be inclined to disobey any orders he didn't like. In turn, the German intelligence services – a mixture of pro-Nazi Arab sympathisers and a very effective army-run signals intercept service operating near the front line – revealed to Rommel just how small the British Western Desert force was.

In fact, it was hardly even British, but a coalition of (eventually) three divisions of Australians and one of New Zealanders; detachments of Greeks and Free French, East and West African troops and British-officered Indian Army units, containing a wide ethnic mix of cultures and languages, Free Poles and the tough 1st South African Division, whilst other South Africans and Rhodesians flew overhead with the Desert Air Force. Elsewhere, Wavell's forces were further distracted by the need to suppress a pro-Nazi rising in Iraq in April 1941, followed

---

*Ironically, Wavell never knew that Rommel carried with him an annotated copy of Wavell's 'Lees Knowles Lectures', delivered at Trinity College, Cambridge, in 1939; published as *Generals and Generalship* it had been translated by the German General Staff.

by a bitter war in June–July against Vichy French forces in Syria and Lebanon, which had supported the coup in Iraq.[22] The net effect was to drain forces away from North Africa that might have stopped Rommel in his tracks.

Rommel was now being actively talked up by the Goebbels propaganda machine and first featured on the front cover of the mass-circulation *Signal* magazine on 2 May 1941. It is also significant that, despite the misgivings in Berlin about his African adventures, in February–March 1941, Rommel's was the only front mounting offensive operations, and Goebbels lost no opportunity in promoting the hero of the 7th Panzer Division. However, he had been forced to pause before Tobruk due to petrol and other logistical shortages, as Paulus knew he would. Such problems would dog the rest of his war in North Africa and dominate the whole campaign over the waterless desert wastes for both sides.

After Bletchley Park intercepted Paulus's damning report back to OKH, Churchill felt confident, ordering Wavell to attack the weakened Fox before his 15th Panzer Division had arrived and deployed. First, Wavell was reinforced by a supply convoy which brought him 238 tanks and 43 Hawker Hurricane fighter planes; then (pressured by Churchill, who in turn was being leant on by the Australian government, whose troops were imprisoned within Tobruk) he launched Operation Brevity on 15 May. It soon became all too evident that Rommel had ignored Paulus's assessment. So too had Hitler, who felt the army's complaints about his favourite general were born of jealousy, which, however valid, in part they were.

Paulus's attention was soon diverted to the attack on Russia, launched on 22 June, which eventually determined his own fate. Rommel, meanwhile, coped logistically by capturing vast dumps of British stores, including fuel. As Jochen Schraepler admitted to his wife, 'the supply of our troops is not in step with our advance. We live almost exclusively on English rationing. It is very good.'[23] Wavell followed Brevity with Operation Battleaxe (15–17 June) to relieve Tobruk and its antipodean garrison. Battleaxe was a disaster for British armour, losing 220 tanks (the Germans lost ten percent of that number). It underlined the superior nature of the panzer formations and their tactic of luring hostile armour onto screens of anti-tank guns and the poor armour–infantry cooperation of the Western Desert Force.

After the failure of Battleaxe, Churchill, ever impatient for victory and blind to the fact that his own meddling had contributed to British setbacks, replaced his unfortunate theatre commander, Archie Wavell (who had also to contend with the abortive operations in Greece and Crete), with General Sir Claude Auchinleck on 1 July 1941. In every respect a nice and honourable man – perhaps too nice for Churchill's needs – Wavell was politely moved to C-in-C India, where he eventually became viceroy and had to contend with the Japanese. Auchinleck, meanwhile, was given express orders from Churchill to launch a counter-offensive as soon as possible.[24]

Tall and athletic, looking every inch a general, 'the Auk' was Monty's senior by three years, and an Indian Army soldier through and through, being commissioned into the 62nd Punjab Rifles in 1904. Awarded a DSO in 1917, he had attended the Indian Staff College at Quetta (where he was later an instructor), *and* attended the Imperial Defence College, had overseen the 1940 Norway expedition, and as GOC of Southern Command in 1940 had clashed continually with his subordinate, Monty. Some of the friction may have stemmed from the early prejudice Monty developed against the languid ways of the Indian Army. In his memoirs he candidly stated of Auchinleck: 'I cannot recall that we ever agreed on anything.'[25] Although others regarded the Auk as 'free from snobbery, able to listen and a keen talent-spotter',[26] Montgomery's operating style seems to have been largely determined by whom he had offended or with whom he disagreed. Much of this appears to have been gratuitous, for on the surface there was no reason why these two talented and distinguished commanders should have been at loggerheads.

On 1 September 1941, the Afrika Korps (essentially 5th Light and 15th Panzer Divisions and corps troops) was officially renamed *Panzergruppe Afrika*, and now included 90th Light Division and two corps of Italian troops. With both sides reinforced, Cunningham's Western Desert Force was renamed 'Eighth Army' in September 1941 (eighth because the Allies had fielded seven armies against the Germans in 1914–18 and it was felt that a sense of continuity was needed in fighting the old enemy). The Eighth Army took as its badge a red cross on a white shield as displayed by Crusader knights who had fought in desert campaigns of old, but when it was realised a red cross might be confused with the internationally recognised symbol for medical personnel, the emblem on the white shield was changed to a yellow cross. The name change was important. The opposition already had their own 'brand name', Afrika Korps, with its logo of a palm tree and a swastika. It was hoped the rebranding might help focus public attention in the newspapers, pride at home, and highlight requests for increased reinforcements and supplies of better quality. And in the arid desert, an environment free from other distractions, men could ponder on the morale-boosting implications of the name change.

Cunningham was absolutely right in expecting the new title, badge and associations would bring an increased sense of identity – but not increased luck. He named his next attack (18 November–30 December 1941) Operation Crusader, hoping it would yield more than Battleaxe had. Initially the attack was successful, but Cunningham seems to have lost his nerve and persisted only until 26 November, having advised an early curtailment due to Rommel's counter-strokes, despite signs of success, and Auchinleck was forced to replace him with Major General Neil Ritchie, Deputy Chief of the General Staff in Cairo.

The affable, pipe-smoking Ritchie (a much younger man than Monty, by ten years) had been commissioned into the Black Watch and won an MC in 1918; he was a brigadier in 1940 and army commander eighteen months later; not without reason, some thought him over-promoted. Crusader was preceded by a spectacular special forces attack on a headquarters complex known to be used by Rommel at Beda Littoria in Libya (Operation Flipper). The Desert Fox was absent and its leader, Lieutenant Colonel Geoffrey Keyes, was killed during the raid but awarded a posthumous VC for his bravery. Keyes's father, an admiral and MP, was at the time Chief of Combined Operations and very close to Churchill. The older Keyes had been a leading anti-appeaser with Churchill and Eden; outspoken in his criticism of Neville Chamberlain, he led the 7 May 1940 debate on British strategy in Norway, when his dramatic address to the House of Commons in full admiral's uniform tipped political support away from Chamberlain, and was partly responsible for the elevation of Churchill as Prime Minister on 10 May. It was one reason why Winston retained such a close interest in the Desert War.

Operation Crusader broke the siege of Tobruk on 4 December 1941, and by the time the offensive ended, it had forced Rommel to withdraw west all the way back to El Agheila. The British also occupied several towns and captured some 25,000 men. Though the British outnumbered Rommel in the Crusader attack, individual German units within the Italo-German army were better led, and possessed much greater initiative and resourcefulness, with the result that the campaign subsided into a confusing series of inconclusive battles between the Egyptian frontier and Tobruk. However, the moment German reinforcements arrived in Tripoli (which British intelligence failed to detect), Rommel struck back. From 21 January 1942 he managed to advance ten miles into the British line with three armoured columns and retook Benghazi on the 29th; his opponents, taken by surprise, fell back. On 23 January, *Reichsminister* Goebbels noted in his diary:

> Rommel's boldly conceived attack in North Africa is extremely gratifying. The English are trying to excuse themselves with weather difficulties. In the course of the day, however, they must nevertheless admit being pushed back quite a distance. Rommel is praised highly by the English press. He is altogether one of our most popular generals. We could well use a few more such big shots.[27]

This was echoed in England by Oliver Harvey, Anthony Eden's Principal Private Secretary, who observed on 27 January:

> Bad news from Libya. Rommel has again broken loose and overrun our people – he is back past Benghazi. What a General! I wish some of ours were half as good. There can be no defence for our failure to hold him. He can't have the resources which we have.[28]

Even Churchill seemed to have fallen under the Rommel spell. Describing the setbacks of January 1942 to the House of Commons, Churchill explained: 'We have a very daring and skilful opponent against us, and, may I say across the havoc of war, a great general' (a tribute he would repeat in 1950 in the fourth volume of his history of the *Second World War*). Goebbels, barely able to disguise his glee at the British reaction, wrote in his diary the same month:

> The English press calls Rommel a rascal who has once again pulled a rabbit out of the hat. They [the English] are making him one of the most popular generals in the entire world. That's perfectly all right with us, for Rommel deserves it.[29]

Worse was to follow. On 1 February 1942, Auchinleck issued an Order of the Day in which he attempted to dispel the magic of the 'Desert Fox', but which only served to have the opposite effect.

There exists a real danger that our friend Rommel is becoming a kind of magician or bogey man to our troops, who are talking far too much about him. He is by no means a superman, although he is undoubtedly very energetic and able. Even if he were a superman, it would still be highly undesirable that our men should credit him with supernatural powers.

I wish you to dispel by all possible means the idea that Rommel represents something more than an ordinary German general. The important thing now is to see to it that we do not always talk of Rommel when we mean the enemy in Libya. We must refer to 'the Germans' or 'the Axis powers' or 'the enemy' and not always keep harping on about Rommel.

Please ensure that this order is put into immediate effect, and impress upon all commanders, that from a psychological point of view, it is a matter of the highest importance.

(Signed) C. J. Auchinleck,

General, Commander-in-Chief, MEF

P.S. I am *not* jealous of Rommel

News of the order soon filtered through to the Germans – and to Rommel: a translated copy of the order was later found amongst his papers.[30]

When Generals O'Connor and Neame were captured on 7 April 1941, their AEC-manufactured Armoured Command Vehicles (ACVs) became Rommel's own. Nicknamed the 'Mammoths', Rommel code-named his (formerly O'Connor's) Max. Leutnant Schmidt recalled how his general would 'usually sit up high on the roof of his Mammoth, dangling his legs through the open doorway', and 'often took over the wheel from his tired driver. His sense of direction was remarkable, and he had an almost uncanny ability to orientate himself by the stars at night.'[31] This was also the origin of one of the most famous Rommel props, found whilst he cleared his new Mammoth of its British gear:

Among the stuff turned out he spotted a pair of large sun-and-sand goggles. He took a fancy to them. He grinned, and said, 'Booty – permissible, I take it, even for a general.' He adjusted the goggles over the gold-braided rim of his cap peak. Those goggles for ever after were to be the distinguishing insignia of the Desert Fox.[32]

Rommel's Mammoth was the undoing of Jochen Schraepler, the loyal ADC who had also served him in the 1940 French campaign; on

9 December 1941 he was accidentally crushed by the heavy vehicle.
Perhaps callously, his death merited only the briefest of belated footnotes, weeks later, in Rommel's Christmas Day letter home to Lucie: 'PS
– I don't think I've told you yet that Schraepler met with a fatal accident (run over by the Mammoth).'[33] This underlines the Desert Fox's
emotional detachment from even those close to him. When Rommel's
command expanded into Panzergruppe Afrika, Schraepler had remained
as adjutant to the DAK's new commander, Ludwig Crüwell, and he
compared the style of the two: 'the position of an adjutant with Crüwell
has a greater importance than with Rommel, who called me only when
he had an order to give'.[34]

Goebbels had been a keen follower of Rommel's progress during the invasion of France (via his man on the ground, Karl Hanke) and quickly spotted
the propaganda potential of North Africa. As the good news from the
Russian front faded with the onset of poor weather in November 1941,
Goebbels turned to Rommel and the Desert War instead. In honour of his
fiftieth birthday, the Nazi party daily newspaper *Völkische Beobachter*
featured on its front page a photograph of the 'Desert Fox' wearing his sun
helmet. It marked the start of a sustained media campaign celebrating
Rommel's virtues and achievements in the desert. From February 1941,
Goebbels had seconded to Rommel another of his trusted subordinates,
Leutnant Alfred Ingemar Berndt, as Karl Hanke's replacement. Just like
Hanke, Berndt was an early party member (from 1924), *SS-Oberführer*,
Propaganda Ministry official and Leutnant in the army reserves. Like Hanke,
part Nazi plant and part conduit between Rommel and Goebbels, Berndt
was also used when 'anything unpleasant needed saying to Adolf Hitler . . .
because he was a brave man'.[35] Leutnant Schmidt recalled of Berndt:

> I got to know Berndt very well in those days. I saw then that he was
> contributing more than most people realised to the growing Rommel
> legend. He took every opportunity of arranging for photographs to be
> taken of the 'Desert Fox' for publication at home and in neutral countries.
> Rommel himself, as war correspondents will testify, readily allowed himself
> to be photographed. I noticed that he often deliberately fell into a pose
> that would make the photographer's task easier and more effective.[36]

Rommel proved a natural showman, giving polished performances to
camera – whether to photographers or movie cameras – in an era when

this was uncommon outside the profession of acting. (This was also a skill Monty would acquire.) At the end of the year a delighted Goebbels wrote in his diary:

> I saw Rommel in a documentary film . . . He talked for more than three quarters of an hour about his successes at the beginning of this year. Without making any gesture he talked in a classic style, practically without correcting himself a single time. What he said and the way he said it, the play of his features and his whole appearance – all gave evidence of the greatness of an outstanding personality.[37]

Along with the photography, Berndt oversaw the adoption of a complement of thumping tunes with syrupy lyrics to add to the Wehrmacht's growing lexicon of marching songs. These often provided the background to Goebbels's weekly newsreel reports – *Die Wochenschau* (*The Weekly Show*) – that German audiences watched in their cinemas. There was already the catchy '*Heia Safari*', the traditional march of General von Lettow-Vorbeck's undefeated East African army of 1914–18, to which was quickly added '*Panzer rollen in Afrika vor*' ('Panzers are rolling forward in Africa'), and '*Unser Rommel*' ('Our Rommel'), with lyrics talking of personal sacrifice and revenge for the African colonies lost after 1919.[38] '*Unser Rommel*' was the only marching song that mentioned a commander by name:

> *We are the German Afrika Korps, the Führer's daring troops*
> *We assault like the Devil*
> *Oversalt the Tommy's soup*
> *We fear neither heat nor desert sand*
> *We resist the thirst and the blazing sun*
> *March to the beat of our drum*
> *Forward, forward, forward with our Rommel!*
>
> *The Brits fear us like the plague*
> *They are on tenterhooks*
> *We revenge German East [Africa] and revenge South-West [Africa]*
> *Which once were dastardly stolen from us*
> *Let Churchill and Roosevelt be inflamed with rage*
> *We beat the enemies in every country*
> *The drum beats 'get ready'*
> *Forward, forward, forward with our Rommel! . . .*

Serving as Rommel's Italian interpreter was the war artist Kurt Kaiser, a German-Italian painter, journalist and comic-book artist whose near-photographic representations of the desert war and its personalities in pencil and charcoal communicated an heroic image of the campaign to a much wider audience, in books and on postcards.[39] Rommel's aide-de-camp, Leutnant Heinz Werner Schmidt, recalled that in the evenings he would spend many hours answering the general's considerable fan mail:

> Many came from hero-worshipping boys, but the majority came from girls and women. They all verged on adoration. Nearly all asked for a photograph. To answer this demand we kept a large carton of postcard portraits taken by Hoffman of Munich, Hitler's official photographer. Replenishments of stock were received regularly, and Rommel personally autographed every photograph I sent out . . . I was always amused when he signed letters to see the tip of the General's tongue protruding and comically following in the air the outline of the unusually florid flourish of his pen when he completed the Bold 'R' in his signature.[40]

Schmidt was also expected to take down all details of Rommel's conferences and verbal orders, as and when Rommel required, and these would form the basis for his future war memoirs: 'Every wish and every order given by the general had to be accurately recorded in writing, and there were endless memoranda on exact times, names, localities, unit strengths, and so on.'[41]

In late July 1941 a unique battle honour was awarded to all German army personnel with two months service in theatre, a small armband worn on the right cuff of uniform jackets, embroidered with the word 'Afrikakorps'. Though by then Rommel's force was technically the Panzerarmee Afrika, a collective term to include the Italians, the Afrika Korps 'brand' had stuck – for the British as well as the Germans. Another of Rommel's aides had recorded on 14 June 1941: 'Members of the Afrika Korps will receive a band with the inscription Deutsches Afrikakorps, which was my idea.'[42] An equivalent of the British Africa Star medal, it was superseded towards the end of the campaign, on 15 January 1943, by another cuffband, embroidered with the single word 'Afrika' between two palm trees. An Italo-German campaign medal was also instituted by the Italian government to recognise the achievements of the Afrika Korps. It was awarded only to German troops, the first presentations taking place in 1942, but after Italy's withdrawal from the war in September 1943, the wearing of all Italian awards was prohibited.

Every German newspaper and magazine carried features on Rommel and the Afrika Korps, far more so than the British featured their Eighth Army, whose own *Crusader* weekly newspaper had commenced in May 1942, followed by the *Eighth Army News*, a four-page daily paper. Rommel was lucky that between February–April 1941, his photogenic advances in Africa were the only combat operations being reported in the German media, so he was able quickly to establish an image, free from competition. The Rommel propaganda effect also spread further afield with the result that *Time Magazine* featured him on the front cover of their 13 July 1942 edition: a head and shoulders portrait hovers over a giant arrow pointing towards the Suez Canal, leaving no doubt about the strategic issues at risk. A similar accolade was not extended to Monty for seven months.

Then there was Wolfgang Willrich, an art student who had served as a *Feldwebel* on the Western Front. Throughout the war Willrich painted famous military leaders and in the Nazi era drew iconic portraits, emphasising the Aryan, Nordic features of his sitters, which appeared as posters and postcards. In September 1939, Wolfgang had written to Rommel, at Hitler's headquarters, suggesting that he be sent to the front as a combat artist. He took part in the campaign in France, where he made much-reproduced portraits of Rommel, Guderian and German airborne troops. It was a poster-sized reproduction of Willrich's famous 1940 portrait of a rugged, greatcoated Erwin Rommel, goggles thrust up over his hat, in the midst of a smoking battlefield, that Montgomery had fixed on the wall of his own caravan in the Western Desert and in Europe.

Auchinleck had spoken in his Order of the Day of Rommel being seen as a 'magician' or 'superman', and the Desert Fox certainly seemed to possess a sixth sense, however confused the battlefield, of knowing where and when to strike, or concentrate his reserves. This built on the earlier mystique of his Ghost Division in France, with its seeming ability to materialise at will behind enemy lines. The growing legend surrounding Rommel was thus a mixture of his own undoubted skill, based on intelligence he received and exploited, but with a hefty dollop of propaganda from Josef Goebbels. Captain Erich Hartmann, who served with Rommel, explained the phenomenon:

We used to say 'Where Rommel is, the front is' ... He was always attempting and bringing off things that no one else would have thought of trying. He seemed to have a sort of *Fingerspitzengefühl*, a sort of sixth sense, an intuition in his fingers ... Hard, yes, though he never asked anyone to do more than he would do himself ... Perhaps officers did not like him as much as the men because he always expected more of them and there were very few who could go his pace. But he was the best of comrades.[43]

Intuition, or something else? Rommel had brought with him to Africa the 621 Strategic Intercept Company, commanded by twenty-six-year-old Leutnant Alfred Seebohm. The company had worked under Rommel in France and throughout 1941–42 would provide the Desert Fox with much valuable signals intelligence. Seebohm and his men were able to monitor British voice transmissions sent between all levels of the army over high-power VHF radios, noting in a report, 'during the battle, as often before, a considerable amount of carelessness in the use of plain language was noticed ... matters relating to Command and Intentions were spoken about with a freedom hardly ever encountered before'.[44] Seebohm's men were highly competent radio operators, with a sophisticated working knowledge of English, and able to transcribe English conversations at speed (although occasionally the British would resort to Urdu or schoolboy French in a lame attempt at security). Leutnant Hertz, Seebohm's second in command, later admitted, 'we don't have to bother much about [translating] ciphers. All we really need are linguists, the sort who were waiters at the Dorchester before the war started.'

Seebohm also had a 'situational awareness officer' who could furnish Rommel or his staff with immediate reports. For example, when on 27 May 1942 during the battle of Gazala, Brigadier A. A. Filose, commander of 3rd Indian Motor Brigade, told General Messervy of 7th Armoured Division that he had 'a whole bloody German armoured division in front of him', Rommel received the message as soon as the brigadier had sent it to his commander. Although 621 Company also passed data back to the *Abwehr**, its primary function was to support Rommel's tactical picture. The young Seebohm became one of Rommel's

*The *Abwehr* was Germany's military intelligence-gathering organisation, based in Berlin, headed by senior naval officers until 1944, when it was abolished and absorbed into the SS.

inner circle of advisers, reporting to him each night. Even though the British had also deployed the 300 men of the 'Y' Service (a tactical intelligence group) since 1941, which collected raw Enigma traffic and passed it back to Bletchley Park for decoding, its primary task was to enable the flow of strategic intelligence in and out of Bletchley, rather than illuminate the tactical picture for the Eighth Army, as Seebohm did for Rommel.

Another, though unwitting, source of intelligence that enabled Rommel to appear 'the magician' was Colonel Bonner Frank Fellers, the US military attaché in Cairo. Fellers sent frequent reports, back to General George C. Marshall in Washington, of British intentions, casualties and unit locations using the US diplomatic code which the Germans had broken. Until his removal in July 1942, his reports proved a source of invaluable material for Rommel (though the security breach apparently did his career no harm: he later served as Military Secretary and the chief of psychological operations under MacArthur in the Pacific). The German *Abwehr* also operated a spy ring in Egypt called the Kondor Mission; it relayed low-grade intelligence back to Rommel's analysts, using the Daphne du Maurier novel *Rebecca* as a cypher key. After an extended hunt, the ring was compromised in August 1942 and thereafter false information fed back to the Afrika Korps.

On 10 July 1942, Seebohm moved 621 Company onto high ground overlooking El Alamein in the hope of gaining good interception and direction-finding for the forthcoming battles, but was only 600 yards behind the front, held by an Italian division. A raid by the opposing Australians punched through the Italians and killed or captured all but sixty-nine of Seebohm's company and he himself was killed. Although 621 Company was reformed in September 1942, it was never as effective; shocked at the implications from the haul of documents captured from 621 Signals Company, immediate security measures were adopted by the Allies – and Rommel's 'magical' insights ceased suddenly.

15

Duel in the Desert II:
The Battles of El Alamein

ALL DESERT VETERANS are united in their recollections of being at war with nature. The heat hit newcomers as a solid wave, but the nights were equally grim when temperatures dropped to freezing and men died of exposure. When the Germans could, they acclimatised their troops, exposing them for several weeks to artificially induced temperatures prior to departure, and sending them to training areas on the Baltic coast in the summer months.[1]

The Italians did likewise: Lieutenant Emilio Pullini of the Folgore Parachute Division remembered that he arrived having received 'very tough training in southern Italy for a couple of months, on difficult ground and [in a] very hot climate . . . when we arrived we were very fit'.[2] With no natural water supplies everything had to be trucked forward; both sides were totally reliant on the fuel, water and food that flowed through the giant umbilical chords that trailed out of the seaports to feed the desert armies.

Rommel certainly had a blind eye to logistics. Although he tended to blame the Italians (and the marauding British) for his shortages – judgements that many historians have unwittingly accepted – new research suggests that Rommel's forces were receiving a useful average of 800 tons per division per day. (In 1944, hungry US armoured divisions had to make do with 600 tons of supplies per day, including fuel.) It was Rommel's own land logistics arrangements that let him down. Because he was so impulsive his supply staff could not plan ahead (in total contrast to Montgomery, when he eventually arrived in theatre). No more than two thirds of the fuel that left the docks in Tripoli (and later Tobruk) ever reached Rommel's front line, for his trucks consumed the rest in making the return journey to deliver it. More fuel was used bringing forward ammunition, water and food. Military logisticians recognise that Rommel soon became overstretched logistically: when he reached the Alamein area in 1942 he was 1,500 miles from Tripoli. In contrast, the Allied logistical bases at Alamein were less than a tenth of that distance, and they could use the efficient coastal railway that ran from Alexandria.

When not fighting each other or the heat, soldiers had to contend with the sand, as fine as talcum powder, that got into every orifice and machine, that formed deceptive crusts on areas of soft sand capable of swallowing small vehicles and people, and occasionally formed into towering sand-storms – the *gibli* – which even the war respected as all fighting ceased and humans hunkered down and waited. Besides the sand, snakes, scorpions and spiders spread their fair share of horror stories amongst front-line soldiers and new recruits alike. But above all it was the flies that veterans recall with most distaste; the insects swarmed uncontrollably, in vast numbers, spreading disease. They appeared from nowhere and settled on everything, on the dead and the living, battling with the troops for every meal, every scrap of ration. The fly-swatter became a *de rigueur* piece of officers' personal equipment; Monty is seen frequently bearing one, like the field marshal's baton he would one day clutch.

The concept of both sides fighting nature in an isolated theatre of war bred mutual respect, underscored by the British adopting the German song, 'Lili Marleen', the type of syrupy love song broadcast from the German-run Radio Belgrade, whose powerful transmitters reached throughout the Mediterranean. The mournful tune, which marked the station signing off every night at 9.55 p.m., evoked memories of home for all soldiers. Though recorded in German by Lale Andersen in 1939, the British soon invented their own lyrics to the tune, and it even became the regimental march of the Staffordshire Yeomanry. Rommel positively admired it, but Monty was invariably asleep before Miss Andersen's nightly crooning began.

This fellow feeling was manifest in many ways, and desert warriors looked back on the North African war, brutal though it was, as chival-rous; as one Eighth Army veteran, Ted Fogg, recalled of his mate, Jack Gadsden:

Often we sent out foraging parties to cannibalise broken down and abandoned vehicles ... Jack was busy dismantling a partially wrecked vehicle ... With his head deep in the engine compartment he was surprised to hear himself being addressed in German ... Expecting the worst, Jack put his hands up and prepared to surrender. The officer demanded his papers, inspected them and gave them back. He then said: 'Go on your way, you silly little Englishman.' Jack got in his Austin and roared off.[3]

In November 1942, Monty entertained to dinner General Ritter von Thoma, acting head of the Afrika Korps, after his capture. Though he retired at his usual 9.30 p.m., Monty noted in his diary of the German general: 'he is a very nice chap and was quite willing to talk about past events ... I doubt if many generals have had the luck to discuss with their opponent the battle that has just been fought.'[4] Churchill's wit sparkled when he heard of the encounter: 'Poor von Thoma,' he quipped, 'I too have dined with Montgomery.'

Even the hard-living Rommel found ways of relaxing in the desert, recording on 10 September 1941: 'I went out shooting last evening with Major von Mellenthin and Lieutenant Schmidt. It was most exciting. Finally I got a running gazelle from the car. We had the liver for dinner and it was most delicious.'[5] Schmidt later recalled of the gazelle hunt with Rommel, he 'flashed out a large hunting knife and finished off the job. He eviscerated the animal expertly, sawed off the horns, and had the carcass loaded.'[6]

For the young Hans von Luck (who commanded Rommel's reconnaissance battalion in 1940), looking back, the fight against the British in the desert was regarded as the most 'sporting' contest of the war; the deep bitterness of the French campaign and the dehumanising ethos of the Russian front were absent. The to-and-fro nature of the struggle meant that both sides got to know each other's units – sometimes each other's personalities – quite intimately. A captured German medical officer might be 'swapped' for a supply of synthetic quinine of which the British were in short supply. Towards the end of the campaign, in a Tunisian bivouac, a Bedouin suddenly came to von Luck's tent and presented him with a letter. It was from the CO of the Royal Dragoons and read:

Dear Major von Luck,

We have had other tasks and so were unable to keep in touch with you. The war in Africa has been decided, I'm glad to say, not in your favour. I should like, therefore, to thank you and all your people, in the name of my officers and men, for the fair play with which we have fought against each other on both sides. I and my battalion hope that all of you will come out of the war safe.'[7]

Despite the filth and detritus of battle, from a distance the desert looked clean, and thus the campaign appeared sanitised in terms of the way the war was fought. In the Mediterranean, organisations like Popski's Private Army, the Long Range Desert Group, Special Boat Service, Special Air Service, Parachute Regiment and Commandos abounded; even the decidedly non-military, almost piratical way they dressed (extensive use of pullovers, corduroy trousers, scarves, crepe-soled shoes and bizarre headgear), exemplified by the eccentricities of dress worn by the cartoonist Jon's Two Types, and copied (or inspired?) by Montgomery himself, underlined the point. Modern historians, however, have concluded that the campaign was not so clean. Rudolf Schneider, then aged eighteen, who served as Rommel's driver with the Desert Fox's own personal protection squad, the *Kampfstaffel*, later recalled:

Rommel enjoyed touring the front lines. We would go deep into the desert to explore. One time we came across fourteen German soldiers who seemed asleep. When we got closer we saw each had his throat cut. Nearby we found a *kukri* – the knife of the British Gurkha soldiers. I still have that knife.[8]

Whilst no Waffen-SS units of any sort served under Rommel in North Africa at any time (probably at Rommel's insistence), the sinister *Sicherheitsdienst* – the security service of the SS – established a Tunis office in July 1942, run by *Obersturmbannführer* Walther Rauff. His detachment was responsible for the detention of over 5,000 Tunisian Jews, many of whom died in a network of SS slave labour camps in Tunisia before the German collapse of May 1943. Nazi documents reveal that Rauff's men took silver, jewels and religious artefacts from their victims, including forty-three kilograms of gold, stolen from the Jewish community on the island of Djerba[9] – although most of the SD's activities took place after Rommel's departure .

Throughout Rommel's military career, there are numerous personal demonstrations of an ethical approach to war, backed up by the moral courage to defy his superiors when he felt it necessary. An important example of this was his decision to burn rather than ratify Hitler's Commando Order, issued to the Führer's generals in secret on 18 October 1942, which required all enemy soldiers caught behind German lines in uniform to be executed immediately. This contravened the 1929 Geneva Convention, ratified by Germany, which protected service personnel captured in uniform during war. After 1945, the wartime conduct of many senior German commanders was often assessed in terms of whether they had obeyed the Commando Order. Most were too terrified of Hitler to disobey, but Rommel (with the exception of his shooting of the French colonel in May 1940) seems to have maintained a front-line soldier's sense of honour towards his opponents.

Though Operation Crusader had forced Rommel to withdraw west all the way back to El Agheila, his counterattack had rolled Ritchie's Eighth Army back east to Gazala, close to Tobruk. Significantly, all the logistical infrastructure that was being built up in preparation for the British advance was captured, to Rommel's delight, and Ritchie foolishly started issuing orders over the heads of his corps commanders, straight to the divisions, increasing tensions within Eighth Army. Recognising, perhaps, that Ritchie was out of his depth, Auchinleck spent time mentoring him at the front, but to no avail.

In January 1942, Rommel's *Panzergruppe Afrika* became the *Panzerarmee Afrika*, with the inclusion of 164th Infantry Division, the Ramcke *Fallschirmjäger* (Parachute) Brigade and three Italian corps. Strictly speaking, the original Afrika Korps (commanded after Rommel, by Ludwig Crüwell until his capture in May 1942, then by Walther Nehring) still existed as a subordinate formation, and simply became part of the ever-expanding German and Italian presence in North Africa.

Within days of Rommel's successful counteroffensive (on 24 January) came the news of his promotion to full *Generaloberst*, reflecting this increased responsibility. Just past his fiftieth birthday, Erwin Rommel was the youngest officer ever to have attained this rank. In mid-February the Desert Fox's achievements brought him Swords to his Oakleaves of the Knight's Cross, only the sixth award made of this decoration at that time.

By contrast, the sort of Churchillian pressure Auchinleck (and subsequently Montgomery) was having to resist is well illustrated by Alan Brooke's diary entry for 2 March 1942:

> Another bad Monday ... Found PM had drafted a bad wire for
> Auchinleck in which he poured abuse on him for not attacking sooner,
> and for sending us an appreciation in which he did not propose to
> attack until June!! ... [Later] much time wasted in hot air, but he finally
> accepted wire we had drafted to replace his.[10]

From February 1942 there was a period when both sides rebuilt their
strength to launch fresh offensives, until Rommel again attacked at Gazala
on 26 May, and with good Italian support eventually broke through after
vicious fighting around the Knightsbridge Box. British tactics at this
stage were not to hold a front continuously, but to concentrate troops
into mutually supporting all-arms 'boxes' that contained armour, artillery,
infantry and anti-tank weapons, behind a screen of wire and minefields.
Ritchie would, alas, discover that obstacles not covered by fire were no
obstacle at all. Although the British outnumbered Rommel in tanks, they
were dribbled into battle in a way that allowed Rommel to destroy many
without sacrificing his own: a mass armoured attack with infantry support
would have yielded different results. The British southern flank at Bir
Hacheim was held by a Free French division under Koenig, who had
fought their way up from Chad, and in causing 3,000 Axis casualties
whilst resisting a sixteen-day siege before escaping provided the first
evidence that de Gaulle's Free French units could play a significant role
in the war.

Rommel's penetration caused Ritchie to order a withdrawal into Egypt
on 13 June, and on the 21st Tobruk fell to a lesser force of investing
Germans, seeming to underline the British failure in North Africa.
Churchill took the defeat badly ('one of the heaviest blows I can recall
during the war'), whilst Rommel was rewarded instantly with his field
marshal's baton. Thanks to Goebbels's near-continuous press coverage
since May 1940, when the final promotion came, Rommel was already
extremely well known. *Signal* magazine again featured Rommel on its
cover, in March 1942, in a sunhat staring into the distance, to accompany a commentary within by its own correspondent, Baron von Esebeck,
former head of 15th Panzer Division.[11] Other photo opportunities arose
which benefited both Axis partners, when Rommel was awarded the
splendid-looking sash of the *Ordine coloniale della Stella d'Italia* (Colonial
Order of the Star of Italy) by the Italian commander General Ettore
Bastico on 28 April 1942.

Although officially neutral throughout 1941, the USA had been supplying economic and military aid to the United Kingdom and Soviet Union which via Roosevelt's carefully-crafted Lend-Lease programme. Whilst the Japanese strike at Pearl Harbor brought America into the war against the Japanese on 7 December 1941, it was Hitler's declaration of war against the USA four days later on 11 December (a strategic error if ever there was one) that brought America into the European war. Within weeks, Churchill had joined Roosevelt in Washington DC for the first-ever summit conference (code name Arcadia), where both leaders agreed to pool their resources strategically to win, first in the Mediterranean and Europe, then in the Pacific. The news of Tobruk's fall in June came at a particularly inopportune moment, for Churchill was staying with Roosevelt at the White House and felt humiliation as well as defeat. They were sitting in the Oval Office when a message was passed to Roosevelt, who 'passed it to me' Churchill recalled, 'without a word'. The American President immediately offered an infantry division to reinforce the Eighth Army, but eventually he and Winston concluded that 300 Sherman tanks and 100 howitzers transported by fast convoy would be better employed.[12]

Tobruk's capture was iconic for the Germans, but not strategic. Both sides, however, magnified its importance. There was no real reason why Hitler should have promoted Rommel to *Generalfeldmarschall* for it, after all he had been promoted to *Generaloberst* only in January, but this was an expression of the Führer's chaotic, illogical, though highly effective and incentivised stick-and-carrot system of ruling; those whom he favoured received disproportionate praise and honour, whilst those on whom displeasure fell could find themselves sacked without warning, and often without reason. Rommel perhaps had realised this sooner than other officers and used Hitler as a conduit to quick promotion and laurels. But who, in fact, was using whom? Rommel's relatively unjustified elevation to the pantheon of German heroes on 21 June 1942 (he would have got there one day, but not for Tobruk) might, however, have been in order for him to be of appropriate rank to receive Mussolini in Egypt. The latter, a noted equestrian, intended to ride his favourite white Arab thoroughbred (already flown into North Africa for the purpose) into Cairo, hard on the heels of the defeated British.

By this stage, Eighth Army morale was suspect, especially between infantry and tank units, and if there was any fighting doctrine at all, it appears to have evolved along divisional lines – fatal for an army trying to fight a coordinated, coherent battle. Auchinleck also suffered his reinforcements being diverted to the Far East, and his logistical tail across the Mediterranean was under severe German and Italian air and U-boat attack. However, Rommel too had outrun his lines of supply, and was surviving only on what had been captured in Benghazi and Tobruk. Both sides had started off pretty evenly matched: Rommel could field around 500 panzers, of which half were his most dangerous medium Mark IIIs and IVs; while the British deployed 850 tanks, including 167 of the American-supplied Grants (described to me by the Knight's Cross-winning German tank commander Oberleutnant Otto Carius in 2004 as 'very tall, and consequently comical', meaning they were easy targets to hit[13]).

Auchinleck took the opportunity to remove Ritchie on 25 June (a move many believed he should have made earlier) and assume command of the Eighth Army himself. But he was still forced to retreat before Rommel, fighting what were essentially delaying actions: first at Mersa Matruh on 26–27 June, then Fuka on 28 June and thence on 7 July to a fortified line on the Alam Halfa ridge, between the railway halt at El Alamein and the Qattara Depression thirty miles to the south. Auchinleck responded with a series of attacks; the first, the Battle of Tell el Eisa (10 July 1942) was an attempt by XXX Corps (under Major General William Ramsden) to turn the German position from the northern end of the El Alamein Line. Ramsden deployed two divisions (1st South African and 9th Australian), both of which secured their immediate objectives. Rommel, however, prevented further penetration and disrupted the overall plan. Though considered by his contemporaries a capable corps commander, Monty would later consider Ramsden, who had commanded a battalion of Hampshires under him in Palestine, over-promoted.

The second Eighth Army attack, the First Battle of Ruweisat (14–16 July 1942), was orchestrated by XIII Corps (under Lieutenant General William Gott). Freyberg's New Zealand and 5th Indian Divisions took on two Italian divisions, the Brescia and Pavia. The Germans counterattacked successfully in support of the Italians and retook lost positions. The Second Battle of Ruweisat (21–23 July 1942) was a night attack, which, although initially successful, was sealed off by a German counterattack. It was during the first clash at Ruweisat that the wounded New Zealander Captain Charles Upham led his infantry company, destroying German tanks and machine-gun positions, until reduced to a strength of six before capture. As a result, Upham was awarded a unique bar to the

Victoria Cross he had already been awarded for bravery on Crete in May 1941.

In both the Ruweisat battles, coordination and cooperation between infantry and armour was ad hoc and inadequate for the task. Unwilling to contemplate further failures and losses, Auchinleck opted to regroup and rebuild his force. This series of lesser battles between the armies' arrival at Alam Halfa and 23 July, collectively known as First Alamein, proved the exhaustion of both forces, and Rommel's inability to advance further without substantial reinforcement.

Both sides paused, spent, but now with shorter logistics lines. Alexandria lay only sixty miles from the front and the British were able to reinforce more quickly. Meanwhile the Allied Middle East commanders were shuffled again, this time by Churchill in person.

Sensing a disaster looming, Winston flew to Egypt, taking long, comfort-less and dangerous flights in the primitive accommodation of a converted four-engined Liberator bomber (no mean feat for a sixty-eight-year-old), and landing (via Gibraltar, central Algeria and southern Libya) in Cairo on the morning of 3 August. He joined Brooke, who had travelled sepa-rately, and left on the evening of the 10th. Churchill was very good at spotting what are known as 'decision points',[14] when choices have to be made, and he had no fear of difficult interviews when it came to relieving generals of their command. Apart from the obvious need to halt Rommel, keep him away from the Middle Eastern oilfields and assist the viability of the upcoming Operation Torch, Churchill's decision point was driven by two separate strategic, but interrelated issues.

First, Churchill needed to deliver a victory to the British people. It had been an awful year so far, with catastrophic losses in the Atlantic, Arctic and Mediterranean convoys and the fall of Tobruk. In July 1942, convoy PQ-17 in the Arctic had just lost all but one of its twenty-five merchant ships. The worst-hit Mediterranean convoy, Operation Pedestal, summed up the maritime experience at that time. The convoy of four-teen fast merchant ships, escorted by up to forty-five Royal Navy warships, was vital to the very survival of Malta, from where Rommel's supply lines were regularly interdicted. Leaving the Clyde on 2 August, Pedestal entered the Mediterranean on the 11th and over the next four days nine merchantmen, the carrier *Eagle*, cruisers *Manchester* and *Cairo* and destroyer *Foresight* were sunk, and most of the other vessels damaged; but four ships and the tanker *Ohio* reached Valetta harbour by 15 August, enabling the island to fight on. A tactical victory for the Axis in the damage inflicted on Britain's Royal and Merchant Navies, Pedestal's arrival in Malta emerged as an operational victory for the Allies in the way it eventually enabled Rommel's logistics to be shut off.[15]

Churchill was also under additional pressure domestically: his National Government had lost a seat in a parliamentary by-election in June and on 2 July he had faced (though survived by a resounding 475–25) a motion of censure in the House of Commons, that 'this House, while paying tribute to the heroism and endurance of the Armed Forces of the Crown in circumstances of exceptional difficulty, has no confidence in the central direction of the war'. Winston could not brush this two-day debate aside lightly: he may have recalled a similar crisis in December 1916, when the sitting Prime Minister, Asquith, lost his job to Lloyd George. Overseas, Churchill would have appreciated the turn of the tide in the Pacific after the battles in the Coral Sea, and especially at Midway, over 4–7 June, and felt the need to provide a British victory in order to claim at seat at the victors' banquet, whenever that feast might occur, alongside the Russians and Americans.

On 5 August, Churchill and Brooke flew to Burg-el-Arab airfield, about twenty miles west of Alexandria, and breakfasted with Auchinleck at Eighth Army headquarters on the Ruweisat Ridge. They were not impressed – either by the fly-ridden location or Auchinleck's briefing to Churchill in his command caravan. Auchinleck was not the best of communicators and his plans required time that Churchill was reluctant to grant. Winston was, however, favourably impressed by 'Strafer' Gott, whom he interviewed in his car en route to Auchinleck's desert HQ.

At RAF headquarters, Air Marshals Tedder and Coningham, commander and deputy of the Desert Air Force, on whom military operations increasingly relied, then entertained Churchill at lunch on the coast, away from the flies, in a style for which airmen around the world are famous: a special luncheon had been brought by van from the famous Shepheard's Hotel in Cairo. Following his tour of the various headquarters, Winston deliberated hard about the game of musical chairs he was about to initiate. He was convinced a major change of direction was necessary, and triggered what became known as the 'Cairo purge'. Auchinleck had apparently already offered his resignation on 24 June, so Churchill knew he had to do something.

Initially, Churchill proposed that Brooke should take over Cairo, with Monty at Eighth Army – an idea Brooke turned down, realising his place

was with Winston. The latter was adamant and countered with the suggestion of Alexander for Cairo.[16] In a flurry of 'Most Secret Cypher' telegrams
to the Deputy Prime Minister, Attlee, on 6 August 1942, Churchill stated
he had made up his mind. He proposed breaking up Auchinleck's Middle
East command into Near East (Egypt, Palestine and Syria), under Alexander,
and Middle East (Persia and Iraq), to be offered to Auchinleck. He also
stated that Generals Corbett (Auchinleck's CGS in Cairo), Ramsden (XXX
Corps) and Dorman-Smith (Auchinleck's personal Chief of Staff) needed
to be relieved and that Gott should take over Eighth Army. In these telegrams,
there is far more weight attached to getting the Near East command appointment right, and the need to avoid impugning Auchinleck's reputation, than
to selecting the new Eighth Army commander. Churchill's fifth and twelfth
points in the main telegram read as follows:

> 5. General Montgomery to succeed Alexander in TORCH. I regret the
> need of moving Alexander from TORCH but Montgomery, though a
> first-rate soldier, would not be suited to the varied duties of the Near
> East Command. He is on the other hand in every way qualified to
> succeed Alexander in TORCH.

> 12. The above constitutes the major simultaneous changes which the
> gravity and urgency of the situation require ... I have no doubt the
> changes will impart a new and vigorous impulse to the Army and restore
> confidence in the Command, which I regret does not exist at the present
> time. Here I must emphasise the need of a new start and vehement
> action to animate the whole of this vast but baffled and somewhat
> unhinged organisation. The War Cabinet will not fail to realise that a
> victory over Rommel in August or September may have a decisive effect
> upon the attitude of the French in North Africa when TORCH begins.[17]

What is interesting here is that in a War Cabinet minute quite specific
reservations were aired over Monty, to which Attlee and the War Cabinet
must have been privy. Monty was seen as suitable to command First
Army in the future Operation Torch, which would have required considerably more tact (in working with the Americans and possibly with the
Vichy French) than he was known to possess. Ever since June, the British
political establishment had been speculating on a change of command
in North Africa. On 24 June, the unusually well-informed Oliver Harvey,
Anthony Eden's Private Secretary, confided to his diary:

The PM, I believe, rather favours Alexander to succeed [Auchinleck]. He is a fine fighting general who did very well in Burma, but AE says he has no armoured warfare experience or local knowledge, no great brain. An alternative, I hear, is Montgomery, a most ruthless man, with pale steel blue eyes who would clean up Cairo and put the fear of God into the staff.[18]

Harvey had observed earlier, on 15 June: 'Desert battle bad, Rommel too much for our Generals again, pushing us about as he likes' – which says much of the stir Rommel had caused already in British military and political circles. Auchinleck duly departed with more honour than Monty would later suggest. On 22 August 1942, *The Times* wrote:

General Auchinleck's farewell message to the Eighth Army will have raised a responsive echo in the hearts of the troops, where he retains their unstinted confidence. The great reputation which he brought with him has not seriously dimmed in the eyes of those who served under him in the Middle East by the misfortunes which attended the recent campaign. Little of the blame for them is considered to rest on his shoulders: on the contrary he is given full credit for changing defeat into partial victory at El Alamein. From the moment he assumed direct command of the Eighth Army his personality made itself felt, as he was everywhere on the battlefield, inspiring fresh confidence in badly rattled commanders by direct instructions and useful bits of information, and encouraging tired troops by stopping his car at the roadside, asking their news, and telling them to continue to do their best . . . Everybody who saw him in those days realised that here was a real leader.[19]

It might almost have been written of Montgomery after the later second battle of El Alamein. *The Times* reflected here the official line of concern to cushion Auchinleck's departure as much as possible, whilst also pointing out what was expected of army leaders in August 1942: proof that others had adopted the Monty style before Monty.

One wonders why Churchill felt that Lieutenant General William Henry Ewart Gott was the right man for the Eighth Army. Gott was a well-connected old Harrovian. Ten years younger than Monty, he'd been commissioned in 1915 and emerged from the First World War as a captain with an MC. He attended Staff College in 1930–31 and by 1939 was commanding 1st King's Royal Rifle Corps in Egypt. A desert warrior, he'd taken part in Dick O'Connor's Operation Compass in 1940 commanding 7th Armoured Division's Support Group (of artillery and motorised infantry), and subsequently led 7th Armoured Division for Crusader from September 1941 (which brought him a DSO), before being given XIII Corps in February 1942. He was considered 'young, energetic, large, religious, relentless, humourless and kind', and he commanded respect and affection.[20] The 'Strafer' soubriquet, a pun on the German First World War slogan 'Gott strafe England' ('May God punish England'), was the product of an army where nearly every regular officer (from Bimbo Dempsey, Windy Gale, Pug Ismay, Jumbo Wilson, Boy Browning, and Jorrocks [Horrocks] to the Auck, Monty and Brookie) had grown up with a nickname.

Gott had also leant heavily on Jock Campbell, who succeeded him in command of both 7th Armoured's Support Group and then the division itself. Campbell was cast in the same heroic mould as Rommel, leading his men into battle clinging to tanks, manning machine guns, laughing and joking. The fearless Campbell won a DSO and bar and the VC in quick succession, but was killed in February 1942. One can see why an old Harrovian (an education Gott shared with Churchill and Alexander) from a smart regiment, who was energetic, popular, relatively young but already a very experienced desert hand, appealed over Brooke's choice of Monty, a decade older, who was widely known only as an excellent trainer of troops, but had no desert experience.

So why was Brooke drawn to the outspoken, awkward Monty? Both were members of the great Anglo-Irish Protestant brotherhood. Apart from Brooke's professional admiration for Bernard's intellect, witnessed over the many times they had worked together at Staff College and elsewhere, and a very similar career path, Brooke may have felt he and Monty shared something else. In April 1925, Brooke's adored wife, Janey, was killed in a car crash when Brooke was at the wheel. He had swerved on a wet road and overturned their car; he escaped with several broken bones, but she died within days, leaving two young children. Thereafter Brooke immersed himself completely in work, concealing a sensitivity that only emerged when birdwatching or fishing – his methods of unwinding. Brooke remarried in 1929, but there were clearly parallels with Monty's life. Maybe Brooke saw in Monty a similar soul who buried his grief in the ruthless pursuit of professionalism.[21] Monty's isolation from his previous life was made almost complete when, in January 1941, a warehouse containing his and Betty's possessions – furniture, family heirlooms, photographs, Betty's paintings – was destroyed in an air raid on Portsmouth. Henceforth, Bernard would be a man without a past, focused only on his work. Thereafter he would devise and accumulate an alternative history – his caravans, portraits, books and archive – to replace that which had been lost.

It is clear that Gott was being talked up in London by his friend Anthony Eden, who held enormous sway over Churchill. Eden and Gott were the same age, had both served in the King's Royal Rifle Corps, and finished the previous war as Captains with an MC. In his memoirs, Eden makes it clear that he was in regular contact with Gott during the inter-war years, visiting him in Egypt in February 1940, and again in October 1940 in Libya, when Eden recalled in his diary: 'Gott has come on tremendously since he has been given his larger command and gave us a valuable appreciation of the Italian fighting value.'[22] Oliver Harvey, Eden's Principal Private Secretary, referred to him positively: 'AE's view is that Gott is the best general we've got and says that this is the opinion of the troops too' (22 July 1942); 'at least the PM seems to be moving in the direction of putting Gott in command' (30 July) and; 'AE has done his best to prime the PM about the generals in Libya and Gott in particular. He has warned him against the line that "Gott is so good but he is really rather tired", which is the line the military opposition may take up'[23] (31 July). Interestingly, Monty in his memoirs echoed *exactly* that objection, stating somewhat ungenerously that:

> it is now clear to me that the appointment of Gott to command the Eighth Army *at that moment* [Monty's emphasis] would have been a mistake. I had never met him; he was clearly a fine soldier and had done splendid work in the desert. But from all accounts he was completely worn out and needed a rest. He himself knew this. He said to a mutual friend: 'I am very tired. Also we have tried every club in the bag and have failed. A new brain is wanted out here on this job; it's an odd job but it needs a new brain. If they want me to do it I will try. But they ought to get someone else, a new man from England.'[24]

That mutual friend was Brooke, who recorded in his diary on 5 August 1942, a rather different version:

Motored south to Gott's HQ, and had a useful talk with him. There is no doubt that a rest home would do him a lot of good, and I do not feel that he would yet be ready to take over the Eighth Army. He requires more experience. However I do not know what opinion the PM formed of him and how much he will press for him instead of Monty.[25]

Indeed there is no convincing evidence that Gott was particularly tired, or stale, above the norm. What is clear is that Gott's appointment was controversial; it seems that politicians in London, led by Eden from his regiment, were pushing the able Gott, whilst Brooke seems to have genuinely felt it was the turn of his protégé, Monty, to have a good field command. He may also have been leaning on an historical precedent. Brooke would have been aware that the arrival of Edmund Allenby on 27 June 1917, to take over the Palestine theatre from the lackadaisical General Archibald Murray, had had an immediate effect and boosted the morale of the ordinary soldier in the Egyptian Expeditionary Force (the 1917 equivalent of Eighth Army). Allenby quickly won the respect of his men by making frequent visits to front-line troops (which Murray rarely did), and by moving GHQ from Cairo to Rafah, much nearer the front lines at Gaza. He organised the hitherto disparate forces into three corps, and supported the leadership of T. E. Lawrence over the Arabs. Allenby was a new face from the Western Front and brought fresh insights to the Desert War with incredibly successful results, rising to field marshal and a viscountcy.[26] Brooke may have been hoping Monty, appointed under similar circumstances in another desert campaign, would deliver in similar fashion, and was not disappointed.

Gott, however, was tired enough to want some leave in Cairo, and duly set off by air on 7 August. According to historian James Holland, who interviewed the pilot in 2006, Gott boarded a twin-engined Bristol Bombay during the afternoon at Burg el Arab airstrip, west of Alexandria for the brief flight to Cairo. A short way out, and flying tactically at fifty feet, the Bombay was 'bounced' by a flight of Messerschmitt 109s from JG–27 which set the engines on fire. Nineteen-year-old Sergeant Pilot 'Jimmy' James skilfully put the aircraft down and got the crew ready to disembark. The Messerschmitts continued to strafe the stricken aircraft. As flames spread from the engines, James exited through the cockpit hatch, but found that only four others had escaped, the rest being trapped because the rear hatch had shot forward and jammed closed. Badly burned (he would spend four months in hospital), James went for help, but on returning found that Gott and the remainder had indeed perished, trapped within the burning aircraft by an unlucky fluke of the jammed rear door – victims apparently of a mid-air assassination.[27]

Strangely, the route from Burg el Arab to Cairo was a routine one, flown most days, and had not been molested before. In later life, Jimmy James was in no doubt that this was an assassination, for in 2005 he met one of the Luftwaffe pilots, *Oberfeldwebel* Emil Clade, who revealed that his squadron had been congratulated on the killing as soon as they landed. There can be little doubt that this was a planned interception, but the target was Churchill, not Gott. German agents and Arab sympathisers in Cairo would have been well aware of Winston's presence – for he was ostentatious by habit, and not given to moving around furtively. Winston had flown exactly the same route on 5–6 August, and in June 1943 the Luftwaffe again shot down an unarmed aircraft over the Atlantic which they thought was carrying Churchill. It seems certain that Winston himself was the unwitting architect of Strafer Gott's untimely death.

On the evening of 7 August, Churchill was told that Gott had been killed. He then, after a little further prevarication, agreed to appoint Montgomery in Gott's place and Kenneth Anderson, who was replacing Monty at South East Command, was given First Army in Operation Torch instead. Monty got the call whilst shaving the next morning and dashed out to Cairo from England, arriving on 12 August, where he immediately met Auchinleck. Oliver Harvey recorded the reaction in London on 10 August 1942:

Today learnt the shocking news that Gott had been killed on Friday in an aeroplane crash. But this is something like a national disaster – our only first-class desert general killed like this, on the eve of his recognition and appointment. What frightful luck pursues us! ... It is now decided that Montgomery is to be the general to take his place. He has the reputation of being an able and ruthless soldier and an unspeakable cad.[28]

Poor old Gott is famous only for the circumstances of his death; it is now apparent that despite Monty's comments about him, he was an outstanding leader, whom history has unfairly overlooked. Monty claimed in his memoirs that at this stage Auchinleck was ready to retreat back to the Nile delta if Rommel attacked again, though Auchinleck always denied this. He did, however, stress the need to keep the army intact.[29]

Perhaps Auchinleck intended to resume the offensive, but his staff in Cairo continued to work out contingency plans for a withdrawal, should that prove necessary. Certainly Monty formed the impression there were too many troops – and staff – in Cairo, together with an air of pessimism. Although not scheduled to take command immediately, Monty in fact virtually appointed himself C-in-C Eighth Army on 13 August – two days early – forcing Auchinleck's premature departure.

One of the able officers Monty inherited that day was the unflappable Director of Military Intelligence, Middle East, Brigadier Francis de Guingand. Born in 1900, 'Freddie' as he was universally known, had been commissioned into the Middlesex Regiment (a background he shared with Brian Horrocks) in 1919. Major Montgomery had already met him in York, 1923–25. De Guingand was a staff officer par excellence, with a quick brain and innate diplomatic skills. Monty approved of him, and after the First Battle of El Alamein he was appointed Eighth Army's Brigadier General Staff (Operations), effectively Monty's Chief of Staff. Promoted major general after the capture of Tunisia in May 1943, Freddie was possibly at his most valuable to Monty in smoothing the ruffled feathers of coalition partners offended by his boss. Prior to the Normandy landings, he formed a useful friendship with Eisenhower's Chief of Staff, Walter Bedell Smith, and – at some personal cost – managed to keep Monty out of trouble (or rescue him) until May 1945. Monty would owe him his career at least once.

That evening Monty assembled his new HQ staff – some fifty to sixty officers in all – and issued a 'no withdrawal order' that subsequently became famous. In fact, Monty's arrival at the Eighth Army's headquarters appears very similar to Rommel's entry into North Africa back in February 1941, in his instant grip of the situation, speed of action and inclination to disobey orders he didn't like. Monty's inspiring address to his newly inherited command staff on 13 August 1942 is fêted by leadership gurus today, and illustrates the breezy confidence and energy he brought to his new command. At this stage he had no desert experience, and had last seen combat over two years earlier. Bernard bade his staff sit in the sand, whilst he addressed them from his caravan, beginning thus:

> You do not know me. I do not know you. But we have got to work together; therefore we must understand each other and we must have confidence in one another. I have only been here a few hours. But from

what I have seen and heard since I arrived, I am prepared to say, here and now, that I have confidence in you. We will then work together as a team, and together we will gain the confidence of this great army and go forward to final victory in Africa.[30]

*

Rommel's second attempt to break through at Alam Halfa, between 31 August and 5 September 1942 (essentially a Second Battle of Alamein), confirmed the growing material ascendancy of the Allies. It was also the first battle fought by Montgomery as Eighth Army commander.

It seems that, despite having denigrated his predecessors, Monty was happy to fight Alam Halfa on the defensive plan conceived by Auchinleck, his Chief of Staff Dorman-Smith and Gott (then commanding XIII Corps), drawn up before Monty's arrival. Rommel was blocked from breaking through by 7th Armoured Division, and under aerial attack and short of fuel, withdrew. Here, Montgomery's operational competence shone through, as did his sense of caution and reluctance to commit his forces unless conditions were overwhelmingly in his favour. Their attack blunted, the Afrika Korps withdrew to regroup. Montgomery had prepared to pursue the Germans but in the afternoon of 2 September he gave corps commander Brian Horrocks clear orders to allow the enemy to retire. This was for two reasons: to preserve his own strength and, secondly, to allow the enemy to observe, and be misled by, the dummy preparations for an attack in the area. Nevertheless, Montgomery was keen to inflict casualties on the enemy and orders were given for the as-yet-inexperienced 2nd New Zealand Division, positioned to the north of the retreating Axis forces, and 7th Armoured Division to attack on 3 September. The attack was repelled, however, by a fierce rearguard action by the 90th Light Division and Montgomery called off further action to avoid casualties.

On 5 September, Rommel was back where he had started, with only heavy losses to show for his efforts: 2,940 casualties, fifty tanks, a similar number of guns and perhaps worst of all 400 trucks, vital for supplies and movement. The British losses, except armour losses of sixty-eight, were much less, further adding to the numerical inferiority of *Panzerarmee Afrika*. Tedder's Desert Air Force inflicted the highest proportion of damage to Rommel's forces, who now realised the war in Africa was unwinnable without more air support, which was impossible since the Luftwaffe was already stretched to breaking point in Russia.

After the stalemate at El Alamein, Rommel hoped to go on the offensive again before massive amounts of men and materiel could reach the Eighth Army. Allied vessels from Malta were, however, intercepting his supplies at sea and the Desert Air Force kept up a relentless campaign against Axis supply bases in Tobruk, Bardia and Mersa Matruh. Most of the supplies reaching the Axis troops still had to be landed at Benghazi and Tripoli and the enormous distances supplies had to travel to reach the forward troops meant that a rapid resupply and reorganisation of the Axis army could not be undertaken. Further hampering Rommel's plans was the fact that the Italian divisions received logistical priority over his, with the Italian authorities shipping materiel for their formations at a much higher rate than for those of Rommel.

Montgomery, firmly in command of a confident force which had tasted victory, waited for his moment. At this juncture, Rommel appears to have surrendered the initiative. This was no doubt connected with his physical state; he exhibited a catalogue of medical ailments, resulting from exhaustion. On 24 August he wrote home to Lucie:

> I was unable to write again yesterday. I'm now well enough to get up occasionally. But I'll still have to go through with the six weeks treatment in Germany. My blood pressure must be got properly right again some time or other. One of the Führer's doctors is supposed to be on his way. I'm certainly not going to leave my post here until I can hand over to my deputy without worrying. It's not yet known who is coming [Guderian was asked but declined]. I'm having another examination today. It's some comfort to know that the damage can probably all be cleared up. At the rate we've been using up generals in Africa – five per division in eighteen months – it's no wonder that I need an overhaul some time or other.[31]

Two days later, Rommel's propaganda chief, Alfred Berndt, also wrote to Lucie:

> Dear Frau Rommel,
> You'll no doubt be surprised at hearing from me in Africa ... The reason for my letter is to inform you about the state of the Marshal's health. Your husband has now been nineteen months in Africa, which is longer than any other officer over forty has stood it so far, and, according to the doctors, an astonishing physical feat ... All this in the

nature of things has not failed to leave its mark, and thus, in addition to all the symptoms of a heavy cold and the digestive disturbances typical of Africa, he has recently shown signs of exhaustion which have caused great anxiety to all of us who were aware of it. True, there is no immediate danger, but unless he can get a thorough rest some time, he might easily suffer an overstrain which could leave some organic damage in its train . . . I ask you, Madam, not to worry. As for his personal safety . . . every one of us, officers and men, would be ready to die for the Marshal.[32]

After the war, Field Marshal Albert Kesselring claimed, startlingly, that Rommel 'had a nervous breakdown in Africa and was hospitalised. He was depressed mainly. At El Alamein he was not the Rommel he had been any more. From then on, it was too much for him.'[33] There is no doubt that Rommel's chaotic model of leadership, rushing here and there all over the battlefield, personally commanding tactical detachments and neglecting to employ the skills of his talented staff officers, contributed to his breakdown in health. He was fit and possessed of enormous energy, but even his stamina was not up to the enormous task of commanding a substantial army on operations in unforgiving terrain for an eighteen-month stint without leave, between February 1941 and August 1942.

Handing over command of his army to Georg Stumme (whom he had succeeded at 7th Panzer Division in 1940), Rommel flew home on 22 September. The blow was softened on the morning of 30 September 1942, when, with Berndt by his side, Rommel strode into Hitler's office in the Reich Chancellery to receive his field marshal's baton personally from the Führer. That evening he was honoured at a rally in the Berliner Sportpalast, where Hitler praised Rommel as a model of 'the next generation of revolutionary officers and generals', that is, those from lower-middle-class backgrounds – the antithesis of the Prussian military caste. With most of Berlin's Nazi elite assembled in Rommel's honour, the extensive newsreel footage of the event suggests it was similar in size and magnificence to a pre-war Nuremberg rally. Perhaps this exact moment was the high tide of the Third Reich – for the battle of Stalingrad had just commenced, and the opening of Alamein was days away. Never again would Germany be as militarily dominant. This was Rommel's fifth promotion since his promotion to *Generalmajor* in August 1939, and in achieving five-star rank, he had become 'an immortal'. With his rank came lifelong privileges, but also problems. In receiving these rewards, Rommel would excite undying envy, but more particularly, enmity, from many army colleagues.

On 31 March 1942, Josef Goebbels had written: 'He is a National Socialist; he is a troop leader with a gift for improvisation, personally courageous and extraordinarily inventive. These are the kinds of soldiers we need. Rommel is the coming Supreme Commander of the Army.'[34] But there were also growing numbers who felt that Rommel was becoming too powerful and needed putting in his place.

From the summer of 1942, Rommel had begun suffering catastrophic losses to the supply convoys sent across the Mediterranean from southern Italy, which impeded his progress, for he could not move without petrol. In August he lost seven out of thirty supply ships – much of the Allied success in ambushing convoys being down to Ultra intelligence gleaned via Bletchley Park. This was something Montgomery was never able to reveal; the Bletchley operation was only made public in 1974, at a stroke rendering most previous campaign histories obsolete.[35]

It was at the first Churchill–Roosevelt conference held in Washington DC (code name Arcadia), of 22 December 1941–14 January 1942, that an Allied 'Europe First' strategy was agreed. From this evolved the Anglo-American amphibious attack against Axis forces in the Mediterranean of November 1942: Operation Torch. Against this strategic background, throughout September and October 1942, Monty built up an over-whelming force of tanks, artillery and men. The new commander set about turning Eighth Army into a confident, capable, aggressive and successful force. This is where he really made his mark, in the nature of his preparation of the Eighth Army for El Alamein, for he was, truly, an unsurpassed trainer of men. Any of the others considered for Eighth Army or even as C-in-C Middle East – Gott, Alexander, even Brooke – could never have rehearsed the new, vastly increased Desert Force with its new equipment to the extent and in the timeframe that Monty did. Whatever his shortcomings as a general – slow in the pursuit, arrogant in victory – he had no peer as the British (and Commonwealth) Army's top trainer, beyond measure.

Monty was heavily reliant on air power. No one in RAF's Desert Air Force had more experience of providing close air support than Air Marshal Sir Arthur Coningham. Known to all as 'Mary' from his earliest flying days (this was a corruption of 'Maori', a reference to his upbringing in New Zealand), Coningham had flown with the RFC in the First World War, commanding a squadron at twenty-three and winning a DSO, DFC and MC. In July 1941, he joined Arthur Tedder, the newly appointed head of Middle East Air Command, as his deputy, to command the Desert Air Force. Thereafter, Tedder and Coningham would work together as a successful team for the rest of the war.

The Desert Air Force worked well enough against the Italians, but the arrival of Rommel meant that the superior aircraft of the Luftwaffe posed a challenge to the RAF's less-advanced machines. Coningham's problems were as much logistical as tactical, with shortages of everything, a higher incidence of sickness than in the UK, challenges of weather and climate and a poor record of cooperation with the Army. Recognition was shaky on both sides: the RAF lost many aircraft to friendly ground fire, whilst pilots mistakenly shot up Allied troops and vehicles. At this time there was no joint doctrine for air–land operations: the Army called the concept 'air support' and the RAF called it 'army cooperation'. The Army and RAF forward headquarters were five miles apart; there was no joint force HQ or joint task force commander. There was no formal establishment; repair and maintenance units had little mobility, which meant that the quality and quantity of the aircraft was fairly irrelevant if they could not be serviced. To add to the challenges, most days and nights the Luftwaffe would shoot up the airfields. Everything worked in a haphazard way, with several units out of the chain of command altogether. Gradually 'Mary' identified these problems and corrected them. By 1943 in Tunisia, for example, Coningham perfected the tactics learned in Egypt and Libya that would be applied in Normandy: that every fighter should also be a fighter-bomber, based not more than twenty miles from the front on forward airstrips. Prior to D-Day, Coningham ensured, in desert fashion, that his aircrew knew how to refuel and rearm their own aircraft, remedy minor mechanical defects and use personal weapons. Ground crews were rehearsed in setting up temporary airstrips within hours and moving them just as quickly to new locations.

In early 1942, Coningham placed his desert command caravan adjacent to Alan Cunningham's, then Eighth Army commander, and requested that both sets of senior officers share the same mess. This boosted confidence as well as generating true cooperation. Although the practice has often been ascribed to Montgomery's arrival in August, it had been Coningham's initiative, several months earlier. But it *was* Monty's arrival in August that changed the nature and pace of air–land cooperation completely. Monty immediately appreciated the importance of air power on the modern battlefield; he had witnessed its successful application by the Germans in the 1940 campaign, and had witnessed the German aerial onslaught during the Battle of Britain.

The cooperation during these first months was profound. Monty and 'Mary' met each evening, when the former explained what he wanted to do, how he expected the enemy to react, and the latter volunteered how he could support. In this way each day's air plan was worked out and the following morning the two would meet again.[36] Crucial to this success was personality, and the recognition that the Desert Air Force was an independent command, not a supporting arm of the army. Monty co-operated with Mary as an equal; Coningham had no qualms about turning down missions he thought were inappropriate or impractical. His basic doctrine was that the strength of air power lay in its flexibility and capacity for rapid concentration. Therefore it followed that control must be concentrated under command of an airman, and therefore air forces must be concentrated in use and not dispersed in penny packets: air–land coordination should be by liaison, not command.

In December 1944, Monty penned 'Some Notes on the Use of Air Power in Support of Land Operations', which distilled all the doctrinal wisdom he had gained from his campaigns, though built chiefly on his Western Desert experience. These are his opening thoughts, which demonstrate how quickly military thinking had matured since 1940:

> Any officer who aspires to hold high command in war must understand clearly certain principles regarding the use of air power . . . The greatest asset of air power is its flexibility. Whereas to shift the weight of effort on the ground from one point to another takes time, the flexibility inherent in Air Forces permits them without change of base to be switched quickly from one objective to another . . . It follows that control of the available air power must be centralised and command must be exercised through Air Force channels. Nothing could be more fatal to successful results than to dissipate the air resources into small packets placed under command of Army formation commanders.[37]

By the autumn of 1942, Coningham had achieved air superiority for the first time, with 420 fighters (including 50 Spitfires) and 120 bombers, enabling Monty to assemble the ground resources he needed and take the offensive. The real challenge for Coningham came during the pursuit, with his squadrons constantly leapfrogging forward to stay within range of the retreating Afrika Korps. Many Axis prisoners referred to the constant air harassment, and the Luftwaffe were notably absent from the skies in October–November 1942, destroying their own aircraft on the ground as they retreated.

But Tedder and Coningham were by now critical of Monty: they accused him of failing to exploit the opportunities offered by Rommel's rapid retreat; Monty's insistence on grabbing the media limelight and omitting to acknowledge the RAF's contribution began to rankle, too. Coningham saw a growing gulf between Monty's deeds on the battle-field and the glowing reports he was sending back home; shortcomings in close air support began to be cited as reasons for failure on the ground (an issue that would come to a head after Operation Goodwood's failure in July 1944). After Tunisia, Monty and Coningham met only occa-sionally. The close working relationship that was so necessary had gone; for this Monty must bear most of the blame.*

In what would become known as the Second Battle of El Alamein, Montgomery anticipated a twelve-day clash, with an attack on the Afrika Korps executed in three phases. To assist him, Monty took a leaf out of Rommel's book and capitalised on the skills of 'A' Force. The outfit had been set up by Wavell in January 1941, and was dedicated to counter-intelligence, camouflage and deception. Its guiding force was a master of illusion, forty-two-year-old Jasper Maskelyne, a music hall conjurer who was serving as a major in the Royal Engineers. Maskelyne and his 'Magic Gang' had already managed to 'conceal' Alexandria harbour and misdirect German night bombers to attack an adjacent bay, where he had recreated the night lights of Alexandria with fake buildings, a light-house and anti-aircraft guns. Deploying considerable skills of illusion in Operation Bertram, prior to Alamein, Maskelyne and 'A' Force, in the northern half of Monty's front, disguised a thousand tanks as trucks with the aid of painted canvas screens on frames. Further south, they created over 2,000 fake tanks out of painted plywood and hessian, supported by illusory artillery pieces, thus deceiving Rommel into

---

* Coningham would retire in 1947 (not having attended a Staff College and with Monty now as an enemy, he was unlikely to make four-star rank). In January 1948 he set out by air across the Atlantic, but the Star Tiger, an Avro Tudor in which he was a passenger, disappeared without trace over what is now known as the Bermuda Triangle.

thinking Monty's main effort would be mounted from the south. Maskelyne's artificial armies were supported by a fake railway line, and a decoy, but incomplete, water pipeline, made out of thousands of tin cans. The Germans assumed when the pipeline was complete then Monty would attack. Every aspect of the illusion worked.[38]

Alamein was the main effort for Britain when it was launched at 9.40 p.m. on 23 October 1942. Some 822 medium and field guns opened an ear-splitting barrage in unison that lasted until each barrel had fired 600 rounds, some five and a half hours later. The Afrika Korps, in contrast, commanded less of Hitler's attention; his focus was firmly on the Russian city on the banks of the Volga that bore the name of his adversary – Stalingrad. In the first stage of Monty's attack, Operation Lightfoot, the break-in, XXX Corps (under Oliver Leese) attacked the Axis defence in its centre, heavily fortified and defended by minefields. Meanwhile, XIII Corps (under Brian Horrocks) attacked in the south, but neither corps was able to break through to exploit the more open country to the rear of the Axis position.

Although Rommel, who had been out of theatre for medical reasons, left for Africa immediately he heard of the Allied attack, it took him two vital days to reach his forward headquarters. Through no fault of his own, he was absent when firm, decisive action was needed; not that he could have challenged Monty's huge logistical back-up. The German defensive plan at El Alamein was more static in nature than Rommel preferred, but his hand was forced by too few motorised units and a shortage of petrol. The defensive line had strong fortifications and was protected with a large minefield, which in turn was covered with machine guns and artillery. This, Rommel hoped, would allow his infantry to hold the line at any point until motorised and armoured units in reserve could move up and counterattack any Allied breaches.

On 1 October, *Panzerarmee Afrika* had been retitled *Deutsch-Italienische Panzerarmee* – a political nod to the fact that the majority of Axis troops in North Africa were, indeed, Italian. No one, of course, expected that Rommel's temporary replacement, General Georg Stumme, would die of a heart attack during the first wave of the fighting. This paralysed the German command functions until Ritter von Thoma took over. Thoma, in command of the subordinate DAK since 17 September, was, in turn, captured just as Rommel returned to the desert. Thus, Rommel was not only confronted with a disrupted command chain, but more seriously his stocks of fuel, already critical when he had left in September, were now disastrously low, whilst counterattacks by the 15th and 21st Panzer Divisions on 24–25 October resulted in heavy tank losses due to the intensity of British artillery and air attacks. Rommel opted for a further counterattack in full force, to throw the British out of his defensive lines, as the only chance of avoiding defeat. The counterattack was launched early on 26 October but the British units that had penetrated his lines held fast on Kidney Ridge. The Allies continued pushing hard with armoured units to force a breakthrough, but defensive fire destroyed many British tanks, leading to doubts among the officers in the British armoured brigades about the chances of clearing a breach.

Then Montgomery's second phase, christened the 'Dog Fight', began in the midst of the Axis position. Between 26–31 October, the Axis fortifications were steadily, and characteristically for Montgomery, reduced by attrition ('crumbling' was his term for this casualty-heavy process); Axis counterattacks were repulsed with the use of air power. The final phase, the breakout, took place between 1–4 November. In Operation Supercharge, the reinforced New Zealand Division drilled through the weakened Axis defensive position, making it possible for X Corps, which had been in reserve, to break out into the Axis rear and achieve a penetration of Rommel's line. There was an immediate counterattack by the remaining panzers in an attempt to encircle the pocket on 2 November, but heavy Allied fire stopped every blow. By this time his *Panzerarmee* had only one third of its initial strength remaining, with only thirty-five tanks operational, and virtually no fuel or ammunition. The British were in complete command of the air, yet their armour had been fought to a standstill, having taken murderous losses with some armoured brigades reporting losses of seventy-five percent.

On 3 November, Montgomery found it impossible to renew his attack, and had to wait for more reinforcements to be brought up. This lull was what Rommel needed for his withdrawal, which had been planned since 29 October, when he assessed his own prospects as being hopeless. But at midday Rommel received a personal missive from Hitler:

The German people and I are following the heroic struggle in Egypt. In the situation which you find yourself, there can be no other thought but to stand fast, yield not a yard of ground and throw every gun and every man into the battle . . . It would not be the first time in history that a strong will has triumphed over the bigger battalions. As to your

troops, you can show them no other road than that to victory or death.
Adolf Hitler.

The instruction demanded the impossible, and virtually ensured the destruction of *Panzerarmee Afrika*. It also served to destroy Rommel's faith in his Führer, with whom he believed he had developed a special understanding.

Rommel had not grasped that Hitler was now preoccupied exclusively with the struggle for Stalingrad, and the latter's 'victory or death' order betrayed a complete unwillingness to understand the situation, or the nature of desert warfare. For the first time, Rommel realised the importance of friends in Berlin (he had only one – his former 'minder' Rudolf Schmundt) and how the General Staff at OKH was ranged against him. Even though he was a field marshal and had enjoyed special access to Hitler, he was now powerless. He began his retreat, against orders, on 4 November. Thereafter his withdrawal westwards was swift, though Montgomery was slow to exploit the pursuit. Nevertheless, by 17 December, the Eighth Army had reached El Agheila, from whence Rommel had set out twenty-one months previously, in March 1941. The Axis supply base of Tripoli fell on 23 January 1943, and although the port was partially wrecked, it started taking shipping within a week.

On hearing the news, Hermione, Countess of Ranfurly, working for SOE in Cairo, confided in her diary:

> Memories slid through my head like a newsreel . . . I thought of General Wavell and how tragic it was that Mr Churchill could never like or understand him – what a combination they would have been; of Auchinleck, so handsome and charming but not quite able to cope with the intricate Desert war; and of Monty whom few of us like or admire – who did not go into action until he was reinforced beyond Wavell's wildest dreams. And I thought of Rommel who, in spite of being our enemy, gained our admiration and respect – almost our affection.[39]

*

The Torch landings at Casablanca, Oran and Algiers in French North Africa to Rommel's rear (which followed Montgomery's victory by a mere four days) altered completely the nature of the campaign. It is possible, therefore, to argue that Second Alamein was 'strategically unnecessary', as Operation Torch would have forced Rommel to withdraw in any case, and that Alamein was fought therefore for political reasons. As we have seen, Churchill desperately needed a victory after a year of setbacks, which had culminated in by-election defeats, the fall of Singapore, the loss of the *Prince of Wales* and the *Repulse*, and the tragic error of the Dieppe raid; Malta just about hanging on and there were prospects of further bad news from the Eastern Front and the Pacific. Churchill could also see that Britain's war effort was about to be subsumed in the larger Anglo-American campaign, and he wanted a final British *hurrah!* before that happened. In fact the Anglo-American effort had already resulted in the delivery of Sherman tanks to Eighth Army which helped tip the balance at Alamein. Monty's 13,560 casualties, though only a fraction of the 195,000 men he had assembled (to Rommel's 116,000), would weigh heavily on his mind ever after, up until his own death in 1976. He would take his post-war returns to the battlefield very seriously indeed.

With Eisenhower as theatre commander, Admiral Sir Bertram Ramsay (responsible for the 'miracle' of the Dunkirk evacuation) oversaw the naval side of Torch, beginning on 8 November 1942. He was Deputy Naval Commander to his exact contemporary Admiral Sir Andrew Cunningham, brother of the ill-fated Alan, Eighth Army's first commander. Ramsay – one of Britain's great admirals, but an unsung hero who died in a plane crash in 1945 – served as deputy to Cunningham at his own request, for though senior he had already retired. Bertram Ramsay would later oversee every naval aspect of Operation Overlord in 1944, and thus twice contributed to great moments in Britain's war – Dunkirk and D-Day. Torch and Husky (the invasion of Sicily) were his stepping stones of expertise between the two.

The scale of troops and vessels required for Torch had not been seen before, even during the First World War: 70,000 British and US troops of Kenneth Anderson's First Army and their supplies had to be landed from 350 transport and landing ships, supported by 200 warships, all sailing directly across the Atlantic and assaulting three different areas. There was some fighting with the Vichy French before Darlan, a senior minister in Pétain's Vichy government, who happened to be in Algiers, ordered a ceasefire. Having secured Oran, Algiers and Casablanca, the Allied First Army advanced eastwards, taking Bone on 12 November before being stopped at Mejez el Bab, just thirty miles south-west of Tunis.

Hitler had reacted swiftly to the invasion, sending his troops into the

Unoccupied Zone of France and reinforcing his troops in North Africa with what was to become Hans-Jürgen von Arnim's Fifth Panzer Army. Amongst the three extra divisions rushed to Tunisia in the aftermath of Torch was 10th Panzer Division. On 11 February 1943, *Oberstleutnant* Claus Schenk, Graf von Stauffenberg, a talented staff-trained officer, arrived in Tunis as the division's Chief of Staff. The thirty-six-year-old colonel, who shared Rommel's birthday of 15 November, first conferred with the Desert Fox on 19 February and was instrumental in launching the attacks at the Kasserine Pass (19–25 February), but was wounded on 7 April in an Allied air attack on his HQ, which cost him his right hand, two fingers of his left hand and the sight in his left eye.[40] Invalided back to the Fatherland and eventually posted to Berlin, Stauffenberg would also be instrumental in launching, on 20 July 1944, the coup that would cost both him and ultimately Rommel, their lives.

Meanwhile, Arnim, anxious to prevent the Allies from cutting him off from Rommel, who was withdrawing before Montgomery, stopped First Army in hard-fought battles at Tebourba and Longstop Hill, and then counterattacked in January 1943, knocking it off balance. When Rommel, who had now fallen back into southern Tunisia, first drew American blood in the Kasserine Pass with great success, the Allies were even more badly rattled. Eisenhower was preoccupied by political concerns and lacked relevant experience, and many of his troops and their commanders were green. Alexander was appointed to command an army group consisting of First and Eighth Armies in order to improve coordination. It was fortunate for the Allies that the Axis forces had problems of their own. Rommel was unhappily under the authority of the Italian *Comando Supremo*, while Arnim was directed by Kesselring, the German C-in-C South-West. On 23 February, Rommel's *Deutsch-Italienische Panzerarmee* was retitled *Heeresgruppe Afrika* (Army Group Africa), acknowledging the arrival of even more German troops – which would have made all the difference before Alamein but by early 1943 would simply be overwhelmed by Allied numerical superiority. Rommel was still dangerous, and planned to use all three of his armoured divisions against Montgomery, who was now approaching the Mareth Line, a pre-war French defensive system which had been designed to prevent the Italians moving from Libya into south-eastern Tunisia.

What proved to be Rommel's last offensive in North Africa was initiated on 6 March 1943, when he attacked Eighth Army at the Battle of Medenine (Operation Capri), employing 10th, 15th, and 21st Panzer

Divisions. By now, his oldest desert warriors, 5th Light Division, had been re-equipped and retitled 21st Panzer. Pre-warned by Ultra intercepts, Montgomery deployed large numbers of anti-tank guns in the path of the offensive. After losing fifty-two tanks, Rommel called off the assault, attributing his failure to leaks by senior Italians on his staff. He then asked Berlin for permission to withdraw to a more tenable position – his command now numbered 346,000 men, and required more logistical effort than the Axis could manage. On receiving Hitler's refusal, something inside Rommel seems to have snapped; his health (already poor, hence his absence at the beginning of Alamein) had not improved and the will to carry on appears to have just evaporated.

Rommel had been retreating continuously since Alamein, four months earlier, and with his back to the coast, squeezed between Anderson's First British Army with an American corps to the west, and his old foe, Monty, to his south, the Desert Fox could foresee the inevitable result of Allied logistics – if not necessarily skill – applied on a vast scale. The Axis were about to lose in North Africa, and no amount of cunning could reverse this. Rommel thereupon embraced the inevitable and decided to retire gracefully on medical grounds. Accordingly, on 9 March, he handed over command of *Armeegruppe Afrika* to General Hans-Jürgen von Arnim and flew out of Sfax, bound for Rome. Whatever his intentions may have been, Erwin Rommel would never stride the sands of North Africa again.

Although Montgomery's frontal attack on the Mareth Line failed, he outflanked it from the south, and the Italian commander, Messe, fell back on Wadi Akarit. By now Arnim's position was worsening daily as the Allied sea and air blockade throttled him. While Eighth Army took Wadi Akarit and advanced northwards to Sousse and Enfidaville, First Army fought its way towards Bizerta and Tunis. Arnim's men fought hard to the end, but Bizerta and Tunis were both captured on 7 May and the last Axis troops surrendered on the 13 May. The Allies took 238,000 prisoners, and had at last won the campaign in North Africa. Monty had triumphed by thorough preparation, innovative deception, superiority in resources of all kinds (including the brand-new Sherman tank), Coningham's air superiority, grim determination and self-belief. After the war, Ritter von Thoma would suggest, somewhat sourly, to the historian Basil Liddell Hart that Monty won his battles because 'in modern mobile warfare the tactics are not the main thing. The decisive factor is the organisation of one's resources to maintain the momentum'.[41]

Though some historians have seen the campaign as a strategic irrelevance for both sides, given the importance of Egypt and the Suez Canal to Britain it is hard to see how the war in the Western Desert could have been averted. If El Alamein, the Torch landings and the advance into Tunisia did not contribute directly to the Allies' main strategic goal – the invasion of north-west Europe – then they provided the Allies, and Monty especially, with invaluable experience. Certainly, given their showing in the Tunisian battles over the winter of 1942–43 it is difficult to avoid the conclusion that the Allies were not yet ready for a cross-Channel invasion. However, the theatre was a low strategic priority for both sides. The British would have fared better had they not diverted troops to Greece and the Far East back in 1941, while far more Germans surrendered in Tunisia than Rommel ever commanded in the Western Desert.

Looking back on Monty's success – for the North African victory was largely ascribed to his generalship – the aspect that stands out the most was his astonishing ability to change an organisation, the Eighth Army, with low morale and haunted by defeat, into an army that took on and beat its much-vaunted foe just two months later.

Monty's personal energy and enthusiasm were largely responsible for altering the Eighth Army's view of itself. He did this with a mixture of personality, and props. The most iconic prop of all – Monty's beret – did not arrive on his own head by accident. Historian Stephen Bungay has related how Bernard realised it was important for his desert warriors (in Monty's eyes, thinking civilians who read newspapers) to be able to recognise him instantly, and not just as a mediaeval overlord, but also as a celebrity, or a mascot – a conscious reaction to the distance he felt surrounded Haig and other senior officers in the First World War. The headgear idea was devised by three men: Captain Warwick Charlton, the editor of the desert newspapers *Eighth Army News* and *Crusader*; Geoffrey Keating, the Head of the Army Film and Photographic Unit (AFPU); and Montgomery's ADC, John Poston. All three had been in the desert for some time and saw the effect of Rommel's cap, goggles, scarf and greatcoat on the soldiers of both nations: instant recognition, which bred infectious inspiration. They hoped to 'use' their new general, a slight man, with a pointy nose, high-pitched voice and white knees, to counter the personality of the Desert Fox.

First, they tried a broad-brimmed slouch hat, which had been presented to Monty when he visited the 9th Australian Division at the Tel el Eisa ridge near Alamein on his first full day in command of the Eighth Army. It soon became festooned with badges, presented to him by each unit he visited, and frankly looked a bit silly when topping his short frame. The hat, in any case, emphasised one nation (Australia), instead of an all-encompassing coalition. Their next attempt was a couple of days before the start of El Alamein, when they still had no iconic image. This time, they tried the black beret of the Royal Tank Regiment – at that time the only unit to wear a beret in the British and Empire forces. With the application of a general's insignia next to the striking RTR badge (depicting a tank, symbolising modernity), and when placed in the turret of his Grant command tank, the picture they were after suddenly came alive. Monty posed with and without binoculars; Keating's shutter snapped away, and soon this image became the stock photo of the victor at Alamein that photo editors around the world needed.

In perhaps the best-known image, apparently filmed under shellfire, Monty has his field glasses at the ready, whilst his ADC John Poston is looking through his, in the background. Monty's new look was reinforced by a press call on the morning of 5 November, when von Thoma, captured the night before, was photographed being entertained by Monty to dinner. Thoma was the first German general to be captured, whilst Bernard's casual dress, grey pullover and the beret, also caused a stir. Monty's beret featured in a *Life Magazine* spread of 5 April 1943 (which claimed 'British Tank Corps beret worn by General Montgomery inspires a popular wartime fashion'). The beret also made it, sitting on Monty's head, onto the front cover of *Time Magazine*, as a surprisingly late counter to Rommel's appearance in the same publication, on 1 February 1943. In half-shadow, Monty sports an enormous beret (incorrectly sporting a single Tank Regiment badge), with a row of US-made Grant tanks firing a barrage in the background. The beret and its owner appeared again on *Time*'s front cover (this time, with the correct pair of badges) on 10 July 1944 in the midst of the Normandy campaign: all proof that the instincts of the AFPU, Charlton, Keating and Poston, were spot on.[42]

The Grant was driven by twenty-two-year-old Private Jim Fraser of the Royal Tank Regiment, who recalled first setting eyes on Montgomery:

I wasn't impressed; to me he didn't look like a general. He was small in stature, thin with hawk-like features and a high-pitched voice. He was wearing a pair of shorts that were far too big and an Aussie bush hat covered by regimental cap badges.[43]

The tank, of course, was supposed to be incognito, unrecognisable as the army commander's 'battle wagon'. However, Monty had his own name painted on the front, his general's flag attached to the wireless aerial and with his head stuck out of the turret and pennant flying, he would tour front-line formations. When visiting his troops, Monty's routine was invariably the same and Fraser observed:

Although Monty was a non-smoker, a large quantity of cigarettes were always carried in the tank. On the move up the line when he spotted a group of squaddies the tank would be brought to a halt and Monty would get out to speak to them. His usual line of approach was 'Good morning, do you know who I am?' Most of the time there was no response to his question. Monty would then pass the fags round telling them he was Monty, their new commander and that we were going to kick this chap Rommel out of the desert for good. The look on the lads' faces was a picture.[44]

Behind the lines, although Monty used the ubiquitous American Jeep, he soon adopted another signature vehicle – 'Old Faithful' a Humber Snipe 4x2 four-seater open tourer, built in Coventry. Camouflaged, with an RAF roundel painted on the bonnet, 'Old Faithful' carried him right through the Western Desert and on to Sicily and Italy until his return to England, when Monty left the car to his successor, Oliver Leese. The Humber formula worked so well that Monty then took a second suitably camouflaged Snipe, christened 'Victory', through France, Belgium, Holland and into Germany.

Whether on his tank or in his Humbers, as his troops crowded round, Fraser observed Monty's opening shot was usually:

'I'm Monty, your new commander. I've studied your track record, there's nothing wrong with you. You are a fine body of men; the problem in the past has been leadership; that will now end. Together I will lead you to victory; there will be no more bellyaching.' The lads loved it . . . One could feel the confidence of the troops getting stronger, they were told what was going to happen and when it was going to happen. I must admit that I felt dead, dead chuffed when driving round the forward unit positions with the lads cheering and shouting 'Good old Monty!'[45]

Although we now know that political pressure, logistics and Ultra intercepts were partly behind the victory at El Alamein, Monty's force of character, stage management and sheer professionalism altered the balance in North Africa overnight. Upon reflection, perhaps, this could only have been achieved by an outsider, fresh from England. In this context, it is difficult to see how an old desert hand, such as Wavell, Auchinleck or even Gott, might have brought about such a fundamental change of fortune within the narrowest of timeframes. Monty's triumph was recognised by an almost immediate knighthood on 11 November 1942 and promotion to full general at the same time – an admirable counter to Rommel's elevation after Tobruk on 22 June.

During the lull before the Sicilian invasion, King George VI flew out to visit his troops overseas, for the first time since 1939, spending two weeks (11–25 June), visiting North Africa and Malta. During a four-day stay in Tripoli, His Majesty knighted Monty in person on 19 June 1943. Monty's ADC, Johnny Henderson, recorded that King George was put up in one of the caravans:

each evening after dinner Monty, as was his wont, would go to bed just after 9 p.m. The King certainly enjoyed sitting up with us after Monty had retired and, as the night drew on, he stammered less and less. There was no doubt that he appreciated the relaxed atmosphere of the TAC HQ.[46]

Significantly, Hitler never seemed to travel as widely as the King or Churchill, overlooking the tonic this could provide his armies, tending to summon commanders to *his* presence in Berlin or Berchtesgaden. When Monty's troops were concentrated in Tunisia for the invasion of Sicily in good weather during the summer of 1943, he launched on a series of much-reported and highly successful pep talks, which fostered the image of being close to his troops. Monty had realised before Alamein that the majority of his men weren't soldiers, but civilians in uniform, and treated them as such. He believed that, even more than their predecessors in 1914–18, he owed them sight of the 'big picture'. Contrary to myth, these talks did not occur much before Alamein; they were simply not possible with the widely dispersed Eight Army. But these informal, highly stage-managed talks were extensively used before the invasions of Sicily, Italy and Normandy and the Rhine crossings – to soldiers and civilians alike. Lieutenant Jack Swaab, a forward observation officer with 51st Highland Division, wrote of Monty addressing his unit on 26 September 1943 after the capture of Sicily:

> Monty was tremendous today. He's a grand little man with an enormous personality. The troops cheered him furiously at intervals . . . The great affection felt for Monty by everybody is a wonderful thing.[47]

*

Technically, the naval aspect of Operation Husky, the Allied invasion of Sicily in July 1943 – again, in the safe hands of Admiral Bertram Ramsay – may have equalled the initial undertaking in Normandy; the invasion involved 1,350 Allied warships, including six battleships, and about 1,850 landing and other craft, escorted by 4,900 aircraft. Commencing on the night of 9–10 July, they deposited 66,000 US and 115,000 British and Canadian troops ashore, on twenty-six separate beaches, whilst 5,000 deployed by air along a 105-mile front. It was certainly an operation from which many valuable lessons were learned. Husky employed many of the innovations that were to feature in June 1944: amphibious six-wheeled trucks (DUKWs) to ferry supplies ashore, and the use of thousands of specialised landing craft that sailed with combat loads from Tunisia (for Torch, all troops had come by large merchant ships and were ferried ashore by smaller vessels).

Though there had been some Vichy-French resistance against Torch, the Sicily landings were the first to meet determined opposition. Ramsay also had to make the final recommendation to go, given the weather and sea state (which were poor) – anticipating his role in June 1944. In Sicily, Monty's Eighth Army employed two army corps, XIII under Dempsey (whom he would take to Normandy) and XXX under Oliver Leese: both would soon command armies of their own. They were accompanied by Patton's US Seventh Army, including Bradley's II Corps. The jigsaw pieces that would be finally assembled for the Normandy campaign were now rapidly coming together: figures like Eisenhower, Tedder, Ramsay, Dempsey and Coningham, Monty's headquarters staff, his command style.

Eisenhower (apart from the brief, cigarette-lighting encounter of

May 1942) first really got to know Monty whilst planning the Sicily invasion in March 1943, during which Ike confided his impressions in a letter
to George C. Marshall in Washington:

> Montgomery is of different calibre from some of the outstanding British
> leaders you have met. He is unquestioningly able, but very conceited.
> For your most secret and confidential information, I will give you my
> opinion, which is that he is so proud of his successes to date that he
> will never willingly make a single move until he is absolutely certain of
> success – in other words, until he has concentrated enough resources
> so that anybody could practically guarantee the outcome. This may be
> somewhat unfair to him, but it is the definite impression I received.
> Unquestionably he is an able tactician and organiser and, provided only
> that Alexander will never let him forget for one second who is the boss,
> he should deliver in good style.[48]

Rommel took no part in the campaign for Sicily, which took thirty-eight
days, during which Monty managed to fall out with Patton. After his
victories at Alamein, in Tunisia and now Sicily, Monty had started to
regard himself as the greatest fighting commander alive, unable to take
criticism from his seniors (the only one he respected was Brooke) or
contemporaries, and ready to lay down the military law to everyone else.
According to one loyal 'servant', (later Brigadier Sir) Edgar 'Bill' Williams,
Monty was:

> most awkward to serve alongside, impossible to serve over, he was an
> excellent man to serve under, especially on his staff . . . He was indul-
> gent to the young aides-de-camp and liaison officers whom he picked,
> trusted and trained. His sense of fairness and especially of truth were
> not as other men's. In his convictions he was ruthless, even baleful, yet
> his insensitively arrogant self-confidence was combined with an indis-
> creet, mischievous, schoolboyish sense of humour which buoyed up the
> spirits of those far from home.[49]

Williams was one of a growing collection of experts and brilliant organisers Monty had assembled around him in North Africa and later took
to Sicily and Normandy. An Oxford don commissioned into the King's
Dragoon Guards, he gravitated towards military intelligence jobs, for
which his quick mind was ideally suited. When de Guingand was

appointed Auchinleck's Director of Military Intelligence he promptly obtained the highly recommended Williams, a junior captain, for his intelligence staff. In August 1942, Montgomery inherited both de Guingand and Williams when he took over Eighth Army. Williams then served through North Africa, Sicily, and Italy, eventually becoming brigadier and chief intelligence officer in 21st Army Group, which invaded France and continued to Berlin. Monty was full of praise for Williams in his memoirs:

> He was the main source of inspiration; intellectually he was far super-ior to myself or anyone on my staff but he never gave that impression. He saw the enemy picture whole and true; he could sift a mass of detailed information and deduce the right answer . . . As time went on he got to know how I worked; he would tell me in ten minutes exactly what I wanted to know, leaving out what he knew I did not want to know.[50]

Although Williams was under no illusions about Montgomery's character (he referred to his 'peacock vanity'), they became close friends, probably because Monty regarded him as an equal, but in another sphere. Monty, however, tended to look down on someone of the calibre of Freddie de Guingand, his long-suffering Chief of Staff, or other subordinate regular officers. Williams also observed Monty's sense of humour and fads, such as appropriating captured enemy stationery and using it, complete with foreign language letterheads, eagles and other insignia, for his correspondence with the War Office – part schoolboy jape and part a throwback to his austere upbringing perhaps. As well as dealing with the demanding and temperamental Montgomery, the Ultra-cleared Williams was required to be a first-class actor; the quality of the intelligence from Bletchley Park was often so good that he would be questioned about its source and had to give plausible replies to officers superior in rank.

Outside the milieu of his headquarters, and illustrative of Monty's continuing inability to empathise with the needs of Allies, in July 1943 the Canadian C-in-C Andrew MacNaughton flew to Sicily to visit his troops, but Monty banned him from landing, saying he was too busy fighting the campaign to see him. Legend has it that Monty threatened his arrest if he set foot on the island; the livid Canadian returned to London and harangued Brooke about Monty's behaviour. The net result when the Canadian government heard was to block the despatch of a further Canadian division to the Mediterranean theatre.

Monty's views of his American allies were equally inflamatory; his dangerously naive thoughts on Ike were penned to Brooke on 4 April 1943:

> Eisenhower came and stayed a night with me on 31 March. He is a very nice chap, I should say probably quite good on the political side. But I can also say, quite definitely, that he knows nothing whatever about how to make war or to fight battles; he should be kept away from all that business if we want to win this war. The American Army will never be any good until we can teach the Generals their stuff.[51]

Monty's comments here betray his failure to appreciate a higher level of the direction of war, something the Germans (originating with Clausewitz) had long understood and taught.

After the Sicily campaign ended on 17 August, operations inevitably shifted onto the Italian mainland, with Dempsey's XIII Corps of Britons and Canadians departing from Messina and landing near the tip of Calabria – Italy's toe – on 3 September (Operation Baytown). Six days later, in Operation Slapstick, British 1st Airborne Division landed by sea at Taranto, and General Mark Clark's US Fifth Army also landed at Salerno on the same day (Operation Avalanche). After a tough fight breaking out of the Salerno beachhead, Clark's Fifth Army advanced up the west coast of Italy, whilst Monty's Eighth followed the Adriatic coast.

Rommel, meanwhile, had been posted to defend northern Italy, with an embryo headquarters staff, Army Group 'B'. The Desert Fox advised that defending Italy south of Rome should not be a strategic priority but defensive preparations should begin along the natural barrier of the Italo-Austrian Alps, also in line with Allied expectations. As a result Kesselring, who was defending southern Italy, had been forbidden to call upon reserves from the Northern Army Group. Eventually, the success of Heinrich von Vietinghoff's Tenth Army in inflicting heavy casualties on the Salerno beachhead and nearly destroying it and Kesselring's strategic arguments persuaded Hitler to alter his strategy and keep the Allies away from German borders by defending forward.* Rommel interpreted the two opposing strategies in personal terms: this was a battle

---

* Such was the nature of the close combat at the Salerno beachhead that General Mark Clark would personally win the Distinguished Service Cross, the second highest US award for bravery.

between himself and Kesselring for Hitler's trust. However on 6 November Rommel and his headquarters were posted to northern France to report on the state of the Atlantic Wall – Hitler's coastal defences in northern Europe – leaving Kesselring in command of the whole of Italy with orders to keep Rome in German hands for as long as possible. Monty and his Eighth Army spent the autumn crawling up Italy's eastern coast. However Montgomery's eyes were largely elsewhere – focused on who would lead the invasion of northern Europe.

The campaign in North Africa was similar in the demands it made of both sides: each had been obliged to prolong their campaigns in the desert unintentionally due to the demands of other theatres which were considered more important by their political masters. Thus the British diverted forces to the Far East and Greece at crucial moments, as did the Germans to Russia. For both sides also the campaign had stressed the tensions of coalition warfare. The British effort from the start had been an Empire-led one, with South Africans, Indians, New Zealanders and Australians fighting under British command, and later joined by Free French, Greeks, and after Operation Torch, Americans. The Germans had to contend with the wildly variable Italian army and a highly political command structure based in Rome. As the Allied and Axis forces in North Africa relied on air cover to protect their Mediterranean shipping convoys and ground troops in the desert, at a strategic level joint (air–land–maritime) cooperation was vital: the Allies mastered this, whilst the Axis failed.

Rommel's continual supply problems were not, as he frequently claimed, the result of intransigence or slacking by the Italians, who handled the trans-shipment of his supplies, but were a result of his aggressive actions in overextending his lines of communication. Staff training might have done much to anticipate and offset some of his logistics challenges. Indeed, the historian Martin van Creveld has calculated that of the twenty-seven ships Rommel lost in September–October 1942 only two were tankers, and that even during the key battle at El Alamein, one third of his fuel stocks remained in Benghazi, suggesting that movement of supplies *in* theatre was more of a problem that sending war materials *to* theatre. Creveld concludes, controversially, that:

> for all of Rommel's tactical brilliance, the problem of supplying an Axis force for an advance into the Middle East was insoluble. Under these

circumstances, Hitler's original decision to send a force to defend a limited area in North Africa was correct. Rommel's repeated defiance of his orders and attempts to advance beyond a reasonable distance from his bases, however, was mistaken and should never have been tolerated.[52]

In a post-war intelligence debriefing the staff-trained, Luftwaffe field marshal Albert Kesselring observed of Rommel: 'He was the best leader of fast-moving troops but only up to army level. Above that level it was too much for him. Rommel was given too much responsibility. He was a good commander for a corps of army but he was too moody, too changeable. One moment he would be enthusiastic, next moment depressed.'[53] Soon, Rommel – perhaps already over-promoted – would be given not one, but two armies to coordinate: in northern France, where he would either flourish or flounder.

PART FOUR

ONCE MORE UNTO THE BREACH

From North Africa to the Rhine there were too many D-Days, and for every one of them we had to get up in the middle of the night.

Robert Capa

16

Two Return to France

I AM IN THE bedroom of my favourite seaside hotel in Normandy, the Hotel de la Marine, on the front at Arromanches-les-Bains, next door to the Overlord Museum; beneath it the Channel laps at the sands and artificial harbour which enabled Monty's 1944 invasion of France – a return to the land of his ancestors – to succeed. A good place from which to commence any battlefield tour, Arromanches was the geographical centre of the landings. East and west lie the invasion beaches, today studded with the remains of decaying concrete bunkers amidst the dunes and seafront buildings which comprised Hitler's Atlantic Wall.

Despite its name, this was never a continuous line of defences, but a series of coastal forts and an obstacle belt scattered along the German-occupied coastline facing Britain, from Norway to southern France. In March 1942, following a British attack on the French port of St Nazaire (Operation Chariot), Hitler ordered fixed defences to be built around naval and submarine bases. The raid on the port of Dieppe in August 1942 seemed to confirm this need. The changed strategic situation from late 1942 onwards extended these defensive positions beyond the harbours to any location where the Allies could conceivably land. *Festung Europa* ('Fortress Europe' – Hitler's collective name for the entire concept) was constructed by the Organisation Todt, the Reich's civil engineering movement, which had built Germany's pre-war autobahns and the Siegfried Line, but now utilised gangs of slave labourers instead of the conscripted German youth it employed prior to 1939.

Plans of selected positions were soon smuggled out to the French resistance and back to London, and the soldiers who landed on 6 June 1944 had surprisingly intimate knowledge of the Atlantic Wall. Millions of tons of cement and steel (vitally needed by the German war machine elsewhere) were used to create artillery bunkers, machine-gun nests, troop shelters and beach obstacles, accompanied by a generous scattering of minefields, both real and fake. Although propaganda photographs made it seem far stronger than it actually was, the wall was incomplete at the time of invasion. Nevertheless it was constantly being improved, thickened and extended, and for each day the Allies delayed their invasion, Hitler's defences grew stronger and more formidable. Employing the same thinking as lay behind the trench lines of the First World War, it was never intended to be a physical barrier to halt an invasion, but a defended obstacle to delay or canalise Allied troops whilst reinforcements were rushed to the threatened sector. In this respect it worked, deterring Allied planners from several options (such as Calais), though it was overwhelmed in Normandy quicker than expected and the reinforcements delayed by air power.

At the end of the Sicilian campaign in August 1943, Montgomery had set down in his diary his reflections on the successful conduct of future military operations. 'What was needed first,' he wrote, 'was a commander who knows what is wanted, who says so quite clearly and who has such prestige and fighting experience that everyone will accept his opinion and get on with it.' One wonders who he had in mind? Monty anticipated the European war would necessitate two major, vicious fights before he entered Berlin. The first would be landing on a hostile enemy shore, D-Day; he then expected an enemy withdrawal to the huge natural terrain barrier that protected the German frontier, the River Rhine, which would frame his second major land operation, an opposed river crossing. The course of events did not at all follow expectations, but there was a certain inevitability about another Monty–Rommel contest, somewhere in northern France, once the pair had left Italy.

Shortly before his fifty-second (and, as it turned out, last) birthday, celebrated on 15 November at his Lake Garda headquarters, Erwin Rommel had been ordered by Hitler to visit the north European coastline threatened by the expected Anglo-American invasion and assess the strength of the much-vaunted 1,600-mile Atlantic Wall. Giving

Kesselring supreme command in Italy had been a real blow to Rommel's ego and prestige, as Hitler knew it would be, and the new assignment was partly an attempt to assuage his loss of face by a Führer still captivated (though admittedly less so than before) by his field marshal's boundless energy. At this stage, there was nothing about commanding in the west, just studying the problem, an issue that had been brought to Berlin's attention by *Generalfeldmarschall* Gerd von Rundstedt himself, Supreme Commander in the west. At the end of October 1943, Rundstedt had submitted a report (drafted by his Chief of Staff, Günther Blumentritt) to Berlin warning that the Atlantic Wall was more propaganda than reality, and Jodl, deputy at the OKW, had seen in the report a solution to the problem of how to employ Rommel. As a prominent and extremely popular figure, he could not be sent on some minor mission or given a transparent sinecure. Rommel needed a decent challenge in public view and the Atlantic Wall study appeared the happy compromise.

Rommel began on 1 December, and after an initial tour of Denmark enjoyed a spot of leave at Herrlingen with Lucie and Manfred, whilst the new house the state had given him (formerly part of a Jewish boarding school) was prepared. Manfred and Lucie had remained in Wiener Neustadt ever since Erwin had been posted there as commandant of the *Kriegsschule* in 1938, but he was keen to move them away from the likelihood of air raids, so found country quarters for them in his native Swabia, in Herrlingen. The villa would be Erwin's last home, but one where he spent almost the most time with his family during the war. The inspection was resumed on 18 December and moved to France – last visited by Rommel under somewhat different circumstances in 1940 – where his team were briefed in Paris by Rundstedt, with whom he had campaigned some four years earlier.

Rundstedt had been commanding all German land forces in France as *Oberbefehlshaber West* (C-in-C West) since March 1942. The senior surviving officer of Wilhelm's Imperial army and perhaps the leading soldier of the Nazi Reich, Rundstedt epitomised the old Prussian military elite, and struggled to reconcile its code of conduct with that of Hitler. Serving in the interwar *Reichswehr*, he had adhered to a code of strict political neutrality, unlike several colleagues, and was thus seen as a safe pair of hands, perpetuating the old values. By 1933, he was a full general, C-in-C of First Army Group (the senior command), but privately (never publicly) reviled the Nazi party. At the Nuremberg trials of 1946, he said he interpreted his military role as executing orders to the best of his ability, but never to moralise to his superiors. This personal inability – or reluctance – to distinguish between right and wrong prevented the German army from putting an effective break on Hitler's military ambitions. Rundstedt had led Army Group A in the invasion of France through the Ardennes to the plan formulated by his own Chief of Staff, Erich von Manstein. The resultant unlikely victory gained him his promotion to *Generalfeldmarschall*. Rundstedt was C-in-C of Army Group South for Operation Barbarossa in June 1941, and remains culpable for transmitting down to his subordinate commands the Commissar Order, ordering all political commissars to be shot out of hand. On 1 December 1941 Hitler sacked him for withdrawing against orders, but it is significant that the Führer later apologised at the behest of SS-General Sepp Dietrich.

In March 1942, Rundstedt was recalled as C-in-C West; and in January 1944 he and Rommel found they disagreed over the strategy of how to defeat an Allied invasion. In practice the sixty-nine-year-old Rundstedt, Hitler's senior field marshal, had little effective control over suggestions made by the sprightly fifty-two-year-old Rommel, the Third Reich's youngest marshal. When they had last met in 1940, Rommel had been a mere divisional commander to Rundstedt who 'owned' forty-five divisions, including seven of panzers. Now they were equals, both field marshals – somewhat unsettling for the older man*. They differed also in energy and attitudes towards personal comfort. Arriving for the briefing in Paris, Rommel found Rundstedt occupying a suite in the Hotel Georges V in the type of luxury the Desert Fox found suffocating – he was not unlike Monty in his preference for austerity.

Though well liked and able, in Paris Rundstedt proved himself too out of touch, too unprepared to cooperate with his subordinates and out of favour with Berlin to influence the forthcoming campaign in a positive fashion. In any case, arguably his post was an additional, superfluous layer of command.[1] Rommel's initial hunch was that the invasion would fall in the Calais area, in Fifteenth Army's sector – not an unreasonable assumption to make – and so on 20 December he lunched with its commander, Hans von Salmuth, in Tourcoing, northern France. Fifteenth Army was the key job in the west and comprised seventeen divisions, trained to repel the expected invasion (which by December 1943, was taken as a 'when', not an 'if'). It had been established in France in 1941,

---

* Rundstedt had actually joined the German army as a seventeen-year-old *Fähnrich* (cadet) eight months before Rommel was born, in March 1892.

when the invasion threat was *against* England; Salmuth's arrival as its chief in August 1943 was recognition that the strategic balance had shifted.

Salmuth had been Chief of Staff of Army Group North for the invasion of Poland, and performed the same task with Army Group B in May 1940 for the invasion of France. These operations had earned him a Knight's Cross in July and promotion to *Generalleutnant* in August. Later on, in Russia, he was responsible for war crimes: he provided assistance to the SS *Einsatzgruppen*, massacring Soviet Jews, and was active in ordering the execution of partisans and hostages (ten for every Axis soldier killed). And yet he disapproved, as many did, of Hitler's Commissar Order and instructed his subordinates to disregard it, whereas many (like Rundstedt) passed it on. Salmuth, promoted *Generaloberst* and commander of Second Army, eventually became trapped at Stalingrad. He broke out, keeping his battered formations intact, but against Hitler's orders – another instance of Hitler's 'no retreat' mentality, which bore no reality to the situation on the ground, which Salmuth would witness in Normandy. For this success, in February 1943, he was relieved of his command and effectively demoted. Thereafter he came to loathe Hitler and the OKW, leaving the Eastern Front altogether that summer. He was nevertheless one of the most successful, experienced and capable Russian front commanders and on 1 August 1943 was posted to Fifteenth Army, stationed in the Pas-de-Calais, though it is significant that no further decorations followed his Knight's Cross of 1940.

Rommel was unusual in that he had no Russian front experience, and those who had (the vast majority of the army, including Rundstedt and Salmuth), may have looked down on him for his African 'holiday'. Perhaps the *Ost*-fronters felt they had earned an easier billet in France, but none of them understood the Allied use of air power as Rommel did. He also experienced considerable command problems: was he subordinate to von Rundstedt or directly under Hitler? At Rundstedt's suggestion, on 15 January 1944, Rommel was given command of the coastal areas most threatened with invasion: those occupied by the Seventh and Fifteenth Armies, deployed between Brittany and the Netherlands. Initially, Rommel's Army Group B was quartered at Fontainebleau, later moving to La Roche-Guyon, about thirty miles north-west of Paris.

Privately, Rommel thought Salmuth idle when the two first met on 20 December 1943, and Erwin was soon back on the road, perturbed by the lack of depth of Fifteenth Army's defences and the great disparity between the much-photographed big artillery bunkers at Cap Gris Nez facing Dover and the almost non-existent beach defences elsewhere. He noted, too, that the average age of Salmuth's enlisted men was thirty-seven, whereas his own Afrika Korps had comprised much younger, fitter men averaging fifteen years younger.[2] Hitler's policy had been to weed out the younger generation of warriors and send them to the mincing machine of the Russian front; Army Group B, Rommel soon realised, chiefly comprised those thought unsuitable for the east. Within a very short space of time Rommel grasped that the true state of the defences comprising the Atlantic Wall was even worse than Rundstedt's October report had suggested. In his view, if an invasion were to arrive within the next few months, then Germany was effectively defenceless.

In retrospect it is surprising that Monty ever got the job of 21st Army Group commander. Two names were in the frame: Alexander and Monty, and Eisenhower had let it be known that he was more than happy to continue with his existing field commander, Alex, having written to Marshall in Washington that he would prefer 'a single ground force commander. If the British give him to me, I would like to have Alexander.' Meanwhile Monty noted that three British generals (himself, Alex and Sir Henry Maitland Jumbo Wilson) were up for three senior commands: Supreme Commander Mediterranean, Army Commander Mediterranean, and Army Commander Western Europe. Every nerve in Monty's body strained for the third option. Writing his private reflections on the Italian campaign on 23 November 1943, Monty's view of his rival for the post was:

> Alexander is a very great friend of mine, and I am very fond of him. But I am under no delusion whatsoever as to his ability to conduct large scale operations in the field; he knows nothing about it; he is not a strong commander and he is incapable of giving firm and clear decisions as to what he wants. In fact no one ever knows what he does want, least of all his own staff; in fact he does not know himself.[3]

Brooke, surely a better judge from afar than his protégé, did not seem to share Monty's thoughts exactly, but confided to his own diary on 11 December:

> Ike's suggested solution was to put Wilson in supreme command [in the Mediterranean], replace Alex by Monty, and take Alex home to

command the land forces for Overlord. This almost fits in with my idea
except I would invert Alex and Monty, but I don't mind much.[4]

Soon after, Brooke visited both Monty and Alex in Italy, and on 14
December wrote: 'Monty strikes me as looking tired, and definitely
wants a rest or a change . . . Frankly I am rather depressed from what
I have seen and heard today. Monty is tired out, and Alex fails to grip
the show!!'[5] The 'tiredness' comment is exactly the phraseology Brooke
had used in August 1942 against choosing Gott for Eighth Army (and
in retrospect it substantially lessens any criticism Brooke may have
had for Gott in preference to Monty in August 1942). Now it appeared
that Monty himself was 'tired', yet on this occasion Monty was Brooke's
preferred option. Or was he? Earlier, Brooke had stated privately on
11 December, 'I don't mind which', so the job was by no means in the
bag for Montgomery.

Churchill, it seems, was also in two minds, but two factors probably
influenced him towards Monty. First, in 1942, Monty had been his
choice (admittedly after Gott), so in a way he had 'made' Monty.
Secondly, Winston was well aware of Montgomery's outspokenness, his
unpopularity, his inability to get along with the Allies, and his arrogance,
but may have assumed that Britain would need an excessively outspoken
representative in planning and leading Operation Overlord; someone
who would not be overwhelmed by the Americans. Alexander was a
gifted diplomat and a natural gent, but if an outspoken and aggressive
flag-waver was needed to protect Britain's interests, then it would have
to be Monty, not Alex. As always, a gifted Chief of Staff helped, but both
were equally well served by Freddie de Guingand and John Harding. The
First US Army commander, Omar Bradley, observed of the two
commanders:

> By nature a restrained, self-effacing, and punctilious soldier, Alexander
> was quite content to leave the curtain calls to his subordinate
> commanders. As a consequence he was soon eclipsed in fame by the
> bereted figure of Bernard Montgomery. But while the latter had emerged
> as a symbol of Britain's comeback in the war, it was Alexander who
> carried the top rating among Allied professionals who knew them both.[6]

It was a pretty poor indictment of Britain's pool of senior officers trained
for high command that the choice was down to just these two. Jumbo

Wilson had age against him in Brooke's eyes (he was sixty-three), though proved a very successful Mediterranean Supreme Commander, with grip and vision. Churchill had already impatiently cast aside two talented commanders, tough, bloodied and bright, who might have been suitable: Field Marshals Wavell and Auchinleck. The Auk might have carried off a very creditable performance in Normandy with the logistics denied him in 1941–42. Fate had conspired against O'Connor and Neame (both captured) and Gott (killed); though O'Connor later returned as a corps commander he lost out on seniority through his months of captivity. Neame, a Cheltenham-educated Sapper, was a year younger than Monty. A high achiever with much to offer he was not employed on his return from Italy in 1943. Churchill had made the shortlist inadvertently smaller than it needed to be and made a rod for his own back in those who were excluded. Kenneth Anderson, who led First Army in Tunisia, was ridiculously sent home and not re-employed operationally after May 1943; 'it is obvious that Anderson is completely unfit to command any army', Monty wrote to Alex in 1943, later describing him as 'a good plain cook', though Eisenhower praised his 'boldness, courage and stamina'. Neil Ritchie's later appearance in Normandy running XII Corps, having been sacked from Eighth Amy in 1942, was proof that operational commanders could be re-employed.

Monty's own 'exclusion' policy heralded the end of some promising careers: the departures of the perfectly capable William Ramsden, Herbert Lumsden and Alex Gatehouse also served to shrink the pool of potential operational commanders, as the army was 'Montyfied'. Andrew 'Bulgy' Thorne, who has sunk into total obscurity, pre-war military attaché in Berlin, was a rising man in 1940, commanding his (48th) Division well in France, but never got another operational command, although only two years older than Monty. Vyvyan Pope is someone we might have heard more of, being promoted to command XXX Corps in August 1941, but died in an air crash two months later. Bill Slim, admittedly busy in Burma, might have proved very adaptable to the Italian or Normandy campaigns, was strong-willed, better suited to coalition operations than Monty and possessed an attractive dose of humility. (Interestingly, Monty, with all his eccentricities, might have contributed as much had he and Slim changed places.)

Brooke privately signalled Monty the news on 23 December, which became official on Christmas Eve. The latter relinquished Eighth Army command in Italy to his own protégé, Lieutenant General Oliver Leese, on 31 December, and inspected the outline invasion plans drawn up by the pre-SHAEF planning staff known as COSSAC (Chief of Staff to the Supreme Allied Commander, Designate). Headed by a Briton, Lieutenant General Frederick Morgan, with an American deputy, Brigadier General Ray Barker, COSSAC had been established in April 1943 to devise the invasion of Western Europe. Back in early 1942, partly in response to Russian pressure for a second front, and partly through American ignorance of the vast logistics required, Churchill and Roosevelt agreed on their strategy for Europe. They would immediately concentrate all available Anglo-American forces in England (Operation Bolero) for a massive cross-Channel assault (Operation Roundup) in the spring of 1943. There was a vague initial plan (Operation Sledgehammer) for an exploratory landing in northern France (either Calais or the Cotentin peninsula) in 1942 should Germany suddenly weaken or the Eastern Front suddenly become critical.

The initial author of these plans was the head of the US Army General Staff's War Plans Division, Brigadier General Eisenhower. With Eisenhower in the Mediterranean, Morgan's COSSAC had been busy devising a plan which envisaged an invasion of Normandy on 1 May 1944 and incorporated some of the lessons learned from previous seaborne assaults. The fiasco of the Dieppe raid of 19 August 1942 had proved that the Allies needed air supremacy, German coastal defences needed softening up beforehand, and that tanks and assault-engineering equipment had to be capable of being landed in the first wave.* The assault would have to be an overwhelming concentration of force at a given point, backed up by a massive logistics chain, and German reinforcements would have to be prevented from reaching the battle zone. The tenacity with which the Germans held Dieppe, and the extent of its defences (subsequently much strengthened) suggested that it was extremely unlikely that the Allies would be able to capture a working port intact, capable of handling the invading force's needs. The very fact that the strongest German defences were concentrated around ports suggested that an attack elsewhere, where defences were somewhat weaker, would stand a greater chance of success. Harbours apart, landing craft

---

* None of the twenty-nine tanks landed at Dieppe was capable of getting off the shingle beach; the Canadian attackers lost 3,670 killed, wounded and taken prisoner – over half the attacking force – and 106 RAF planes were destroyed. The Germans lost 500 men and forty-six aircraft.

design also had to be improved, and British shipyards had no spare capacity.

Before arriving in England, Monty flew to Marrakesh, where Churchill was recovering from pneumonia, landing on the evening of 31 December. Winston gave Bernard the latest version of the COSSAC plan for invading France to read overnight. Churchill was in the midst of what would prove to be a monumental sixty-seven days out of the country (unthinkable for a modern PM), which included the 28 November–1 December Tehran Conference (code name Eureka) – the first meeting of the 'Big Three', Stalin, Roosevelt and Churchill – at which plans for Normandy were discussed with post-war policy for Germany. It was here that the Soviet delegation was presented with (and dropped) a jewelled sword for the defenders of Stalingrad, and where Stalin suggested that after the war the senior 50,000 German officers should be shot. Churchill and Roosevelt felt understandable and irresistible pressure to open up the Second Front, and had already appointed Eisenhower as Supreme Commander.

Winston's health – he was seventy – had deteriorated rapidly, to the extent that his personal physician, Lord Moran, had felt he might die – hence an eighteen-day recuperation in what Winston termed his 'Shangri-La', Marrakesh.[7] Churchill had fallen in love with the place ('the loveliest spot in the world') with its views of the Atlas Mountains on his first visit in 1936. Some of his best paintings are of the range's warm, vibrant colours: Monty's own favourite was a bright, vivid landscape painted by Churchill in Marrakesh and presented to Bernard after the war. In January 1943, following the Casablanca Conference, Winston encouraged Roosevelt to accompany him to what he described as the 'Paris of the Sahara'. They stayed in the Villa Taylor (placed at their disposal by an expat American) rather than Churchill's favourite watering hole, the Mamounia Hotel, and soon after their arrival two of Winston's valets carried the polio-stricken US President up to the rooftop. Reclining on a divan, Roosevelt was so taken by the scene that stretching his arm out he said to Churchill: 'I feel like a sultan, you may kiss my hand, my dear!'

Churchill felt safe and cosseted in Marrakesh, so it was a good location to explore the merits of the COSSAC plan, to which Monty took immediate objection. Monty, typically, tried to get out of the New Year's Eve celebrations with the Prime Minister (probably the only person in the known world who would have actively avoided an evening of bonhomie in Churchill's company), read the plan and wrote a comprehensive four-point memo. Churchill's private secretary, John Colville, was present in Marrakesh and recalled Mrs Churchill's ability to subdue Monty over the New Year's Eve guest list:

> On this occasion, when it was time to have a bath before dinner she turned to [Monty's] ADC, Noel Chavasse, and said she looked forward to seeing him in half an hour. 'My ADCs don't dine with the Prime Minister,' said Monty tartly.
>
> Mrs Churchill gave him a withering look. 'In my house, General Montgomery, I invite who I wish and I don't require your advice.'
>
> Noel Chavasse dined.[8]

Colville then recalled that the party saw in the New Year early,

> so that General M. could go to bed. Punch was brewed, the PM made a little speech, the clerks, typists and some of the servants appeared, and we formed a circle to sing 'Auld Land Syne'. I was linked arm in arm with General Montgomery and the American barman.[9]

In his memorandum to Winston, Monty stressed that he had not seen the plan before but felt immediately that it was too modest, landing too few troops on too narrow a front, with inadequate logistics arrangements – no provision for capturing ports had been made – and that the air battle needed to be won before commencement. His knowledge of Sicily and Salerno (particularly) persuaded him that ports were vital and the British and American armies (lacking any inter-operability) should be kept separate, each with their logistics chain and own harbour. The later landings at Anzio from 22 January (Operation Shingle) would support his conclusions. Eisenhower had also seen the plans on his appointment, on 27–28 October in Algiers, coming to a similar, private conclusion. As Ike was fortuitously passing through Marrakesh at the time, Monty repeated his objections to him, and got authority to revise everything when he arrived in London.

Churchill and Monty spent New Year's Day 1944 discussing the plans whilst walking in the foothills of the Atlas Mountains. Winston recalled Monty saying:

'This will not do. I must have more in the initial punch . . .' Presently I pushed forward into the mountains and our cars zigzagged slowly up the road to a viewpoint I knew. But the General would have none of this. He got out of the car and walked straight up the hill, 'to keep himself in training', as he put it. I warned him not to waste his vigour, considering what was coming . . . These admonitions were in vain. The General was in the highest spirits; he leaped about the rocks like an antelope, and I felt a strong reassurance that all would be well.[10]

This was Monty showing off – in another round of the 'I'm 200 percent fit' game of one-upmanship that had begun when the pair first met in 1940; Monty's 1914 wounds still troubled him (stiff left knee, slightly breathless), so his 'antelope' performance amidst the Atlas foothills would have been at some personal cost.

In London, 21st Army Group's headquarters was established, by pleasing coincidence, in Monty's old school, St Paul's (Monty inevitably occupying the headmaster's study). On 3 January 1944, the COSSAC planners presented their vision to him of the future invasion. Monty caused immediate consternation by stating the ideas were too puny and inadequate. Part of the COSSAC plan had evolved from the time when General Frederick Morgan had commanded a defensive sector of Britain in 1940 and war-gamed a projected German cross-Channel invasion, to test his troops. The scale of planning to mount a successful operation became apparent to him and he was under no illusions as to the complexity of his task, but had limited staff, few resources and no idea of the facilities, technology or numbers that would be made available in 1944. Morgan, whose COSSAC headquarters was in Norfolk House, St James's Square, was a very bright gunner who had passed through Staff College in Quetta before Monty's time, and had been Chief of Staff of 3rd Division when Bernard was commanding 9th Brigade (at the time of his wife's death). There may, therefore, have been 'history' between them. In many ways, running COSSAC had been a thankless task for Morgan, who worked hard, to the extent of sleeping in a camp bed in his office, but was hampered by not having a supreme commander to deputise for, thus could not pull any sort of weight until Eisenhower's appointment.

The COSSAC plan involved dropping one airborne division south of Caen, and landing three infantry divisions on three beaches (the eventual Gold, Juno and Sword) with two infantry divisions in reserve.

It wasn't a bad plan, it was just limited in its ambitions and would have come unstuck logistically. Instead of thanking Morgan and his team for their hard work on limited resources, Monty denigrated their efforts, making an enemy of Morgan, who subsequently joined the SHAEF planning team, where he became Deputy Chief of Staff and was knighted in 1944. (In his memoirs, Monty needlessly criticised Morgan for being pro-American, pro-Eisenhower and for maintaining a vendetta against him.)

At St Paul's, Monty's instant grip was felt as he spent several days challenging the planners to look at landing on the west coast of the Cotentin peninsula, or western Brittany as alternatives, but eventually settled on Admiral Ramsay's advice for the fifty-mile-wide beachhead of the Seine Bay, between Cherbourg and Le Havre. On his return from the United States, on 21 January, Eisenhower accepted the revised plan, and on 13 February COSSAC was formally absorbed into SHAEF, its job done. Monty foresaw the importance of capturing a port early and the inclusion of Utah Beach on the eastern Cotentin was precisely an attempt to make a stab at Cherbourg in the hope of seizing the harbour before the Germans demolished it. Recent scholarship has revealed the Utah landings and the insertion of two airborne divisions behind it were based largely on the second Sledgehammer option of 1942, to capture Cherbourg, seal off the Cotentin and use that as a springboard for offensive operations.[11] Whilst this would not have worked in 1942, it would succeed spectacularly in 1944.

Because of the changes to the COSSAC plan, some of the US assault formations only set sail for Britain in late January 1944, and it was only whilst training in south Devon and Cornwall that some green American soldiers had enough live ammunition to get to know their weapons ('That's where my mortar platoon became expert, most of the men had never fired it,' recalled one[12]). American troops were transported across the Atlantic by troopship, including the liner *Queen Mary*. On one occasion, crammed with 16,000 US troops, she was hit broadside-on by a rogue wave of approximately a hundred feet high; the ship tilted fifty-two degrees, and would have capsized had she rolled another three degrees (the incident subsequently inspired Paul Gallico to write his story, *The Poseidon Adventure*).

Morgan came up with the solution to the harbour problem: building on an earlier idea of Churchill's, he suggested that the Allies build their own ports and float them to the invasion area, hence the two prefabricated Mulberry harbours. These were conceived as only a temporary expedient until a fixed harbour could be captured and operated. The idea of a floating harbour can be traced to Churchill's memo of 30 May 1942 to Lord Mountbatten: 'Piers for use on beaches. They must float up and down with the tide. The anchor problem must be mastered. Let me have the best solution worked out. Don't argue the matter. The difficulties will argue for themselves.'[13] Progress at first was slow as discussions on competing ideas by the many interested parties were considered. Churchill was irritated by the apparent lack of progress and penned a number of increasingly irate messages, culminating on 10 May 1943: 'This matter is being much neglected. Dilatory experiments with varying types and patterns have resulted in us having nothing. It is now nearly six months since I urged the construction of several miles of pier.'

By the end of January 1944, Montgomery had reaffirmed the choice of Normandy: the invasion had to be mounted between the Scheldt estuary to the east and the Cotentin (Cherbourg) peninsula to the west, because that was the effective flying range of Allied fighters which were required in massive numbers for close air support. The Calais area was too heavily defended, and was expected by the Germans to be the object of a seaborne assault, and much of the coastline further west featured unassailable cliffs. Substantial reserves were to follow on, each beach becoming the focus for a whole army corps. Monty insisted that substantially more forces be deployed; the initial assault force was eventually eight full divisions (three airborne and five infantry) attacking five beaches along the fifty-mile front, to be supported by a massive aerial and sea bombardment. Leigh-Mallory, the air component commander, initially resisted the deployment of the two extra US airborne divisions, claiming that he had little extra lift capacity, and that he expected up to seventy percent casualties amongst the glider troops and at least fifty percent amongst the paratroopers. As late as May, at Monty's prompting, Ike had to insist that the airborne operation was essential to Overlord, as he recalled in *Crusade in Europe*:

> On May 30 he [Leigh-Mallory] came to me to protest once more against what he termed the *futile slaughter* of two fine divisions. He believed that the combination of unsuitable landing grounds and anticipated resistance was too great a hazard to overcome. This dangerous combination was not present in the area on the left where the British airborne division would be dropped ... To protect him in case his advice was

disregarded, I instructed the air commander to put his recommendation in a letter . . . I telephoned him that the attack would go ahead as planned and that I would confirm this at once in writing. When, later, the attack was successful he was the first to call me to voice his delight and to express his regret that he found it necessary to add to my personal burdens during the final tense days before D-Day.[14]

Churchill was fascinated with code names and personally selected the names for all major operations.[15] The British beach names, Gold, Jelly and Sword – all types of fish – were the original choices, but Churchill disapproved of the word Jelly for a beach on which many men might die, so he had the word changed to Juno. Omaha and the later Utah were named after the birthplaces of the US V and VII Corps commanders 'Gee' Gerow and 'Lightning Joe' Collins. There was also a little-known sixth beach, code-named Band, with signified the coastline to the east of the Orne river, and was possibly for use had a disaster occurred elsewhere or the need suddenly arose for a maritime flanking movement along the coast, as had been the case during the Sicilian campaign. Borrowing First World War German practice, whose code names he knew well from writing his multi-volume Great War history, Churchill saw the names of culturally significant figures as useful sources of operational code words:

Proper names are good in this field. The heroes of antiquity, figures from Greek and Roman mythology, the constellations and stars, famous racehorses, names of British and American war heroes, could be used, provided they fall within the rules above.[16]

Churchill's hand also is certainly evident in the names for Normandy; the plan for the 1944 invasion was originally Roundhammer an amalgam of the code names for invasions planned in previous years, Sledgehammer (1942) and Roundup (1943).[17] The fact that Churchill changed it with remarkable inspiration to Overlord[18] reveals much about his own view of the previous name.*

The drawback of Normandy was its hinterland: tactically, the *bocage* of thick hedgerows, which limited visibility and manoeuvre, could restrict the Allies to their beachhead, enabling the Germans to counterattack and push them into the sea. Operationally, the terrain and road network did not lead the Allies towards Germany (whereas the roads out of Calais did) and the Anglo-American forces would have to fight against the grain of the land and infrastructure. The *bocage* would also inhibit the Allies from constructing temporary airstrips for their fighters. By 30 June (D+24) within the bridgehead, the British and Americans had built ten airstrips each, a figure increased to an astonishing eighty-nine temporary airfields in use by D+90 (4 September), underlining the importance of air power. The runways were paved with steel mesh tracking and per-forated steel plate (which can still be seen today fencing the farms of Normandy). From the air, flying in a modern jetliner at 38,000 feet, the whole theatre is visible in a glance and one is struck by the patchwork-quilt appearance of the landscape, even more so than in England. The *bocage* remains today, though less obviously, around population centres like Caen and St Lô. Even at such heights, the landscape twinkles – the result of the sun being mirrored by quite small pieces of glass or metal on the landscape below. It was so much more the case in 1944, when the slightest reflection from a windscreen, binocular lens or mess tin could bring down devastating hostile fire within seconds, from either side.

Originally the strategic plan was to launch a simultaneous landing in the south of France – Anvil – but in April it was realised that there were too few landing craft to mount both. Assault landing craft were the main factor that drove Allied grand strategic planning, for the Pacific campaign needed landing craft too; it was a busy war for landing-craft crews, for there was only one fleet available for all theatres. Anvil was eventually delayed until 15 August 1944 and renamed Dragoon; Churchill was violently opposed to Anvil, wanting to maintain the pressure on Italy instead.[19] By the time of Normandy, in addition to Dieppe, the SHAEF planning staff had gained invaluable experience of having seen the progress of seaborne invasions on hostile shores in North Africa (Operation Torch), Sicily (Husky), at Salerno (Avalanche) and Anzio (Shingle).

Montgomery's policy was for each of the five assault divisions to consist of as many battle-hardened troops as possible, withdrawn from other campaigns, and specially trained for the invasion over the preceding months. Thus, he replaced 49th Division with 50th (who had fought under him in the Eighth Army) for the assault on Gold Beach in January 1944, much to the former's annoyance. The 3rd Division (largely unbloodied), which assaulted Sword, was Montgomery's old command from May 1940. On the US beaches, only 1st Infantry Division were combat veterans, whilst 4th (Utah) and 29th (who also landed on Omaha) were green.

Monty quite ruthlessly shuffled battalions and brigades around until he had the mix he required. Usually, he did not want to be overwhelmed by data and had detail summarised and packaged, but as the date for invasion neared Monty interviewed the commander of every British unit in the assault wave, down to lieutenant colonel – in excess of a hundred officers – replacing some as he did so. In some respects, this was a good example of a leader keeping in touch with his foot soldiers, but equally this activity could be criticised as being at the risk of losing strategic vision. Certainly it was not part of an army group commander's remit; it was the sort of issue a division or, at most a corps, commander should have been managing – not that Britain routinely trained any of its officers to operate at army group level. Yet, as we have seen, Monty (and Rommel) always took a close interest in those serving under him, perhaps in reaction to the distance of many First World War commanders. The one-eyed Brigadier Sir Alexander Stanier recalled:

> Monty came down to inspect 183rd Brigade ... he asked me a lot of searching questions to which I felt I'd given many wrong answers. He was a bully and at the end of the day I was thoroughly gloomy. To my great surprise I received a telegram the next day offering me command of 231st Brigade. The brigade, which was part of 50th (Northumbrian) Division contained no Geordies at all, but was entirely made up of south coast regiments, who had just come back from a long stint in the Middle East.[20]

The veteran troops of 50th, 51st Highland and 7th Armoured Divisions, all ex-Eighth Army formations, who were earmarked for Normandy, did not appreciate Monty's compliment. Already worn out from years of desert warfare, in the bitter fighting ahead they suffered some of the highest numbers of casualties from nervous exhaustion.

A total of thirty-nine divisions was allocated to participate in the invasion: twenty American, fourteen British, three Canadian, one French and one Polish. The number of US Army troops based in Great Britain doubled in the first six months of 1944, rising from 774,000 at the beginning of the year to 1,537,000 in the week preceding the final assault. More than 16 million tons of supplies were stockpiled; also 137,000 jeeps, trucks, and half-tracks; 4,217 tanks and fully tracked vehicles and 3,500 artillery pieces.

It may be that Monty's greatest contribution to the success in Normandy was not his conduct of it, but his planning for it.[21] Omar Bradley observed:

> Monty's incomparable talent for the 'set' battle – the meticulously planned offensive – made him invaluable in the Overlord assault. For the Channel crossing was patterned to a rigid plan; nothing was left to chance or improvisation in command. Until we gained a beachhead we were to put our trust in The Plan.[22]

Monty's instant grip of the COSSAC organisation and forceful denigration of its plan nevertheless resulted in the eventual recipe for success; it is difficult to envisage the original proposals succeeding against Rommel. Although Monty was forceful, his arrival was not initially accompanied by mass sackings; he kept much of the existing team in place, utilising their best talents to the full; he even remained on good terms with the Americans for a while. However, Alan Brooke (CIGS) noted in his diary on 10 March that Bernard was 'making good headway in making plans', but was 'equally successful in making enemies as far as I can see! I have to spend a great deal of my time soothing off some of these troubles'.[23] Writing of his command philosophy, Montgomery said that a C-in-C

> must spend a great deal of time in quiet thought and reflection. He will refuse to sit up late at night conducting the affairs of his Army; he will be well advised to withdraw to his tent or caravan after dinner at night.[24]

Compare this insight of Monty with Rommel at the same time – who was chasing around France, Belgium and Holland on multiple inspections, covering great mileages, with little time for the suggested 'quiet thought and contemplation' – and we have two very contrasting styles. The fact of Monty's appointment was important in itself; General Omar Bradley, who had gained promotion fighting alongside Monty in Sicily and was not a noted Monty supporter, nevertheless observed:

> Psychologically the choice of Montgomery as British commander for the Overlord assault came as a stimulant to us all. For the thin, bony, ascetic face that stared from an unmilitary turtle-neck sweater had, in little over a year, become a symbol of victory in the eyes of the Allied

world. Nothing becomes a general more than success in battle, and Montgomery wore success with such chipper faith in the arms of Britain that he was cherished by a British people wearied of valorous setbacks.[25]

While another commentator, a former actor who would become Monty's double, captured the mood of the country at the time:

The whole nation was on tenterhooks about the coming invasion which might easily become a blood-bath, and even a shattering defeat. And one of the exceptions to this general nervousness was the man who had most cause to worry, the man who was in charge of it.[26]

Monty selected the forty-seven-year-old Miles 'Bimbo' Dempsey, who had led a corps under him in Sicily, as his Second Army commander. Unfortunately, Dempsey has become as obscure as Monty has become famous; his identity remained opaque even to his own Second Army troops in Normandy. Major Johnny Langdon of 3/RTR, spoke for many when he told me that his unit to a man knew of Montgomery and their corps commander (O'Connor), but few had ever heard of Dempsey.[27] Tall, slim and always perfectly turned out, Bimbo had commanded an armoured division and took over XIII Corps in December 1942 after Horrocks was wounded six weeks after Alamein. He had served in the Italian campaign and was an outstanding tactician. Monty certainly tried to ensure that only his own nominees commanded under him in Normandy, and sought to exclude his old rival Dick O'Connor (recently escaped from an Italian POW camp) from the campaign. In a War Office compromise, Oliver Leese was left in command of Eighth Army in Italy, O'Connor was appointed to XIII Corps, Dempsey's old command, whilst Bimbo (the origins of the nickname are lost to history) got Second Army, as the more experienced corps commander.[28] He was very much Monty's protégé; indeed so close was the Monty–Bimbo relationship that 'Dempsey claimed after the war that Montgomery's operational directives to him were only written up for the record after the plans had been settled by mutual agreement between them.'[29]

Brigadier Hubert Essame, who commanded 214th Brigade of 43rd Wessex Division in 1944–45 and therefore saw Dempsey at first hand, painted this picture of his army commander: 'a friendly, unassuming man, Dempsey had a thorough knowledge of all aspects of staff work and plenty of common sense. Absolutely loyal to his chief, he admirably filled the role of a modest first violin in an orchestra playing a piece conducted by the composer, Montgomery.'[30] In wartime photographs, Dempsey certainly stands to the rear, almost lost in the background when pictured with his boss. Dempsey declined to write his memoirs and 'even ordered that his diaries be burned rather than they be used to stir up inter-Allied bickering'. Perhaps Monty's obvious insecurity can be understood in this light; his fear was not so much from future historians, but his contemporaries. Dempsey's diaries were not apparently destroyed until his death; the fact that they were consigned to the flames suggests they contained real dynamite: only Dempsey and Monty can have known what. To what extent Miles Dempsey's obscurity was the result of his own shyness or Monty's determination to hog the limelight is difficult to tell, but it did underline Bernard's inability to acknowledge the team of which he was captain.

Despite Monty's attitude towards the COSSAC plans and their author, General Frederick Morgan's shortcomings were not so much of vision but of means: without a high-profile boss to lobby for support or resources, Morgan's team could achieve little; their valuable contribution was to identify Normandy as the beachhead. Other alternative commanders (particularly with experience of the Mediterranean, like Alexander) would also have been able to see the shortcomings of the original plan, but all would have been hampered, as Morgan was, by the lack of resources until 1944.

The day of the invasion would go remarkably well, with around 9,000 casualties (a fraction of the 30,000 expected; at one stage Allied planners anticipated and were prepared to accept sixty percent casualties on D-Day). In preparation for the D-Day landings on Utah and Omaha Beaches, the US forces had conducted a series of exercises on Slapton Sands in south Devon*. From a temporary training area comprising around 30,000 acres, a total of 3,000 people (750 families) and 180 farms with livestock were evacuated. This was completed in six weeks. On 27–28 April 1944, a dress rehearsal of the 4th US Infantry Division landing at Slapton (Exercise Tiger) went disastrously wrong. By chance, nine German E-boats from Cherbourg intercepted a fully loaded and manned assault convoy practising in the Channel. Two of the landing craft, LST 507 and LST 531, were sunk and others damaged. On board the two

*Monty had exercised his 9th Infantry Brigade here in 1938.

landing ships the casualties were severe: 638 men killed (197 sailors and 441 soldiers) and hundreds injured. This was ten times greater than the casualties sustained in the real D-Day assault on Utah Beach (forty-three killed, sixty-three wounded), and reinforced the military training maxim, 'Train hard, fight easy.' Altogether, including casualties from other ships and those killed by friendly fire on shore, a total of 946 Americans died during Exercise Tiger. News of this disaster was kept a closely guarded secret; it was feared that the operation had been compromised, making the planners and commanders even more nervous – but the Germans remained unaware of the significance of the convoy, or their success.*

By 1944, Ike's Americans had established a different 'way of war' to that of Monty's Britons and the doctrine of the two nations never quite fused. In the 1940s, the USA was easily the world's most affluent nation with an industrial capacity that rivalled that of Europe. America placed great value on the lives of its citizens: they had the highest average standard of living in the world and as a result had huge expectations from life. Then, as now, US politicians and generals came to use all their great resources and wealth to minimise casualties, by fighting war with technology, machines and firepower. In some ways, British aspirations were similar, because of the 'Shadow of the Somme', where Monty and most other senior commanders had served. However, the British had fewer resources with which to protect the lives of their servicemen and women, and so established a tradition of 'institutionalised amateurism' – or 'muddling through'. Having also experienced the Blitz, which had resulted in high casualties on the home front (51,000 civilians died ultimately from aerial bombing), the British population were far more tolerant of casualties from fighting overseas than their US cousins.

In spite of all precautions taken to protect the secrets of D-Day, some officers still engaged in careless talk. One such case was that of US Major General Henry Miller, chief supply officer of the US Ninth Air Force, who, during a cocktail party at London's Claridges Hotel, talked freely about the difficulties he was having in obtaining supplies. He added that things would ease after D-Day, declaring that would be before 15 June. When Eisenhower learned of this indiscretion, he ordered that Miller (whom he knew personally, having been at West Point together) be

---

*The disaster was kept secret long after the war. In 1984 a Sherman tank was raised from the seabed and stands sentinel over Slapton Sands as a fitting memorial to those lost on Exercise Tiger.

reduced to the rank of colonel and sent back to the US, where shortly after he retired from the service.

Historian Keith Simpson has suggested that in May–June 1944 there was great Allied nervousness of disaster: 'Churchill, Brooke, Eisenhower, and to a lesser extent, Montgomery, had severe doubts about the success of the landings, with the British being particularly fearful of failure – memories of Gallipoli, Dunkirk and Dieppe.'[31] Indeed, Eisenhower's aide Harry C. Butcher recorded post-war that Ike, with great humility, actually penned the following note on 5 June, anticipating failure:

> Our landings in the Cherbourg and Havre area have failed to gain a satisfactory foothold and I have withdrawn the troops. My decision to attack at this time and place was based upon the best information available. The troops, the air and the navy did all that bravery and devotion to duty could do. If any blame or fault attaches to the attempt, it is mine.[32]

At the time, until the breakout from Normandy in August, the invasion and campaign seemed to the Allies a close-run thing, while more recent evaluations of the campaign by military historians have come to the same conclusion. Whilst D-Day itself was crucial for establishing the bridgehead, the subsequent ten days determined whether the Allies would get beyond Normandy, or the Germans would contain them, and force an attritional campaign, pushing the Allies into the sea; the odds were against the Germans because of the Allies' pre-invasion planning, deployment and campaign plan. The fact that Hitler decreed a 'stand-and-fight' policy in Normandy, frequently issuing 'no retreat' orders, meant that the battle for France would be decided in Normandy, as few troops would be permitted to escape.

The Gallipoli-Dardanelles campaign of 1915 and Normandy 1944 make for an interesting comparison, mounted twenty-nine years apart. The British and their German–Turkish opponents used the same bolt-action Lee Enfield and Mauser rifles as Monty and Rommel's troops would use in Normandy and whilst naval gunfire support in 1915 was provided by the guns of the new dreadnought HMS *Queen Elizabeth*; her sister ship, HMS *Warspite*, fulfilled the same role in 1944. But there were clear differences. On 25 April 1915, the British and colonial troops who landed were rank amateurs who had yet to develop the battle tactics that would beat the Germans in 1918. Infantry units had no light machine guns in 1915 (the Lewis Light Machine Gun was to appear a year later), and no sophisticated landing ships had been developed, unlike in 1944. The Turks anticipated an Allied landing in 1915, whereas Overlord achieved complete operational surprise. Gallipoli was the Allies' first major amphibious assault in the First World War, but Normandy had been preceded by several other (successful) seaborne invasions. While the Allied commanders were right to worry in 1944, their soldiers were in top form and supported by a host of specialised weapons and technology, developed specifically to make D-Day a success.[33]

Overlord did, however, suffer from a lack of clear doctrine, and was an unhappy fusion of the British and US traditions of amphibious warfare. No doctrine was developed for amphibious operations in the interwar period by any of the Allies (or the Germans for that matter). Many interwar theorists thought that Gallipoli represented the future of amphibious war, and its failure was 'fought all over again, in printer's ink and at Staff Colleges in Britain, the US and Australia . . . Daylight assaults against a defended shore were considered a folly . . . money was tight and what there was, had to be spent on more obviously urgent experiments than amphibious operations'.[34] Zeebrugge, St Nazaire and Dieppe represented the British tradition of using amphibious operations for limited raids, nothing more, where inter-service cooperation was marginal. The Americans, by contrast, evolved a different doctrine from their experience in the Pacific. In that theatre, the US Marine Corps had its own air, naval and ground assets, so that inter-service rivalry was not a factor. The British determined that tactical surprise was absolutely essential for a successful amphibious assault, which usually meant a night-time or pre-dawn operation and sacrificing the naval gunfire/aerial bombing support that would destroy or suppress enemy positions. Such timings and tactics had been adopted for Dieppe, North Africa, Sicily, Salerno and Anzio – the daylight assault with minimal fire support in Normandy would be a first. The US Marine Corps believed the exact opposite – that firepower was more important than surprise and their island-hopping experiences in the Pacific confirmed this: a massive pre-invasion naval and aerial bombardment lasting days could destroy over fifty percent of the enemy fixed defences before the troops went in. At Tarawa (November 1943) the pre-invasion bombardment lasted three hours; at Iwo Jima (February 1945) it was three days, and at Okinawa (April 1945), it lasted seven days. On 6 June, Utah and Omaha were to receive just thirty minutes each, and Gold, Juno and Sword only an hour more, from a smaller naval fleet, whilst most aerial bombs would miss altogether.

There were also vast differences in the British and US approaches to command. Bradley recalled:

> Weeks after the Sicilian campaign was ended, Patton visited Monty at the latter's CP [command post]. During their conversation George [Patton] complained at the injustice of an order issued by Alexander's Army Group HQ. Monty looked at him with amusement. 'George,' he said, 'let me give you some advice. If you get an order from Army Group that you don't like, why not just ignore it. That's what I do.'[35]

In some ways, it was a curious piece of advice, for Monty would never have tolerated such behaviour within his own 21st Army Group. Thus, the Americans took the view that the British considered an order as a basis for discussion between commanders, and if a difference of opinion emerged, it would be ironed out and the order amended. Within the US forces, differences were ironed out before the order was issued. The Americans also observed that 'the British in higher echelon prescribe details, which in normal US practice are left to responsible commanders in the lower echelons',[36] suggesting that US forces had already, in 1944, adopted 'mission command', whereas the British had not.

This perhaps underscored the flaws in Monty's military education – his not having attended the Imperial Defence College before the war, and having little detailed knowledge of the air force's and navy's requirements and capabilities, let alone those of his allies. Knowledge of the doctrine of one's allies is taken for granted amongst today's senior officers within NATO, but it was non-existent on all sides in 1944: to a certain extent this produced mutual ignorance. Monty's solution was to try and impose his will, forcefully, on all Britain's allies, whether appropriate or not. Such a defensive and insular attitude towards allies had already been obstructive in the Mediterranean campaign and would prove so again in Normandy.

Overlord and the subsequent campaign also suffered by having no overall tri-service commander. Ike fulfilled that function at a strategic level, but below him were just the single-service chiefs – Admiral Bertram Ramsay, Monty, and Air Chief Marshal Trafford Leigh-Mallory. Thus the Allies had no real (in modern terminology) Joint Task Force Commander; Montgomery fulfilled the function of Ground Commander but in reality was commanding the Anglo-Canadian 21st Army Group and 'advising' Bradley's First US Army. Monty irritated Omar Bradley no end, and the latter's aide Chester Hansen wrote mockingly of Montgomery 'striding in with his corduroy trousers, his enormous loose fitting gabardine coat and his beret like a poorly tailored Bohemian painter'.[37]

Monty would lose his Ground Commander's role in July with the activation of Patton's Third US Army and the resultant creation of the US 12th Army Group (under the elevated Bradley), who thus became his equal, rather than a subordinate. With two Allied army groups, there was no one above directing them, for Eisenhower and his deputy Air Chief Marshal Arthur Tedder were turning political aspirations into military reality at the grand strategic level, barely coordinating, much less directing or commanding, the Allied armies. The Allied navy and army now appear to have suffered because of poor relations with the RAF and USAAF: in particular the strategic bomber commanders (Arthur T. Harris and Carl 'Tooey' Spaatz), who were reluctant to abandon bombing Germany and shift their resources to Normandy. Bradley later wrote that these two 'sincerely believed that Germany could be defeated by bombers alone, and that Overlord was not even necessary'.

In their final form, the D-Day plans authored by Monty (and they were his, in the sense that he firmly insisted his 21st Army Group planners follow his explicit wishes) called for first-wave attacks consisting of two brigade groups, or US Army Regimental Combat Teams (RCTs) attacking Omaha, Gold and Juno, and one each at Utah and Sword. Second and third waves would follow, ensuring that more than one complete infantry division, plus armour (well over 20,000 men) had landed by the end of the day on each beach. The D-Day objectives were to seize a bridgehead up to eight miles inland (keeping the landing beaches beyond the range of hostile artillery) ringed by the towns of St Mère Eglise, Carentan, Isigny, Bayeux and Caen. In reality, few of these tasks were actually realised.

The first echelon of each brigade group was to be preceded by frogmen who, thanks to the work of Captain Jacques Cousteau in the development of underwater breathing apparatus, were able to defuse many of the obstacles that aerial photo-reconnaissance cameras and special forces beach reconnaissance parties had revealed. Infantry battalions emerging from landing craft supported by amphibious Sherman tanks would then hit the beaches with assault engineers, whose task was to clear lanes through the obstacles and enable passage off the sands. Whilst the armour suppressed enemy bunkers, divisional artillery would fire from landing craft and DUKWs bring in supplies from offshore. Each beach would also witness fire support from special landing craft fitted with rockets as well as extensive naval gunfire support.

British assault engineers would land with specialised armour, including flail tanks to clear mines, and armoured bulldozers; other British specialised armour included Churchill engineer tanks (AVREs) with fascines to bridge obstacle belts, bridgelayers, recovery tanks, road-layers (for the soft-sanded beaches) and tanks armed with heavy mortars to deal with blockhouses. Although the specialised armour belonging to Percy Hobart's 79th Armoured Division – 'the Funnies' (so-named because of the odd-looking adaptations to standard Churchill and Sherman tank hulls) – was offered to the Americans, they declined to use much of this equipment. The prevailing American view was that the Brits had over-engineered a solution to a challenge that American brawn and guts would overcome. The exception was the amphibious Shermans (equipped with canvas side screens supported by compressed air, and driven by propellers) which they used at Utah and Omaha. The lack of specialised armour certainly contributed to the American problems at 'Bloody Omaha'.

Percy Hobart was Monty's brother-in-law, Betty Carver's brother. He had had a 'good' First World War, largely in Mesopotamia, where he had earned a DSO, OBE and MC, and preceded Monty at the Camberley Staff College in 1919. Very much a rising star in the 1930s, Hobart had commanded 2nd Battalion of the Tank Corps in 1931–33, was the Tank Corps' Inspector 1933–36 and Director of Military Training at the War Office in 1937. He raised and trained the force that became 7th Armoured Division in the Middle East, but fell out with Wavell and Jumbo Wilson whilst there. Regarded as a loose cannon and not a team player, Hobart (often portrayed as the visionary surrounded by reactionaries) could be impatient and tactless, and seems to have been his own worst enemy. (These quirks of the cavalier loner, understood and tolerated by his sister Betty, may have been what drew her to a similar eccentric genius, Monty.) By late 1940, Hobart was out of work and back in England, serving as a lance corporal in the Home Guard, which is where Churchill – ever on the lookout for eccentric geniuses – rather than his brother-in-law Monty, rescued him from obscurity.

Hobart's comeback was meteoric: commander of first 11th, then 79th Armoured Divisions, and knighted in 1943, his Funnies made a significant contribution to the success of D-Day and the subsequent fighting. But not all the Funnies were as successful as is often made out. Each beach assault had a battalion of swimming tanks allocated to it; on Omaha, most of the tanks sank in the rough weather taking their crews straight to the bottom; on the British beaches, many landing craft coxswains took the tanks all the way in, than risk them foundering in the poor sea conditions, while the few swimming tanks that made it ashore rapidly became magnets for enemy fire. They were used with greater success in the calmer conditions of the Rhine crossings on March 1945, but have not been imitated since.

Although he may not have appreciated it at the time, in early 1944 Bernard Montgomery was at the height of his power and influence; he was popular and instantly recognisable at home and abroad, while much of the world identified him as the author of a string of victories in the desert and Mediterranean. His clashes with other senior commanders were not public knowledge and were tolerated (though not welcomed) by his colleagues and peers. His force of personality had pushed through necessary revisions to the Overlord plan, without which D-Day might have failed. Monty's commitment to the training and morale of his men – Americans, Canadians and Britons – was without parallel and perhaps made the period of January–June 1944 his 'finest hour'.

17

Defending Normandy

EISENHOWER AND MONTY knew that opposing them was nearly one fifth of Hitler's field army, fifty-eight divisions concentrated in the west under von Rundstedt. The latter complained, somewhat grumpily, that the only troops under his direct command were the guards outside his headquarters at St Germain-en-Laye, near Paris. Rundstedt's command was eventually subdivided into two army groups: Rommel's *Heersgruppe* (Army Group) B – and *Heersgruppe* G created on 26 April, under the anti-Nazi *Generaloberst* Johannes Blaskowitz. Army Group B contained the two armies of *Generaloberst* Friedrich Dollmann (Seventh, which covered Normandy and Brittany), and Salmuth's (Fifteenth, occupying the Channel coast from the Seine valley to Holland. The First and Nineteenth Armies of Army Group G covered the less-vulnerable Biscay coast and the Riviera with seventeen divisions.

Unlike General Morgan, Rommel never formally war-gamed out an Allied cross-Channel invasion, otherwise he too might have settled on Normandy. The German defences suffered due to a much-reported difference of opinion between Rommel and Rundstedt. Rommel was determined to hit the Allies as they landed, when they were at their most vulnerable: disorganised, wet, cold, seasick and disorientated; he was well aware of the damage Allied air interdiction could do to his reinforcements moving into the combat zone, therefore the closer his reserves were to the main effort, the better. Rundstedt, however, believed that the invaders should be held by a crust of static defenders, supported by nearby infantry, and then counterattacked by as many of his ten panzer divisions (totalling about 1,400 tanks) as he could assemble in the vicinity.

Rundstedt knew that the efforts of the coastal divisions (of which there were thirty-seven by June 1944), defending from their 16,000 bunkers, could not prevent a landing, and was convinced that if he could forge the panzers into a powerful, fully mobile reserve he would be able to destroy any beachhead before it became too powerful. He knew the Atlantic Wall could never be a physical, continuous obstruction – the most it could do would be to canalise opponents onto ground of the Germans' choosing, or buy time (whilst the obstacle belt was demolished and crossed) for appropriate reinforcements to be rushed to the correct spot. Ironically it appeared that for Berlin the Atlantic Wall was assuming the same sort of mythical impregnable status that the Maginot Line had represented for the French in 1940, until its defeat by the Germans; now the same psychological process was beginning to obsess Hitler. To coordinate the training and deployment of the armoured divisions collectively, at Rundstedt's behest, Hitler created *PanzerGruppe West* (later renamed Fifth Panzer Army) under Geyr von Schweppenburg. Placed under von Rundstedt on 24 January, it was arguably the only organisation with the ability to destroy the Allied beachhead.

Leo Dietrich Franz, Freiherr Geyr von Schweppenburg (hereon referred to as Geyr) was born into the Prussian military aristocracy and descended from a family that produced two Prussian field marshals.[1] He was raised in Potsdam, the home of the Prussian army, and had a natural fluency with languages which led to his appointment as military attaché to Great Britain, Belgium and the Netherlands with residence in London in April 1933. This was an important posting at a key moment, for his time in London coincided with the start of Hitler's rule. His military assessments of Allied military preparedness were a vital element of the decision to reoccupy the Rhineland (March 1936) and absorb Austria and Czechoslovakia (although he left in September 1937 before the Munich crisis erupted a year later). He seems to have had a breezy, debonair self-confidence – which helped him on the diplomatic circuit but made him no friends amongst the Nazi hierarchy in Berlin, especially when he argued for caution, and was openly opposed to Hitler's bold foreign policies.

In July 1943 he joined *PanzerGruppe West*, a position created for him by Guderian (who had become Inspector General of Armoured Troops in March), giving Geyr command of ten armoured divisions.

These were the *Wehrmacht*'s elite and there were frequent squabbles within the German hierarchy as to who should control them. Geyr, therefore, was Guderian's watchman in the west, keeping the panzers safe from Waffen-SS and Luftwaffe (there was already a Hermann Goering Panzer Division) interference. He utilised many of the staff from *PanzerGruppe Afrika*, dissolved the previous year, and was charged with training and deploying all panzer units in the west, including those of the Waffen-SS.

Based near Paris, fifty-eight-year-old Geyr trained his command very well. His ideas have a very modern ring to them: there were night exercises three times a week and extensive studies and experiments in hedgerow fighting, with demonstration battalions schooled in British doctrine engaging in simulated combat against German units. Once a week was *Fliegertag* (aviation day), set aside for training against air attack, but Geyr had not appreciated the Allies' ability to dominate the skies, which had been demonstrated clearly to Rommel in the Mediterranean. Geyr was one of the earliest panzer commanders and may have seen Rommel as a panzer 'upstart', as had many; but both Geyr and Rundstedt were out of touch. Neither appreciated the extent of Allied air superiority by 1944 or the ability of Allied close air support to interdict troops and tanks moving forward to the combat zone; the Soviets had not yet developed this capability when Rundstedt or Geyr had last served in the east. Although Rommel and Rundstedt disagreed over the placement of the panzers, the two field marshals appear to have got on quite well; the acrimony over the tanks centred on Rommel and Geyr. Rundstedt and Hitler grew tired of the endless bickering between Rommel – who wanted control of all the armour in the west and to have it stationed near the coast – and Geyr, supported by Guderian, who wanted it further to the rear.

On 26 April, the Führer announced the formation of Army Group G to oversee the First and Nineteenth Armies in southern France (effective on 12 May). Attached to Army Group G would be three of Geyr's panzer divisions, the 9th, 11th and 2nd SS; at the same time three were put under Rommel's tactical control (2nd, 21st and 116th Panzer Divisions), leaving only the remaining four armoured divisions (three SS – the 1st, 12th and 17th – and General Fritz Bayerlein's excellent Panzer Lehr) with Geyr as a strategic reserve. Additionally, the deployment of any panzers was to be cleared with Berlin first. Hitler's compromise of course satisfied no one, and because of their dispositions and

Allied air power, few of the formations would be thrown against the Allies immediately. Consequently, although Geyr had trained his men well, his formations were drip-fed into the invasion front. The Führer's compromise on armour ensured Allied success on 6 June because he removed from Rundstedt the authority of the man on the ground to deploy armour as events dictated. The sad reality was that *PanzerGruppe West* was also superfluous: it added an extra layer to an already over-complicated command structure for the Germans and ensured that any response to an invasion was bound to be slow and interrupted by petty jealousies.

The Rommel–Geyr disagreement mirrored an identical debate fought out in British high command circles in 1940, for on 22 June 1940 Monty had been promoted to command V Corps within Auchinleck's Southern Command. Monty had wanted a completely mobile defence, using commandeered buses to race to a trouble spot, whereas Auchinleck decreed a more rigid defence of likely beaches, to repel any invaders as they landed. In the event, Monty had got his way, going over Auchinleck's head to the War Office and Churchill – but in view of the similarity to the Rommel–Geyr debate, and the way the invasion unfolded in Normandy, history suggests that the shoreline defence advocated by Auchinleck in 1940 and Rommel in 1944 was more likely to succeed than the mobile reserve (harried by air power and constrained by time) policy advocated by Monty in 1940 and Geyr/Rundstedt in 1944.

All the evidence suggests that Geyr had the self-confidence to be a dangerous enemy in Normandy, but he also had the fearlessness to squabble with those above him and the misfortune to lose his newly established headquarters near Caen (destroyed in an RAF air attack on 10 June 1944) just as his moment of glory arrived. Consequently, he left the stage having failed to influence the campaign at all. However, the panzer divisions trained by him owe Geyr a debt of gratitude for the preparation they received, without which they might have been over-whelmed far sooner. Both Rommel and Geyr were talented leaders of men who understood their enemy; however, they were both headstrong and to a certain extent they worked against one another. Here we see again one of Rommel's failings: that he could become obstinate and petty when he encountered others as headstrong as himself. He had sent Karl Hanke packing and despatched an aide to Berlin to cancel his Knight's Cross in 1940; he relieved several senior commanders in North Africa abruptly, making enemies in the process; and he had fallen out with

Kesselring, Guderian and then Geyr. Rommel knew that his personal access to Hitler would protect him against his rivals, but if he lost that access, he would become acutely vulnerable.

Geyr would be sacked by Hitler on 2 July, along with Rundstedt, as both had concluded that a strategic withdrawal from Normandy was needed and said as much to Berlin. Hitler may also have recalled his cautious reports from London before the war; David Fraser, recalled

> Geyr was a large, handsome, heavily built man, a cavalry officer and a distinguished horseman ... In July 1934, immediately after the Night of the Long Knives, Geyr came to my father's Brussels office in a distressed state. He burst into tears. Weeping, he implored my father not to believe that these gangster methods were the real Germany ... The figure of Geyr von Schweppenburg is clear and sympathetic in my memory. He was a gentleman.[2]

Rommel knew he could do little to correct the massive imbalance in the air – in any case the Luftwaffe's shortcomings were Goering's responsibility. Instead, he assessed that success against an invasion rested on three inter-related factors: trying to straighten out the upper echelons of the convoluted German chain of command; gaining control of the *Wehrmacht*'s best weapon, the panzers; and raising the morale of his men. He was aware that the tactical picture on the ground was less than reassuring. German units in Normandy ranged from the inept to the formidable; the size of divisions ranged from 8,000 – the 700-series *Bodenständige* (static) coastal defence units, which were undermanned and poorly equipped – to the nearly 20,000 men of the 12th SS. Stationed south of Paris, the 12th SS Hitlerjugend Panzer Division was formed from a cadre of 1st SS Panzer (*Leibstandarte Adolf Hitler*) veterans in June 1943, and posted to France in April 1944. Most of its soldiers were young, ex-Hitler Youth members and fervent Nazis. A snapshot of one infantry company reveals that 115 out of 169 (sixty-nine percent) were eighteen, born in 1926. Its company commander was the oldest at age twenty-nine.[3] Normandy would be their first battle honour. By contrast, most of the non-SS soldiers belonged to much older-aged conscript groups and were not physically fit. The younger troops had been 'combed out' for the Eastern Front, and in return the static units received frost-bite cases, those with stomach ailments, heart trouble and other medical problems, whilst their officers occasionally possessed artificial limbs.

Generalleutnant Wilhelm Richter's 716th Infantry Division caused Rommel most concern, being the weakest formation with a strength of barely 8,000 (of whom 6,261 would become casualties by 11 July), and its transport consisted of bicycles. One of its battalions comprised *Osttruppen* – Polish, Czech and Russian volunteers, who appeared to show little spirit in defence. Their officers and NCOs were German but the non-German rank and file reflected the fact that most had volunteered when Germany was in the ascendant and had long since lost any enthusiasm. Most of the defending German divisions included a few battalions of Eastern Europeans, who soon melted away; one of their commanders had remarked that deploying them to an invasion front was inviting trouble, observing, 'We are asking rather a lot if we expect Russians to fight in France for Germany against Americans.'[4] Fifty-two-year-old Richter commanded from a series of bunkers in a quarry on the outskirts of Caen.*

Stationed behind Omaha and Gold Beaches was *Generalleutnant* Dietrich Kraiss's 352nd Division, with a strength of 13,000. Frequently described as a crack unit; it was, in fact, formed at St Lô in September 1943 from divisions which had been decimated on the Eastern Front. It may have contained a few veterans (although most of its recruits were born between 1924–26), but it was not itself a veteran division. Lack of ammunition meant that few had conducted live firing of their weapons before 6 June; between March and June each soldier had spent two thirds of his day helping to build the Atlantic Wall, rather than training. By 25 July, it would have suffered 8,583 casualties, or sixty-six percent of its 6 June strength, whilst Kraiss himself died of wounds on 6 August.

With such shortcomings, Rommel understood that he could increase the effectiveness of his troops with morale-boosting visits, where he would inspect their training, weapons and bunkers and at the same time offer a pep talk. He had learned from the desert that German soldiers loved a sing-song around a camp fire, and a hallmark of his inspections was often the gift of an accordion. Largely due to Rommel's efforts, German morale in Normandy and elsewhere was high, and on 6 June it is fair to argue that his men largely believed that they could defeat an Allied landing in France, with Rommel at the helm. They thought (with

---

* The quarry is now filled in and the site is appropriately occupied by *Le Mémorial de Caen*, an excellent state-funded museum which tells the story of the Normandy campaign specifically and of wars and genocides generally.

good reason) that an Allied failure might buy the Third Reich several years before another attempt was made. After all, Dieppe (in August 1942) had been a spectacular disaster and previous landings at both Salerno (September 1943) and Anzio (January 1944) had nearly ended in Allied defeat. From their experiences in Tunisia and Italy, Berlin did not rate American infantry good soldiers and dismissed British equipment as inferior.

Fortunately for Montgomery, the nearest armoured unit to the invasion coast was Rommel's least effective one – 21st Panzer Division under *Generalmajor* Edgar Feuchtinger. Rommel's former student and subordinate from 7th Panzer Division Hans von Luck remembered Feuchtinger as 'an artilleryman, with no combat experience, and none of panzers'. Feuchtinger was an opportunist Nazi, who owed his position more to political zeal than military competence, and had helped stage the Nuremberg rallies. His HQ was in Falaise, though some units were stationed north of Caen, and Feuchtinger himself spent much of his time in the fleshpots of Paris rather than preparing for an invasion. The division was equipped with a mixture of formidable Panzer IVs and many outmoded French light tanks captured in 1940, but never managed to concentrate as a formation on 6–7 June.[5]

If Feuchtinger was a disgrace, Rommel was also badly let down by his Seventh Army commander *Generaloberst* Friedrich Dollmann, who was older than most of his comrades – sixty-two at the time of the invasion. Dollmann had been stationed in Munich (the cradle of Nazism) continuously from 1923–33 and thus saw the way the political wind was blowing. He went out of his way to promote National Socialist ideals within the army and the speed of his subsequent promotions certainly indicates political favouritism from Hitler. By 1 March 1939 he was the seventh most senior officer in the German army; he had no experience of modern war in Poland, France or Russia, and had commanded the Seventh Army on occupation duties in northern France since June 1940.

The invasion would catch Dollmann by surprise and out of his depth. Bizarrely, his last combat experience had been in 1918, before the time of high-tempo armoured engagements and when close air support was in its infancy. He had not flown in an aircraft since 1916; Dollmann was so dangerously out of date that he ordered Panzer Lehr and 12th SS forward in daylight and in radio silence on 6 June, which contributed to the disorganisation of both and the loss of many vehicles. He was increasingly bypassed by more active subordinate commanders and overtaken by events. He comes across as weak and indecisive, ordering counterattacks then cancelling them and diverting forces across the battlefield on a whim. In the end, the capture of Cherbourg on 26 June proved the last straw; Hitler ordered an enquiry (although the port had been comprehensively sabotaged) but on the morning of 29 June Dollmann was discovered dead in his bed. Given his lamentable lack of experience, Dollmann, who always appear anxious or uncomfortable in photographs, was the worst possible choice anyone could have made for senior command on a potential invasion coast. Even given the large proportion of weak *Osttruppen*, a Seventh Army with almost anyone else at its helm would have given the Allies a much harder fight on 6 June 1944.

The headquarters that Rommel's Army Group B chose was the thousand-year-old castle complex of La Roche-Guyon, overlooking a once-important crossing point of the Seine. Consisting originally of a stone tower built atop a rocky outcrop with a commanding view, a fortified manor was later constructed at the base of the cliff. Manor and castle were linked by a passage cut into the rock, and over time other rooms were chiselled out of the stone. The manor at ground level was considerably enlarged in the eighteenth century, and the resultant glorious château offered Rommel a suitably generous number of rooms for his staff, stabling for horses and vehicles, pleasure gardens and bomb-proof shelters in the rock; concrete bunkers also enhanced the protection. Still the home of the Rochefoucauld family today, it is a quirky mixture of ornate château and chilling mediaeval castle.* Its chief advantage for Rommel lay in its location, being within reach of his two army commanders (Salmuth and Dollmann), and Rundstedt in Paris.

Whenever Rommel was away, La Roche-Guyon was presided over by *Generalleutnant* Dr Hans Speidel. Bright, bespectacled and crafty, Speidel was a rare combination of soldier and scholar, who had gained a PhD before the war. He was an exceptional staff officer and, like Rommel, a Württemberger. He was commissioned in 1915 into the Württemberg infantry and first arrived at Army Group B headquarters on 15 April 1944, as Chief of Staff to Rommel, whom he had first met in 1915; they had also served together in the same regiment briefly in the interwar period. Speidel was fluent in English and French and thus the success

---

* Rommel would not be the first inhabitant to suffer misfortune at the hands of the English: a previous owner lost his life at Agincourt in 1415.

of the Allied deception campaign, Operation Fortitude, was magnified at Army Group B because Speidel was able to translate many of the 'intercepted' (in reality, planted) Allied radio transmissions himself. Speidel (formerly on the staff of the military governor of Paris) is very important to understanding the Rommel story, for after the war he became the acceptable face of German militarism, spokesman for the Rommel family, historian of the Normandy campaign, one of the creators of the *Bundeswehr* in 1955 and ultimately held the key NATO position of COMLANDCENT, Commander of Land Forces in Central Europe, 1957–63. It is Speidel's tenure at Army Group B (he stayed on for two months after Rommel's death, serving under Kluge and Model) that fascinates historians, for the softly spoken intellectual was deeply implicated in the Stauffenberg plot against Hitler and through him, Rommel came to learn something of the plans for a 'regime change', but quite possibly no more than that. By July 1944, Speidel's motivation must have been waning as he was hoping the forthcoming Stauffenberg coup would bring an end to the fighting in the west.

Rommel certainly came to rely on the efficient, meticulous Speidel a great deal, who ran things at La Roche-Guyon like clockwork, allowing Rommel to roam about on his endless tours of inspection. To what extent Rommel believed an invasion would happen in Normandy is debatable; Speidel asserts that he did, but that Hitler did not and overruled him, yet evidence also exists to the contrary, that Hitler's hunch was Normandy and Rommel's was Calais. In either case, Rommel considered a cross-Channel invasion to be not unlike a First World War battle. The Channel was a wide obstacle, a no-man's land to be traversed before an assault on his trenches. With his ability for field sketching, Rommel was able to illustrate to his subordinate commanders how he envisaged German defences (many of which he personally devised) would 'fix' the Allies, whilst machine guns and artillery chewed up their infantry and armour. Anti-tank and anti-personnel mines, and over half a million wood or steel obstacles, were strewn between the high- and low-water marks, and inland on likely glider and parachute drop zones. This was, in effect, creating the deepest possible obstacle belt, like the barbed wire entanglements of the Western Front, known so well to Rommel's generation.

In September 1942, Rundstedt had received as his Chief of Staff at OB West a very capable staff officer in the person of Günther Blumentritt, who recalled that during 1944

Rommel did not smoke, and ate and drank but little . . . In the evening[s] . . . there was an unmistakeable feeling that Rommel was far away with his thoughts. His brain worked incessantly on new ideas. No landscape, no historic building, interested him; he was just a soldier. During a meal he would often take his pencil and a sheet of paper and sketch some new technical idea. Then he would hand it to his engineer general with the request that he would give his views on it the next morning before starting out. Engineers, artillerymen and sailors were for the most part pressed into service as attendants, and Rommel was always requiring new proposals from them. He went to bed very early in order to be able to start off fresh again in the morning. As a rule he had one or two keepers of war records with him who took photographs on every possible occasion.[6]

Friedrich von Mellenthin, his staff officer in Africa, observed that Rommel shared Monty's battlefield austerity;

[his] lunch consisted of a few sandwiches eaten in the car, with a mouthful of tea from a bottle. Dinner in the evening was no less Spartan; Rommel usually dined by himself or in the company of a few of his closest staff officers. During dinner he allowed himself one glass of wine. For himself and his staff Rommel insisted on the same rations as the troops.[7]

Heinz Werner Schmidt was also quite specific about the wine: 'he permitted himself a glass of wine only when a special occasion called for a show of sociability. He never smoked.'[8] His naval adviser, *Vize-Admiral* Friedrich Ruge, remembered Rommel saying to a cameraman: 'You may do with me what you like if it only leads to postponing the invasion for a week.'[9] Here we see a Rommel similar in his personal habits to Monty – few outside interests, early nights and abstemious with food and drink. Yet whilst Monty appears to have 'switched off' at night, Rommel did not, doodling with his ideas over dinner. His personal attention to detail was, like Monty's, fastidious, and in one sense over-indulgent, yet it gave him an exact and detailed knowledge of those under his command, their weapons and equipment. Hitler worked in a similar way, and many German officers remembered the photographic memory of their Führer and his ability to regurgitate huge quantities of accurate statistics, to counter and disarm any argument of a military adviser with whom he disagreed – an approach Rommel would have witnessed during 1939–40, which he perhaps imitated.

Rommel's technical solutions included steel girders welded together – 'hedgehogs' – which were embedded in the sand to pierce the hulls of landing craft. Wooden stakes were driven into the ground with mines attached to detonate on contact with seaborne assault craft, or gliders landing by air. Collectively these were known by those who had to build them as 'Rommel's asparagus', as Erwin's sketches of his obstacles made the beaches look like vegetable gardens. He also hit on the idea of flooding low-lying areas behind the coast and turning them into inaccessible marsh. Blumentritt recalled:

> Rommel had particular interest in the Engineers and used to declare with satisfaction that as a cadet at the War Academy [Danzig Kriegsschule], engineering was what attracted him most. For this reason, besides Admiral Ruge as naval adviser, he always chose the Engineer-General of Army Group B, General Dr Meise, to accompany him on his tours of inspection . . . Naturally time, labour and materials were inadequate . . . Rommel could be very unpleasant if in his opinion units had not done enough work, and he permitted no excuses.[10]

In the rush to complete the beach defences, *Wehrmacht* soldiers toiled alongside slave labourers of the Organisation Todt in the construction work. From the autumn of 1943, most military training had ground to a halt as all German combat troops were involved in erecting beach defences, siting mines and uncoiling miles of barbed wire. After the war, Geyr, who wrote historical monographs on the campaign for the US military, claimed 'the time and numerical strength of the [German] troops was wasted on the construction of fortifications at the expense of training'.[11]

Overlooking the Normandy beach defences, gunners of 1716th Artillery Regiment manned sixteen coastal gun emplacements and camouflaged positions inland, totalling sixty heavy cannon of various calibres. The gunners, whose headquarters was in Crépon, had little mobility, while their weapons and equipment were towed by horses or even oxen. One of their batteries was on the cliff tops at Longues-sur-Mer, west of Arromanches-les-Bains, and incorporated guns from a decommissioned warship which had a range of 25,000 yards (fourteen miles). Preserved today, this site is known locally as Le Chaos, after the offshore reefs which have traditionally deterred large ships from the Normandy coast. Each coastal battery operated different calibre guns (there were twenty-eight different calibres of artillery in use in Normandy alone), evidence of the huge amount of war materiel the Wehrmacht had absorbed from conquered nations and a quartermasters' procurement nightmare. The technically curious Rommel would have been gratified that his training with an artillery regiment back in 1914 still came to his assistance thirty years later, when overseeing the Atlantic Wall.

Building the defences was never-ending: a series of missives from Berlin in succession ordered that all bunkers eventually had to be capable of self-defence, have their own independent water supply, be gas proof, be able to communicate with other bunkers by armoured field telephone wire, have an electricity dynamo, and so on. Much was robbed from the Maginot Line and other inland fortress systems, but the Atlantic Wall fell far short of Hitler's dreams in June 1944. Though heavily bombed, most positions were nevertheless operational on 6 June; this is perhaps testimony to the inability of the Allies in 1944 to pinpoint and destroy specific targets. RAF Bomber Command's definition of accurate bombing in 1944 was 'within five miles of the aim point'. Along the seafront other concrete bunkers were built, housing 88mm and smaller weapons firing in enfilade, which overlooked the variety of beach obstacles, barbed wire and mines that had been sown beforehand.

Behind Sword Beach was a depth position, *Widerstandnest* (Resistance Strongpoint) No. 17, which comprised twelve underground bunkers linked by trenches, and the headquarters of Oberst Ludwig Krug, commanding 736th Infantry Regiment. Krug's bunker complex, code-named Hillman by the British, is currently being excavated by the community of Colleville-Montgomery and tells us much about the German defence plans. It is cleverly sited, extremely well protected by mines, wire and machine guns – and was almost complete by D–Day. Had the invasion been delayed, the Germans would have had time to build many more bunker complexes like Hillman, providing Montgomery with an infinitely more difficult challenge. Of all the brigade-sized German regiments in Normandy, Krug's would be the most tested on D-Day, with his battalions covering Juno and Sword Beaches, and the area of British 6th Airborne Division's parachute and glider descent. I have walked the ground of Hillman – where Krug would hold out until the morning of 7 June before surrendering – with his son Hans and grandson Christian. They related to me that Ludwig not only survived, but his defiant defence was recognised by promotion whilst in subsequent British captivity (surely a rare event) from *Oberst* to *Generalmajor*.

Rommel hoped to be able to destroy an invasion first at sea, but in June 1944 only a few scattered surface units of the *Kriegsmarine*, mostly E-boats (motor torpedo boats), minesweepers and destroyers, were stationed in Cherbourg and Le Havre. *Vize-Admiral* Friedrich Ruge had also been Rommel's naval adviser initially at Lake Garda from November 1943. He recalled that Rommel expected the invasion in one of three possible areas: the Scheldt, the Somme and the western part of the Bay of the Seine –where the landings actually took place.

> At first Rommel thought that the coast on both sides of the mouth of the Somme the most probable location because our defences there were especially weak. Later the Calvados and Cotentin seemed to him most likely, because of Allied air activity in May 1944 . . . He thought, besides the main invasion, there possibly would be minor operations such as cutting off the north-west part of Brittany by taking Brest, or cutting off the Cotentin or Le Havre.[12]

Although Rommel's initial hunch was the Calais area, it can be seen here that he really wasn't sure and prevaricated until the last moment as to the likeliest site. It was never the case, as many historians allege, that Rommel was inclined to Normandy and Hitler remained fixed on Calais as the probable invasion area. Although he only controlled land forces, Rommel's defensive capability was strongest on land, weak at sea and almost negligible in the air. Goering's once-powerful Luftwaffe was represented by pathetically few fighter squadrons, widely scattered to avoid being bombed, and therefore unable to concentrate and co-ordinate their efforts. *Generalfeldmarschall* Hugo Sperrle's *Luftflotte* 3, covering France and the Low Countries, could only field 890 aircraft of all types, of which 497 (or fifty-six percent) were operational on 6 June.[13] With General Wolfgang Pickert's powerful III Flak Corps of anti-aircraft artillery, the Luftwaffe nevertheless commanded 384,579 uniformed personnel (including 16,109 uniformed women auxiliaries, *Helferinnen*). According to aviation historian James S. Corum, the principal reason for the Luftwaffe's weakness in the west was that all the better squadrons had been pulled back for the air defence of the Reich – thus indirectly justifying the Allied strategic bombing offensive; given the Allied superiority of pilots and machines, the Luftwaffe 'had no hope of winning even local or temporary air superiority over the Normandy beachhead',[14] but it could still have fought a more effective campaign than it did. The failure of the Luftwaffe to contest the skies would have grave consequences; in March 1946, *Generalleutnant* Friedrich Dihm, special artillery adviser on Rommel's staff, wrote in captivity of his belief that 'had it not been for Allied air power, it would have been possible, in my opinion, to prevent a successful invasion during the first few days after the initial assault. These were the most critical days for the Allies.'[15]

Montgomery and Eisenhower were gifted with an elaborate picture of the German defences, which came to them in four forms. Aerial photographs were taken by Photo Reconnaissance Unit (PRU) Spitfires and Mosquitos based at RAF Benson, which enabled an accurate, new mapping survey of the invasion coast, incorporating every German position. Pictures were taken every few weeks to note any changes, but similar photographic work had to be mounted all along the French coast to avoid focusing undue attention on Normandy. Secondly, special forces (mostly British Special Boat Service) took terrain samples from likely assault beaches back to England for analysis to determine where tanks could operate, and spied out the beach obstacles.* Third, tactical intelligence-gathering was undertaken by the 7,000 very brave men and women of the French resistance. Directed from London via field offices in Paris and Caen, the Century network of resistance workers created a living map of Normandy. Each square mile became the responsibility of separate cells; clandestine photos of positions, stolen blueprints of bunkers and details leaked by drunken Germans all formed part of the overall picture, although some 1,500 *résistants* were betrayed by their countrymen or intercepted by the Gestapo. When the chief of the *Milice* (pro-Nazi French militia) in Caen became too successful against Century, he was assassinated. Agents included the Grandcamp café proprietor André Farine, who recorded the barbed-wire fences of the Point du Hoc battery with binoculars from a nearby church tower, supplemented by titbits overheard in his café. The Port-en-Bessin music teacher Arthur Poitevin was allowed to walk the cliff tops because he was blind; he used the opportunity to pace out every defensive perimeter, committing the figures to his sharpened memory. Jacques Sustendal used his position as

---

*It was during one of these that a Commando Officer, George Lane, was captured and subsequently interviewed by Rommel in person.

the doctor of Luc-sur-Mer to visit perfectly fit patients and record every trench and machine gun in his sector. Others smuggled out exhaustive details of concrete thickness, gun calibres, minefields and leave rosters.[16]

Finally, strategic intelligence-gathering by the Allies at Bletchley Park not only produced the exact German order of battle, but gave evidence of the Rommel–Geyr clash over the armour, and Hitler's resultant compromise. The success of the Allied deception plan, Operation Fortitude, was also measured at Bletchley Park. Bogus signal traffic was used to create the impression of a fictitious First US Army Group, supposedly commanded by the legendary and dashing George S. Patton, waiting in Kent to cross the Channel and invade the Calais area – which was exactly what the Germans expected. Dummy landing craft, tanks and camps reinforced this idea for probing Luftwaffe reconnaissance aircraft, and double agents fed reports back to the *Abwehr*.* This deception was instrumental in delaying the switch to Normandy of Salmuth's Fifteenth Army until too late. Once the invasion was under way, the Allies fed a battle plan piecemeal to the Germans that Normandy was only a feint, and that a further assault would take place six weeks later on the Pas de Calais, and a smaller operation against Norway (Operation Fortitude North). As late as 17 July, Berlin still expected a further attack on Calais and held back reserves accordingly.

Also part of the deception plan, the officer-actor Lieutenant Colonel David Niven, working for British intelligence, discovered a lieutenant in the Pay Corps who bore an uncanny resemblance to Montgomery. Fortunately, Lieutenant M. E. Clifton James had been an actor before the war, and duly briefed by MI5 and having studied his subject's mannerisms and voice, was deployed to Gibraltar and Algiers in an attempt to distract German attention away from the obvious target of northern France. We now know that this subterfuge worked. Before assuming the part of his subject, James, who had fought in the First World War, watched Monty with the detachment of a professional, recalling:

> He strode along dominating the scene, but never interfering unnecessarily. Every now and then he stopped and fired questions at officers,

---

*New research has revealed that Rommel initiated exactly the same deceptions on 14 May 1944. Under Operation *Landgraf* replica tanks were to be distributed throughout northern France and eight fictitious divisions were to be created by bogus radio broadcasts.

NCOs and privates – checking up, offering advice, issuing orders . . . What personality he had! . . . This man was what we should call a 'natural'. The moment he appeared, before even he spoke, his personality hit people bang between the eyes. He would have made a fortune on stage, I thought . . . It was obvious that he would tolerate no second-rate performers in his 'cast', and I noticed his habit of turning suddenly on a man and fixing him with those piercing eyes of his as if he could read his innermost thoughts.[17]

Clifton James's close study and imitation of Montgomery gave him many insights: 'I began to realise that he had himself under control in a way that I have never seen paralleled', and wrote of Monty's 'iron self-control and detachment'. Brigadier Brian Richardson (chief planner at 21st Army Group) echoed this in observing that Monty's confidence 'was but an outward sign of a self-imposed regime, designed to ensure that the army's morale did not deteriorate'.[18]

In addition to 'gripping' the plans and planners and enthusing them with his energy and sense of purpose, in the months leading up to disembarkation Monty got round Britain to visit as many of his troops as possible. As his conviction was that 21st Army Group were temporary warriors under arms, he shared with them as much information about the forthcoming battle as he dared. His old Eighth Army, he had reasoned, 'consisted in the main of civilians in uniform not of professional soldiers. And they were, of course, to a man, civilians who read newspapers.'[19] So, too, his new command; travelling by a special train put aside for his use, named the *Rapier*, he

inspected two, and often three, parades a day, each of 10,000 men or more . . . It was essential I gained their confidence . . . I explained how necessary it was that we should know each other, what lay ahead and how, together, we would handle the job. I told them what the German soldier was like in battle and how he could be defeated . . . [that] I had absolute confidence in them, and I hoped they could feel the same about me.[20]

Monty built on the pep talks he had given to very large gatherings of troops prior to the Sicily invasion, addressing troops by microphone from jeeps and his staff car. John Ford, an artillery observer with 50th (Northumbrian) Division, remembered how his complete division had formed up on three sides of a hollow square on a sloping hill, with Monty's jeep parked on the fourth – and highest – side. The troops were drawn up in ranks about twenty deep; Monty arrived late (a noted trick of many celebrities – building the tension), and ordered the first ten ranks to step forward then about-turn so he could walk round the entire division, through the midst of the troops, pausing here and there to speak to soldiers – often picking out those who wore the Africa Star ribbon. He then climbed onto the bonnet of his jeep and shouted 'Break ranks' and 'Gather round' – the soldiers ran to the vehicle; then 'Everybody sit down.' Almost in the manner of a fireside chat, but delivered with punch and enthusiasm, he asked rhetorically: 'What's your most important possession? – It's your life. And I'm going to save it for you.' His talk was full of confidence and concluded with: 'You and I together will see this thing through.'[21]

Later, this showmanship would revert to exactly that: post-war addresses to troops would invariably involve Monty arriving in his Rolls-Royce wearing a cap and transferring to a jeep, wearing his trademark beret, at the last minute. Alan Moorhead described how Monty during these months would also talk to war workers of the home front: 'it was growing very like an election campaign . . . railway workers, the miners, the stevedores. He held huge mass meetings in the factories . . . posters began to appear in the streets; pictures of that lean, intent face under the beret, a personal message underneath.'[22] To the consternation of politicians (who instructed him to stop – instructions he characteristically ignored), he built up an enormous following. Clifton James recalled an incident in Scotland in the spring of 1944:

> We came across a stone building standing by the road, and above the hum of our engines I could hear the sound of children's voices singing. It was a village school. At once Monty ordered a halt. He got out, crossed the small playground in front of the building and went in through the open door. The singing stopped abruptly. Then we heard frantic cheering which presently died away. I couldn't hear what Monty said, but I imagine it was the sort of homely talk you sometimes hear on speech days at school . . . Now the singing began again, 'Oh God, our help in ages past', the piping voices sang, and Monty came out looking happy.[23]

Proof of Monty's claim to have personally inspected 'well over a million men' came on 22 June in a charming letter from Walter Beddell Smith,

Eisenhower's Chief of Staff at SHAEF, which also demonstrated how Monty had bridged the nationality gaps of the coalition nations.

> Dear General,
>
> I have just received from a most reliable and intelligent source a report on attitude and state of mind of American troops in action. The writer is completely unbiased, and his report contains the following paragraph, which I hope will give you as much pleasure as it has given me:
>
> > 'Confidence in the high command is absolutely without parallel. Literally dozens of embarking troops talked about General Montgomery with actual hero-worship in every inflection. And unanimously what appealed to them – beyond his friendliness, and genuineness, and lack of pomp – was the story (or, for all I know, the myth) that the General "visited every one of us outfits going over and told us he was more anxious than any of us to get this thing over and get home". This left a warm and indelible impression.'
>
> The above is an exact quotation. Having spent my life with American soldiers, and knowing only too well their innate distrust of everything foreign, I can appreciate far better than you can what a triumph of leadership you accomplished in inspiring such feeling and confidence.[24]

There were two major briefings of the Overlord plan: on 7 April (Good Friday) and 15 May at Monty's old school, St Paul's, in Chiswick, south-west London. At the first, Monty, in the presence of Eisenhower, presented the detailed plan for the ground assault against the beaches: Ike recalled that 'an entire day was spent in presentation, examination and coordination of detail'. According to Bradley:

> A relief map of Normandy the width of a city street had been spread on the floor of a large room in St Paul's School. With rare skill, Monty traced his 21st Army Group plan of manoeuvre as he tramped about like a giant through Lilliputian France ... During our battle for Normandy, the British and Canadian armies were to decoy the enemy reserves and draw them to their front on the extreme eastern edge of the Allied beachhead. Thus, while Monty taunted the enemy at Caen, we were to make our break on the long roundabout road toward Paris.[25]

Bradley's recall of the eventual strategy is important because the conduct of the Normandy campaign would provoke huge debate, with allegations that the British were slow and had been 'fixed' by the Germans at Caen, when they intended to break out. Bradley went on:

> when reckoned in terms of national pride, this British decoy mission became a sacrificial one, for while we tramped around the outside flank, the British were to sit in place and pin down Germans. Yet strategically it fitted into a logical division of labors, for it was toward Caen that the enemy reserves would race once the alarm was sounded.

It was a long session, with Monty introducing the day and speaking for ninety minutes. Ramsay and Leigh-Mallory's presentations concluded the morning. The afternoon saw Bradley and his V and VII Corps commanders (Gerow and Collins) present the American forecast of their battle, followed by Second Army's General Sir Miles Dempsey with Lieutenant Generals John Crocker and 'Gerry' Bucknall (I and XXX Corps). General Sir John Kennedy, Director of Military Operations at the War Office, observed:

> When the conference started in the morning, Montgomery has asked us not to smoke . . . Later on he announced that he had had a message from Winston to say that he would join us after tea. Monty added that, as the Prime Minister would undoubtedly arrive with a large cigar, smoking would be allowed after tea. He made the announcement in such a puckish way that there was a great roar of laughter.[26]

So Monty the non-smoker could laugh at himself. The final conference at St Paul's on Monday 15 May was attended by the King, British Chiefs of Staff, the War Cabinet and scores of Allied generals. 'During the whole war I attended no other conference so packed with rank as this one,' remembered Eisenhower;

> This meeting gave us an opportunity to hear a word from both the King and the Prime Minister. The latter made one of his typical fighting speeches, in the course of which he used an expression that struck many of us, particularly the Americans, with peculiar force. He said 'Gentlemen, I am hardening toward this enterprise', meaning to us that, though he had long doubted its feasibility and had previously advocated its further

postponement in favour of operations elsewhere, he had finally, at this late date, come to believe with the rest of us that this was the true course of action in order to achieve the victory.[27]

Wartime austerity abounded and General Kennedy remembered: 'It was very cold. Winston and [Field Marshal] Smuts came in overcoats, and kept them on for luncheon . . . I got hold of a blanket and shared it with an Admiral and an Air Marshal; we were glad to have it on our knees.' The King's biographer noted that the hall was a big panelled room:

> The King and Mr Churchill accorded the privilege of armchairs, but the rest of the company sat on school forms facing a large map of the invasion area which hung above a low dais. After a brief introduction by General Eisenhower, each commander demonstrated his own particular role and task in the invasion . . . When the last . . . had concluded his statement on this portentous blueprint of the shape of things to come, the King rose and, to the surprise of all, stepped on to the platform. He had not been expected to speak and he did so without notes.[28]

George VI was by nature rather shy and not a natural public speaker, a situation aggravated by a pronounced stammer (which had been tentatively 'cured' by an Australian speech therapist), of which many of his audience were aware. Consequently it took personal courage on his part to mount the dais and make his short speech, which John Kennedy observed 'was perfect for the occasion, and created an excellent impression on the Americans, as well as on us. I met Alan Lascelles [the King's Private Secretary] at dinner . . . and he told me he had no idea the King would speak until he got onto the platform.'[29]

In later life some adults revisit the places of their childhood to restructure memories and reinvent their own imperfect youth in their own minds. It is impossible not to believe there was an element of this whilst Monty was based at St Paul's, his school from 1901–07. Monty had not chosen St Paul's: by chance the school had been requisitioned by 21st Army Group before it was known that Monty would be its boss (the smart money would then have been on Alexander, an old Harrovian). St Paul's was the land force headquarters, while Eisenhower's SHAEF had opened at Camp Griffiss, Bushey Park, Teddington; Ramsay's naval HQ was at Southwick House, Portsmouth; and Leigh Mallory was based at RAF Bentley Priory near Stanmore (from where the King, Churchill and Eisenhower would monitor the initial landing reports on D-Day). From the moment Monty arrived at St Paul's, he made the old school his territory. Consequently for the two key briefings Monty could not resist the idea of playing to his captive audience: the great and the good of Britain and America, who were seated on uncomfortable wooden benches facing the stage. There was no better way for Monty to reassure himself that he had made considerably more of his life than had been expected when he left the school thirty-seven years earlier than with the entire Allied top brass being lectured to by their 'headmaster', Monty, in this bizarre, austere setting.

General Hastings 'Pug' Ismay, Churchill's chief military assistant during the war (and Monty's exact contemporary in age) was at the 15 May conference and noted:

> The administrative arrangements were explained in considerable detail by General Humfrey Gale, the Chief Administrative Officer. In his desire to give an idea of the magnitude of the undertaking, he revealed that the number of vehicles to be landed within the first twenty days was not far off 200,000. The Prime Minister winced. He had been much amused, but shocked, by the tale – probably apocryphal – that in the Torch operation, twenty dental chairs had been landed with the first flight in from Algiers . . . the first instruction he gave me as we drove away from St Paul's was to write to Montgomery and tell him that the Prime Minister was concerned about the large number of non-combatants and non-fighting vehicles which were to be shipped across the Channel in the early stages of Overlord.[30]

Churchill certainly had an obsession about the administrative tail of 21st Army Group and was prepared to push the point sufficiently to want to quiz Monty's staff over what was absolutely necessary – a common tendency to become absorbed in the detail at moments of high stress (Churchill was very worried about D-Day, not least because he had overseen the Gallipoli operation in 1915 and it had made his position in the Cabinet untenable). Despite his lifelong enthusiasm for military affairs, such interference betrayed also Winston's lack of Staff College training. Montgomery took the 'long screwdriver' of Winston's interference personally – so much so that, when on 19 May Churchill visited Monty's own temporary headquarters at Broomfield House, in Southwick, Hampshire, the Prime Minister was led by Bernard into his study and behind closed doors the two strong personalities came head to head. No one else was present, but it was clearly a spirited debate, with Monty displaying the moral courage to stand up to his PM (correctly, as it happens, on this occasion). Monty glossed over the issue lightly in his memoirs, but it was a formative moment for both. Monty sat Winston down, then said:

> I understand, sir, that you want to discuss with my staff the proportion of soldiers to vehicles landing on the beaches in the first flights. I cannot allow you to do so. My staff advise me and I give the final decision; they then do what I tell them . . . I consider what we have done is right; that will be proved on D-Day. If you think it is wrong, that can only mean you have lost confidence in me.[31]

Monty played his cards very well, for he knew Winston could not replace him at such short notice. It was a charged atmosphere and both parties clearly became very emotional: in later life Churchill apparently threatened to sue Monty's first biographer Alan Moorhead if he revealed that he, Churchill, had broken down and wept. Eventually the doors opened and Churchill was introduced to Monty's waiting staff officers, and said 'with a twinkle in his eye', 'I wasn't allowed to have any discussion with you gentlemen.'[32] If it had proved difficult to keep Monty in check up to this point, henceforth it would prove impossible, for he had lectured his King in a cold classroom and now had browbeaten his Prime Minister into submission.

Southwick Park – a Regency-period country house near Portsmouth commandeered by the Navy (and still a military base today) – became the focus for senior commanders as the date of invasion neared. Ramsay established his HQ within the main building and had a custom-built huge map of Normandy installed in the former library. The map, commissioned from the Chad Valley toy company and made of plywood, included terrain from Norway to Spain, with ports, rivers and other details painted on. Of course only the Normandy pieces, which alone covered an entire wall of a large, high-ceilinged room, were relevant; the rest of the map, ordered to deceive curious minds, was burnt and the civilian workmen who assembled it, in becoming privy to a secret of world significance, found themselves detained until the invasion was under way. Churchill parked his personal train nearby, Eisenhower set up a trailer in the grounds and Monty lurked in nearby Broomfield House. The village pub, The Golden Lion, found itself serving drinks to an august range of celebrities that summer: apparently half-pints of bitter for Ike and grapefruit juice for Monty.

From late 1943 the British assault divisions had been in Scotland, training for their role, and in January 1944 moved to East Anglia. Dempsey's Second Army HQ was opened at Didlington Hall in Norfolk, before moving to the south coast, and then Normandy.[33] When they transferred to coastal ports in late May, troops were 'sealed' into quayside transit camps. Brigadier 'Sammy' Stanier recalled:

Maps were issued and everyone received 200 French Francs [the equivalent of £1]. Special ration packs, sea-sickness tablets and bags were issued. At a memorable church parade our padre blessed our home-made regimental flag . . . and the stirring personal message from Monty our C-in-C was read out.[34]

The 'stirring personal message' was a continuation of the Orders of the Day, 'to be read out to all troops', which Monty had pioneered with great success in the Mediterranean. He was not the first: Haig and his subordinates had issued them occasionally in the First World War (his 'Backs to the wall' of 1918 was the most famous) and Auchinleck had also used them ('Our friend Rommel' of 1 February 1942) – but Monty's were more frequent and more effective. This one was the eighteenth since his first, issued pre-Alamein on 23 October 1942, of an eventual thirty-one. Eisenhower also produced one on the eve of D-Day, and it is interesting to compare the two, for it tells us much about the men. Both orders were neatly printed on small pieces of paper about the size of a paperback book, and signed with facsimile signatures. Monty's first succinct message of the north-west Europe campaign began:

Personal Message from the C-in-C. To be read out to all Troops

1. The time has come to deal the enemy a terrific blow in Western Europe. The blow will be struck by the combined sea, land and air forces of the Allies – together constituting one great Allied team, under the supreme command of General Eisenhower.

2. On the eve of this great adventure I send my best wishes to every soldier in the Allied team. To us is given the honour of striking a blow for freedom which will live in history, and in the better days that lie

ahead, men will speak with pride of our doings. We have a great and righteous cause. Let us pray that 'the Lord Mighty in Battle' will go forth with our armies, and that His special providence will aid us in the struggle.

3. I want every soldier to know that I have complete confidence in the successful outcome of the operations that we are now about to begin. With stout hearts and with enthusiasm for the contest, let us go forward to victory.

4. And, as we enter the battle, let us recall the words of a soldier spoken many years ago: –

> *He either fears his fate too much,*
> *Or his deserts are small,*
> *That dares not put it to the touch*
> *To gain or lose it all.*[35]

5. Good luck to each one of you. And good hunting on the mainland of Europe.

Monty's message is replete with biblical and hunting terminology, appropriate for a bishop's son and a good horseman. Although he had last used the hunting phrase before invading Sicily in a 10 July 1943 message ('Good luck and good hunting in the home country of Italy'), it was probably a world away from the soldiers he was about to commit to battle. Monty's real appeal was reflected in his first and third paragraphs: the positive 'deal . . . a terrific blow' – the equivalent of 'we're going to hit the enemy for six' that he had written in his first message before Alamein. The 'I have complete confidence' of his third paragraph was a phrase he deployed often in his pep talks to soldiers, and with great success. If Monty's principal contribution to victory was training, his second was certainly the raising and maintenance of his troops' morale.

Eisenhower's altogether different message communicated a political vision, with talk of 'the Great Crusade' and the 'United Nations' and our 'Home Fronts', and reflected the strategic level at which he was commanding and the many different audiences he was addressing; no accident, therefore that he would one day become a great President. He too – like Monty – deployed a prayer:

Soldiers, Sailors and Airmen of the Allied Expeditionary Force!

You are about to embark upon the Great Crusade, toward which we have striven these many months. The eyes of the world are upon you. The hopes and prayers of liberty-loving people everywhere march with you. In company with our brave Allies and brothers-in-arms on other Fronts, you will bring about the destruction of the German war machine, the elimination of Nazi tyranny over the oppressed peoples of Europe, and security to yourselves in a free world.

Your task will not be an easy one. Your enemy is well trained, well equipped and battle-hardened. He will fight savagely.

But this is the year 1944! Much has happened since the Nazi triumphs of 1940–41. The United Nations have inflicted upon the Germans great defeats, in open battle, man-to-man. Our air offensive has seriously reduced their strength in the air and their capacity to wage war on the ground. Our Home Fronts have given us an overwhelming superiority in weapons and munitions of war, and placed at our disposal great reserves of trained fighting men. The tide has turned! The free men of the world are marching together to Victory!

I have full confidence in your courage, devotion to duty and skill in battle. We will accept nothing less than full Victory!

Good Luck! And let us all beseech the blessing of Almighty God upon this great and noble undertaking.

18

Britain's Last *Hurrah!*

L ONG AFTER THE event, the 1944 campaign in Normandy remains a battle of awe-inspiring proportions. The simple facts of D-Day itself still conjure visions of wonder in an era used to telephone-number salaries, international travel and lightening-speed communication, and represented an unprecedented feat of military planning: 195,700 seamen manned 6,939 ships, including 1,213 warships. There were 4,126 landing vessels of all types, including 1,073 tank landing craft, as well as 864 merchant ships, which by the evening of D-Day had deposited 132,715 troops and 20,000 vehicles directly onto the beaches, with another 23,490 parachutists and glider-borne troops dropped by the Allied air forces. The aerial armada supported the landings with 11,590 aircraft, which flew 14,674 sorties. In the previous nine weeks 197,000 sorties had been flown (at a cost of 1,251 aircraft and 12,000 aircrew) and 195,000 tons of bombs had been dropped on German military and communications targets.[1]

The overwhelming majority of the warships present in the waters off Normandy were from Britain's Royal Navy (892 RN and 200 US Navy, the balance coming from other nations); most of the landing craft were British (3,261) as opposed to American (865); a significantly higher number of Britons arrived by sea than Americans; many US infantrymen and Rangers landed from craft crewed by British sailors; more pilots and aircrew flying overhead were from Britain and her Empire forces than hailed from the United States. Whilst Dwight David Eisenhower, an extremely likeable and anglophile American may have been in charge, his deputy, Arthur Tedder, and the three land, sea and air commanders – Monty, Bertram Ramsay and Trafford Leigh-Mallory – were all British. At no further stage during the Second World War, or since, has Britain managed to match, let alone outnumber, her American partner in military personnel or resources deployed. British-equipped and uniformed Free French commandos also fought in the campaign under Monty's command; so too would an armoured division and an airborne brigade recruited exclusively from Free Poles, and Major General Liška's 1st Czechoslovak Armoured Brigade, the Piron Brigade of Belgians and Luxembourgers, and the Dutch Prinses Irene Brigade. Thus, 6 June 1944 was Britain's final day (with Monty at the helm) as a superpower. Thereafter, throughout the Normandy campaign, the nation was increasingly overshadowed by the more numerous and prosperous United States. As such D-Day constitutes, for the Empire and imperial influence, a final *Hurrah!*

From a purely European point of view, Allied cooperation for the campaign in which Monty's role as the land force commander was pivotal, was the antecedent of NATO and politically anticipated the European Union. American, British, Canadian, Free French, Belgian, Dutch, Luxembourger, Polish, Czech, Greek, Danish and Norwegian naval and military forces all took part, or flew overhead with the vast air armada. Australians, South Africans and New Zealanders also participated, and even the Russians had observers present. Many soldiers from the Irish Republic fought with British units, whilst anti-Nazi Austrians and German Jews were involved in the Allied deception and intelligence war. The *Wehrmacht*'s ranks were equally diverse, and included Russians, Romanians, Italians, Poles, Czechs, Ukrainians and others fighting (with varying degrees of enthusiasm) for Hitler.

The Normandy campaign lasted for seventy-seven days (as against the ninety days predicted by Montgomery and his planners), and resulted in the destruction of the German Seventh Army and Fifth Panzer Army in the Falaise Pocket by 21 August; 209,672 Allied soldiers would be killed, wounded or posted missing; 16,714 Allied aircrew lost their lives, whilst estimated German losses were 250,000.[2] Additionally, 2,483 Normans connected with the French resistance were executed by the Germans before or during the campaign, whilst possibly as many as 35,000 civilians died (the lowest estimate is 15,000) and 60,000 wounded in the liberation. This averages out at 6,674 casualties *per day* for the entire campaign. Such a figure, of course, doesn't reflect the reality of any single day, but provides a useful benchmark against which to measure other campaigns. This daily average actually *exceeds* the Great War daily casualty rates of Verdun 1916 (2,389 per day over 299 days), the Somme 1916 (6,444/day over 142 days), or Passchendaele 1917 (4,601/day over 113 days).[3] By mid-July of 1944, the region was more heavily populated with troops than an equivalent-sized sector of the Western Front in the First World War: there were 2,052,299 Allied troops and nearly half a million German soldiers in Normandy by the end of that summer of 1944.[4]

Ironically, given all Bernard Montgomery's experience of the Great War and concerns to avoid unnecessary bloodshed, these casualty statistics suggest that Normandy was more attritional than the worst battles of the First World War. And though we may argue perhaps over the exact accuracy of some of these figures – no two sets of statistics for big battles are ever the same – the fact remains that Normandy was far more costly than the Allies had anticipated – and that is perhaps the measure of the generalship of Monty and Rommel in June–July 1944. And yet, for all modern Western armies, Normandy remains the defining example of innovative leadership on both sides set against a backdrop of an attritional slog between well-matched and motivated opponents.

As the planned day for invasion approached, the weather in the English Channel became stormy, the worst weather in the Channel for twenty years. Heavy winds, a five-foot swell at sea, and lowering skies compelled Eisenhower – after a series of nail-biting conferences at Southwick Park, to where all the senior commanders had gravitated – to postpone the assault from Monday 5 June, by twenty-four hours. Conditions remained poor, but SHAEF's chief meteorologist, Group Captain James Stagg, advised that there would be a window of thirty-six hours of adequate weather in which to launch a seaborne invasion, something his German opposite numbers failed to foresee. Ike's decision to go was made at Southwick House with his senior commanders: Monty was bullish and urged an immediate start; Leigh-Mallory was pessimistic, for high winds would scatter the airborne troops and possibly ground some of his aircraft. They went, but a further wave of bad weather resulted in a worse storm between 19–22 June, which sank or beached 800 Allied ships, mostly landing craft, and severed the submarine cable that enabled Monty to talk to Southwick. As a result, Stagg later wrote to Monty observing that had he, on 5 June, advised a further postponement to 19 June, as he might, the result would have been disastrous: the Allies either might have just landed but, isolated from reinforcements and with air cover grounded, have been destroyed by Rommel, or would have been about to land and defeated by the storm, like the Spanish Armada. We often overlook the fact, therefore, from the comfort of the twenty-first century, that D-Day itself was an incredibly fortuitous, close-run affair.

Though he later agreed 'only Monty could have got us across the Channel', the ultimate burden of responsibility for Overlord lay with Dwight David Eisenhower. Initially, Ike is a puzzle: the overall commander in Europe during 1944–45, yet possessing no combat experience whatsoever. Almost without exception, his military and political contemporaries from all nations had seen action in both the First World War and the early years of the Second World War. What were Eisenhower's special qualities to presume to command so great an enterprise? Born of poor parents in 1890, he was commissioned into the infantry from West Point, in the Class of 1915 (famous for producing 59 generals out of 164 graduates – Omar Bradley was a contemporary) but missed being posted overseas during the First World War. His aptitude for staff work was demonstrated in 1926, when he graduated first out of 275 at the US Army's Command and General Staff School, Fort Leavenworth, and again in 1928, when he attended the Army War College. Lieutenant Colonel Eisenhower's career changed course when he attracted the patronage of the US Army's Chief of Staff, George C. Marshall, in late 1941 whilst serving with the Third Army as Chief of Staff.

After Pearl Harbor, Ike was appointed Brigadier General to head the Army War Plans Division, and later Major General of the Operations Division in Washington DC. This concept of a professional staff officer is anathema to the British, who intersperse staff work with regimental duty, whilst the Germans followed more closely the US model. In June 1942, Marshall promoted him over 366 senior officers to be commander of US troops in Europe, a radical move which was more than justified, but unthinkable in some other armies. That it worked was due to Eisenhower's personal charm and tact, where a lesser man would have excited envy or bitterness. This personal quality, as well as Marshall's patronage, is surely the key to Ike's success. He was the ultimate 'team player', able to reconcile different national interests, as well as being a first-class staff officer, whilst possessing the inner confidence of his exact British contemporary in age, the patrician Irish Guardsman Harold Alexander.

From his experience of overseeing the landings on Sicily and at Salerno, it is not now difficult to see why Eisenhower was appointed to lead the Normandy invasion – though still only a substantive lieutenant colonel. Crucially, he had the support of military contemporaries and the backing of his political masters. It is easy with hindsight to assume that Normandy was always going to succeed, but the responsibility of launching the huge force in bad weather weighed heavily; the success of the landings was a vindication of Monty's choice of the site and the time to go. At times in Normandy and later on, Ike found himself as much keeping the peace between Monty, Patton and Bradley as fighting the Germans, which is precisely why he had been selected Supreme Commander, to manage his team, all of whom were brilliant, in their own ways. In reaction to Monty's elevation to the rank of field marshal on 1 September 1944 (five-star rank, which the US armed forces did not have), Roosevelt initiated a Congressional Act which created US five-star ranks on 12 December. Eisenhower was one of the first recipients, receiving his promotion to General of the Army on 20 December 1944.

After the early days in the desert, Monty rarely got on well with airmen. He fell out with 'Mary' Coningham, Tedder's deputy and later commander of the Desert Air Force in the Mediterranean, and by Normandy would only deal with Coningham's deputy, Air Vice Marshal Harry Broadhurst. Much of the acrimony seems to have been generated by Monty refusing to give due acknowledgement to the air arm for its part in the campaigns. By Normandy, Tedder (another airman he came to detest) had become Ike's deputy at SHAEF, and was technically Monty's superior, which caused much tension. Finally, Trafford Leigh-Mallory (younger brother of the noted climber George Mallory, who died on Everest in 1924), the air component commander, was notoriously prickly with everybody, which led to much inter-service bickering, notably with Monty. Leigh-Mallory had grown increasingly hostile to the use of the three airborne divisions, until Eisenhower demanded they be deployed. With Leigh-Mallory's expectations of very high casualties among the airborne forces ringing in his ears, Ike bid farewell to the 101st Airborne Division before their gliders carried them off to battle from Newbury; a journalist recording the scene later told friends he had seen the Supreme Commander's eyes water. Monty, by contrast, turned in for an early night at Broomwood House, near Southwick. Meanwhile, field hospitals in southern England were readied to process a minimum of 30,000 casualties from D-Day. On 12 February, Monty's 21st Army Group HQ had sent an estimate of British and Canadian D-Day casualties to the War Office: it arrived at the grim conclusion that out of a landing force of 70,000 there would be 9,250 casualties, including 3,000 men drowned.[5] This planning figure escalated dramatically when the invasion force was effectively doubled, and US troops were included.

The poor weather actually worked to the Allies' advantage. When the BBC broadcast some lines of the French nineteenth-century poet, Paul Verlaine at 9.15 p.m. on 5 June announcing to the resistance that the invasion was imminent ('*Blessent mon coeur d'une longueur monotone* – 'Wound my heart with monotonous languor'), Oberst Wilhelm Meyer-Detring, chief intelligence officer at Rundstedt's headquarters, realised their significance. He persuaded Salmuth's Fifteenth Army HQ on the northern edge of Tourcoing to go onto a higher invasion alert status, but Dollmann's Seventh Army HQ at Le Mans and Speidel at La Roche-Guyon did not follow suit, surmising that an invasion force would not put to sea in such conditions. Indeed, many divisional commanders in Seventh Army's area had already left for Rennes to participate in a war-games exercise designed, ironically, to simulate an Allied landing in Normandy. Rommel himself had left at 6 a.m. on 4 June for his home in Herrlingen, near Ulm, for his wife's birthday on the 6th, and also to visit Hitler at Berchtesgaden to ask for reinforcements.[6] It is ironic that for all Rommel's efforts with the Atlantic Wall he was absent at the precise moment the landing craft ramps touched down. It is impossible not to believe that Rommel's personal decisiveness (and moral courage to disobey even Hitler if the circumstances demanded) would have caused a more violent German reaction to D-Day than was the case. He might not ultimately have won in Normandy, but the Allies would have had a much harder time getting ashore. Twice Rommel had the misfortune to be away from his headquarters at a turning point of the war: earlier he had missed the start of El Alamein.

As planned, airborne units led the invasion, with the aim of seizing key bridges beyond the landing beaches, which were natural choke points. Pathfinders landed first by parachute, then paratroopers, followed by glider-borne troops later in the day. Leaving aerodromes, including Harwell, Fairford and Brize Norton, shortly after midnight, Major General Richard ('Windy') Gale's British 6th Airborne Division dropped north-east of Caen, near the mouth of the Orne River, in Operation Tonga, where it anchored the British eastern flank by securing crossings over the Orne river (Horsa Bridge) and the Caen canal (Pegasus Bridge), and attacking the Merville coastal defence battery. Taking off from Tarrant Ruston near Blandford, Major John Howard led 'D' Company, 2/Oxfordshire and Buckinghamshire Light Infantry in a textbook assault on his two bridges,[7] but Lieutenant Colonel Terence Otway's 9/Parachute Battalion was widely scattered and only 150 out of 550 attacked Merville. Instead of deadly 150mm guns, the battery was found to contain obsolete 100mm cannon, and Merville was retaken by the Germans (it changed hands four times in the next twenty-four hours), underlining just how hazardous night-time airborne operations really are.*

Amongst the very first soldiers to jump into Normandy were 22nd Independent Parachute Company – Pathfinders, charged with marking 6th Airborne Division's drop zones and setting up Eureka ground-to-air radio beacons to attract the incoming transport aircraft.[8] To the west, the US 101st and 82nd Airborne Divisions dropped near St Mère-Eglise and Carentan to secure road junctions and beach exits from which it was planned that US VII Corps would move quickly to secure the key port of Cherbourg. Some of the US airborne troops landed near their objectives, but most were scattered over a wide area. Many of the 'missing' drowned in the marshes behind Utah, deliberately flooded as part of the German defences. In Operation Detroit, the three parachute infantry regiments of Major General Matthew B. Ridgway's 82nd ('All-American') Airborne landed on drop zones west of St Mère-Eglise, severing all German ability to communicate with Cherbourg. Whilst colleagues cleared their sectors, 3/Battalion, 505th Parachute Infantry Regiment took the town itself, and Private John Steele caught his parachute on the steeple of the town's church. The oldest US airborne soldier to have landed on D-Day appears to have been forty-two-year-old J. Strom Thurmond, a glider-borne lieutenant colonel with 82nd Division, and later senator of South Carolina for fifty years, who died aged 100 in 2003.

American paratroopers flew by twin-engined C-47 Dakota transports to Normandy, flying first south, and then turning east over the Channel Islands, to approach from the west. The 101st flew from Exeter, eighteen to a plane, at a height of 1,500 feet, and jumped into battle at 700 feet. Flying in several waves, the division left England at 11.50 p.m., arriving over their target two hours later. It was a lonely night for all the Allied

---

* Since D-Day no major army has attempted a large-scale night-time parachute drop; the concept is seen as simply too hazardous to repay the investment.

paratroopers during the early hours of 6 June. Many, lost amongst the *bocage* hedgerows, entangled in trees, or dropped by disorientated aircrew, failed to reach their assembly areas, but their very presence sowed confusion in German-held Calvados. When reports trickled back to German HQs of parachutists, there was little indication as to their size or objectives; commanders were initially reluctant to commit forces against what might prove to be a diversion or a coastal raid, like Dieppe. The Allies added to the confusion by parachuting dummies, luring German units away from the landing zones and the coast, where their presence might have done considerable damage to the attackers.

Operation Chicago saw paratroopers of Major General Maxwell D. Taylor's 101st ('Screaming Eagles') Division land on drop zones behind Utah and between Carentan and St Marie-du-Mont, ready to clear the way for the seaborne troops. The paratroopers of the 501st, 502nd and 506th Parachute Infantry Regiments were more scattered than their colleagues in 82nd Airborne Division, due to the high winds and accurate anti-aircraft fire, but had been issued with a toy 'cricket' to click for recognition. Taylor landed completely alone, later recalling that 'any order he might have given would have been received only by a circle of curious Normandy cows'. Later in the morning, more airborne troops landed by glider, but fewer than half the flimsy, plywood and canvas gliders assigned to 82nd Airborne reached their assigned landing zones intact, the rest catching anti-landing stakes, or landing in the flooded areas. By mid-morning as many as 4,000 men of the 82nd were unaccounted for, whilst the 101st had lost their assistant divisional commander, Brigadier General Pratt, killed sitting in an armoured jeep in his glider (an incident that featured in the movie *Saving Private Ryan*). Both US airborne divisions suffered heavily, the 82nd losing 1,259, whilst the 101st suffered 1,240 casualties. A further large-scale (or so it seemed) airborne drop was created by Operation Titanic, when 500 dummies were parachuted into Normandy with SAS soldiers at four sites. On landing, recordings of gunfire were played by the SAS; the ruse was completely swallowed by the Germans who moved troops to counter the phantom army. Meanwhile, Operation Glimmer saw RAF aircraft drop quantities of silver foil strips ('window') over the Channel, which was known to jam German radar. Rommel had been at the receiving end of Monty's deceptions before Alamein and was no stranger to using such stratagems himself. One is left wondering if he might have seen through any of the Allied ruses, had he been physically present in Normandy that night.

As dawn neared, medium bombers began to strike up and down the coast, flying the first of what would become, by the end of the day, more than 11,000 sorties against enemy batteries, headquarters, railways and troop concentrations. The gliders Eisenhower had watched departing from Newbury also arrived. Unlike the parachutists, the glider-borne troops of 6th (British) and 101st (US) Airborne Divisions landed mostly on target, though 82nd Airborne were less fortunate, and as much as sixty percent of the equipment they carried was lost or damaged. A second, daylight, lift fed in airborne reinforcements to the three divisions. At 7 pm on D-Day, in Operation Mallard, Halifax bombers towed Horsa and tank-carrying Hamilcar gliders across to Normandy, containing 6th Air Landing Brigade, equipped with Tetrarch tanks, Bren-gun carriers, 25-pounder field guns, scout cars and even Bailey bridge pontoons.

By first light at 5.30 a.m. on 6 June 1944, the entire horizon off Normandy between Ouistreham and Vierville-sur-Mer was filled with a huge seaborne armada. The naval bombardment began at 5.58 a.m., detonating minefields and destroying many blockhouses and artillery positions. Emerging from behind a screen of smoke, three E-boats attempted briefly to contest the attack, but only one inflicted any damage, sinking the warship *Svenner*, a British S-class destroyer sailing under the Free Norwegian flag. This was the *Kriegsmarine*'s sole contribution to the fight that day. The surface fleet was always too small to make a difference and the U-boat arm had already been defeated by superior technology in the Battle of the Atlantic. Likewise, the failure of the Luftwaffe to appear over the D-Day beaches caused no end of bitterness and imposed an operational effect – the Allies had learned their lessons well from Dieppe and the Mediterranean landings on Sicily and at Salerno. In Normandy the Luftwaffe managed no more than 4–500 sorties a day, even fewer on D-Day itself, and their greatest success would come from sowing sea mines in the Seine Bay later in the campaign. By contrast the Allies flew a hundred times that amount – 49,000 sorties in the first six days. As the Allied naval bombardment ended, the Germans had a brief respite before the landing craft beached and the troops waded ashore. The rough sea was a challenge for the smaller landing craft assigned to British I and XXX Corps as they headed for Gold, Juno and Sword Beaches, aiming to touch down at 7.25 a.m., and for US VII and V Corps, which began to disembark on Utah and Omaha at the earlier time, due to tidal variations, of 6.30 a.m.

A detailed look at the attack on Sword Beach, as conceived by Monty, will give a flavour of 6 June 1944. Sword, the coast between Lion-sur-Mer and La Brêche, and subdivided by the planners into four sectors (following the phonetic alphabet of the day – Oboe, Peter, Queen and Roger), was assaulted at low tide by Monty's old 3rd Division. Two battalions led the attack on Queen Beach, 2/East Yorkshires and 1/South Lancashires, both of 8th Brigade, supported by swimming Sherman tanks of 13/18th Hussars and 5th Assault Regiment, RE, in armoured engineer tanks. Some swimming tanks were launched into the rough seas and soon sank, whilst others were delivered by landing craft directly onto the beaches – a total of eighteen made it ashore. Within the first few minutes five engineer tanks, eight Sherman minesweeping tanks and two armoured bulldozers were knocked out, but soon the engineers had cleared eight lanes through the obstacles.

Shoreline opposition came from individual bunkers and two local strongpoints code-named Cod and Trout, which were overwhelmed rapidly, and by 9.30 a.m. 8th Brigade had captured Hermanville but eventually came up against artillery on the Périers Ridge. Brigadier the Lord Lovat's 1st Special Service Brigade (the abbreviation of SS for Special Service was soon dropped for obvious reasons) of four Commando battalions – preceded by Lovat's personal piper, Bill Millin – also landed on the eastern edge of Sword and assaulted Ouistreham from the west, then marched to Pegasus Bridge to link up at about 1.30 p.m. with 6th Airborne Division. The kilt-wearing Millin remarked to me, years later, that (wearing nothing underneath his kilt) the coldness of the water took his breath away. The onshore wind drove the tide in so fast that the engineers had insufficient time to clear more obstacles before the next formation arrived from about 10.30 a.m., in observance of Monty's strict timetable. Amongst the second-wave units at Sword was 2/Royal Warwicks, a battalion of Monty's old regiment. One of its platoon commanders, Lieutenant Kingston 'Tiny' Adams, recollected how in the half-light

> it was misty and we could just make out the houses behind the beach which were getting heavily bombed. All the men were packed below deck and it was pretty grim for them. There was a lot of seasickness and they did not get much sleep. I had all my men to worry about, so this took my mind off what was about to happen – but the privates only had their own concerns to dwell on; it was harder for them.

His platoon struggled ashore under the weight of 70-pound (32kg) equipment loads, the shock of the cold water momentarily stunning them. Tiny had joined his regiment on leaving Shrewsbury School with my father and had spent his war years training for D-Day. But the next afternoon a bullet ignited a phosphorus grenade in his webbing, burning his hands and legs. Evacuated by medics, he was so badly burned that it took him eleven months to recover, and he became one of the few non-RAF members of the Guinea Pig Club of burns victims treated by the pioneer of plastic surgery Archibald McIndoe during the war. After years of training, Tiny's war had lasted a little over a day.[9]

The three battalions of 185th Brigade, including 2/Royal Warwicks and their supporting armour of the Staffordshire Yeomanry, were soon stuck in traffic jams for at least an hour, queuing to pass through the few lanes cut through Rommel's obstacle belt, whilst harassed by mortar and artillery fire; 9th Brigade – the divisional reserve – did not land until after midday. Nevertheless, by nightfall some 28,845 personnel had passed through Sword, a slightly higher total that those who had landed on Juno or Gold. Whilst 3rd Division secured their beachhead and linked up with the division on their left (6th Airborne), they failed to reach 3rd Canadian Division on their right; although 21st Panzer Division started to exploit this gap in the inter-divisional boundary, they were unaware of its significance and soon withdrew.

Far more importantly, 3rd Division failed to seize Caen as planned (which has been interpreted since as a strategic setback of the highest order), for two reasons: partly because the deployment of 21st Panzer Division distracted the advance on Caen, and also because of Oberst Ludwig Krug's resolute defence of the strongpoint code-named Hillman. This complex of bunkers on high ground south of Colleville was surrounded by wire entanglements and minefields, housed machine guns and heavier artillery, and proved a far tougher nut to crack than anticipated. The inability to capture Hillman throughout 6 June delayed the entire 3rd Division's advance from Sword. Krug, commanding 736th Infantry Regiment (of Wilhelm Richter's 716th Division), was disgusted that a nearby artillery position, code-named Morris, had fallen to the attacking Lieutenant Colonel Dick Goodwin's 1/Suffolks at about 1.30 p.m.: apparently the garrison of sixty-seven surrendered with suitcases packed, having spent the preceding months running a Calvados distillery and sausage-meat factory, rather than troubling themselves with military training.

Lulled into a false sense of security, 1/Suffolks may have anticipated the early surrender also of Krug's position, but soon came to grief: their intelligence summaries prior to D-Day show that Hillman's hardened steel cupolas and deep concrete shelters had been overlooked by aerial photo analysts. It needed two attacks to take Hillman (which should have been a brigade task) before it was subdued by about 9 p.m. – the Suffolks and 1/Norfolks suffering almost 200 casualties between them in the process. Though he was personally unaware of the significance of his actions during the afternoon of 6 June, Krug at Hillman managed to draw the attention of two infantry battalions, a tank squadron and an armoured engineer squadron – at that moment a huge proportion of 3rd Division's available combat power. But it was not until 6.45 a.m. on 7 June that Krug emerged with three of his officers and seventy surviving other ranks from an underground bunker that had been overlooked the previous night, having almost single-handedly sabotaged beyond recovery Montgomery's plan to take Caen on D-Day.[10]

Sword Beach was one of the invasion sites of Frederick Morgan's original COSSAC plan, approved in outline by Churchill and Roosevelt at the Quebec Conference (code name Quadrant) in August 1943. Morgan and his team had envisaged D-Day as a three-division seaborne attack, and the attacking division at Sword (eventually chosen as Monty's old 3rd) was to have assaulted with two of its three infantry brigades forward (as at Gold, Juno and Omaha). But on Monty's revision of the COSSAC plan in January 1944, the force destined to begin the attack on Sword Beach was revised downwards to a single brigade, largely because of the offshore reefs constraining the available sea space and passage to the shore. Whilst this made logistical sense, and, in keeping with Monty's anxieties over casualties, would have minimised risk to the attackers, the revision did not sit easily with the task.

The primary D-Day objective of 3rd Division was to capture or dominate (the phrase used was actually 'mask') Caen, Normandy's route centre, which offered roads out of the region onto the Caen-Falaise plain (good for deploying the Allied surfeit of armour) – and beyond. Initially Morgan had advocated a break-out from Normandy in the east, but Monty revised the concept to that of an Anglo-Canadian force in the east attracting German reserves – particularly of armour – to the Caen area, whilst the Americans initiated a break-out in the west. Major General Tom Rennie's 3rd Division also had secondary missions on D-Day, which included linking up with 6th Airborne Division to the east, 3rd Canadian Division to the west, securing the beachhead and creating an armoured reserve to deploy against the expected enemy tank counterattack. Beyond doubt this was far too much for a single division. In addition, because 3rd Division was trickling its strength onto the battlefield – landing brigades sequentially, not simultaneously – it could not deploy its full strength until mid-afternoon and the most that Rennie could spare to throw at Caen that morning were a couple of infantry battalions and a regiment of sixty tanks.

At about 1.00 p.m. a mortar bomb landed on a 9th Brigade conference, seriously wounding the commander, Brigadier Cunningham, and some of his staff. Earlier the CO of 5/Assault Regiment, RE had been killed shortly after H-hour (the hour of assault), as had the commander of 5/Beach Group. On the beaches, the CO of the South Lancashires had been killed by a sniper and that of the East Yorkshires wounded by mortar fire. Incidents like these also slowed down 3rd Division's response to Krug's rugged defence at Hillman and the urgency of reaching Caen. The assault on Sword Beach, then, is best envisaged as a funnel into which the mass of 3rd Division with its armour and supporting units was tipped at the top, but relatively little dribbled out at the bottom, and that, slowly.

Whilst 3rd Division did extraordinarily well, at a cost of around 650 casualties, it is clear with hindsight that deploying a second brigade in the assault wave onto Sword Beach (as on three of the other beaches) might have enabled 3rd Division to reach Caen on 6 June, which would have changed the course of the Normandy campaign. Therefore, whilst Montgomery's revisions to the original COSSAC plan altered the organisation and sequence of the attacking troops, 3rd Division's tasks were not reassessed; if they had been, then the first conclusion would have been that Major General Rennie had too few with which to achieve too much, and that he needed at least two brigades ashore initially, or an even stronger attacking force: the Americans, for example, planned to land elements of two divisions (1st and 29th) ashore on Omaha Beach. Rennie, who commanded a brigade under Monty in North Africa, does not appear to have objected, but neither did his corps commander, John Crocker. In Montgomery's army, complainers (Monty called them 'belly-achers') were not tolerated – so there might have been severe reluctance to challenge the plan; everyone wanted 'in' on the big day. The fact that no one intervened illustrated a flaw in the planning cycle, and suggests that Monty should have thought of revising his plan sufficiently to enable

3rd Division to have a fighting chance of seizing Caen, whose northern outskirts went undefended for most of D-Day*. With rich hindsight, Brigadier Nigel Poett of 6th Airborne Brigade later observed that one of his companies could have taken Caen in the morning; by the afternoon it would have needed a battalion, and at nightfall such a task would have required a focused divisional effort. Whilst Monty had been busy interviewing a hundred battalion and brigade commanders, he should perhaps have been casting one final, critical eye over his plan instead.[11]

Also under Monty's watchful eye, along the five-mile stretch of Juno Beach, Major General Rod Keller's 3rd Canadian Division deployed 8th Canadian Brigade to St Aubin and Bernières at 7.45 a.m., whilst to their right, colleagues of 7th Canadian Brigade had attacked Courseulles ten minutes earlier. They were supported by 2nd Canadian Armoured Brigade, equipped, trained and clothed exactly as their British counterparts. In the centre of Second British Army's sector, the Canadians' multiple tasks were to establish a beachhead, capture the three small seaside towns, and advance ten miles inland. Their objective was to cut the Caen–Bayeux road, seize the Carpiquet aerodrome west of Caen, and form a link between Sword and Gold Beaches. The all-volunteer Canadian Army would later also deploy its own 2nd Infantry and 4th Armoured Divisions through Juno. Of the 21,400 Canadian and attached British troops who landed at Juno on 6 June, losses totalled 1,204, but by nightfall the Canadians were further into France than any other division. The opposition they had faced was stronger than that of any other beach save Omaha; this was indeed revenge for 2nd Canadian Division at Dieppe and an accomplishment in which the whole nation could take considerable pride.[12]

Gold Beach (itself subdivided into Jig and King sectors) was wide enough for two brigades of Major General Douglas Graham's 50th Division to land side by side, with H-hour at 7.25 a.m.. The formation was known as the Tyne-Tees division due to their divisional flash featuring two 'Ts', the two principal rivers in their recruiting area; and a trace of this badge remains painted on a house wall in Ver-sur-Mer. In the western sector of Gold, Brigadier 'Sammy' Stanier's 231st Brigade landed at le Hamel (Jig Beach), and to the east 69th Brigade assaulted La Rivière (King Beach). It was here that Company Sergeant Major Stan Hollis of

---

*Modern armies have started to use mentors (usually retired senior officers) to provide an impartial sanity check during key military operations of this type.

'D' Company, 6/Green Howards, almost single-handedly subdued the German artillery battery at Mont Fleury overlooking King Beach and won the only Victoria Cross of D-Day. Australia's only warrior to land on D-Day, Major Jo Gullett, was also attached to 6/Green Howards, and the battalion later advanced to Crépon that afternoon, where today a fine bronze sculpture of a 1944 Tommy stands as the Green Howards' memorial in the village square, the face being a likeness of Hollis. Further west, it took 231st Brigade until late in the afternoon of D-Day to capture the bunker complex at Le Hamel, whilst in the east 69th Brigade had lost ninety-four men killed capturing La Rivière. By about 11.00 a.m., General Dietrich Kraiss, commanding the local 352nd Division, was so convinced that the American attack on Omaha Beach had failed that he actually redirected his reserves to the Gold Beach area, where the British appeared to pose a greater threat against his 914th Regiment. They might have made all the difference at Omaha but made not the slightest impression at Gold, where 50th Division was already striking far inland. By nightfall, 24,970 men had landed on Gold, for a loss of 413 killed, wounded or missing, and a link-up had been made with 3rd Canadian Division from Juno, on the left flank. On the right flank of Gold, a link with the Americans on Omaha was only made on 8 June (D+2), after 47 Royal Marine Commando had taken Port-en-Bessin, for 200 casualties.

As under Monty in the desert, British troops had not yet appreciated the need to fully integrate infantry and tank units on the battlefield. Consequently, each British and Canadian infantry division that splashed ashore on D-Day was followed by its heavy metal – a distinctly separate armoured brigade – which included three armoured regiments. Regiments were equipped with thirty-ton M4 Shermans, in three squadrons, each of four troops of four. By this date, every fourth Sherman carried a 17-pounder gun capable of destroying a Tiger or Panther; the remaining Shermans would mount a 75mm gun until the availability of 17 pounder guns increased. Armoured brigades also possessed a motorised infantry battalion, accompanying them in Bren carriers and lorries; proper protection in the form of armoured half-tracks was not yet available in sufficient quantities to be allocated to infantry, although the casualties of the next month would soon correct this flaw in doctrine. This motorised battalion was the only formation trained to work with tanks, or who could keep pace with an armoured advance. Each British tank regiment supported an infantry brigade, and each armoured squadron (roughly twenty tanks), a battalion. Frequently the cooperation extended down to troops of four tanks being assigned to infantry companies, but none of these arrangements was permanent and British doctrine was that the armour departed at last light to league, refuel and re-arm, which left the infantry vulnerable, bitter and often jealous of their cavalier colleagues.

By 1944 Rommel's troops had adopted a superior system of infantry – armour cooperation – a fact overlooked by Montgomery, the doctrine writer and trainer. One of the outstanding features of German armoured divisions was that their mobile infantry (called Panzer Grenadiers) could keep pace with their tanks by riding into battle in armoured half-track vehicles, which had cross-country mobility. In British armoured divisions at this time, the infantry support generally rode in unarmoured lorries which were confined to roads, thus tanks had no infantry support when manoeuvring cross-country. This doctrinal flaw also contributed towards the failure to seize Caen: had an armoured brigade with infantry in half-tracks landed at H-hour and set off immediately with the specific task of seizing Caen they would certainly have reached the town and might have been able to hold it, given the paucity of local German reserves.

The popular images of the D-Day invasion we have today are over-whelmingly of US equipment and soldiers, despite the fact that it was a minority (albeit substantial) of Americans who landed in the first hours, compared with British and Canadians. The absence of images of the distinctive soup-bowl-shaped Tommy helmets and woollen khaki battle-dress is largely down to the ludicrous suspicion the British War Office had (and their successors at the MOD retain) of photographers and journalists. Monty, the arch publicity hound, did nothing to correct this major historical mistake. Nearly two hundred civilian journalists and photographers roamed the American sector on D-Day; not one was allowed into the British zone. One of the best-known US photographers, *Life Magazine*'s Robert Capa, landed on Omaha Beach with Company 'E' of the 16th Regimental Combat Team, and struggled ashore, dodging mortar bombs, taking photographs all the way. Recalling the day in his memoirs, Capa observed:

the flat bottom of our barge hit the earth of France. The boatswain lowered the steel-covered barge front, and there, between the grotesque designs of steel obstacles sticking out of the water, was a thin line of land covered with smoke – our Europe, the 'Easy Red' beach. My beautiful France looked sordid and uninviting, and a German machine gun, spitting bullets around the barge, fully spoiled my return . . . with the invasion obstacles and the smoking beach in the background . . . I paused for a moment on the gangplank to take my first real picture of the invasion . . . The water was cold, and the beach still more than a hundred yards away. The bullets tore holes in the water around me, and I made for the nearest steel obstacle. A soldier got there at the same time, and for a few minutes we shared its cover. . . . The sound of his rifle gave him enough courage to move forward, and he left the obstacle to me. It was a foot larger now, and I felt safe enough to take pictures of the other guys hiding just like I was.[13]

Capa made for a disabled Sherman tank, feeling 'a new kind of fear shaking my body from toe to hair, and twisting my face'. With trembling hands he reloaded his camera and used all of his three rolls of film, exposing 106 frames. Witnesses remember him wading about, holding his cameras high above his head to keep them from getting wet. Eventually he followed a medic team carrying a stretcher to a waiting craft; ahead of him most of the medics were cut down by machine-gun fire but he was hauled out of harm's way by the crew. Full of wounded, the slow-moving boat eventually reached England on the evening of Wednesday 7 June, when Capa committed his films to a waiting courier who sped them by train to the *Life* darkroom in London's Piccadilly for developing. Capa then returned to Normandy. The darkroom assistant was so anxious to see the invasion images that, in his haste, he turned up the heat to dry the film too quickly and the emulsion ran. Out of 106 exposures, only eight ('the magnificent eight') blurred, almost surreal shots, alive with drama, were recovered.[14]

Coverage of the British and Canadian landings was left to the skills of a mere fifteen men from 5th Army Film and Photographic Unit. Ingeniously, the Canadians also fixed movie cameras inside some assault landing craft, which were triggered automatically by the action of lowering the bow ramp (the resultant footage, shot from behind, of helmeted men scrambling out of rocking landing craft, gripping weapons and ladders, has become part of the core imagery of many TV documentaries). Eight AFPU men landed on Sword, two were wounded on D-Day and two killed soon after; among them Sergeant George Laws who landed with No. 4 Commando on Sword Beach:

I took several shots of the other assault craft on the way and as we neared the beach everybody was ordered to crouch down during the final run-in . . . On landing I found my ciné camera had been switched on . . . and about a third of my 100ft roll of film had been wasted. I wound up the clockwork and chased after the Commandos to their assembly area. In that rush up the beach, they had suffered forty casualties including their CO. After they had assaulted the battery in Ouistreham, and we were moving inland to support 6th Airborne, a German plane, flying very low, strafed us and bullets splattered along the road – no time to take cover. Later, when mortar and shell-fire caught us, I dived into a ditch and broke the spring on my camera's motor. What an anti-climax to all my training![15]

Despite Monty's courting of the press in the Mediterranean theatre, they were kept at arm's length throughout the north-west Europe campaign, and especially on D-Day. This call was certainly Monty's to make, both as commander of two national contingents (Canadian and British) within 21st Army Group, and as overall land component commander. It seems he was good at reacting to any opportunities the press offered him, but was not proactive in his relationship with them. Press conferences were held, but were rare; there was no attempt to shape the news, as today. Ultimately, the press of 1944 were a breed that Monty did not understand and he failed to obtain the maximum value from their presence on the battlefield. Crucially for future generations, the relative paucity of British media presence in Normandy has created a false impression for future generations of American dominance in the campaign.

19

Where is Rommel?

THE FIRST INKLING amongst Rommel's men that anything was wrong came when two British paratroopers from 6th Airborne Division overshot their drop zone and landed on the front lawn of *Generalmajor* Josef Reichert's 711th Divisional headquarters at Cabourg, east of the Orne river, where the general was playing cards with his staff at about 1 a.m. At the same time, 709th Infantry Division reported capturing airborne troops from the US 101st Division. General Erich Marcks, commanding LXXXIV Corps, immediately deduced from these two reports, received at about 1.45 a.m. on 6 June, that with paratroopers of separate divisions (and nationalities) dropping to the east and west, the Calvados coast was about to witness the invasion. Incredibly, the earlier invasion warning from Oberst Meyer-Detring at Rundstedt's HQ had not been taken seriously by either Speidel at La Roche-Guyon or Dollmann's Seventh Army, and had not been passed down to subordinate commands, so this was the first Marcks knew of a possible assault.

General der Artillerie Erich Marcks was considered one of the brightest General Staff officers of his generation and one of Rommel's most talented subordinates in Normandy. He left behind a reputation both of an efficient practitioner of war, but also a cultured man with a love of the arts and sciences (as chief of staff of Eighteenth Army, with whom he invaded Belgium and France in 1940, he took steps to ensure the preservation of Bruges and the Seine bridges in Paris[1]). In the autumn of 1940, Marcks was involved in drawing up plans for Army Group North's eventual invasion of northern Russia. He was given command of 101st *Jäger* Division, and was wounded in June 1941 four days into Operation Barbarossa, which resulted in the loss of his right leg. After a long period of convalescence, Marcks was given 337th Division in March 1942, based on the Loire, then St Malo in Brittany. Promotion to LXXXIV Corps followed in August 1943 and Marcks' command spanned the future invasion front.[2] Before the invasion, Rommel had proposed that Marcks be given command of Seventh Army (which might have had a decisive impact on the Allies), but Hitler rejected this, preferring the more politically reliable – though inept – Dollmann, with disastrous consequences. Perhaps Rommel also saw in Marcks the acceptable face of a non-Nazi German army.

The 6th of June saw Marcks celebrating his fifty-third birthday (he was portrayed in the film *The Longest Day* limping with his false leg and cutting his birthday cake). Six days later, on 12 June, he insisted on carrying out his daily round of inspection of the invasion front, despite the protests of his staff. Within fifteen minutes of leaving his HQ his car was caught by Allied fighters on a main road near Hébécrevon as the morning fog lifted. The general died during the encounter in a roadside ditch. Whatever Hitler's reservations about Marcks' political reliability, his military skills were undoubted and the Führer arranged for a posthumous award of Oakleaves to his existing Knight's Cross.

From the two reports of paratroopers and others that began to arrive soon after, Marcks had concluded the invasion was beginning and persuaded Seventh Army to issue an alert at 2.00 a.m. on the 6th. This quality of quick thinking and decisive action by Marcks and Rommel mark them out as quality commanders, compared with the indolence of Dollmann (who had left his Seventh Army headquarters for the war games in Rennes) and others. Alert and monitoring the command frequencies, Rundstedt's headquarters actually ordered Panzer Lehr and 12th SS Divisions ready to move at 4.00 a.m., which would have seen them in the beachhead area on D-Day. But because of the Rommel–Geyr acrimony over control of the panzers, all movements of armour were subject to Berlin's approval, which was not forthcoming (for fear that Normandy was a feint) until 4 p.m. that afternoon. When, on Rundstedt's request, General Alfred Jodl, Deputy Chief of Operations Staff at Armed Forces High Command (OKW) refused to release the panzer divisions at about 7.30 a.m. (his boss, Wilhelm Keitel, and Hitler were still asleep and unaware of developments), Rundstedt reacted with exceptional bitterness at what he perceived as a slight to the Third Reich's senior field marshal. Urged by his Chief of Staff, Blumentritt, to phone Hitler direct, Rundstedt let pride overcome him and refused to beg before 'that Bohemian Corporal'.

Meanwhile, Speidel appears to have dithered also, probably expecting an invasion – if one was to appear at all – to materialise somewhere in Fifteenth Army's area. The evening before had seen Speidel hosting a dinner party at La Roche-Guyon in Rommel's absence, which may have been a gathering of anti-Hitler plotters, for the guests included the celebrated anti-Nazi writer Ernst Jünger (the ex-stormtrooper who had been Hitler's favourite author). Speidel's initial sluggishness of response may have been due to a hangover, but was talked up after the war as highlighting his anti-Nazi credentials. He was already aware of reported paratroop drops in Normandy when, at 6.15 a.m., Dollmann's Seventh Army Chief of Staff, General Max Pemsel, reported a naval bombardment but conceded he had no reports of seaborne landings. At about this time, Speidel took the precaution of phoning the news through to Rommel in Herrlingen (awake at that early hour preparing his wife's birthday presents) that something was afoot, but did nothing more.

The phone logs at Seventh Army HQ and La Roche-Guyon show that Pemsel called Speidel back at 9.10 a.m. to report actual landings on the Calvados beaches, and the two discussed countermeasures for about thirty minutes. Meanwhile, Allied journalists had been summoned to London University's Macmillan Hall for an 8.30 a.m. press conference, where Colonel Ernest Depuy of SHAEF's Public Relations Division read a brief communiqué, prepared days earlier, announcing to the world that the invasion was in progress. With this confirmation, it fell to one of Hitler's staff – actually Rommel's friend, Rudolf Schmundt – to wake his Führer and tell him the news shortly after 10 a.m.

Rommel took a second call from Speidel at 10.15 a.m., whilst tending his wife's roses in the garden at Herrlingen. (According to Cornelius Ryan, Montgomery was busy pruning roses too on D-Day, at Southwick Park, as he had little else to do.) Only at this stage did Rommel realise that a major landing was being attempted; apparently the colour drained from his face as he listened to Speidel's report. Soon Rommel was on the road, racing towards La Roche-Guyon, but with a journey of over 500 miles ahead of him. Thus the panzers were not forthcoming with the timeliness or in the quantity required because of an over-complicated command network and petty jealousies within the German army and

Nazi party. In the meantime, Berlin even reprimanded Rundstedt for issuing preliminary orders to the panzer units without prior approval.

It took Rommel all day to reach La Roche-Guyon, his powerful Horch staff car, driven by *Gefreiter* Daniel, speeding the hundreds of miles through towns, scattering pedestrians, its horn blaring. It paused briefly in Rheims at 6 p.m., where Rommel spoke to Speidel, before continuing. Hauptmann Helmuth Lang, Rommel's aide-de-camp, was with his boss and noted that Rommel remained quiet and grim-faced throughout. At about 10.30 p.m. the Horch's tyres finally threw up the gravel in the darkness at the front entrance of the La Roche-Guyon château.

By landing at low tide, the Allies had circumvented all the obstacles of 'Rommel's asparagus', for it stretched only between the high- and low-water marks, though following units were caught as the tide ebbed in again. Rommel realised that only two assets could prevent the Allies from gaining a foothold in Europe: the Luftwaffe and the panzer divisions. Unlike those whose service had been primarily in Russia, Rommel had seen the Allied air forces at work against his Afrika Korps and had witnessed the technological might and ingenuity of the British Eighth and First Armies. He fully expected the same, but greater, to be deployed against northern France. On 17 May, he had told Fritz Bayerlein, commander of the Panzer Lehr Division:

> Our friends from the east [those who had served on the Russian front] cannot imagine what they're in for here. It's not a matter of fanatical hordes to be driven forward in masses against our line, with no regard for casualties and little recourse to tactical craft; here we're facing an enemy who applies all his native intelligence to the use of his many technical resources, who spares no expenditure of material and whose every operation goes its course as though it had been the subject of repeated rehearsal. Dash and doggedness alone no longer make a soldier, Bayerlein; he must have sufficient intelligence to enable him to get the most out of his fighting machine. And that's what these people can do; we found that out in Africa.[3]

Though Rommel already feared the Luftwaffe to be a spent force, he was far more optimistic about the German armoured forces. Fortunately for the Allies, all day on 6 June he was too far away to have any influence over the deployment of the panzers. Nevertheless, late in the afternoon of D-Day, Oberst Hermann Oppeln-Bronikowski's armoured brigade of 21st

Panzer Division, the nearest armoured unit to the beaches, which had been ready to move since the first alerts, began to threaten the outskirts of the beachhead. Oppeln, a Prussian aristocrat from a long line of distinguished generals, was a talented tank commander whom Rommel had almost sacked when he discovered him drunk on duty during a lightning inspection earlier in the year. After the war he would coach a riding team in the 1964 Olympics, but D-Day required precision and excellence of a different sort. His boss, *Generalmajor* Edgar Feuchtinger – hauled out of a Parisian nightclub in the small hours – had been reluctant to deploy his division without Berlin's sanction. But at 6.30 a.m., when he saw it would not be forthcoming, he ordered his tanks to move anyway, by which time it was already far, far too late.

Oppeln's two tank battalions had been initially ordered to attack 6th Airborne *east* of the Caen canal; had they moved earlier, his tanks would have massacred the lightly armed paratroopers, who had virtually no anti-tank weapons. However, Marcks, the local corps commander, soon perceived that the real threat was from the sea, and the panzers were needed back along the coast, not rounding up scattered parachutists. En route, the tank battalions were ordered to turn around and attack *west* of the canal instead. With relatively few miles to go, Oppeln nonetheless had to travel far inland to avoid being ambushed and dodge the attentions of Allied aircraft in order to reach his start line. Thus it was only mid-afternoon when Oppeln's Panzer IVs deployed to where they needed to have been by mid-morning. As they approached the Sword bridgehead from the south, beyond Caen, at about 4 p.m., they were ambushed broadside-on by Sherman tanks of Colonel Jim Eadie's Staffordshire Yeomanry, who accounted for seventeen within minutes. This was exactly Rommel's fear: that if his tanks were stationed too far away, the seaborne attackers would have consolidated their positions by the time the panzer counterattacks began.

Colonel Eadie, a Territorial soldier who had served under Montgomery at El Alamein, had anticipated such a move and had his Shermans stationed hull-down at the ready. His Yeomen wore a red triangle behind their regimental badge, for before the war Eadie had been a director of Bass breweries in Burton-on-Trent, whose corporate logo, the Red Triangle, instantly marked out his company's beer in pubs throughout the Midlands. In the undergrowth nearby lurked the infantry of the 2nd King's Shropshire Light Infantry, who were by then supposed to be in Caen. Leaving burning and crippled panzers in the fields, the remnants of Oppeln's armoured spearhead (the *spitz*) kept going, accompanied by some of 21st Panzer's armoured infantry – the panzer grenadiers of 192nd Regiment. Led by the one-legged Marcks, standing erect in his staff car, they and six of Oppeln's tanks appear to have punched a narrow corridor through to the coast at Luc-sur-Mer, between Sword and Juno Beaches at about 8 p.m.

When, later still, a formation of planes overflew them, discharging parachutes to their rear, nerve failed and General Feuchtinger commanding 21st Panzer Division, in radio contact and monitoring the operation, ordered a withdrawal – thus losing the chance to surround and isolate two beachheads. (Ironically the parachutes were a supply drop that had gone off course, not a counterattack against the tanks. If Feuchtinger had even tried to assess the situation calmly and rationally, he would have realised that the aircraft had taken off from southern England two hours earlier and couldn't possibly be reacting to his presence on the battlefield.) For Oppeln's panzer men, this had been their only hope of defeating the invasion – by preventing the separate beachheads from merging, then surrounding each one in turn and destroying them.

The leadership of *Generalleutnant* Feuchtinger, commander of 21st Panzer Division, would not have been worthy of note, except that it was so poor and illustrates the contradictions of the Nazi regime.[4] Feuchtinger had not been at his divisional HQ on 6 June 1944, but in a nightclub in Paris. Though his presence would not have made a great deal of difference, the point was he was absent when needed. Despite his division relying on captured French armour (it was the only panzer division rated unfit for service in Russia), elements of 21st Panzer performed very creditably against overwhelming odds in the Normandy campaign, largely because of its able regimental and battalion commanders. This brought Feuchtinger a Knight's Cross and promotion to *Generalleutnant*, despite the observation of Rundstedt's deputy chief of staff that 'Feuchtinger took to his heels' when he aborted the armoured thrust to Luc-sur-Mer on D-Day. But on 5 January 1945, Feuchtinger was arrested and imprisoned at Torgau Fortress to face a court martial over his absence from divisional headquarters on the night of 5–6 June 1944. Apparently, when the Nazi official investigating his case arrived at 21st Panzer's HQ on Christmas Eve 1944 (as the division was engaged in heavy combat on the Saar), Feuchtinger was again absent without leave. He was found guilty, sentenced to death (though later pardoned) and reduced to the rank of *Kanonier* (artilleryman), his rank in 1914.[5]

It seems incredible that Feuchtinger was trusted with any divisional command at all, let alone an armoured one in what was likely to become a key theatre – a decision for which Hitler and the Nazi regime were to blame. Perhaps Rommel tried to replace him but found him too well protected politically.

Americans have always had a reluctance to allow their forces to be placed under command of a foreign national. For D-Day, they relented and permitted Monty, as overall Ground Commander (and subject to Eisenhower's scrutiny), control of all US land forces.[6] Diplomatically, Monty was obliged to *advise* rather than *command*, a task which fell to Omar Bradley, commanding the First US Army. Concerned that Hitler would respond to the invasion with extreme violence, and might even resort to poison gas, Eisenhower's Chief Surgeon, Major General Albert W. Kenner, and the Chief Surgeon of the US Army's European Theatre of Operations, Major General Paul R. Hawley, had prepared their staffs to process at least 12,000 killed and wounded on D-Day in the First US Army alone. In the British sector, owing to reefs and geological obstacles, troops had landed on a flood tide an hour to an hour and a half after their American counterparts at Utah and Omaha. 'This gift of time,' wrote Bradley afterwards, with a hint of bitterness,

> enabled the warships of the Royal Navy to deliver to Monty's beaches a two-hour daylight bombardment, nearly four times the length of the bombardment at Utah and Omaha. To this was added a massive attack by British heavy bombers. The combined air and sea attacks in the British sector were far more effective than those in the American sector.[7]

The Americans insisted on landing as soon after first light as possible to preserve tactical surprise. At Utah, German artillery managed to sink the destroyer USS *Corry* and swift currents carried the landing craft of 4th Division well to the south of their intended beach, ironically to a stretch that was very lightly defended. One of the strongpoints attacked was WN5, commanded by Leutnant Arthur Jahnke, whose position had been inspected personally by Rommel on 11 May. Despite the fact that Jahnke sported a Knight's Cross, won in Russia, and was clearly made of 'the right stuff', Rommel asked him to remove his gloves and hold out his hands; on seeing they were covered with scratches from barbed wire, demonstrating he was working as hard as his men, Rommel grunted his approval and moved on.

Within three hours, any resistance from Jahnke and his colleagues had been overcome and US troops of the 4th Division were moving inland, linking up with the airborne forces. The latter had ambushed and killed *Generalleutnant* Wilhelm Falley, commander of 91st *LuftLande* (Anti-Airlanding) Division, who was caught in his staff car, racing to his command post. One is struck by the paralysis of the German command network at critical moments throughout the Normandy campaign: Falley was one of several key commanders killed or wounded on D-Day, blunting the German operational response. A report from Rommel at the end of June 1944 enumerated losses of twenty-eight generals and 345 commanding officers. Despite the determination of some formations to counterattack, their efforts were undermined by the confused and even contradictory German chain of command. The paralysis inflicted upon the defenders' commanders by the violence and surprise of the Allied assault ensured that no coordinated activity was possible against the invasion at operational level.

By nightfall, some 23,000 men had landed on Utah, at a cost of 197 casualties among the ground troops. The 4th Division's assistant division commander, Brigadier General Theodore Roosevelt (son of President Theodore Roosevelt and a distant cousin of Franklin Delano), landed with the first wave and had to decide whether to continue landing on the wrong beach, with the implied logistics nightmare, or shift to the planned beach. In an excellent example of 'mission command', he directed the landing to continue where he was ('We're going to start the war from here'), having achieved his commanders' intent of a successful landing. Dominating Utah were two powerful gun emplacements that neither sea nor aerial bombardment suppressed. Whilst the Azeville Battery (with four 105mm guns) surrendered on 8 June, the Crisbecq Battery, housing four powerful 210mm guns, held out until 12 June, all the while harassing the landing operations.

If the attack was going well on Utah, the situation was quite different on Omaha. There, the beach is five miles long, overlooked by sandy cliffs ('bluffs') approximately thirty metres high, and when first chosen by the COSSAC planners was undefended. A night-time beach reconnaissance had been carried out in January 1944 by Captain Logan Scott-Bowden of the Royal Engineers and Sergeant Ogden-Smith of the SBS, who had swum ashore from a midget submarine, each armed with only a commando knife and a Colt .45 automatic, and returned with soil and sand samples. Immediately on their return, Scott-Bowden (who would also witness D-Day itself) was whisked off to COSSAC headquarters at Norfolk House in St James's Square, to be debriefed by Omar Bradley on the nature of the beach; his work brought him an immediate DSO.[8]

Prompted by Rommel's inspections and energy, it was only in March 1944 (surprisingly late) that Seventh Army started sowing Omaha with beach obstacles and cliff-top bunkers, making its assault a very different proposition. Five valleys ('draws') wind off the beach inland, of which only two were suitable for armour or wheeled transport. German strongpoints dominated these exits, and the slightly concave shape of the coast meant that German guns could enfilade the length of the whole beach. At low tide, the depth of the exposed sands was about 460 metres; above the high-tide mark was a shelving shingle bank, offering minimal cover for troops and an impediment to vehicles. On the western portion of Omaha Beach, just behind the shingle, was a seawall constructed of wood and concrete. Between the seawall and the cliffs, a sandy shelf ran for a depth of approximately 200 metres and over this area the Germans placed concertina wire and minefields; the final physical barriers were the cliffs. Overlooking Omaha were fifteen miniature fortresses (WN Nos 60–74), containing artillery and machine guns targeted on the beach itself. They were mutually supporting and covered every scrap of the terrain; interconnected by trenches to allow deployment of troops between defensive positions, they were protected by barbed wire and minefields. The catalogue of Omaha's defences included eight artillery casemates, thirty-five pillboxes, six mortar pits, eighteen anti-tank guns, approximately forty rocket-launcher sites, eighty-five machine-gun nests and six tank turrets, nearly every one set in a thick, reinforced concrete position. Additionally, there were four artillery batteries on call in the vicinity

The 1st and 29th Infantry Divisions of Leonard T. Gerow's V Corps experienced the worst conditions encountered of any formation during the invasion. Known as the 'Big Red One' after its divisional emblem, 1st US Infantry Division was to deploy its 16th Regimental Combat Team to capture the western beach exit and link-up with troops from Utah, whilst 116th RCT (of the 29th US Division) was to take the eastern exit, and make contact with the British advancing from Gold. Bradley deliberately used elements of two divisions because the Omaha troops had to plug a gap of twenty-five miles between Gold and Utah, and he expected both complete divisions to be ashore before nightfall. High seas swamped many landing craft during their ten-mile run in, survivors reaching the beaches seasick and weak. Of two companies (twenty-nine tanks) from 741st Tank Battalion approaching the eastern half of the beach, all but two swimming Shermans foundered in the heavy seas they had not been designed to withstand.[9] With an H-hour of 6.30 a.m., there was only half an hour of daylight for supporting fire from the eighteen warships and 450 B-24 Liberator bombers assigned to soften enemy positions. Blinded by thick cloud, pilots overshot their targets and dropped their loads up to three miles inland.

The assaulting troops, however, had not anticipated strong opposi-tion, as only a battalion of the 716th Division was expected to be manning the coastal defences. The appearance of 352nd Division (not picked up by Ultra or the French resistance) came as a nasty shock; worse still, its 916th Regiment were manning many of the fortifications on an anti-invasion exercise, alert and with full scales of ammunition. US soldiers had been led to expect an effective destruction of the German defences by the navy and airforce, but in the half-hour prior to H-hour, little had been accomplished. In a narrative distressingly similar to the briefing for British troops before 1 July 1916 on the Somme, the historian of the second-wave 115th Infantry Regiment wrote:

it came as quite a shock to many when, just prior to going ashore, the men . . . heard that they might have to land fighting. Briefing had stressed the fact that the landing itself would be relatively simple; that troops would merely walk ashore, make for the high ground, and then walk until the objective was reached.[10]

Omaha itself was a tangle of obstacles: logs tilted towards the sea with mines lashed to their tips; steel rails welded together at angles to smash in the keels of landing craft, dominated by concrete bunkers and machine-gun nests. Most of the radios required by artillery spotters ashore went down on landing craft that never reached the beach, so the effort to destroy the beach obstacles was only marginally successful. In the vanguard of the attack, members of the 6th Engineer Special Brigade and naval demolition units attempted to cut avenues of approach to the beach by blasting lanes through the obstructions. Despite mishaps (of sixteen bulldozers, for example, only three survived enemy fire, and one was prevented from moving by American riflemen who sought shelter

Montgomery was not the first choice to command the Eighth Army; Lieutenant General William Gott shared a school (Harrow) with Churchill and a regiment (KRRC) with Eden and had unrivalled desert experience, but was felt by Brooke to be 'tired'. His death in a mid-air assassination on 7 August 1942 cleared the way for Monty.

It was Air Marshal Arthur ('Mary') Coningham who first co-located his headquarters caravan with those of the ground commander in the desert. Initially he and Monty cooperated very successfully, but later the relationship soured.

The fellow feeling between Allied and Axis soldiers was prompted by a sense of being at war with the desert itself. All had to contend with flies, snakes, scorpions, spiders – and the acute shortage of water.

The successful management of logistics was the key to victory in the desert. Food, fuel, weapons and ammunition – like these used shell cases during the Battle of El Alamein – were all at the end of a long logistics 'tail'; ultimately too long for the Afrika Korps.

Tunisia, June 1943: King George VI reviews his air force accompanied by Air Marshal Coningham (*back seat*) and Monty (*front seat*) in the latter's staff car, 'Old Faithful'. The truth behind this picture is that Monty and Coningham were barely on speaking terms and the P-40 Kittyhawks in the background had been massacred in huge numbers by the Luftwaffe.

Montgomery hands
them out

OUR
FIGHTING MEN

They long for a Smoke

JUST ten shillings will send 500 good Cigarettes, duty free, to the Fighting Men—not forgetting the Wounded in ★ Casualty Clearing Stations and Base Hospitals overseas, and in Hospitals in this country—and help to cheer them up.

£1 will send 1,000.

Postcards will be enclosed with every parcel, bearing your own name and address, to bring back to you grateful thanks from the men themselves !

Don't they deserve it ?

★ *A Commanding Officer writes :—* " Many thanks for your cigarettes. They have gone to the Wounded at the Salerno Beaches."

MOS/175 ⓢ P.T.O.

Monty had enjoyed tobacco as a young man but became a notorious anti-smoker after his 1914 lung wound. Nevertheless he used cigarettes as a means of boosting his troops' morale. This promotional bookmark exhorts the wartime British public to support the smoking needs of their troops.

GENERAL MONTGOMERY

Bernard had a difficult and distant relationship with his son, David, an only child whom he saw infrequently. Here Monty relaxes with him and David's guardians, Major and Mrs Tom Reynolds on a rare visit home in 1943.

Americans never knew what to make of Monty: those who knew him found him abrasive, but to the US public, he was the opposite of British stuffiness, with his signature beret, pullover and flying jacket. A Brit on the front cover of *Life* (here on 15 May 1944) was a rare event indeed and a tribute to the closeness of Anglo-American cooperation.

Monty's pep talks to his troops were innovative and hugely popular within the field army. He initiated these as a conscious reaction to the perceived distance in the First World War between senior commanders and their men.

The chateau at La Roche-Guyon was part mansion and part medieval castle. Overlooking the Seine, it was chosen as the HQ of Rommel's Army Group B in 1944, being midway between his two subordinate army commands and von Runstedt in Paris. The cliffs (*right*) offered additional protection for his extensive staff.

Rommel greets Wehrmacht nurses at a field hospital in early 1944. He may have been hero-worshipped in the Fatherland, but after his affair with Walburga Stemmer in 1912, and birth of their daughter Gertrud, he remained loyal to Lucie and resisted the attentions of female admirers.

Rommel has just presented an accordion to troops defending the French coast. He saw his role in 1944 as much to boost morale as to inspect and improve defences of the Atlantic Wall.

Major General Edgar Feuchtinger owed his command of 21st Panzer Division to Nazi zeal, rather than military competence. Almost anyone else would have posed a dangerous threat to the Allied invasion, but Feuchtinger had to be hauled out of a Paris nightclub to be told of the landings.

THE SECOND ROUND

Reflecting widespread public opinion, the British satirical magazine *Punch* saw the 1944 Normandy campaign as an extension of the desert duel between the two commanders. Within five days of this 12 July cartoon, Rommel would be rendered *hors de combat* by Allied air power.

Field Marshal Gunther van Kluge, who replaced first von Rundstedt, then Rommel after the latter's injury. Kluge arrived in Normandy full of Hitlerian enthusiasm but soon came round to Rommel's view that the campaign in the west was being grossly mishandled.

(*Above*) History commends Eisenhower to us as an easy-going, friendly Supreme Commander, who Monty nevertheless managed to offend on several occasions during and after the war. Monty (correctly) saw the need for an overall Land Component Commander in the European campaign; but (incorrectly) saw himself as the ideal and only candidate. (*Right*) By the time this photograph had appeared in the *Illustrated London News*, Monty had just been promoted to field marshal. On his left sits Miles Dempsey, British Second Army commander, who loathed publicity as much as Monty loved it.

THE ILLUSTRATED LONDON NEWS

SATURDAY, SEPTEMBER 2, 1944

18 October 1944: The charade of Rommel's state funeral in Ulm. It was here that the SS leader Heinrich Himmler sent word to Frau Rommel that he had taken no part in the events of her husband's forced suicide.

Monty's reputation was sustained post-war by Eighth Army reunions, held annually in London's Royal Albert Hall.

Monty never had an easy relationship with his king. After the war Monty spent an unhappy time as CIGS (Chief of the Imperial General Staff – his role in this 1948 photograph) but subsequently was hugely influential in the establishment of NATO as a credible military force in Europe.

Monty rarely relaxed in public. An appearance with a pretty woman on his arm – in this case the actress Constance Moore – was even more unusual.

After the war Fritz Bayerlein (*left*) and Rommel's son, Manfred (*right*), helped restore the Desert Fox's reputation, building on Desmond Young's bestselling biography of 1950.

Monty the author, then in his seventies, displaying tattoos including his regimental badge of an Antelope first acquired in India before the First World War. During the Second World War he usually wore his sleeves rolled down – even in the desert – to conceal them.

around it), they succeeded in opening six complete gaps. Even so, they could mark only one, because all the buoys and poles had been destroyed. Casualties in both 5th and 6th Engineer Special Brigades were about forty percent, mostly incurred in the first half-hour. British combat engineers were, of course, equipped with specialised armour, which would have made all the difference here.

The story of *Feldwebel* Hein Severloh was typical of the experience of many of the German defenders. After firing over 12,000 rounds from his MG-42 machine gun, Severloh's bunker took a direct hit at about midday. Wounded, and with no more ammunition – and no prospect of a resupply – he withdrew to nearby Colleville. Eighteen-year-old Franz Gockel was another defender in WN 62, whose position had been inspected by Rommel on 30 March 1944. Of the invasion day itself, he remembered:

> The alarm call into the bunker woke us from a deep sleep. A comrade stood in the entrance and continued to shout the alarm, to dispel any doubt, and urged us to hurry. We had so often been shaken to our feet by this call in the past weeks that we no longer took the alarms seriously, and some of the men rolled over in their bunks and attempted to sleep. An NCO appeared in the entranceway behind our comrade and brought us to our feet with the words 'Guys, this time it's for real. They're coming!'[11]

The artillery bunkers at WN 62 ran short of 75mm shells at this time and soon afterwards WN 61 to the east also fell silent. Whilst Severloh, Gockel and their friends stayed in their bunkers gunning down their opponents until their ammunition was exhausted, no attempt was made to launch any form of counterattack. At Omaha Beach this would have been decisive and swept the demoralised Americans from the beaches. Unaware that his defences were beginning to crumble, Oberst Ernest Goth, commanding 916th Regiment, after watching the massacre on the beach below him, sent this report to his divisional HQ:

> At the water's edge the enemy is in search of cover behind the coastal-zone obstacles. A great many motor vehicles, among them ten tanks, stand burning on the beach. The obstacle demolition squads have given up their activities. Disembarkation from the landing boats has ceased . . . the boats keep further out to sea. The fire of our battle positions and

artillery is well placed and has inflicted considerable casualties on the enemy. A great many wounded and dead lie on the beach.[12]

This was precisely in line with Rommel's expectations, and what he had instilled his men along the coast to expect. More wreckage accumulated at the water's edge as timetables slipped and landing craft became hopelessly entangled on the beach obstructions, all the while under German mortar and artillery fire: less than a third of the first-wave troops even reached dry land. Lacking most of their heavy weapons, these survivors had little choice but to huddle behind sand dunes or in the lee of a small seawall. Captain Richard F. Bush of 16th Infantry Regiment recorded of his men:

> they lay there motionless and staring into space. They were so thoroughly shocked that they had no consciousness of what went on. Many had forgotten they had firearms to use. Others who had lost their arms didn't seem to see that there were weapons lying all around them. Some could not hold a weapon after it was forced into their hands. Others, when told to start cleaning a rifle, simply stared as if they had never heard such an order before. Their nerves were spent and nothing could be done about them. The fire continued to search for them, and if they were hit, they slumped lower into the sands and did not even call out for an aid man.[13]

This underlines the utility of tough, realistic training, to prepare men for the realities of war; seemingly V Corps's training was perhaps not hard enough. The inability to retreat back into the sea and the seeming reluctance to advance the distance across Omaha's sands to the seawall, stalled soldiers and encouraged the onset of battleshock. Some troops needed time to adjust to the situation before they started to act, but this varied from individual to individual. This period of adjustment could have been shortened by the presence of strong leaders, but most company and platoon commanders were killed or wounded when exiting their landing craft (1st Infantry Division Standing Orders specified that 'an officer be the first man to go off the boat'). Unit cohesion had all but disintegrated among the troops on Omaha, especially within the inexperienced 29th (National Guard) Division. Eventually, the weight of Allied firepower (from off-shore destroyers) told and more landing craft pushed their way to the beach, bringing fresh troops, heavy weapons, radios, ammunition and most importantly leaders.

Natural leaders also emerged from the survivors of the first wave, who gradually formed small groups and fought their way off the beach. Leaders of all ranks asserted themselves. 'Two kinds of people are staying on this beach,' the commander of 1st Division's 16th RCT, Colonel George A. Taylor, told his men: 'the dead and those who are going to die. Now let's get the hell out of here.' The NCOs were especially abrupt, and resorting to curses they had accumulated over the years, they exhorted their men to get off the beach. The assistant divisional commander of the 29th, Brigadier General Norman 'Dutch' Cota[14] was particularly effective as he ignored enemy fire and moved up and down the shore-line among his troops, cajoling, rallying and urging them forward. By nightfall, 34,000 men were ashore on Omaha, but had sustained losses of around 2,000. The beach itself was a shambles of burning and disabled vehicles, but almost all of the coastal villages located inland were in Allied hands.

The accusation is often made that the Americans were deliberately given the most difficult beaches, implying Montgomery's spite. Nothing could be further from the truth; the landing order reflected dispositions in southern England. It made military and logistical sense for US troops, based in Devon and Cornwall, to assault those beaches opposite them, whilst Canadians and the British, based in Hampshire and south-east England took on the eastern beaches of Normandy. There was no logic in complicated criss-crossing movements in mid-Channel to put the Americans ashore in the east. The landing order also had implications that remain into the twenty-first century, for when the Allies broke out of Normandy, with the Americans on the right, this translated in 1945 into US forces approaching and occupying southern Germany, where their descendants are still based, whilst the British, on the left, ended up garrisoning northern Germany throughout and after the Cold War.

20

Exploiting the Beachhead

B^{Y} 1944 MONTY had perfected a highly personalised system of commanding in the field via two headquarters: his small tactical headquarters (TAC) forward, close to the battle zone, and a main HQ (MAIN), run by his chief of staff, Freddie de Guingand, with most of the staff, further back. This had worked well in the desert and Italy for Eighth Army and he repeated the method for his elevated command of 21st Army Group. This physical division of headquarters, which is still widely practised today, would have benefited Rommel considerably, who instead rushed 'home' each evening to La Roche-Guyon.

Leaving his main headquarters under the watchful eye of Freddie de Guingand near Southwick House, Montgomery sent his tactical headquarters on ahead aboard three American-crewed landing craft. They arrived in France on 7 June, and landed with the minor mishap of 'the three-ton truck carrying the officers' and sergeants' mess kit, together with all the Scotch, slid off the ramp and disappeared into the sea'.[1] They set up camp with all the vehicles and personnel in fields at St Croix-sur-Mer, three miles inland, south-west of Courseulles. Montgomery, meanwhile, had set sail on D-Day evening at 10.15 p.m., aboard the destroyer HMS *Faulknor*, and arrived off France the following morning. He first found Bradley aboard his command ship, the cruiser USS *Augusta*, then Eisenhower aboard the fast mine-laying cruiser HMS *Apollo*. Ike recalled:

During the course of the day [7 June] I made a tour all along the beaches, finding opportunities to confer with principal commanders, including Montgomery. Toward evening and while proceeding at high speed along the coast, our destroyer ran aground and was so badly damaged that we had to change to another ship for the return to Portsmouth.[2]

Here Ike mistook his initial ship (HMS *Apollo*) for a destroyer: it was a destroyer, HMS *Undaunted*, which gave him and Admiral Ramsay a fast passage back to Portsmouth that evening. His flag, signed by Eisenhower using an indelible pencil dipped in whisky, became a wardroom trophy for many years. Monty, still aboard HMS *Faulknor* (which, much to the embarrassment of her captain, ran aground also), and conferring through the day with other commanders afloat, was retrieved by his TAC staff early on 8 June by DUKW, and went straight to the new headquarters, where his caravans awaited him. It was immediately obvious to all that the choice of St Croix was

an unsuitable if not disastrous one. It was too close to the enemy lines, and under regular shellfire, while the chaos as more and more troops poured into the cramped confines of the bridgehead made its operation almost impossible . . . Monty decided to move as soon as possible. Fortunately [his Canadian liaison officer] Trumbull Warren had located the grounds of a lovely old château.[3]

What is clear here is that all the senior Allied commanders were in Normandy within twenty-four hours; admittedly Eisenhower and Ramsay did not stay, but the important point was that they viewed the ground and situation, and gained a personal feel for the success of the operation. Monty's new venue was the eighteenth-century château at Creullet, which his TAC occupied until 22 June, though typically he slept in his caravan set up in the grounds rather than in the house. Albert 'Bert' Williams, who served with Montgomery in the TAC HQ at Creullet, and later Blay, recalled that Monty 'was a quiet man', who

liked to distance his own caravan, even from the rest of the HQ. He would surround himself with farm animals though. He found their company relaxing.[4]

It is clear that Monty liked his solitude; staff officers provided him with data, intelligence and assessments, but he appreciated the physical and mental space to arrive at solutions – hence the lonely settings for his TAC HQ, first in the desert, then Europe. His early officer training required him to be a competent horseman, and on campaign he acquired a menagerie, chiefly birds and dogs – not unlike Rommel, who also liked dogs and horses. Indeed, one of Monty's small dogs was named Rommel; so, too, was one of his horses, acquired after the war and apparently named after its former owner! Animals are often calming to the human soul – they react, but do not answer back. Monty was not by nature a committee man and the records do not suggest him sitting through time-consuming meetings discussing options. He did, however, relax over dinner with his young liaison officers before turning in early each night.

On 12 June, Monty welcomed Winston Churchill; the latter wrote of the visit:

> We lunched in a tent looking towards the enemy. The General [Monty] was in the highest spirits. I asked him how far away was the actual front. He said about three miles. I asked him if we had a continuous line. He said 'No.' 'What is there to prevent an incursion of German armour breaking up our luncheon?' He said he did not think they would come.[5]

Here was the Churchill–Monty game of one-upmanship again, with the seventy-year-old Winston still pining for combat, but with Bernard definitely having the upper hand.

The village of Creully, beyond, was liberated on 7 June and the BBC were directed to set up their first Field Broadcasting Unit in the nearby twelfth-century castle which dominates the area. (Perhaps Monty felt safer with newsmen whom he could lock up in a castle when he chose.) For their benefit, Monty held his first press conferences at Creullet. On 14 June he hosted General de Gaulle, and two days later King George VI, on a day-long tour of Normandy, sped out and returned in the cruiser HMS *Arethusa*, writing in his diary how he had lunched with Monty and decorated some troops (including Major John Poston, one of Monty's liaison officers, who received an MC, underlining the dangers of working at TAC), then moved to Monty's map caravan, where he received an explanation as to how the battle was going. An apocryphal tale (at least I have not been able to track its source) of Monty's relationship with his sovereign went along the lines of Eisenhower remarking to His Majesty in a light-hearted moment that he felt Montgomery was after his job as Supreme Commander, at which the King immediately retorted: 'You relieve me greatly; I thought he was after mine.'[6]

It was at Creullet that Monty established the routine for his TAC HQ that would last for the rest of the war: TAC was his own personal fiefdom, and after the war he retained his three caravans, first keeping them at his home in Hampshire, then leaving them to the Imperial War Museum. Monty's TAC was highly mobile and could disassemble everything (tentage, signals masts, wires) and move within about forty-five minutes and re-establish itself elsewhere within a similar timeframe. TAC had a couple of unique components: one was the 1st GHQ Liaison Regiment, known as Phantom, which was a secret intelligence and communications unit. Monty's requirements for a constant, real-time picture of the battlefield were serviced by Phantom, which provided patrols – teams of specialist radio operators, drivers, despatch riders under an officer – attached to every headquarters of every division and corps within 21st Army Group, and all American corps headquarters. Patrols passed back tactical summaries direct to Monty via 'bomb lines', transmitting using the regiment's own cipher system, innovative radio equipment and an endless allocation of frequencies. Its officers, including the film actor David Niven and a bizarre number of future public figures including Hugh Fraser, Maurice Macmillan, Robert Mark, Christopher Mayhew and Peregrine Worsthorne, were trained to collect unembellished information about the tactical situation at the front and about future operational plans, and to concisely, regularly and quickly report it to Monty.[7]

The second unique aspect to Monty's style of command were his liaison officers, who performed a similar function, but were personally selected and directed daily by Monty himself. They included Majors John Poston (formerly his ADC in the desert), Carol Mather and Dick Harden, and went out by jeep each morning with a list of tasks – people and headquarters to visit, a tactical picture to form – and delivered their oral reports each evening to Monty in person, usually in his map caravan, upon return. The role was dangerous: Poston would be killed in his jeep by Nazi stragglers on 21 April 1945, two weeks before the war's end, whilst Mather and Harden were badly wounded in their spotter plane in early 1945. Mather's brother William also worked at TAC; their parents had been friends of Betty Carver's, before her marriage to Monty.

One of Monty's two stepsons, Dick Carver, had also worked in TAC. On his arrival in 1942, Monty had written: 'I am devoted to him and it will be delightful to have him with me.' After El Alamein, Carver was captured by an Afrika Korps patrol, who never realised their prisoner's relationship to their arch-opponent; he was imprisoned in Italy. Carver subsequently escaped and rejoined his stepfather, who welcomed him back with mock severity:'Where the hell have you been?' Thus TAC was, in many ways, a happy family.

The liaison officers' activities formed part of the TAC day-to-day rhythm, for in the evenings they would often dine with Monty and entertain their chief: shop talk (of military affairs) was rarely permitted over dinner, which was a time for all concerned to unwind from the stresses of their lives. The liaison officers tended to be young, strong-willed, well connected and most became devoted to their boss. They needed to be, for sometimes they were regarded with suspicion by other headquarters as 'Monty's spies'. They therefore had great powers to make or break quite senior officers, who, given Bernard's impatient nature, would often find themselves fired unceremoniously by Monty on a snap visit: nearly every British division in the Normandy campaign lost at least one brigadier in this fashion. This worked both ways, however; whilst intrusive, they gave subordinate commanders (brigade and battalion commanders) a direct channel through to their commander-in-chief. This was one of Monty's conscious reactions to the First World War – that commanders at the front should never feel that headquarters in the rear was out of touch and committing lives on false intelligence.

On 23 June, Monty's TAC headquarters moved to an open hillside at Blay, liberated on 9 June and just within the American sector, six miles west of Bayeux. Monty would remain based at Blay for six weeks, which was regarded as one of their best sites by TAC's headquarters personnel – a tranquil oasis, seemingly far removed from the war, but whose reminders were ever present. Monty had two captured tanks, a Tiger and a Panther, placed either side of the entrance to the TAC compound. This was for morale purposes and inspired by growing Allied fear of the German heavy tanks: Monty's gesture was his attempt to demonstrate their impotence. In contrast to the tanks of Rommel's 7th Panzer Division in 1940, five-man Panthers carried thick, sloping armour in imitation of the Russian T-34, weighed on average forty-five tons and carried a 75mm gun. Tigers were an altogether different matter: the five-man beasts were heavy (weighing an average of sixty tons), thickly armoured but slow (with a maximum speed of a little over 20mph) and guzzled fuel (they needed refuelling every fifty miles). Reflecting the fact that Germany had switched over to a defensive war, Tigers were effectively mobile pill-boxes, with little need for speed or range, and mounted the deadly 88mm gun, comfortably capable of tank kills at up to 1,500 metres, and often far beyond. They were regarded as formidable opponents and most reports from Allied infantry when confronted by German tanks referred to Tigers, even though only 1,347 were manufactured and most deployed to Russia and some to Tunisia and Italy, so very few – probably around 150 – ever surfaced in Normandy. Although many at the time, and since, have disparaged Allied armour in 1944, the five-man, thirty-ton Shermans had many things in their favour, chief amongst which was numbers (over 50,000 were manufactured); every German tank was irreplaceable, but every Allied tank damaged or destroyed could be replaced from giant tank parks of spare vehicles in southern England.[8]

Despite near-total air supremacy, the going for the Allies was slow; the suffocatingly close country of hedgerows, sunken lanes, and stone cottages favoured the defenders, who resisted savagely; the terrain generally slowed the movement and neutralised the usefulness of Allied armour. This was echoed by the experiences of tank crews more acclimatised to desert fighting, where tank commanders who had always fought with their heads out, were sniped and killed as they drove along an apparently peaceful lane. Crews, who had been accustomed to engage the enemy at half a mile or more in North Africa, found themselves facing hostile anti-tank guns at ranges of 50 yards, or worse still, dealing with boarding parties who leapt on their tanks from the cover of hedgerows. It has been suggested that it was Allied doctrine – of using tanks on their own – that failed here, not the vehicles themselves, and that by the end of the campaign, once armour–infantry cooperation and flexibility had been established, the solution had been found to the Tiger threat.[9]

With Rommel back at La Roche-Guyon and in control, the two panzer divisions which should have been deployed first thing on the morning of D-Day now arrived on the scene. Kurt 'Panzer' Meyer, CO of 25th SS-Panzer Grenadier Regiment, had actually dragged the leading elements of 12th SS Hitlerjugend Panzer Division up to Caen by the evening of 6 June, and attacked the Canadians early on the 7th. Rommel's old Chief of Staff from the desert, Fritz Bayerlein, who had been promoted to command an elite tank division formed from all the German army's tank training schools and demonstration units – the Panzer Lehr Division – was also rushed up to the invasion area, but appeared only on 8 June. The division lost eighty-five armoured vehicles, five tanks and 123 trucks, including eighty petrol tankers, during its hundred-mile march. The tardy arrival of these two important armoured formations, amongst others, more than justified Rommel's plea for the panzers to be stationed nearer to the coast. Meanwhile, the Norman underground cut road and rail links sufficiently for the 275th Infantry Division to take five days to cover 125 miles to the coast, arriving only on 11 June.

'Panzer' Meyer moved into the area west of Caen and established his headquarters in the Ardennes abbey; from the battlements he could just see (as you still can today) the coast, and looking west, he discovered the 9th Canadian Brigade Group advancing toward Carpiquet airport. He threw his Grenadiers at the exposed Canadian left flank with two battalions supported by tanks. Striking with great force in vicious close-quarter battles, the Canadians were forced out of Authie and Buron with heavy losses. In defence, the Canadian infantry proved as stubborn as the Germans, especially once they were able to bring their artillery to bear: in Normandy, as in the trenches of the First World War, artillery was the key killer on both sides, causing three out of every four wounds. Benefiting from naval gunfire support, 9th Brigade retook Buron, but was decimated – the North Nova Scotias lost eighty-four killed, thirty wounded and 128 captured, whilst the Sherbrooke Fusiliers lost twenty-eight tanks. At dawn on 8 June, Meyer's sister unit (26th Regiment) attacked the Canadian 7th Brigade in Putot-en-Bessin with two battalions and surrounded the Winnipeg Rifles. According to Meyer's memoirs, the fighting was so fierce that just as he arrived to witness the battle, his battalion commander's head was removed by a tank shell. A third German battalion attacked Bretteville, where the Regina Rifles stubbornly defended the town and the battle raged all day, but by dawn the next morning the 12th SS retreated after suffering heavy losses. Alas, to stop the counterattack, 3rd Canadian Division paid a high price: the Winnipeg Rifles lost 256 men including 105 killed and the Canadian Scottish 125.

Inexplicably, Meyer's men executed forty-five Canadian prisoners at Ardennes abbey on 8 June, whilst twenty-three others had been shot the previous day, thus beginning a mutual loathing between the Canadians and 12th SS Division. This exposed the different cultures of Rommel's army and the Nazi-indoctrinated Waffen-SS, whom Rommel despised.

When his son Manfred expressed a desire to join the SS in 1943, Rommel explicitly forbade him; Manfred joined a Luftwaffe unit instead, though at this time one of Rommel's nephews was commanding an SS unit in Italy. On 14 June, when Meyer's divisional commander, Fritz Witt, was killed, Meyer was promoted in his place – at thirty-three years, the youngest divisional commander on either side. After the war Meyer was held responsible for the Canadian executions and sentenced to death, a sentence later commuted to life imprisonment.

To coordinate the panzer divisions, Geyr moved his Panzer Group West forward to a site south of Caen, as Rommel had wanted all along. In a spectacular example of the value of signals intelligence (SIGINT), analysis of German wireless traffic revealed that Geyr's Panzer Group had made its headquarters in the Château La Caine, twelve miles south-west of Caen. This was a potentially harmful development for the Allies, because coordination of the panzer divisions in Normandy by Geyr would amount to an operational-level threat. An attack on 10 June by the 2nd Tactical Air Force caught Geyr's headquarters completely by surprise and killed his Chief of Staff, *Generalmajor* Ritter und Edler von Dawans, seventeen staff officers, and wounded many more. With seventy-five percent of its communications equipment and many vehicles destroyed, the HQ was neutralised as an effective command centre for crucial days at a key stage of the battle. The move of Panzer Group West to La Caine had also been confirmed by Ultra, and there was always the tension as to how much tactical use to make of Ultra intelligence without compromising this valuable strategic asset. Bill Williams, Monty's intelligence chief, made the reasonable deduction that Panzer Group West's location could have been revealed by poor German radio discipline, so the raid went ahead.

Encouraged by Rommel – who appeared everywhere along the front, shoring up morale and getting the tactical picture – and drawing on their experiences from the Eastern Front, a German defensive doctrine quickly evolved to create a main defensive belt deep in the *bocage*, around mutually supporting strongpoints, equipped with mortars and MG-42s. Fields were laced with mines, booby traps and covered by snipers, automatic fire and mortars. Positions were well dug in to hedgerows, camouflaged, with generous arcs of fire. The many sunken lanes provided ideal ambush sites and excellent routes for quick counterattacks, which slowed the Allies down, whilst well-directed German artillery administered the coup de grâce. Whilst the Germans used the Normandy terrain to their advantage in defence, when they in turn attempted offensive operations they were faced with the same difficulties. Overwhelming Allied firepower destroyed every German counterattack as soon as troops left their defensive positions.

Monty's preferred offensive tactics of fire and movement were useless in this environment, where hedgerows compartmentalised forces, masked fields of fire and prevented armour from supporting the infantry. When the attackers were forced into small groups by the terrain, it deprived men of a common knowledge of the battle and the physical comfort of adjacent soldiers, and reduced units' fighting power. Individuals had to assume greater responsibilities for which they were untrained; the dispersion heightened stress and deepened the fear of isolation, and casualties further broke down unit cohesion. The US 90th Infantry Division reported that when faced with hostile fire from an invisible enemy, the unit's commanders found the principles of fire and manoeuvre nearly impossible to apply in the hedgerows, especially the precept that an attacking force should move just behind its artillery in order to confront the enemy while he is still off balance.

In the campaign, artillery and mortars caused most damage. Each German division had, theoretically, a battalion of about twenty six-barrelled 150mm mortars, *Nebelwerfers* ('Moaning Minnies'), mounted on wheeled carriages. Like the Stuka earlier in the war, the *Nebelwerfer* was as much a psychological weapon as a physical one. They sounded like this: 'the howling and wailing grew until it filled the sky, rising in pitch as it approached, and ending in a series of shattering explosions all round us . . . then more squeals, the same horrible wail and another batch exploded astride us, so that the pressure [wave] came first from one side, then the other'.[10] To avoid the launcher being tipped over by the blast, the rockets fired in a set sequence; the crew had to be at least fourteen metres away from the launcher, in a trench, to avoid the blast from the exhaust, and a trained crew could load and fire six rounds in ninety seconds to a maximum range of about four miles.

Thus the Germans managed to fight the Allies to a standstill: if they broke through, the Germans simply fell back to the next hedgerow. Ironically, the ground in Devon and Cornwall, where US units had trained beforehand, was remarkably similar to Normandy – small fields and orchards, bounded by high banks or hedgerows and stone walls – but tactics had not been developed for a *bocage* war. Allied planners had underestimated the Germans' use of the terrain because they hoped to be beyond the hedgerows in a matter of days. When the operational requirements of the campaign changed, Allied tactical doctrine was slow to respond. Allied infantry–armour doctrine was poor, hampered by the lack of tank–infantry radios; external telephones mounted on the rear of tanks for the infantry proved difficult to use in combat, as the handsets invariably ended up being dragged along the ground or crushed.[11] As the Allies, led by Montgomery, had been combating the Germans in mainland Italy since September 1943, one might have expected useful doctrine to have influenced the Normandy campaign: principles of infantry–armour cooperation, the use of close air support, the shortcomings of strategic air support (as at Monte Casino), but this did not happen. Normandy soon became a reactive campaign to German defensive techniques.

Having failed to secure the key route centre of Caen on D-Day, Second Army's commander, Miles Dempsey, proposed to encircle it with Bucknall's XXX Corps advancing from the west via Villers-Bocage, and using 51st Highland and 6th Airborne Divisions in the east, moving round from their bridgehead, astride the river Orne: thus Caen would fall to a giant pincer movement. Originally 1st Airborne Division (who ultimately took no part in the Normandy campaign, but dropped into Arnhem instead over 17–27 September) was to be dropped south of Caen to effect this envelopment. This plan proved too ambitious, but a right hook round Caen, Operation Perch, was ordered instead. Thus, on 11 June, XXX Corps' spearhead, Major General 'Bobby' Erskine's 7th Armoured Division (old hands from North Africa), was ordered to move south, bypass strong German positions in Tilly-sur-Selles, and take Villers-Bocage. Exploiting a gap in the German line, Brigadier 'Looney' Hinde's 22nd Armoured Brigade of three tank battalions and a motorised infantry battalion reached Livry on 12 June, and attacked Villers-Bocage the following day.

In a confused morning's combat, SS-*Hauptsturmführer* Michael Wittmann of 101st SS-*Schwere-Panzer-Abteilung* (Heavy Tank Battalion) ambushed the leading armoured column of 22nd Brigade on high ground east of Villers-Bocage. In one of the most famous actions in the history of armoured warfare, 4th County of London Yeomanry (a Territorial Army tank battalion) and its attached infantry battalion were all but destroyed, losing twelve tanks, thirteen half-tracks and Bren carriers and two anti-tank guns in the first ten minutes, with most crews killed or captured, all despatched by Wittmann and his crew. Entering a British-occupied town initially in a lone Tiger tank without supporting infantry, Wittmann, a thirty-year-old tank ace, helped to check the advance of a British armoured brigade. The subsequent British withdrawal (much criticised since as a lost opportunity) was the result of Wittmann's colleagues, who later arrived and destroyed twelve more tanks, losing five of their own. Villers-Bocage was the low point of British fortunes in Normandy. Hinde, Erskine and Bucknall were later sacked, whilst Wittmann died in his Tiger on 8 August near Cintheaux, with 138 tanks, 132 anti-tank guns and countless soft-skinned vehicles to his credit. It is thought that eight Shermans from the 1/Northamptonshire Yeomanry ambushed him, three of which he destroyed before his own death.[12]

By 12 June, 326,000 troops had been landed on the beaches of Normandy, plus 54,000 vehicles; by 2 July, another 929,000 men and 177,000 vehicles were put ashore. Such logistics triumphs came at a cost: in the ten days after D-Day (6–16 June) a total of 5,287 Allied soldiers were killed. Despite the best efforts of Bernard Montgomery to avoid a repetition of the First World War, and keep his casualties to a minimum, between D-Day and the end of the war (the German surrender was signed 337 days after the landings, on 8 May 1945) British losses alone would total 30,280 dead and 96,670 wounded.

Allied logisticians had landed seventy-three percent of their predicted requirements by 18 June. As a result of a Channel storm, the worst of the twentieth century, by 22 June this figure had fallen to fifty-seven percent. The percentage of vehicle landings and personnel arriving similarly fell short of the planned target. Despite the loss of a second artificial Mulberry harbour, Omaha averaged 13,500 tons of supplies per day by the end of June, or 115 percent of planned capacity. Utah achieved 7,000 tons per day, or 125 percent of its target. Most of these American resources were landed directly onto beaches by landing craft, bypassing the remaining Mulberry at Arromanches altogether.[13] Conducting a group of foreign attachés to meet Monty in November 1944, the writer-soldier Major Anthony Powell wrote of first seeing the great harbour:

It was difficult to know what we were regarding. In the foreground lay a kind of inland sea, or rather two large lagoons, the further enclosed by moles and piers that seemed exterior and afloat; the inner and nearer, with fixed breakwaters formed of concrete blocks, from which, here and there, rose tall chimneys, rows of cranes, drawbridges ... What, one wondered, could this great maritime undertaking be? Was it planned to build a new Venice here on the water?[14]

The size of the Allied build-up and the success of the Mulberry harbours only became apparent to Berlin after an air photo sortie on 2 August. That afternoon, *Oberstleutnant* Erich Sommer, one of the Luftwaffe's most experienced and well-respected test pilots, flew the world's first jet reconnaissance mission. Sommer's top-secret, single-seat Arado-234 Blitz, a revolutionary jet bomber of incredible speed (nearly 500mph), took him along the Normandy coast. 'My task was just to take pictures,' he recalled later; 'inside the cockpit you felt like being in a glass tunnel. It was a fantastic aircraft. I could out-fly any enemy plane. You knew nobody could touch you because your speed was superior to any enemy machine's speed.'[15] It took twelve photo interpreters two days to analyse Sommer's photographs. It was only then – after seeing how much materiel had been landed in Normandy – that Hitler and the OKW comprehended for the first time the full extent of D-Day and realised that this was the true main effort – and that no second assault was about to be sprung against Calais.

By 18 June, Lieutenant General 'Lightning Joe' Collins' US VII Corps had cut its way across the Cotentin peninsula and severed all roads leading to Cherbourg. Under orders from Hitler 'to defend the last bunker and leave to the enemy not a harbour but a field of ruins' the remnants of five infantry divisions defended the port stubbornly. Cherbourg eventually fell on 25 June, to psychological warfare. Surrender leaflets dropped by air, which emphasised the dwindling food supplies within the city and offering instant hot meals to the defenders, won over the starving troops. Yet the fall of Cherbourg, whose sabotaged port was left unusable for months, offered only temporary consolation to the Allies. Monty had counted on being able to use Cherbourg or Le Havre at least by the end of the Normandy campaign, and in fact remained reliant logistically on Arromanches until the year's end, so complete was the demolition job on Cherbourg, which had an invaluable sheltered outer harbour and docks for transatlantic shipping. By 1 July the Allies had established a beachhead seventy miles wide and had brought hundreds of thousands of men and 177,000 vehicles ashore, yet, except around Cherbourg, their lodgement was in no place more than twenty-five miles deep, and in most areas it extended little more than five miles inland. The beachhead was becoming congested, when on 5 July, the millionth Allied soldier stepped ashore in Normandy.

Rommel now planned a measured counterattack, intending the newly arrived II SS Panzer Corps (9th and 10th SS Panzer Divisions) to slice through the Allied bridgehead towards Bayeux, reaching the coast at Arromanches. Meanwhile Monty, too, was planning an attack. Unable to take Caen, which had been a D-Day objective, and gain access to a road and rail network out of Normandy, Montgomery launched Operation Epsom on 26 June, momentarily delayed by the Channel storm. Its aim was for XXX Corps to sweep round to the west and south of Caen and reach the main Caen–Falaise road. This would almost encircle the German defenders around Caen, particularly those at the Carpiquet aerodrome, who were preventing any further progress on Monty's left flank. Despite his claims after the war that his intention was to 'fix', i.e. hold, the enemy armour in the east whilst the Americans broke out from the beachhead in the west (and this was without doubt the pre-invasion plan – Bradley's memoirs of the briefing at St Paul's School are quite explicit about this), we can now say that Operation Epsom appears to have been designed to deliver Caen to the British and facilitate a breakthrough.

The XXX Corps plan was to advance on two axes over difficult country but it soon bogged down in the face of determined resistance from 12th SS, who destroyed sixty British tanks. The attack was launched at 7.30 a.m. with a tremendous artillery barrage, rather like a First World War set-piece attack. Amongst the indirect fire support assets were three Royal Navy cruisers anchored offshore and firing shells accurately fifteen miles inland. The first division (15th Scottish) followed the bursting shells, stepping forward into the shoulder-high cornfields, in platoon and section groups, not dissimilar in appearance and method to an attack of 1917–18; success or failure would rest on the determination of quite young leaders who did not know the terrain, and on the equally determined German defenders, who did.

On 28 June, unable to stem the Allies and close to despair, the Seventh Army's commander-in-chief, the tall and gangly *Generaloberst* Friedrich Dollmann, committed the entire II SS Panzer Corps against Epsom, thus spoiling Rommel's planned counterattack. Denying II SS Corps the chance to prepare a proper attack against the British salient from both sides, Dollmann insisted the counterattack begin straight away. Dollmann was also past caring: that night he allegedly committed suicide (Hitler was told of a 'heart attack') on account of losing Cherbourg and having to explain this to the gentlemen of the Gestapo in Berlin. At this moment both Rundstedt and Rommel were also absent, having been summoned to Berchtesgaden in Bavaria for a meeting with the Führer. SS General Paul Hausser was immediately appointed to command Seventh Army, relinquishing command of II SS Panzer Corps, which was in the process of trying to sever the 'Scottish corridor' (as the Epsom thrust became known, on account of 15th Scottish Division's initial assault) from its base. The commander of 9th SS Division, Willi Bittrich, was promoted to command II Panzer Corps in place of Hausser, and one of his regimental commanders took over the division. A battalion commander stepped in to lead the regiment. Thus, Dollmann's death caused a tsunami across every level of command amongst the formations trying to counterattack Operation Epsom. The timing could not have been worse.

As it happened, there was a similar reshuffle in the British camp. Monty sacked the commander of 11th Armoured Division's infantry brigade (159th Brigade) for not driving his brigade hard enough. This was a not uncommon reaction of Montgomery's in the Normandy campaign: 49th Division also lost two brigadiers in this way, whilst the axe later fell on Major General 'Bobby' Erskine, GOC of 7th Armoured Division; the commander of his 22nd Armoured Brigade (Brigadier 'Looney' Hinde); Major General Bullen-Smith of 51st Highland Division, and 'Gerry' Bucknall, of XXX Corps. Enemy action removed other commanders, including Major General Tom Rennie of 3rd Division, who was wounded on 12 June.*

The civilians were suffering too. Most villagers in the path of Operation Epsom and around Hill 112 fled to Bayeux during the Normandy battles and returned to find their houses shattered, livestock killed and crops destroyed by the fighting; mines and booby traps laced the fields and derelict buildings; flies and vermin hovered around the human and animal corpses that littered the battlefield, spreading disease in their turn. Patrick Hennessy, a tank commander with the 13/18th Hussars, recalled: 'We became used to the sight of dead cows and the sickly stench of their bloated and rotting carcasses. This smell became an enduring

---

* Rennie, whose jeep had run over a British mine, suffered a broken arm, but later returned to take over 51st Highland Division. The only British divisional commander to die in action in Europe, the unfortunate Rennie died in March 1945 during the Rhine crossings.

memory of Normandy in the summer of 1944.'[16] Whereas the invasion of 1940 had caused little damage, the early post-war editions of the Michelin guidebook stated, that as a result of 1944, 'more than 200,000 buildings were destroyed or damaged, and nearly all the great towns suffered – Rouen, Le Havre, Caen, Lisieux. Of the 3,400 towns and villages of Normandy, 586 have been totally reconstructed to new plans.'[17] Some personal anger at this damage, caused by the Allies in the name of freedom, seems to have lingered, as one ageing Norman told the *Guardian* in 1994:

> I'm sorry to say that in many towns and villages, there is resentment for American arrogance and hostility for British callousness ... This isn't aimed at individuals, but against distant generals and politicians who seemed to mistake us for the real enemy ... We helped British soldiers but suffered for months because of homelessness and lack of food.[18]

Operation Epsom meant that all the panzer reinforcements were being fed piecemeal into the front, denying the Germans a strategic reserve with which to combat an Allied breakthrough. For example, during Epsom, 12th SS Panzer Division lost 2,662 casualties (twenty-two percent of its fighting strength) – an irreplaceable elite, and gained no replacements. Following Epsom, another Allied attempt was made to gain Caen. Preceded by another strategic air assault, Operation Charnwood was led by 3rd British, 51st Highland and 3rd Canadian Divisions, involving 115,000 men. After 460 bombers had dropped over 2,600 tons of high explosive, the Canadians entered a city on 9 July that had been eighty percent destroyed, for a cost of eighty tanks and 3,500 casualties. In the two days of desperate fighting that followed, the Germans (Hitlerjugend from Meyer's 12th SS Panzer Division) resisted viciously, aided by the rubble thoughtfully created by the RAF. Monty's forces entered Caen and took half the city but moved no farther. Casualty rates during the battle were appalling: already-weakened British and Canadian infantry battalions sustained losses of twenty-five percent.

The capture of Caen had achieved little. Montgomery therefore launched, on 10–11 July, Operation Jupiter, to try and break out towards Falaise, via Hill 112. This again failed because of the skill of German armour, the determination of the defenders and the inability of Allied air power to influence events in the *bocage* below. Flying at over 300mph, Allied pilots were often unable to pinpoint targets in the thick tree cover, and reluctant to engage when the opposing forces were intermingled. Hill 112 remained in German hands despite heavy fighting, and became a no-man's land between the two armies.

On 18 July the most ambitious and controversial of Montgomery's battles, Operation Goodwood, was launched. Its aims have become obscured in a haze of memoirs, as generals have sought to justify their role in what was basically a failed operation. In his memoirs Monty wrote:

> My master plan for the land battle in Normandy . . . was to stage and conduct operations so that we drew the main enemy strength on to the front of the Second British Army on our eastern flank, in order that we might more easily gain territory in the west and make the ultimate break out on that flank – using the First American Army for the purpose . . . On the eastern flank . . . the need *there* was by hard fighting to make the enemy commit his reserves, so that the American forces would meet less opposition in their advances to gain the territory which was vital on the west . . . I never once had cause or reason to alter my master plan.[19]

Whilst his basic contention – fix in the east and break out in the west – is broadly what happened, and is supported by the eyewitness testimony of the final briefing at St Paul's School of 15 May, it is certainly not the case that Monty had neither 'cause' nor 'reason' to alter his master plan. Monty had envisaged reaching the Seine by D+90 (4 September) and actually reached it a fortnight earlier, but none of the major battles – Epsom, Goodwood, or the Falaise Pocket – were foreseen, and until the end of July (D+55) any kind of break-out looked very distant and unattainable by the due date. Monty's timetable was impossibly adrift, but suddenly and miraculously came back on target because the German front collapsed very quickly.

Goodwood brings to the fore the heart of the debate about British and American intentions for Normandy. During the campaign, the American press claimed the British had stalled in front of Caen ever since D-Day, when they failed to capture the city; that subsequently they had gone over to the defensive in that sector; and that Montgomery had run out of ideas, whilst in the west it was the Americans who were doing all the fighting, and the dying. Even the British media had observed that British progress was slow (running headlines containing words like

'Stalemate'), compared with the Americans. Subsequent memoirs, still widely accepted, suggested that the *real* Allied plan cooked up by Montgomery had been to seize Caen on D-Day and then effect a break-out via the road network there, but that British failure had required the Americans to work harder and eventually break-out in the west.

The St Paul's briefing, and Bradley's recollection of it, made it clear that the plan was always to hold at Caen and break through in the west. On 15 July, Monty wrote to Dick O'Connor, then commanding VIII Corps, stating:

> Object of this operation. To engage the German armour in battle, and to 'write it down' to such an extent that it is of no further value to the Germans as a basis for battle. To gain a bridgehead over the Orne through Caen and thus to improve our positions on the Eastern flank. Generally to destroy German equipment and personnel.[20]

If this was Monty's aim – to distract all the remaining German armour away from the west, where the Americans were preparing for their break-out – then why did Monty deploy three *armoured* divisions at Caen for Operation Goodwood, if all he wanted to do was distract the Germans' attention? Specifically, Monty wrote that he wanted VIII Corps to reach the high ground of the Bourgébus Ridge and the Caen–Falaise road, but there is plenty of evidence to suggest that Monty and Miles Dempsey then wanted to attempt *their own* break-out south, via Falaise. One inter-pretation, therefore, is that Goodwood was a genuine attempt to break out, via a left hook around Caen, and not just an attempt to fix the enemy (which appears to be a later justification, once the break-out had failed). The choice of Epsom and Goodwood – both racecourses – as code names, also suggests the fingerprints of Dempsey, a keen race-goer and an outstanding horseman, in the plan.

Monty certainly exaggerated Goodwood to Air Chief Marshal Tedder, Ike's deputy, to enable him to 'borrow' massive aerial support from Bomber Command. Cooperation between air and ground commanders at this crucial stage of the war was appalling,[21] and the airmen were given to understand that Goodwood was a giant left pincer that, with Cobra (the forthcoming American break-out attempt in the west) as an equally large right pincer, would envelop the German forces in Normandy. Had such a masterful stratagem ever been Allied policy, then this would have been an admirable imitation of Russian deep operations, but the sad truth is that the Western allies did not conceive of warfare in such terms in 1944.

In assessing Dick O'Connor's VIII Corps and their opposing German units during Operation Goodwood, it is obvious – with hindsight – that here was an opportunity for an Allied deep operation, but no evidence of such planning in 21st Army Group headquarters can be detected. Biographer Alun Chalfont suggests that Cobra and Goodwood were linked in Monty's mind, as the former's start date was originally intended to be 19 July – which would have given an almost simultaneous left, then a right hook to the Germans. But if that was the intention and Cobra was delayed (eventually to 25 July), then why not postpone Goodwood also, so the two operations remained simultaneous? Today we might reflect that Monty's weapons could have been used more effectively to shape the battlefield for a subsequent manoeuvre, but the contention is that had the British broken through, Monty intended a pursuit, not envelopment.

In this sense, Goodwood was not coordinated in any way with the American Cobra, so the envelopment argument does not hold water. In connection with Cobra, Operation Goodwood might also be interpreted as a *rival*, not a complementary, operation – it would have been a matter of Monty's pride that the British should break out before their American colleagues. Brooke may be partly at fault here, for he knew the *bocage* from boyhood holidays, and thought privately that the Americans 'will never get through it'. He may have encouraged Monty to consider his own break-out. Certainly the line-up for Goodwood suggested a break-through and speedy pursuit, rather than an attempt merely to draw the Germans into an attritional tank battle.

Dick O'Connor's VIII Corps fielded three British armoured divisions – 7th, 11th and Guards, totalling some 75,000 men and 1,300 tanks – who were to blast their way through German lines, substituting their firepower for the already greatly depleted and hard-pressed infantry. Whilst British tanks were the best weapons with which to kill enemy armour, they needed infantry support to remove the threat from German anti-tank weapons. For an anticipated fixing operation – where British armour would be thrown, naked and vulnerable, against the German defenders in a close-quarters fight – there was remarkably little infantry deployed forward. This may have been because the British were already running short of infantrymen; but it may have been because the fifty-five-year-old O'Connor, charming and affable, had spent two and a half years as a POW in Italy following his capture in the Western Desert in April 1941, and was out of touch with recent military knowledge. O'Connor needed to know that the Germans reacted with astonishing speed to operational crises and their combined-arms fighting at a tactical level was frequently the key to their battlefield successes.

Very much a contemporary of Montgomery's, and probably seen by Monty as a rival, the fact that Dick O'Connor was immediately offered a corps command on his escape from captivity suggests the esteem with which he was held by Brooke and Churchill. So another inter-pretation is that the disproportionate presence of armour in VIII Corps was designed to replace British infantry, of which there was a famine. Dempsey, however, was clearly hedging his bets, and did relocate his headquarters next to O'Connor's, presumably to seize the moment if the German front happened to crack wide open. It is significant that Brooke chose this moment to fly over and spend 19 July with Monty. Apparently Montgomery had earlier told Churchill he was not 'at home' to any visitors, even his Prime Minister, who had wanted to visit him at his TAC headquarters. Winston was incensed and Brooke, real-ising Monty was in danger of becoming a liability (he had annoyed the Canadians in exactly the same way during the Sicilian campaign), felt it necessary to fly out to Blay, and during 'a long talk with Monty' forced him to write a personal note to the Prime Minister, inviting him to visit at any time, which Brooke personally delivered to Downing Street later that same evening. Was there another reason for Brooke's presence in Normandy at just the moment when Monty might have broken out?

Another reason for the preponderance of British armour may have been that O'Connor *was* directed to deploy his tanks forward in the hope of a breakthrough – either for the military reasons just described – or possibly due to political pressure from Churchill for an altogether different reason. During the night of 13–14 June, the V-1 flying bomb attacks on Britain had started, and these renewed German attacks on London were shaking public morale at home; eventually a total of 10,500 missiles would be launched, of which 3,957 were destroyed by anti-aircraft defences and fighters, but 3,531 reached England and 2,353 fell on London. The death toll from this last blitz would eventually total 6,184 killed and 17,981 seriously injured.[22] On 6 July, the *Evening News* headline read, 'Flying Bombs Kill 2,752, Injure 8,000: Very High Toll in

London – Premier'. In view of this last, random blitz on southern England, a break-out from Normandy would enable Monty to overrun the launch sites of these morale-sapping weapons. Brooke's diary refers frequently to the flying bomb menace on 18–19 July, with the observation that: 'The tendency is of course to try and affect our strategy in France and to direct it definitely against rocket sites. This will want watching carefully.'[23]

Goodwood began on Tuesday 18 July with a nod to the barrages that had preceded the Somme, Passchendaele and El Alamein: 2,500 heavy and medium bombers carpet-bombed the Germans first, followed by a bombardment from 750 Allied artillery pieces and naval gunfire support from warships anchored off the coast. The Shermans of Pip Roberts's 11th Armoured Division then started rolling at 7.45 a.m., flanked by 3rd Canadian Division to their right and 3rd British Division to the left; lorry-borne infantry followed the armour, rather than escorting it, restricted to roads and tracks. Operation Goodwood's flaw was that VIII Corps' advance was on so narrow a front that each division had to follow each other in turn. The forward German zone of defended villages was penetrated successfully but Allied intelligence had underestimated the strength of the German defences, which were almost ten miles deep, spread over a gentle forward slope. The forward areas, garrisoned by 16th Luftwaffe Division, were largely destroyed by the bombing; but behind lay elements of Feuchtinger's 21st Panzer Division, and beyond that, strung along the Bourgébus Ridge itself, was Luftwaffe *General der FlakArtillerie* Wolfgang Pickert's III Flak Corps with eighty 88mm guns firing from concealed positions. The aerial bombing was extraordinarily effective; Freiherr von Rosen, commanding a company of Tiger tanks, later noted that 'some of the sixty-two-ton machines lay upside down in bomb craters thirty feet across; they had been spun through the air like playing cards'.[24]

Oberst Hans von Luck, nominally commander of 125th Panzer Grenadier Regiment (of 21st Panzer Division), but in fact leading an all-arms battle group based in Cagny, had literally just returned from Paris, and was unaware the British had launched a major attack until he found one of his battalion commanders in a state of shock and all contact with forward units had been lost. Luck dashed to his command tank and drove forward. He found the area north of Cagny in ruins and, where one of his battalions should have been deployed the whole area was dotted with British tanks, which were slowly rolling south, against no opposition. 'My God,' I thought . . . How could I plug the gap? . . . As I was driving past the church of Cagny, which lay in the undamaged part of the village, I saw to my surprise a Luftwaffe battery with its four 8.8cm anti-aircraft guns, all pointing to the sky . . . A young captain came up to me . . .

'My God, what are you doing here? Have you any idea what's happening over there to the left of you?'

'Major, my concern is enemy planes, fighting tanks is your job. I'm Luftwaffe.'

He was about to turn away. At that I went up to him, drew my pistol . . . levelled it at him and said: 'Either you're a dead man or you can earn yourself a medal.'

The young captain realised that I was in earnest. 'I bow to force. What must I do?'[25]

Luck galvanised his defence and that of his neighbouring defenders, and soon the cornfields ahead of him were full of blazing Sherman tanks, belonging to 3/Royal Tank Regiment, the Fife and Forfar Yeomanry and the 23rd Hussars.[26]

At this stage, Monty held a press conference to announce the progress of the operation and did so in terms of suggesting a breakthrough (his biographer, Nigel Hamilton, maintains this was part of Monty's ongoing deception, to distract attention from the forthcoming Cobra). When the advance stopped, the media – and SHAEF and the RAF – interpreted it in terms of a failure. Monty's own interpretation in his memoirs was quite different:

Many people thought that when Operation Goodwood was staged, it was the beginning of the plan to break out from the eastern flank towards Paris, and that, because I did not do so, the battle had been a failure. But let me make the point again at the risk of being wearisome. There was never *at any time* any intention of making the break out from the bridgehead on the eastern flank.[27]

However – with Dempsey moving his TAC forward and Brooke flying out to join him, Monty would have been neglecting his duty if the front had cracked open and he did not take advantage of the opportunity. Surely the greatest military commanders are the most flexible, not the most rigid? The most generous interpretation we can put on Goodwood is that Monty, Dempsey (and possibly Brooke) were hedging their bets. On one hand, they were trying to attract the last trickles of German armour to their front, which is exactly what they achieved; ultimately the arrival of Sepp Dietrich's I SS Panzer Corps (1st and 12th SS Panzer Divisions) on the Bourgébus Ridge late on 18 July prevented a break-through. On the other, they were hoping to break through and had assembled the necessary resources with which to exploit the moment, master plan or no master plan. So much for Allied cooperation!

Goodwood continued into Wednesday 19 July, when the armour and infantry of both sides fought a series of battles for the villages south of Caen, overlooked by the Bourgébus Ridge, but the following day poor weather effectively brought the exhausting combat to a close. With hind-sight it has been suggested that Monty could have extended Goodwood and pushed on, had he been prepared to accept the increased casualties, but refrained from doing so. Whether this was a practical option is debatable, but in some eyes Goodwood was the second occasion when Monty had called off an operation (the first was at Villers-Bocage in June) that might have yielded more.

By this time the British had lost 413 tanks (many of which were salvageable), and sustained 5,537 casualties in just two days (though only 521 were tank crewmen) – over a third of Second Army's tank strength. For Monty's tank crewmen, the end was often brutal, but mercifully swift. One Sherman tank commander in the East Riding Yeomanry (27th Armoured Brigade) later wrote of:

> An enormous bell sounded with a clangour . . . The loud gong-like noise was followed by a colossal roar, and from . . . [the] tank soared a monstrous column of white smoke that formed a huge ring in the air . . . The tank, a moment ago home and shelter of five men, now looked a terrible sight, with flames belching out of its front hatches and engine louvres, and a pillar of fire roaring out of the turret, colossal gouts of smoke billowing across the field in curled torrents. The 88 had struck home! . . . tremendous internal explosions thrust fierce gasses from every aperture . . . no one within could have possibly survived. A rim of smoke and flame like a lace collar squeezed in jets from where the turret joined the hull. The painted metal began to grow black . . . and eyes of white flame looked out from the holes where the armour piercing shot had entered.[28]

Eisenhower later wrote that Operation Goodwood was an advance of seven miles at a rate of a thousand tons of aerial bombs per mile. Perhaps a less appreciative supreme commander would have sacked Montgomery at this stage, as Tedder requested at the time (though who would have been the replacement is a moot point). Monty certainly had himself to blame for any misunderstandings: he could be secretive and was seen eventually as unwilling to confide in anyone other than his trusted collaborators. He was lucky with the press on this occasion. Having brazenly primed the press with stories of a breakthrough at his 18 July press conference, which duly appeared as 19 July's headlines ('Second Army Breaks Through' in *The Times*), he faced major derision when that penetration failed to materialise. He was, in fact, saved by the actions of a one-eyed German colonel, whose activities in East Prussia dominated the following days' headlines and deflected attention from Monty. In the *News Chronicle* on 21 July, the story 'Allies Storm Through 9 Normandy Villages' was totally eclipsed by the banner headlines proclaiming: 'Hitler: Assassination Attempted at his Headquarters'. Misleading the press over Goodwood had been foolish; but with his own Main HQ elsewhere under Freddie de Guingand (it would relocate close to Monty only in mid-August), and Eisenhower and SHAEF headquarters still across the Channel in Portsmouth and London, there would remain plenty of scope for Monty to mishandle people and events.

21

Plots and Breakouts

For most of the forty-one days that Rommel fought the Normandy campaign against Monty he was on the back foot, reacting to the moves of his opponents, never gaining enough time or reserves with which to launch a surprise of his own. Twice only did Rommel try and shape the campaign in strategic terms for Germany. On 17 June he and Rundstedt met Hitler at the reserve *Führerhauptquartier* in France, built originally to oversee the invasion of Britain, at Neuville-sur-Margival, near Soissons.[1] Even by this stage, eleven days after D-Day, Hitler still chose to believe Normandy to be a feint and (swallowing the Fortitude deception) was waiting for a second landing elsewhere – probably in Fifteenth Army's area, around Calais. He flatly rejected the requests of his field marshals to pull back the defenders beyond the limits of Allied naval gunfire, so as to consolidate their defence. His 'no retreat' policy, ruthlessly imposed in the Russian winter of 1941–2 may have been appropriate at that time and place, but had no applicability elsewhere – as Rommel had found to his cost after Alamein, and was now experiencing again in Normandy.

Realising the military situation could lead only to one result, Rommel played another hand. When the stenographers had been dismissed, he hinted that Germany needed to make some sort of peace with the west in order to carry on the fight in the east. Hitler would have none of it, not even from his former security adviser, and brushed Rommel's concerns aside with an angry retort that Erwin should leave the politics of the situation to him. The Allies were to find the battle for Normandy so costly precisely because Hitler ordered this rigid defence of every scrap of ground; it was a policy born of no military logic, but then Hitler rarely acted in a militarily logical way.

On a second occasion, on 29 June, Hitler summoned both his field marshals to Berchtesgaden to discuss the campaign, in the aftermath of the fall of Cherbourg. Rommel, more convinced than ever that the front was about to collapse and that 'political action' was needed in realigning Germany's strategy, took the risk of running his thoughts past Goebbels and Himmler before again confronting Hitler. Shrewdly neither reacted, keeping their own counsel. Again Rommel and Rundstedt briefed Hitler; again Hitler refused permission for any withdrawal. Instead, just as at Margival, the meeting turned into a 'Führer monologue' about the new vengeance weapons and jet fighters that were going to suddenly turn the tide. Rommel tried to steer the conference towards the strategic, boldly stating: '*Mein Führer*, on behalf of the German people, we should begin with the political situation.' Curtly, Hitler invited his field marshal to 'stick to the military situation'. Rommel tried again: 'History demands that we should realise greater issues are at issue,' but Hitler, impatiently demanded that his Desert Fox 'deal only with the military situation in the west, nothing else'.

Later on, during a pause, Rommel interjected: '*Mein Führer*, I must speak openly of the future of Germany.' This time the remarkably calm but menacing Führer-headmaster ordered his Rommel from the room: 'Field Marshal, please leave now before you say something you regret.' Perhaps – almost biblically – Hitler felt he had been thrice-denied by his once-favourite general. At any rate, Rommel quietly left the great hall at Berchtesgaden at 9.45 p.m., never to see Hitler again. Rundstedt – who had just received news of Dollmann's death – was amazed as he then heard Hitler order a thousand Luftwaffe fighters be sent to the Normandy front. From where, he wondered? Shortly after the 29 June conference, Keitel phoned from Berlin reacting to more bad news from Normandy; 'What should we do now?' he mused to Rundstedt. The sixty-nine-year-old field marshal, never a supporter of Keitel's, let alone Hitler's, shouted down the phone: 'Make peace, you fools!' He had experienced quite enough and was not surprised to receive news of his dismissal on 2 July, accompanied by a 'Führer-sweetener', a small leather-bound case containing silver Oakleaves to affix to his Knight's Cross.

Whilst the Germans had reacted very effectively at a tactical level to

Goodwood, they were losing the campaign operationally, and tanks and fresh troops that should have formed an operational reserve for future counterattacks were being used to plug gaps in their front. Rundstedt's replacement, Hans von Kluge, arrived in France believing Rommel to be a defeatist, and the two had sharp words initially, but he rapidly came round to Rommel's viewpoint. Kluge and Rommel had met previously in July 1943, just after the Allied invasion of Sicily. The encounter at Hitler's HQ in East Prussia was recorded by the ADC of a third *Generalfeldmarschall*, Erich von Manstein, who was also present. After a meeting with Hitler the three marshals relaxed and chatted about the course of the war. Kluge said: 'Manstein, the end will be bad.' Rommel then agreed the 'end of the war would be a total catastrophe'. Manstein, architect of the 1940 victory against France, remained more optimistic. Later Rommel told the ADC: 'Your Field Marshal [Manstein] is a strategist of genius, I admire him, but he is cherishing an illusion.'[2] From the evidence of this earlier meeting, it appears that Kluge, too, may have concluded that the war was already lost.

In July 1944, perhaps Rommel and Kluge picked up the threads of their conversation of exactly a year earlier, and discussed possible ways of ending the struggle in Normandy on their own, with some sort of overture to the western Allies. At any rate, Rommel decided to send a last telegram to Berlin outlining the political ramifications of the battle in the west that Hitler had refused to hear on 17 and 29 June. In it, he stated that Army Group B was losing men and material at a catastrophic rate; that adequate reinforcements were not forthcoming, and that the 'unequal struggle is approaching its end. It is urgently necessary for the proper political conclusions to be drawn from this situation.'[3] Rommel despatched this missive to Berlin on Saturday 15 July. It was definitely an unwise move, and underlined perhaps Rommel's political naivety and surely desperation. It was doubly unfortunate in view of what would happen to him two days later.

Rommel, continuing the chaotic (if inspirational) command style he had developed in the Western Desert, was driving all over Normandy, assessing the situation, shoring up morale, visiting his subordinates and returning most evenings to La Roche-Guyon, presided over by the seemingly unflappable Hans Speidel. It was an exhausting time. Although Army Group B had an alternate headquarters at Vernon, further down the River Seine, it was La Roche-Guyon that remained the principal centre of activity. Rommel needed to be centrally located in the command chain: La Roche-Guyon was equidistant between Seventh Army in Le Mans and Fifteenth Army in Tourcoing, who were awaiting another landing in their sector, long after D-Day. Rommel also felt he needed to be close to the OB West headquarters at St Germain-en-Laye, near Paris (Rundstedt's command, then that of his successor, Hans von Kluge), which was a little over an hour's drive away. He would have done much better to have emulated Monty, and operated from a TAC HQ forward, leaving Speidel at the Main HQ in La Roche-Guyon. Nevertheless, whether earlier in Africa, or later in Normandy, Rommel could not resist the comradeship of the front, and loved touring his troops' forward positions, as Friedrich von Mellenthin (his intelligence officer, later his operations officer in the Afrika Korps), recalled:

> During his visits to the front he saw everything and nothing escaped him. When a gun was inadequately camouflaged, when mines were laid in insufficient number ... Rommel would see to it. Everywhere he convinced himself personally that his orders were being carried out. While very popular with young soldiers and NCOs with whom he cracked many a joke, he could become most outspoken and very offensive to senior commanders of troops if he did not approve of their measures.[4]

One might observe that none of this should be a commander-in-chief's function; Rommel's presence may have been inspirational on the battlefield, but much of this 'eyes forward' work should have been done by trained staff officers (as required by Monty), whilst attending to the camouflage of guns hints at an unnecessary obsession with detail that bedevils some generals, such as Gort in 1940. This unstructured approach was also his undoing. Monday 17 July found Rommel visiting 277th and 276th Infantry Divisions, followed by a conference with Sepp Dietrich, commander of I SS Panzer Corps (1st and 12th SS Divisions) at St Pierre-sur-Dives; their business done, Dietrich warned Rommel to be careful on his journey home – his own staff car had been intercepted by fighters and totally destroyed just days earlier.

At about 4 p.m., Rommel, with *Gefreiter* Daniel at the wheel of his Horch staff car, set off accompanied by Hauptmann Lang, Major Neuhaus, and *Feldwebel* Holke acting as aerial observer – necessary for all road journeys in Normandy. Their route took them through Livarot, which they could see was under aerial attack. Often navigation was a matter of taking minor roads and tracking across country when necessary: forced to avoid Livarot, they diverted south on the highway to Vimoutiers, with the intention of heading east later, where, according to Helmuth Lang:

> Holke, our spotter, warned us that two aircraft were flying along the road in our direction. The driver, Daniel, was told to put on speed and turn off onto a little side road to the right, about 300 yards ahead of us, which would give us some cover. Before we could reach it, the enemy aircraft, flying at great speed, came up to within 500 yards of us and the first one opened fire. Marshal Rommel was looking back at that moment. The left-hand side of the car was hit by the first burst. A cannon-shell shattered Daniel's left shoulder and left arm. Marshal Rommel was wounded in the face by broken glass and received a blow on the left temple and cheekbone which caused a triple fracture of the skull and made him lose consciousness immediately . . . Major Neuhaus was struck on the holster of his revolver and the force of the blow broke his pelvis . . . Daniel, the driver, lost control of the car. It struck the stump of a tree, skidded over to the left of the road and then turned over in a ditch on the right . . . Marshal Rommel, who at the start of the attack, had hold of the handle of the door, was thrown out, unconscious, when the car turned over and lay stretched out on the road about twenty yards behind it.[5]

I last met the late Alain Roudeix in April 2001 near Falaise; short and stocky, with a tweed bonnet always glued to his scalp, he had spent much of the post-war era clearing the 1944 battlefields of munitions and had accumulated quite an interesting collection of 'scrap' as he termed his growing pile of steel helmets and other military detritus. In July 1944 he had been mending fences near the Livarot–Vimoutiers road when he heard an aircraft shooting up 'something on the road'; this was at about 6 p.m. He remembered heading home soon afterwards, his work done, when he saw the wrecked staff car in a ditch and a group of Germans, one with his arm blown off (Daniel, the driver) by a gatekeeper's lodge at the roadside. Shouting and agitated, they asked him where the nearest hospital was. Roudeix recommended a doctor in Livarot, and then went on home; air attacks were daily occurrences and like all Frenchmen he was keen to see the back of the Germans.[6] Rommel's wounds were dressed by the doctor suggested by Roudeix, before he and Daniel, both still unconscious and close to death, were moved to a Luftwaffe hospital at

Bernay. The site of Rommel's accident on the Livarot–Vimoutiers road is best remembered by the name of the hamlet opposite where it happened: St Foy de Montgommery – Bernard's ancestors were having their last laugh.

The news of Rommel's injury was suppressed for some time, for fear of the effect on morale amongst his troops, and in Germany. Though many Allied fighters were in the air that day and several attacked lone vehicles, two Allied pilots seem to have a better claim than most to be the cause of Rommel's accident. One was Squadron Leader Chris le Roux, a South African Spitfire pilot, flying with his 602 (City of Glasgow) Squadron. Flying from an airstrip at Longues-sur-Mer, Le Roux claimed an attack on a German staff car near Livarot, though two other members of his squadron also claimed staff cars the same afternoon. An ace with eighteen victories to his credit, Le Roux died unaware of any impact he might have had, killed whilst flying his Spitfire on 29 August 1944. The other pilot was Spitfire pilot Flight Lieutenant Charles W. Fox of the Royal Canadian Air Force, credited with destroying or damaging 153 enemy vehicles and trains, and nine aircraft. Flying with 412 Squadron south-east of Caen, on 17 July he spotted a staff car, attacked it and forced it off the road. It does not perhaps matter who pulled the trigger, but the controversy does illustrate how full the skies were of Allied aircraft at this time.

Rommel would have been hovering in and out of consciousness when Montgomery struck with Operation Goodwood on the very next day. He was still close to death when Oberst Claus, Graf von Stauffenberg made an assassination attempt on Hitler on 20 July.

Generaloberst Ludwig Beck, the Army Chief of Staff until August 1938, was initially a supporter of Hitler's – going so far as to defend three officers accused of Nazi party membership in a court martial when any political activity was illegal, but turned against the Führer when it became apparent that Germany was being pushed into war before she was ready, and that the SS and Nazi party was eclipsing the traditional power of the army within Germany. The crunch had come over Hitler's proposed invasion of Czechoslovakia: on 29 July 1938 Beck had written a secret memorandum stating the German army had the duty to prepare for possible wars with foreign enemies and 'for an internal conflict which need only take place in Berlin' – one of the first documented cases of meaningful opposition to the Nazi regime. Although Neville

Chamberlain let Beck down with his appeasement strategy, Beck there-after started to recruit plotters against the regime. Initially this was difficult when Germany was in the ascendant in 1939–42, but became far easier after the reverse at Stalingrad, though Admiral Wilhem Canaris, head of the *Abwehr*, and his deputy, General Hans Oster, were early sympathisers.

Some, like the diplomat Ulrich von Hassell, author Ernst Jünger, Dr Carl Goerdeler, a former mayor of Stuttgart, and his successor as mayor, Dr Karl Stroelin (who had served with Rommel in the First World War), were moved to resist by knowledge of the 'final solution' or other atroci-ties committed in the name of the German people. Others, like retired *Generalfeldmarschall* Erwin von Witzleben, Alexander von Falkenhausen, Governor of Belgium, Carl-Heinrich von Stülpnagel, Governor of Paris, or Dr Hans Speidel, Rommel's own Chief of Staff, could see where the deteriorating military situation was leading, and wanted to salvage honour for Germany before it was too late. Stülpnagel's personal aide was the Luftwaffe *Oberstleutnant* Caesar von Hofacker, a cousin of Stauffenberg, whose father, the much-decorated *Generalleutnant* Eberhard von Hofacker, had been Rommel's first patron and hero until his death in 1928. Young Hofacker seems to have been very much the link man, travelling between headquarters; he had attempted (but failed) to win over Rommel in a visit to La Roche-Guyon on 9 July. There were more, like Helmuth von Moltke, son of another famous family, who saw the Nazis destroying what they saw as the worthy traditions and honour of the old state.

To suggest these individuals acted with one voice or wanted the same outcome from a coup would be to suggest too much; most were united by personal antipathy towards Hitler and the SS – like Wolf von Helldorf, Police Commissioner for Berlin or General Erich Hoepner, sacked for refusing to obey orders in the east. They had no clear idea of the look of a post-Hitler Germany. Some wanted Hitler imprisoned and tried; others, like the highly motivated Claus von Stauffenberg, who had briefly served under Rommel in Tunisia (as chief operations officer of 10th Panzer Division, where he was severely wounded and disabled by an Allied air attack), wanted Hitler dead and were prepared to act. The bright thirty-seven-year-old aristocrat, from an ancient and distinguished Catholic family, who shared his birthday with Rommel, was evidently a figure of great charisma. He had originally welcomed the Nazi accession to power in 1933 and, like Ludwig Beck, had turned against the regime early on, using his *Kriegsakademie* staff training for the benefit of the conspirators. Most had agreed that Hitler and his gang (usually Himmler and Goebbels were specified) needed somehow to be removed and negotiations opened with the Western powers for peace; but always there remained the unrealistic determination to carry on the fight in the east. The Allied unconditional surrender policy made this unlikely, and with the situation in Normandy deteriorating every day, the plotters knew they needed to move quickly before military events overtook their plans. They also had to contend with the highly effective Allied bombing campaign against German cities, which tended to unite people behind Hitler; direct action was seen by many as 'siding with the enemy'.

As the most serious plotters canvassed the German army for activists, they came across many sympathisers who were anti-Nazi by inclination: aristocrats like Gerhard, Graf von Schwerin and Heinrich, Freiherr von Lüttwitz, both commanding panzer divisions in Normandy, who would not initiate anything but who would not oppose it either. Due to the nature of the conspiracy, little was written down, but after the war many claimed they had been involved, knew beforehand of Stauffenberg's proposed actions, or had engaged in 'meaningful conversations' with some of the plotters. One of these was Sepp Dietrich, Hitler's former bodyguard, commander of I SS Panzer Corps, with whom (apparently) Rommel speculatively explored the idea of closing down the Western Front in order to concentrate on the east. It is claimed Rommel said: 'Something has to happen! The war in the west has to be ended!', to which Dietrich responded 'You're the boss, *Herr Feldmarschall*. I obey only you – whatever you're planning.'[7]

It is very difficult to know how to interpret such off-the-cuff, second-hand comments, much reported by historians. Such testimony is not necessarily incriminating, and could have referred to the military situation; on the other hand, Dietrich was a survivor and a realist, had already grown disillusioned with Hitler, and understood the reality of the military situation in the west – in his mind, here may have been an opportunity for someone with blood on his hands to make amends by changing sides at exactly the right, finely timed moment. On the one hand, therefore, we are left with an alleged vast web of plotters and sympathisers right across Germany and on the Eastern and Western Fronts who were happy to see the back of the old regime, for different reasons, but had unrealistic expectations for its successor. On the other, we see a small group of furtive activists, unsure who to trust, bound together by old friendships, blood ties and common interests, led by Beck and Stauffenberg – but certainly not with a wide network of prearranged support.

In the event, after an estimated sixteen failed assassination attempts, on Thursday 20 July Stauffenberg left a bomb under the briefing table at Hitler's headquarters in East Prussia, and flew to Berlin without waiting to check that Hitler had died in the blast. Stauffenberg was unaware the Führer had survived but initiated anyway the pre-planned coup, built around using the 'Replacement Army', of which he was Chief of Staff, to suppress a supposed takeover by the SS. Support was limited, for the plotters in Berlin had to lie in order to mobilise troops to overthrow the regime, stating that an SS coup was in progress when such was not the case. The turning point came when one of the commanders mobilised to suppress the supposed coup, Major Otto-Ernst Remer, CO of Hitler's Berlin bodyguard battalion stationed in Berlin (ironically, the successor to the *Führerbegleitbattalion* Rommel commanded 1939–40), tried to arrest the Propaganda Minister. Goebbels wisely connected Remer by telephone to the Führer's *Wolfsschanze* (Wolf's Lair) headquarters in East Prussia, scene of the earlier bomb blast, and Remer found himself speaking to the instantly recognisable voice of Hitler, who promoted him on the spot to *Oberst* and ordered him to crush the rebellion.

After that point the coup was doomed. Beck, Stauffenberg and many others were rounded up and either committed suicide, were executed or tried before a travesty of justice called the *Volksgerichtshof* (People's Court), which condemned around 3,000 to a variety of unpleasant deaths. The Gestapo interviewed many more, and the total number of deaths associated with 20 July – some of which were opportunistic, where the regime chose to rid itself of anyone it suspected of dissent – may have been around 5,000. Hitler's regime executed an astonishing eighty-four generals, the majority of these in the wake of the Stauffenberg coup.[8] Many more who survived were nevertheless put into concentration camps, some in all innocence: *Generaloberst* Franz Halder, for instance, the former Chief of the Army General Staff whom Hitler had sacked in 1942 and had no connection to the coup, was incarcerated on the grounds that he might – in view of his past status – become a focus for future anti-Hitler opposition.

In France, the coup (Operation Valkyrie) almost succeeded. Carl-Heinrich von Stülpnagel, the Military Governor, managed to disarm the SD and SS in Paris, and imprison most of their leadership. He then travelled to Günther von Kluge's OB West headquarters to enlist his support and get him to contact the Allies. Another key conspirator was *Generalmajor* Henning von Tresckow, author of several failed anti-Hitler plots, who had been Kluge's Chief of Staff at Army Group Centre in Russia. All day the spineless Kluge hedged and prevaricated, then discovered that Hitler was alive. Realising the plot had failed, Stülpnagel then released the SS and Gestapo men in Paris, coming to the gentlemanly agreement with their commander, SS-*Obergruppenführer* Karl-Albrecht Oberg, a close comrade from the First World War, that they had been 'taken into protective custody for their own safety'. Summoned back to Berlin for questioning, Stülpnagel attempted suicide at Verdun with a pistol, but failed. When he regained consciousness in hospital, he was apparently heard to mutter one name over and over again – Rommel's. Kluge, too, would be recalled to Berlin, and because he knew he would be considered guilty (though never an activist, he hadn't opposed or reported the plot), also chose suicide by poison at Verdun, and succeeded. One is struck here by how much their First World War experiences mattered to Oberg, Stülpnagel, Kluge and many other old warriors.

Rommel knew many of the conspirators personally; he had been approached by Caesar von Hofacker; he had conducted conversations with – according to their memoirs – Speidel, Dietrich, Blumentritt (Rundstedt's Chief of Staff) and his naval adviser Friedrich Ruge, which, if not treasonable, were at least 'nuanced'. He would have known that some sort of plot was afoot; his name appeared on some of the conspirators' lists, variously as a head of state, or Commander-in-Chief of the Army, but none of this amounted to the guilt it implied. Paul von Hindenburg, Germany's foremost commander in the First World War, had been appointed German President from 1925–34 as a figure of strength, honour and national unity, though he was no politician and had not courted the office. Some of the plotters, who seem to have put together several different lists, may have felt Rommel answered the call for a similar popular figure of national unity at a time of crisis.

There is nothing concrete to tell us how close to the conspiracy Rommel really was, though some plotters used his name as a magnet to recruit others – Carl Goerdeler, Speidel and Hofacker certainly so. Although Rommel was certainly disillusioned, he was also a product of the Hitler regime; he owed his later promotions to Hitler personally and his fame to Goebbels. He had lived through the uncertainties of 1918–19 and had no wish to cause revolution in Germany again. According to Lucie, he opposed the idea of assassinating Hitler, fearing his death could create martyrdom and trigger a lasting civil war. Instead, Rommel wanted Hitler arrested and brought to trial for his many crimes. To ordinary German civilians and soldiers, Rommel may have personified at least the 'nicer' side of the Third Reich. In the absence of the firm evidence of corroborated interviews,[9] I remain unconvinced that Rommel was a leading conspirator. Switching to the 'dark' side and taking an active part in a conspiracy determined on Hitler's death and a denial of most of one's lifetime achievements would have been a huge step for an officer never previously involved in politics. Nevertheless, 20 July put the barely conscious Desert Fox in the tightest corner of his battle-scarred life: however implicated, he almost certainly had no foreknowledge of the timing or details of Stauffenberg's plot. The news would have come as a great shock, but the net result was that he was implicated, and soon put under Gestapo surveillance.

For Siegfried Knappe, who we last met graduating from Rommel's classes at Potsdam in 1938, the aftermath of the July plot was worrying. He had just been appointed to Kesselring's staff and recalled:

> some of the names being mentioned were familiar to me. Everybody knew someone who had been arrested. And with the Nazis' system of holding a man's family responsible if he did something wrong, even families were arrested and sent to concentration camps. Rommel's death was announced, but it was attributed to injuries he had received during a strafing attack he had received in France; because he was so popular with the German people, his death was not connected with the assassination attempt in the news ... After [the July plot], the army was forced to adopt the Nazi salute. [It] had been used indoors, when we were not wearing a hat, since I had joined the Army in 1936. Now, however, we were forced to abandon the normal military salute and adopt the Nazi salute exclusively. Political officers were also forced on the Army ... placed in a position to watch commanding officers.[10]

Though the most high-profile aspect of the German resistance (the *Widerstand*) was the army revolt led by Beck and Stauffenberg, there were other unconnected networks, centred on politicians, Communists and the Church, all of whom suffered after 20 July 1944. In this respect, although united in their antipathy towards Hitler, the various strands of the *Widerstand* never managed to coordinate their activities. Furthermore, Allied intelligence had made very little effort to contact or support any of the anti-Hitler plotters, though aware of their existence. German plotters recorded their frustration at being rebuffed by their former British friends, but the political solution agreed by Churchill, Roosevelt and Stalin was to crush Germany militarily so as not to leave a state capable of rebirth, as had happened after 1919. Hence the unconditional surrender policy agreed at Casablanca in January 1943, which saw 'collective guilt' being imposed upon the whole German people for warmongering. It is now obvious that overtures to the plotters might have shortened the European war and saved countless lives, (between 20 July 1944 and 8 May 1945, some 4.8 million Germans died), but Stalin – ever paranoid of an Anglo-American deal behind his back – would never have permitted such a course of action. Instead, the Germans were going to be defeated in the west by the democracies fighting in Normandy, led by Montgomery.

On 25 July, three miles west of St Lô, Bradley launched his long-awaited Operation Cobra to break south, out of the Cotentin peninsula, which would eventually envelop the remaining German forces en route. Monty was aware of the plan, but not intimately involved in its creation or enaction. Cobra, like Goodwood before it, was to be preceded by the aerial carpet-bombing of opposing troops, using the strategic bomber force. This was to be followed by Lightning Joe Collins's VII Corps (70,000 men, 600 tanks and forty-three battalions of artillery) punching their way through on a very narrow, five-mile front. The attack was originally ordered for 24 July but was postponed by bad weather for twenty-four hours. Unfortunately word did not reach some aircraft, who hit Allied forward positions, causing over 300 casualties.

Just like the awful cooperation between Monty and the RAF before Goodwood that required Monty to lie in order to obtain strategic air support, so Bradley had an equally acrimonious relationship with the US Eighth Air Force. For Cobra, Bradley understood that the bombers would fly along the front, rather than at right angles to it, to minimise friendly casualties. USAAF records indicate no such understanding, underlining the poor US Army–Air Force relationship, the lack of a doctrine for such operations, and that basically ground and air commanders simply failed to understand what each other was trying to do. On 25 July, 1,500 heavy and 380 medium bombers appeared again and dropped 4,700 tons (the equivalent of a small tactical nuclear weapon), causing over 600 US casualties, including Lieutenant General Lesley McNair, America's highest-ranking wartime casualty, who now lies in the Omaha Beach cemetery. This experience of 'friendly' air support so unnerved ground troops that when Collins later wanted to call in an air strike to support the US 30th Infantry Division, its commander radioed back, 'if we have any more of the same, then our troops are finished'.

For the Germans the effect was devastating: when ordered by *Generalfeldmarschall* von Kluge to hold the line at all costs, Rommel's old Chief of Staff, Fritz Bayerlein, now commanding the Panzer Lehr Division, replied angrily over the telephone:

> Out in front every one is holding out. Every one. My grenadiers and my engineers and my tank crews, they are all holding their ground. Not a single man is leaving his post. They are lying silent in their foxholes, for they are all DEAD.[11]

The strategic bomber force largely obliterated Panzer Lehr and the other German units opposite Collins's VII Corps and, with the help of Rhino hedge-cutters attached to Sherman tanks, the *bocage* was overcome gradually and a break-out achieved. Rhinos were pointed metal horns (cut down from Rommel's beach defences) welded to the hulls of tanks, which then acted like bulldozers, ripping paths through the hedgerows. They had been devised by US Army Sergeant Curtis G. Culin and demonstrated to Bradley on 14 July, who immediately ordered Rommel's steel beach obstacles cut up and fitted to as many tanks as possible. (Rhinos were a good example of the need to innovate rapidly that the Normandy battlefield imposed on all its visitors.) Behind Collins's infantry rumbled tanks of the 2nd and 3rd US Armoured Divisions, the former exploiting the inter-corps boundary of German Seventh Army, and truly breaking through. Recent detailed analysis of the battle by James Jay Carafano has concluded that the US 30th, 4th and 9th Infantry Divisions had a bitter five-day struggle to overwhelm their opponents; and that 'armoured column cover' (air cover provided by P-47 Thunderbolts, controlled by an experienced pilot who rode in the leading tank of each US armoured column) was the most significant contribution to the battle. Between 27–30 July, 9th Tactical Air Command flew 5,105 sorties in support of VII Corps.[12]

As the Germans crumbled, Cobra revealed operational opportunities far beyond the plan's original intent. With Troy Middleton's US VIII Corps operating on the right and Collins's VII Corps on the left, Bradley's First Army had pushed the twelve miles to Coutances by 28 July, and Avranches on the west coast of the Cherbourg peninsula, which fell on 30 July. The next day, Patton's Third US Army was activated, and Bradley rose to command 12th US Army Group, whilst Courtney H. Hodges took over First US Army. Henceforth, Bradley was effectively an equal to Monty, with an army group of his own, rather than a subordinate. Quickly, George Patton exploited the gap rent in the German lines, and with nothing to stop him, pushed his Third Army formations south and east towards Le Mans.

Operation Cobra illustrated that armour was effective only after the infantry had done their job of breaking through and creating movement corridors for the tanks, covered by close air support, which is why Goodwood had failed the week before. The real achievement of Cobra was that it proved that the US Army had matured very quickly since 6 June, arguably quicker than Monty's British or Canadians, and fast enough to learn the necessity of the all-arms battle and so beat the Germans in close combat. Now out of the *bocage*, the Allies were able to gain the operational flexibility that had been denied them in the constricting terrain of Normandy's three *départements* of Calvados, Manche and Orne.

A panzer counterattack at Mortain on 7 August (Operation Lüttich), personally ordered by Hitler, actually helped to ensnare the German Seventh Army in a trap of its own making, as the Americans raced east, and then hooked north to cut off the Germans who had advanced into their left flank. With the lesser-known Operation Bluecoat – the important break-out from hedgerow country by British XXX Corps via Mont-Pinçon (captured on 7 August) – the German Seventh Army started to be enveloped on its north-west flank. There had been concern that the Germans would be able to use the ideal defensive terrain of the Suisse Normande, particularly Mont Pinçon, as the key to a new defensive position, or as a firm base prior to withdrawal. At Monty's behest, Brian Horrocks's (Bucknall's replacement) XXX Corps therefore launched Bluecoat to prevent this, and fed in first 11th Armoured and then the Guards Armoured Divisions to the hilly terrain. Most senior NCOs and officers within the Guards Armoured Division were experienced infantrymen before converting to armour and the division was the first formation to successfully apply the lessons of combined infantry and armour working together. After the lessons learnt from Operations Epsom and Goodwood, infantry and armoured units were paired off to provide an especially close level of cooperation.

Operation Totalise, on 8 August, was a Canadian armoured dash to Falaise, astride the main Caen–Falaise road, preceded by aerial bombing on a Goodwood scale. It was designed to get General Guy Simond's II Canadian Corps (three Canadian divisions and 1st Polish Armoured, under General Stanislaw Maczek) to Falaise the same day by frontal armoured assault, operating with infantry in Kangaroos – improvised personnel carriers.[13] Totalise contained many hallmarks of Canadian ingenuity – made at night, illumination was provided by reflecting search-light beams off overhead clouds (known as 'Monty's Moonlight'), whilst the left and right of arcs of the corps axis of advance were shown by firing 40mm anti-aircraft tracer rounds. Such ability to improvise was crucial to the success of the campaign, but Monty's hidebound British forces appear to have maintained a system that stifled such initiative – for few startling innovations emerged from it. Whilst technological inno-vation appeared in the run-up to D-Day (swimming tanks, Hobart's Funnies and the Mulberry harbours, for instance), and was important for its success, tactical innovations on Monty's battlefields were few. This might have been an institutional flaw within the British Army, but Monty's system of discouraging 'bellyaching' may also have suppressed many good ideas.

Totalise tested their opponents, the veteran 12th SS Division, to the extreme, as narrow columns of armoured vehicles drove through the German defences at night without a preliminary artillery barrage, but with heavy bombing from the air to seal the flanks. Once on their objec-tives, the infantry exited their Kangaroos and cleared out the defenders. It was only the quick thinking of the local German commander, Kurt Meyer, which prevented disaster, as he described in his memoirs:

> before me, making their way down the Caen–Falaise road in a disorderly rabble, were the panic-stricken troops of the [German] 89th Infantry Division . . . I realised that something had to be done to send them back into the line and fight. I lit a cigar, stood in the middle of the road and in a loud voice asked them if they were going to leave me alone to cope with the enemy. Having a divisional commander address them in this way, they stopped, hesitated, and then returned to their positions.[14]

One suspects the fact that the lone figure was a senior SS officer with a bloody reputation also acted as an incentive to turn about! Having rallied the frightened soldiers of the 89th, Meyer sent armour and anti-tank guns to their positions at Cintheaux before directing his two battle groups to counterattack to the north of the village. Stiffening their resistance against the continued pressure, German anti-tank gunners halted the Canadians after an advance of three miles. Over the next two days, the effects of this action and the continuous grind of counterattacks reduced the German division to little more than a reinforced battle group. Simonds tried to bomb his way through, but the Germans had captured a scout car on 13 August with a copy of the plan for the attack, and Meyer moved his men back in time. In the end, Meyer's front was held only by 500 panzer grenadiers and fifteen tanks, until forced to withdraw when the 2nd Canadian Division broke through on their right flank.

By the time Falaise, birthplace of William the Conqueror, fell into Allied hands on 16 August it had, like Caen, been reduced to brick dust. Six days after Totalise, Operation Tractable abandoned the head-on thrust in favour of a flanking move eastwards towards Trun. 4th Canadian Armoured and 1st Polish Armoured Divisions were to advance from the north-east, whilst the British 53rd Welsh and 59th Staffordshire Divisions attacked from the north-west, in an effort to catch the Germans in a ring of steel, surrounded by eleven Allied divisions, with the Americans to the west and south.

Generalfeldmarschall Günther von Kluge, Rommel's replacement at Army Group B, had spent 15 August visiting the front, but had been out of contact with his HQ and a frantic Hitler for the whole day. Kluge claimed his staff car had been forced off the road many times by Allied fighter-bombers (not unreasonable; Rommel had of course been wounded this way). However, suspicious of his generals after the 20 July attempt on his life, Hitler concluded that Kluge was plotting to make peace with the Allies in Normandy and relieved him of command on 16 August. Although there is not a shred of evidence to support this, conspiracy theorists comment that some senior US commanders were expecting overtures from a 'senior German commander' at this time. If Kluge had been unable to make up his mind whether to join the Stauffenberg conspiracy, it is unlikely that his temperament would have permitted him to initiate any peace overtures on his own. A dedicated Nazi, Kluge had arrived in Normandy full of enthusiasm, taking over first OB West from Rundstedt, then Army Group B from the wounded Rommel (and subsequently merged the two commands, something that OKW should have contemplated before 6 June), but became quickly disillusioned and was considered defeatist by Hitler. It was after being implicated in the 20 July plot and summoned back to Berlin that he committed suicide. His place in the rapidly shifting cast of senior German commanders was taken by the steady-handed *Generalfeldmarschall* Walter Model, who had been summoned in haste from the Eastern Front.

Whatever the truth, the unconditional surrender clause agreed by Churchill and Roosevelt at the Casablanca Conference (14–24 January 1943) would have limited Eisenhower's ability to accommodate any deal proposed by Kluge, or anyone else. But what was Allied strategy? Was it to bring an end to the fighting in Europe and stop the consequent bloodshed as soon as possible, or was it to reach Berlin and crush the Nazi regime? Stalin feared the former might happen without the latter, and the unconditional surrender policy was proposed by Roosevelt as a gesture for Stalin's benefit, who felt he was personally unable to leave his own country and join Churchill and Roosevelt in Morocco.

With hindsight we might observe that the unconditional surrender clause was counter-productive and merely pushed the Germans into a corner, giving them no room to negotiate any sort of peace deal, and possibly prolonged the war in the west. Reflecting popular post-war American opinion of the policy, one later commentator observed:

> The political and diplomatic folly of our wildly proclaimed 'unconditional surrender' policy gave such solidarity to Goebbels's malevolent propaganda and fastened Hitler's grip on the *Wehrmacht* so firmly that it was impossible to subjugate Germany within six months after cracking *Festung Europa* in Normandy.[15]

It would have been an interesting test of the west's political resolve had Kluge, or Rommel before him, or a regime installed by Stauffenberg's actions, offered the Western powers a negotiated peace. Given the excessive Allied losses suffered in Normandy, the prospect of worse to come by invading Germany and a hint of the duplicitous Communist approach towards Poland, the west may indeed have been tempted to agree to a separate peace and weather the storms of rage from Stalin. For, starting on 1 August 1944, the Polish Home Army had risen up and captured

Warsaw from the Germans, with the aim of handing their self-liberated city over to the advancing Soviet army; they had also foreseen the need to emphasise Polish sovereignty and install Free-Polish governance before the Soviet-dominated Polish Committee of National Liberation could assume control. Unhappy at this expression of anti-Russian democracy, Stalin ordered the Red Army to halt along the River Vistula and let the Germans crush the Poles, after a sixty-three-day resistance, on 2 October, before entering Warsaw. This was at a time when Monty was also casti-gating the Free Polish Airborne Brigade (those who had managed to escape from Arnhem) for not 'fighting hard enough'.

In any case, the Anglo-American armies under Montgomery were not remotely equipped to deal with a German peace overture. Today all modern military commanders have political advisers (POLADs) in their headquarters, advising on the political implications of their military activ-ities. Monty had no equivalent individual at his TAC HQ back in 1944, and although Eisenhower's SHAEF headquarters was equipped to under-stand this military/political interface, the Main SHAEF HQ remained at Bushey Park near Teddington, its home since 5 March, until its 8,000 officers and 21,000 men (in personnel terms, the size of a division) relo-cated to Versailles on 20 September. Unsurprisingly, French journalists began to refer to SHAEF as the *Société des Hôteliers Américains en France*. On 7 August SHAEF's Forward HQ opened at Tournières, near Bayeux, then moved to Jullouville on the west coast of the Cotentin, near Granville, at the end of that month. This then became Eisenhower's operational HQ when he assumed direct command of 12th and 21st Army Groups on 1 September. Neither was Monty equipped to deal with de Gaulle. Inspired by the Poles in Warsaw, in Paris the *Forces Françaises de l'Intérieur* (FFI) took over their city in armed insurrection. They held on until their liberation by (in a magnanimous Allied gesture) the only French forma-tion in Normandy, Philippe Leclerc's 2nd French Armoured Division on 25 August. At this point, de Gaulle appeared in Paris and installed himself as (unelected) leader of France.[16]

Meanwhile, by 18 August, the German forces in the Falaise Pocket had been squeezed into an area six miles long by seven miles wide. Allied attempts to close the German escape route between Trun and Chambois had been hampered by bad flying weather on 17 August, but during the 18th over 3,000 sorties had been flown by Spitfires, Typhoons, Mustangs, Lightnings and Thunderbolts of the 2nd Tactical Air Force against the retreating Germans. The congestion in the skies was such that one pilot likened the traffic to a 'bank-holiday rush', reporting that 'no results observed owing to the number of aircraft in the area'. With so many Allied troops converging on the Germans, incidents of friendly fire were numerous: the 1st Polish Armoured Division suffered 263 killed and wounded by Allied aircraft. Retreating German soldiers tried to hitch lifts on passing vehicles and horse-drawn columns, whilst the roads were under fire from tanks, artillery and fighter-bombers. By 19 August, their final escape route had been compressed into a narrow corridor two miles wide between St Lambert-sur-Dives and Chambois. Running between the two small towns is the River Dives, whose waters have cut deeply into the Norman soil over successive generations. With banks of eight to ten feet, the little Dives was an effective obstacle to all transport desperate to cross. Two narrow stone bridges at St Lambert and the ford at Moissy created bottlenecks for the Germans, whose retreat eastwards had degenerated into a disorganised rout. The route to and from the crossing places was under constant Allied aerial and artillery attack, and became known to the Germans as *Das Korridor des Todes* – the Corridor of Death.

In the summer of 2008, I returned to Moissy ford with Herr Nicolas Frank, a radio operator on one of 116th Panzer Division's few remaining Panther tanks, who recalled crossing the ford under fire on 20 August 1944. There he met an SS officer, pistol in hand, who ordered his tank to bulldoze aside an ambulance full of charred bodies that had been hit. Frank's tank had been holed in its fuel tank, but the crew had plugged the damage with German black bread, which, he said, 'set like concrete'. Frank reckons he saved his own life by deserting five days later. Meanwhile, a British artillery observer recounted how:

the floor of the [River Dives] valley was seen to be alive with . . . men marching, cycling and running, columns of horse-drawn transport, vehicles . . . and as the sun got up, so more targets came to life . . . It was a gunner's paradise and everybody took advantage of it . . . Away on our left was the famous killing ground, and all day the roar of Typhoons went on and fresh columns of smoke obscured the horizon . . . We could just see . . . the whole miniature picture of an army in rout. First a squad of men running, being overtaken by men on bicycles, followed by a limber at full gallop, and . . . being overtaken by a Panther tank crowded with men doing well over 30mph, all with the main idea of getting away as fast as they could.[17]

On 19 August, Monty sent a battle group of 2,000 men and eighty-seven tanks from 1st Polish Armoured Division to occupy Hill 262 at Mont-Ormel, as the cork in the bottle that contained the retreating German army. This area became known as *Maczuga*, the Polish for 'mace', which was the shape of the Polish perimeter that August. Withstanding repeated attacks by the Germans desperate to break through, the Poles on Maczuga suffered 1,441 casualties until the Falaise Pocket was closed. One Polish eyewitness recorded:

> the attack soon started up again. Our losses kept mounting . . . I could not believe my eyes and ears: the Huns were marching towards us singing '*Deutschland, Deutschland über alles*'. We let them approach to fifty metres, then we mowed them down, then we ran out of ammunition . . . one wounded soldier, very near me, looked like a mere child: I read his date of birth in his paybook: April 1931! He was thirteen years old . . . we took some prisoners. A few members of the *Wehrmacht* were of Polish origin. We asked them if they wanted to join up with us. When they said yes, they were given a gun and the uniform of a dead colleague. These unexpected reinforcements were invaluable to us. As for the SS and those whose papers showed that they had been in Poland in '39, we showed them no mercy.[18]

Major Wladyslaw Zgorzelski, whose fighting group consisted of the 10th Dragoons, 24/Lancers and two anti-tank batteries, captured the village of Chambois and was ordered by General Maczek 'to hold it at all costs'. Zgorzelski wrote in his diary:

> The weather created particular difficulties on that battlefield. Battledress proved very uncomfortable in the day's heat under the blazing sun. Clouds of dust, raised by hundreds of tracked and wheeled vehicles from dry soil, covered the countryside and penetrated into the eyes and parched throats, while drinking water was in short supply. The most pitiful sight was that of the dispatch riders covered in dust, with black faces, swollen eyelids and reddened eyes. There was no water, so local cider was tried but found to be a poor substitute. The most burdensome thing one had to endure was the stench of the swollen German corpses decomposing quickly under the blazing sun. Their bodies were scattered everywhere on the fields, in the hedges and amongst the buildings. Continuous fighting left no time for burying the dead.[19]

When the fighting died down on the morning of 22 August, the Normandy campaign was effectively over. More than 70,000 German soldiers had been killed or captured in the Falaise Pocket alone – the remnants of nineteen divisions that had arrived in Normandy at full strength, of whom perhaps some 20,000 escaped. The campaign as a whole is reckoned to have cost the Germans 1,500 tanks, 3,500 guns, 20,000 vehicles and nearly 500,000 men, most of whom could have been pulled back across the River Seine to fight another day.

Immediately afterwards, the British No. 2 Operational Research Section found 187 tanks and self-propelled guns, 157 other armoured vehicles, 1,778 lorries, 669 cars and 252 artillery pieces within the small area bounded by the towns of St Lambert, Trun and Chambois.[20] No count was made of the horse-drawn transport, the stench of rotting flesh in the hot August weather being so overpowering that investigators had to hurry past.[21] Thousands of military horses and farm animals perished in the same area, which was quarantined for the two months after mid-August because of the risk of disease. The carnage left an impression on all who witnessed it: Sergeant Norman Kirby of Monty's TAC staff recalled 'there was hardly room for my vehicle to pass between the gruesome piles of bodies of men and horses ... the heat was intense and the smell nauseating'.[22] Even today, just as on the Somme, most weeks the local newspapers report the discovery of human remains during ploughing or building work, along with rusting helmets, twisted weapons and hand grenades, which still kill and injure local farmers and livestock. In 2004, the last resting place of a German NCO was found in a thick *bocage* hedgerow in the St Lô area, where he had crawled and died in 1944. A Tiger tank, abandoned in August 1944, and now displayed beside the Lisieux to Alençon road at Vimoutiers, is the last relic of the hundreds of armoured fighting vehicles the Germans left behind in the Falaise Pocket. Their defeat was total and comprehensive, and to some German generals Falaise was the 'Stalingrad in Normandy'.

Whilst total defeat threatened in Normandy, Rommel remained in Germany, recuperating. He was a very fit fifty-three-year-old and this undoubtedly aided his recovery; within a week, 24 July, he was able to write to Lucie and explain his injuries: he had throbbing headaches from his skull fractures, and his left eye was swollen shut. Despite the verdict of surgeons that 'no one could have survived injuries like these', he returned home to Herrlingen on 8 August. In hospital and at home,

Rommel received a train of visitors, keeping the restless field marshal abreast of military developments and passing on sinister gossip in the aftermath of the Stauffenberg plot. Rommel's fellow Swabian Hans Speidel visited his old boss when he could. His last visit was whilst suspended from his duties at Army Group B on 6 September. The next day the Gestapo took him in for interrogation, underlining for Rommel the precarious position he himself was in. Field marshals, he realised, were not sacrosanct; three had gone already: von Rundstedt had been dismissed on 2 July; the retired Erwin von Witzleben had been found guilty and hanged by the odious People's Court on 8 August, and Hans von Kluge had committed suicide on 15 August. Impacting on the campaign in the west, the bomb plot had cut a swathe through Germany's senior military commanders, removing much irreplaceable institutional wisdom and expertise. It was at this stage, whilst recovering at home, that Rommel destroyed any incriminating evidence relating to his command in Normandy, or the nature of his relationship with any of the plotters, which leaves historians so much in the dark as to the depth of his involvement. He now felt threatened by Berlin and realised his past conversations with Hitler of 17 and 29 June and his telegram of 15 July could be taken in a very different light.

Unfortunately, in his dealings with the Army High Command Rommel had made many enemies: those who were plainly jealous and spiteful, and those he had rudely brushed off in the past, when he was head-strong in Africa and believed he was so close to Hitler that nothing could stop him. He had maintained one friend in Hitler's entourage, General Rudolf Schmundt, Chief of the Army Personnel Department. Ironically, one of the victims of the July bomb plot was Schmundt, who was caught in the blast and died of complications on 1 October. Without Schmundt, Rommel knew his chances of survival, much less a military comeback, were minimal. Rommel was first summoned to Berlin on 7 October, but refused to travel, fearing for his life, and provided medical proof he had still not recovered.

On Friday 13 October Rommel remained sufficiently concerned for his life – convinced that his murder would suit the regime very well – to discuss with his long-standing aide-de-camp Hermann Aldinger, who had served under Rommel in both wars, where he wished to be buried should the worst occur. The gloom was not lifted when a telegram was delivered that afternoon announcing that Schmundt's successor, General William Burgdorf (with whom he had served in the 1930s), and the head of the Legal Section of the Army Personnel Branch, General Ernst Maisel, would visit Rommel at home on the following day. Hitler knew it would cause a major scandal if it were revealed that Rommel had been in any way connected with Stauffenberg's plot, given his enduring popularity with the German people. It might legitimise the plotters in some peoples' eyes and provoke more dissent. The Führer would have asked his various army generals and others in his entourage – those he could trust – their views on Rommel; no one would have condemned him outright, but no one seems to have spoken up on his behalf.

Erwin had crossed swords with the Nazi hierarchy ever since the 1936 Nuremberg rally, when he had blocked their access to Hitler. When commanding the Führer's bodyguard in 1939, he had disagreed violently with Martin Bormann, subsequently head of the Party Chancellery after 1941, and who thus controlled access to Hitler. Heinz Guderian, Army Chief of Staff (OKH) from 21 July 1944, was no supporter after the acrimony surrounding control of the panzer divisions earlier in the year. Keitel and Jodl at OKW remained publicly contemptuous and privately jealous of Rommel's activities in Africa. Kesselring had become a competitor in the Mediterranean and the two had fallen out. Whatever Josef Goebbels felt privately about the legend of the Desert Fox that he himself had helped create, he kept his own counsel. It seemed to be payback time and Rommel's enemies in Berlin outweighed his few friends: amongst the generals who sat on the Officers' Court of Honour to decide Hans Speidel's fate on 4 October 1944 were Heinz Guderian and *Generalleutnant* Heinrich Kirchheim, whom Rommel had dismissed from North Africa in 1941. Speidel agreed he had gleaned information about the plot, but claimed he had passed that knowledge on to Rommel and had assumed Rommel forwarded that information on to Berlin. Ralf Georg Reuth has documented how Kirchheim and Guderian seem to have gone out of their way to exonerate Speidel, thus keeping him away from Judge Roland Freisler's court, but in so doing they concluded that Rommel was at least in possession of 'guilty knowledge'.[23]

A solution to Rommel's 'guilt' was presented to Erwin by the two generals, Burgdorf and Maisel, when they called on him at midday on Saturday 14 October 1944. Rommel was offered a trial at the People's Court, preceded by a formal expulsion from the army, which – he was assured – would happen. Rommel was aware of the injustices meted out by Judge Freisler, and the humiliations: when Field Marshal von Witzleben had been in the dock, he was given baggy trousers to wear without a belt or braces; Freisler berated him 'stop fiddling with your trousers, you dirty old man', and had him hung. Or Rommel could take the cyanide capsule offered, which Burgdorf had brought with him; it would act straight away, there would be no recriminations against his family; Lucie would receive his field marshal's pension during her lifetime and he would be given a state funeral. Rommel had only ten minutes in which to make his choice. The 'men' in the house – his fifteen-year-old son Manfred and his aide-de-camp Hermann Aldinger – favoured shooting their way out, until it was seen that truckloads of SS were ringing the house.

So Erwin Rommel departed as the Desert Fox, in his Afrika Korps uniform that he loved the most, wearing the gilt-crossed batons of the field marshal's insignia awarded to him by Hitler in June 1942 and his *Pour le Mérite*, which he had worn ever since receiving it from his Kaiser in December 1917. He said first goodbye to Lucie, then Manfred and finally Aldinger, his close comrade from both world wars. To combat the slight chill of that October Saturday, he donned his favourite field-grey greatcoat, trimmed with the red facings of a general and gold buttons, and tucked under his arm the field marshal's baton, given to him on 30 September 1942 by the same Hitler who had just condemned him to death, straightened his gold-embroidered cap and strode to Burgdorf's waiting Opel staff car. He shook hands with Manfred and Aldinger and, as the driver Heinrich Doose stood to attention, and Burgdorf and Maisel saluted, *Generalfeldmarschall* Erwin Johannes Eugen Rommel climbed into the car and was gone. He didn't look back. Manfred noted the driver belonged to the SS, an organisation he had once wanted to join, but his father had forbidden it.

At 1.25 p.m. came the call from Ulm hospital that Rommel had been brought in with a suspected heart attack, but was dead on arrival. The official story was altered subsequently to claim his death had been brought about by the wounds he had suffered on 17 July. The Nazis were masters of the charade: Rommel lay in state at Ulm town hall and received a lavish state funeral three days later, on 18 October, to which Hitler and the party hierarchy sent wreaths and condolences. To those in the know, Hitler's vengeance shocked even some senior Nazis. Himmler, sent word via an emissary that he wanted Lucie to know he 'had no part in this'; the emissary was none other than Rommel's aide-de-camp and propaganda expert from North Africa, Alfred Berndt. To make sure that no evidence of foul play could ever surface, the Desert Fox was cremated.

His ashes lie under a simple, much-visited Iron Cross grave marker, carved in wood, in the town cemetery at Herrlingen. A short distance away another plaque set in a stone cairn reads:

Hier wurde Gfm. Erwin Rommel am 14.10.1944 zum Freitod gezwungen! *Er nahm den Giftbecher und opferte sich, um das Leben seiner Familie* *vor den Schergen Hitlers zu retten.* [Here Generalfeldmarschall Erwin Rommel on 14.10.1944 was forced to commit suicide. He took the poison and sacrificed himself, to save the lives of his family from Hitler's henchmen.]

It marks the spot where Heinrich Doose pulled over his staff car on a Saturday in October 1944 and left the field marshal to his own affairs with General Burgdorf, whilst he walked away for a cigarette.

22

Beyond the *Bocage*

THE HEDGEROWS AND orchards of Normandy were, then, Rommel and Montgomery's last battleground. It was not a happy campaign for either man: it cast a shadow over Bernard Montgomery's reputation and cost Erwin Rommel his life. All in all the invasion should have succeeded, given that it was the product of mind-boggling resources and over a year's planning, involving some of the finest military minds on the Allied side. But, as many senior commanders have since admitted, whilst overwhelming attention was devoted to the act of storming ashore, little thought was given to the difficulties of fighting in the Normandy terrain, for it was assumed the Germans would cave in and retreat if the landing was successful, whilst momentum carried the Allies forward. Monty's own reputation – as far as the Americans and Britain's own RAF were concerned – came to rest on his performance in Operation Goodwood, regarded as a tactical defeat, which seemed to epitomise his indifference towards Allied strategy and the notion of acting as part of a team. It is significant that when Monty was ennobled with a viscountcy, he did not take the title Montgomery of Normandy, Caen or Falaise, as he might have done, in homage to his ancestors; but he chose a far-off little railway station where he had been happiest in command, with no allies to offend – El Alamein. In the public mind, too, Rommel would always be the Desert Fox.

In July 1944 that bastion of English satire *Punch Magazine* hit on the Monty–Rommel contest in a cartoon depicting the pair – though neither of them very recognisable – as eighteenth-century dandies, fighting a duel, rapiers clashing. And in some ways they were both military dandies, in their friendships with the media, their self-conscious tailoring of their images, their peacock vanity, and in their fiery relationships with contemporaries.

As the north-west Europe campaign progressed, Monty's TAC HQ moved first from Creullet to Blay, then he shifted his TAC in early August to the Forêt de Cerisy, where he entertained an incongruous visitor. This was the Independent Member of Parliament A. P. Herbert, known chiefly to Bernard for being a close friend of his late wife. Since Betty's death in 1937, Monty had shut out any hint of the happy family life they had enjoyed for ten years (later in life, Monty would snap at the historian Basil Liddell Hart to 'never mention her name again'). Here was proof that Bernard's mourning was buried but not forgotten. In July 1944, Monty was not even at home to Churchill, yet in August – in the midst of a challenging campaign – he invited the eccentric backbencher over for a few days, and even sent his personal Dakota to collect him. Even more remarkably 'APH', who turned up on 9 August in a naval petty officer's uniform, was allowed to smoke his pipe. Herbert was enchanted by Monty's TAC HQ amidst the ferns and tall trees, regarding it as straight from the pages of *A Midsummer Night's Dream*. Perhaps his visit and the surroundings were Monty's conscious antidotes to the stress or tension of the Normandy battle.[1] On the 19th the TAC moved again to Condé-sur-Noireau, overlooking the village of Proussy, where the next day diarist Oliver Harvey visited Monty with Anthony Eden.

[Monty's HQ] was on the side of the road on a sloping hill looking across a wide valley. Here were the general's caravan and two or three tents ... we went into dinner in another open tent. 8 of us, AE[den] and I on each side of M[onty], de G[uingand], Dawnay and ADC.[2] The Gen. sat up like a little bird with his head on one side, sharp as a needle and with very bright eyes. He was in most genial mood and kept putting questions to AE, such as 'what are the statesmen going to do when the soldiers have done?' 'Is there going to be an election?' and 'What are you going to do with Germany?'[3]

Although Monty had little understanding of politics, he at least was looking ahead and framing the right questions, for, in truth, Eden's colleagues had yet to agree policy on any of Monty's three questions.

Harvey, in the same entry, concluded: 'I daresay he is pretty ruthless with his generals . . . I told AE . . . [that] M would have great influence after the war and I felt AE could probably guide him. M is a bit *naïf* in political matters.'[4]

Whatever Monty's predilections about receiving visitors in the field, he was increasingly obliged to entertain the great and the good. From 11–16 October, King George VI again stayed with Bernard, this time when the TAC HQ was camped in a central Eindhoven park, but both Monty and his monarch spent most of the time at MAIN, now suitably ensconced at the Residence Palace in Brussels. From TAC, the King visited troops and presented more medals. Two days later, Monty's long-suffering Chief of Staff, Major General Freddie de Guingand, was knighted (he received a KBE) at Main HQ.

As the weather grew colder, Monty sent home for warm clothing, acknowledging that the war was likely to drag on through autumn and winter into 1945; after Eindhoven, he abandoned his caravans (for the first time since June) and lived indoors. Major Anthony Powell visited Monty's TAC HQ in November with his gaggle of foreign liaison officers, whom Bernard inspected. They posed for photographs and were briefed by Monty. Powell transferred this event into his 1968 novel *The Military Philosophers*, where he presented this penetrating sketch, which echoed the actor Clifton-James's opinion of Monty's extraordinary self-control:

> . . . the Field Marshal's outward personality offered . . . willpower, not so much natural, as developed to altogether exceptional lengths . . . It was an immense, wiry, calculated, insistent hardness, rather than a force like champagne bursting from the bottle . . . their synthesis seemed to offer dependability in utter self-reliance and resilience . . . The eyes were deepset and icy cold . . . oddly sustained by the voice. It was essentially an army voice, but precise, controlled, almost mincing, when not uttering some awful warning.[5]

With the end of the Normandy campaign Monty was forced to disband 59th (Staffordshire) Division, and redistribute its infantry, as he had run so short of manpower. The supply of reinforcements from Britain had dried up, and henceforth the 21st Army Group would have to plunder the field army. By the year's end, 50th Northumberland Division was also broken up, and artillery, sappers and pioneers were being re-roled as infantry. Britain raised forty-eight divisions during the Second World War, the Americans ninety and Canadians five, yet because of the Normandy casualties, each nation was running out of men, having to furnish personnel for the Mediterranean and Pacific theatres as well.[6] Knowing he had a finite supply of troops, Monty squirrelled away eight armoured brigades (1,400 tanks), six Royal Artillery brigades (700 guns) and six Royal Engineer groups to provide a firepower reserve equivalent to six additional divisions. As he knew he was likely to run short of manpower, Monty used firepower to compensate for the lack of men: his 'metal not flesh' approach was also to offset personal fears associated with the First World War casualties he and his contemporaries had witnessed. The 'Shadow of the Somme' always hung over Monty and his generation: he, his five corps commanders, and all but one of his seventeen divisional commanders had seen service in the trenches – and strove to avoid the bloodletting of that war.[7] It is ironic therefore that Normandy became a campaign as attritional as Passchendaele. In some ways, Monty's attempts (in vain as it turned out) to avoid attrition in Normandy probably prolonged the battle and increased the casualties.

Once out of Normandy, it transpired there was no agreed Allied policy for how to advance into Germany, which evolved into a slanging match between Monty, favouring a narrow thrust (led by himself), and Eisenhower, advocating a broad (and multinational) front. Britain and America had approved, and never changed, Eisenhower's basic broad-front war strategy, but Monty continuously badgered him to abandon this policy and attack with a single, overpowering rapier-like thrust into Germany. Finally, Ike agreed to let Monty try his single-thrust theory and approved his plan to drive sixty miles straight through Holland and enter Germany via an airborne carpet of paratroopers.

The outcome was Operation Market Garden of 17–25 September. Eisenhower halted the advance elsewhere whilst Monty attacked; the scheme was very ambitious and, for many reasons, failed. Of the operation, Montgomery stated:

> In my – prejudiced – view, if the operation had been properly backed from its inception, and given the aircraft, ground forces, and administrative resources necessary for the job – it would have succeeded *in spite of* my mistakes, or the adverse weather, or the presence of the 2nd SS Panzer Corps in the Arnhem area. I remain Market Garden's unrepentant advocate.[8]

While Omar Bradley later praised Monty for the general idea of the thrust to Arnhem:

> Had the pious teetotalling Montgomery wobbled into SHAEF with a hangover, I could not have been more astonished than I was by the daring venture he proposed. For in contrast to the conservative tactics Montgomery ordinarily chose, the Arnhem attack was to be made over a sixty-mile carpet of airborne troops. Although I never reconciled myself to the venture, I nevertheless freely concede that Monty's plan for Arnhem was one of the most imaginative of the war.[9]

Market Garden would always remain a thorn in Monty's side: when he described the operation as 'ninety percent successful', Prince Bernhard of the Netherlands observed: 'My country can never again afford the luxury of another Montgomery success.' After Arnhem, Montgomery antagonised the Free Polish Parachute Brigade, led by the (admittedly) prickly Major General Stanislaw Sosabowski, by making them a scapegoat for Market Garden's failure. In a monstrously unjust slur on their fighting skills, he wrote to Brooke, reporting (incorrectly) that the Poles had fought 'very badly and the men showed no keenness to fight', declaring he did not want them under his command and suggesting they be sent to join General Anders' Polish Corps in Italy. The fact that he wanted to palm them off on another theatre supports the contention that the Free Poles were fine fighters, and that Monty's intemperance was founded – yet again – on a clash of personalities. This, at a time when he was critically short of manpower.

By contrast, when TAC was at Eindhoven, Monty was uncharacteristically sympathetic in welcoming the defeated Major General 'Roy' Urquhart, whose 1st Airborne Division had just been crushed at Arnhem, losing 8,000 casualties. Urquhart later recalled he was given one of Monty's caravans to rest in – a privilege hitherto reserved only for the King or Churchill – and that Monty's debrief was gentle; he later ensured that decorations were set aside for the deserving in German POW camps, and that 'a moving congratulatory letter be passed to all survivors'.[10] Whilst TAC was in an earlier location, near Leopoldburg, Monty's signals staff discovered that freak atmospherics gave them direct reception of the division's battle in Arnhem – and sometimes a better tactical picture of the unfolding disaster that was Market Garden than Urquhart himself possessed in the midst of the battle. Monty's behaviour after Market

Garden may have heralded a personal change of mood, as well as revised expectations of the north-west European campaign, for whilst there were all the characteristic flashes of arrogance and impatience (with the Poles), humility seems to have come to the fore in his dealings with Urquhart.

From 14 July 1944, with the establishment of Bradley's 12th US Army Group (US First and Third Armies), Monty had had to face up to the loss of his position as overall Ground Commander and reversion to command of (merely) 21st Army Group. Compensation was arranged on 1 September 1944 in the form of a field marshal's baton. Ostensibly, this was in recognition of his generalship to date, but as Oliver Harvey, Eden's Private Secretary, noted in his diary that same evening, the promotion was also:

> some consolation for having the command of the whole force taken
> from him in the middle of the rout [after Falaise]. Eisenhower and the
> press have published explanations of this decision which could not be
> more laboured or unconvincing. The truth is that Americans in an
> election year must be commanded by an American general so that it
> can be an American victory.[11]

The US presidential, Senate and Congress elections duly took place on 6–7 November 1944 and confirmed Roosevelt in office for a unique fourth term.

Monty collided with the Americans again when, on 16 December 1944, the Germans launched a surprise attack into the American-held forests of the Ardennes, along the borders of Belgium and Luxembourg. The campaign quickly gained the popular name of 'the Battle of the Bulge', after the fist-shaped tranche of territory temporarily retaken by the Germans. Incurring 89,000 US casualties (including over 20,000 POWs), the campaign would prove to be the single largest and bloodiest battle the Americans fought in the Second World War. Under the nominal leadership of von Rundstedt, recalled yet again to service as OB West, Operation *Wacht am Rhein* ('Guard on the Rhine') saw the *Wehrmacht* deploy half a million men, 1,800 carefully husbanded tanks and 1,900 artillery pieces in an attempt to split the Allied armies in two by crossing the River Meuse and reaching the port of Antwerp. It was no coincidence than the attack (a personal whim of Hitler's) was launched over precisely the terrain than heralded the 1940 victory over France. If ever evidence was needed that Hitler in late 1944 was grasping at straws for his strategy, here he was trying to refight 1940 over the same ground, but under conditions of 1944, where the enemy were American, not French, and the Luftwaffe no longer controlled the skies. The furthest point of the German advance was Dinant on the River Meuse, successfully taken by Rommel in May 1940, but a river too far for the 2nd Panzer Division in December 1944. At Dinant Monty's own 3rd Royal Tank Regiment (of 11th Armoured Division) was instrumental in halting the panzers, making the defence of the Bulge (if most of the fighting was done by the Americans) a truly coalition effort.[12]

Had Rommel somehow survived, he would without doubt – as Germany's most distinguished panzer general – have played a major role in *Wacht am Rhein*, yet it is impossible to believe than even his tactical abilities and skills of improvisation would have been able to overcome the Allies' material superiority. As the Ardennes crisis developed, on 20 December, Eisenhower asked Monty to take temporary command of the northern shoulder of the 'bulge', including the US First and Ninth Armies. No one questioned Monty's abilities in first shoring up the American front, then counterattacking, but on 7 January he held a press conference at Zonhoven, Belgium, in which he stated:

> the situation began to deteriorate, but the whole Allied team rallied to meet the danger; national considerations were thrown overboard; General Eisenhower placed me in command of the whole northern front. I employed the whole available power of the British group of armies . . . The battle has been most interesting; I think possibly one of the most interesting and tricky battles I have ever handled . . .[13]

The initial sentences sounded smug and glib – and implied that Monty had been largely responsible for rescuing the Americans by deploying a huge British army, which was certainly not the case. The belittling tone offended the ears of all the senior American commanders, particularly Eisenhower, Bradley and Patton. Monty's inclination during questions with the journalists to take the most of the credit for the campaign's successful conclusion proved another irritant. Monty failed to mention a single American general by name except Ike, and overlooked the fact that the actual British contribution was negligible: for every British soldier there were a hundred Americans in the campaign. Montgomery did go on to acknowledge:

I shall always feel that Rundstedt was really beaten by the good fighting qualities of the American soldier, and by the teamwork of the Allies ... I have spent my military career with the British soldier and I have come to love him with a great love; and I have now formed a very great affection and admiration for the American soldier; I salute the brave fighting men of America ...[14]

However, the whole press conference was seen as self-promoting, focusing exclusively on Monty's own generalship and massively exaggerating the contribution of his own army group. Even Monty subsequently recognised his error and later wrote that he should never have held the conference.

Monty had also been lobbying for a unified ground commander (and saw himself as the only candidate) since the Normandy campaign, which was never Allied policy, and sent another ill-timed missive to Eisenhower on the subject just before the press conference on 31 December. Ike's patience snapped. He concluded, encouraged by Arthur Tedder, that he needed to sack Monty. Freddie de Guingand sensed something was amiss and flew to SHAEF headquarters at Versailles, where he learned that Ike had already drafted and circulated a letter he was about to send to the Joint Chiefs of Staff in Washington requesting Monty's removal from office (actually it asked them to choose between himself or Monty). De Guingand realised how vulnerable Monty was. Montgomery only realised the gravity of the situation when told that Ike was considering replacing him with Alexander – a sensible and viable alternative. To Freddie de Guingand's dictation, Monty penned a grovelling letter which ended:

Very distressed that my letter may have upset you and I would ask you to tear it up.

Your very devoted subordinate,
Monty

Eisenhower relented and did not forward his missive; the crisis passed, but the memory remained. The rest of Monty's statement at the 7 January press conference thus needs to be seen in this context, addressed as it was not to the journalists present, but to Eisenhower personally:

It is teamwork that pulls you through dangerous times; it is teamwork that wins battles ... The captain of our team is Eisenhower. I am absolutely devoted to Ike; we are the greatest of friends ... But let us have done with destructive criticism; that aims a blow at Allied solidarity, that tends to break up our team spirit, and that therefore helps the enemy.[15]

Every senior Allied commander would have known that any 'destructive criticism' or 'blows' aimed at Allied solidarity had not only come from Monty, but had been delivered with persistence – and relish.

On 23 March 1945, Monty's 21st Army Group crossed the Rhine watched by himself, Churchill and Brooke. His war ended on Lüneburg Heath, at 6.30 p.m. on 4 May 1945, when a German delegation consisting of Admiral von Friedeburg, General Kinzel and Major Friedl surrendered all the German armed forces in Holland, north-west Germany and Denmark to him. Unrepentant to the last, Montgomery concluded in his war diary:

> and so the campaign in North-West Europe is finished. I am glad; it has been a tough business . . . When I review the campaign as a whole I am amazed at the mistakes we made. The Supreme Commander had no firm ideas as to how to conduct the war and was 'blown about by the wind' all over the place . . . the staff at SHAEF were completely out of their depth all the time. The point to understand is if we had run the show properly the war could have been finished by Christmas 1944. The blame for this must rest with the Americans. To balance this it is merely necessary to say one thing, i.e. if the Americans had not come along and lent a hand, we would never have won the war at all.[16]

The last *Wehrmacht* soldiers to surrender in Western Europe were an infantry company on the tiny Channel Island of Les Minquiers. When a French fishing boat anchored nearby, a German soldier called from the shore for help: 'We've been forgotten by the British. Perhaps no one on Jersey told them we were here. I want you to take us over to England, we want to surrender.' The date was 23 May 1945, two weeks after the war in Europe had ended on 8 May.

PART FIVE

THE FINAL DUEL: REPUTATIONS

History will be kind to me, for I intend to write it
                                                    Winston S. Churchill

23

How Will History Judge Me?

WITH THE ADVENT of peace it was obvious to anyone in the know that Monty was going to irritate the world by talking up, to the best of his ability, his role in single-handedly winning the Second World War. Rommel, despite his wartime fame, on the other hand, was 'just another dead general'[1] in a wrecked country that had comprehensively lost the greatest war it had ever fought; his prospects for ascending to the worldwide status he has since acquired were not promising.

Montgomery's profile had already been raised on the home front with the release in March 1943 of the Ministry of Information's first full-length documentary film, *Desert Victory*, to great public acclaim. It portrayed Monty versus the Desert Fox – the most skilful, best-known (and well-liked) German general in the world, which made Monty's toppling of this contemporary icon all the more impressive, rather in the manner of Wellington and Napoleon. The film made a major impact as it relied heavily on dramatic newsreel footage, filmed by both British and German combat cameramen, at a time when the inclusion of Nazi film footage in Allied films was usually avoided.[2] Directed by Major David MacDonald of the Army Film and Photographic Unit, of the twenty-six cameramen who contributed to *Desert Victory*, four were killed, six wounded and seven captured. The timing of the film's release, in the wake of El Alamein (which itself had already elevated the figure of Montgomery and his Eighth Army), and two months before the final conquest of Tunisia, gave it a very contemporary value, as if history was unfolding before the audience's eyes. At the beginning a rolling title of credits included a dedication to:

> The men of the Eighth Army ... who, on 23 October 1942, left the holes they had scratched for themselves in the rock and sand of the desert, and moved forward to destroy the myth of Rommel's invincibility.

There was also a second dedication to 'the workers of Great Britain and the United States ... without whose efforts victory could not have been achieved'. The film's emphasis on the workers of the home front having a stake in the result stitched the audience into the story of the Monty and the Desert War, and thus the public identified far more with this campaign than others nearer home. Little wonder it won its director an Oscar.[3]

Montgomery and the Eighth Army also became well known in the war years through a sequence of Ministry of Information paperback pamphlets. Essentially official propaganda, the story of the North Africa campaign was told first in the 1941 publication, *The Destruction of an Army*, which dealt with Wavell's triumph over the Italians, followed by *They Sought Out Rommel*, which portrayed forty-six days of the 1942 campaign, and *The Battle of Egypt*, covering the period from June to November 1942. Finally, published in July 1943 and priced at a shilling, came *The Eighth Army*, a substantial paperback of 104 pages, which had the largest print run of all the wartime government publications, chronicled the events of September 1941 to January 1943, and featured good photographs and maps on most pages – which the earlier titles lacked.[4]

Another way in which Monty and the Eighth Army became household names was through the business of medals. On 3 August 1943, Churchill had announced to the House of Commons that the King had authorised two war medals, the 1939–43 Star and the Africa Star, in the wake of the victory in Tunisia. The medals would be made and issued after the war, he stated, whilst the ribbons would be despatched as soon as available. One of three clasps could adorn the Africa Star ribbon: a figure '8' denoting service with Eighth Army (23 October 1942–12 May 1943), the figure '1' denoting First Army (8 November 1942–12 May 1943) or a rosette for Navy, Merchant Navy or Air Force service. Whilst the ribbon was reputedly designed by George VI himself (buff background for sand, thin dark-blue stripe for the navy, thick red stripe for the army and narrow light-blue stripe for the RAF), the Eighth Army '8' was the specific result of Monty's lobbying.

Montgomery had wanted a medal to reflect *only* Eighth Army's triumph, specifically at Alamein, and refused to accept the existence of First Army's efforts at all. The '1' and '8' clasps were the War Office compromise, recognising the potential difficulties if First Army was excluded. However, Montgomery was adamant that only *his* soldiers would get to wear the '8', thus excluding his Eighth Army predecessors, specifically Auchinleck. On this occasion Churchill and the War Office caved in: whilst the qualifying date for the Africa Star commenced on 10 June 1940, and Eighth Army dated back to September 1941, only those serving under Monty from 23 October 1942 got the '8'. Monty had simply demanded an Eighth Army medal; Churchill masked the impact by announcing the 1939–43 (subsequently 1939–45) Star and Africa clasps at the same time.[5] The identity fostered by Monty for 'his' army thus continued well after the war, with an annual Alamein reunion held in the Albert Hall which Monty usually attended, and occasionally Churchill.[6]

As soon as the European war was over, Monty initiated a secret battle aimed at guarding his reputation, which he guessed would soon be under attack. In August 1945, Sir Bernard Montgomery's 21st Army Group was renamed the British Army of the Rhine (the BAOR), after the post-First World War force stationed in Germany between 1919–29, with whom Monty had served under Sir Wully Robertson back in 1918–19. In January 1946 Monty was ennobled as Viscount Montgomery of El Alamein and the same month his 158-page campaign history, *El Alamein to the River Sangro* appeared, published first in hardback by the Printing and Stationery Services, British Army of the Rhine, ostensibly as a BAOR handbook. The opening page read: 'Published for private circulation in the British Army of the Rhine. NOT to be reproduced. Copyright reserved'. The copyright turned out to be Montgomery's and his Foreword betrayed his concern to get his version of history in print before anyone else:

> I am anxious to place on record an authentic and factual account of the activities of the Eighth Army during the period that I commanded the Army . . . This book contains the story of those days and I have based it on my personal diary.

It was an open secret that the research and most – if not all – of the writing had been undertaken by Major General David Belchem, Monty's wartime Chief of Operations, whose surviving papers make it quite plain the work he was doing.[7] In correspondence with Major Edward Budd, who was the officer in charge of printing at Monty's HQ, Belchem later claimed sole authorship of two volumes: Monty certainly had no time to write *El Alamein*, or his second history, the better-known *From Normandy to the Baltic*, the first editions of which were likewise published in hardback by the BAOR in 1946 and marked 'Restricted', although there was nothing remotely restricted about their contents, which duly appeared under the Hutchinson imprint the following year and sold extremely well. Throughout the late 1940s and 1950s, Monty tended to present people with signed copies of either of these two works (just as Rommel had with *Infantry Attacks*). Hutchinson also published two volumes, one in 1945, containing facsimiles of his wartime Orders of the Day to his troops entitled *Forward To Victory* and a subsequent volume of speeches, *Forward From Victory*, in 1948.

In addition to campaign narratives, Montgomery found another, more discreet, way to shore up his image. As commander of the new BAOR, Montgomery (and his successors[8]) had, until the late 1950s, the challenge of building an identity for the Rhine Army, whose ranks were beginning to fill with National Servicemen who had not seen action during the late war, yet needed an esprit de corps in the face of the Warsaw Pact. There was also a need to keep young men busy in occupied Germany. In December 1945, Monty's engineer-in-chief, Major General Sir John Inglis, devised for junior officers the idea of a battlefield tour of the Royal Engineers' achievements, from the assault landing on D-Day to the bridging of the River Rhine. A hallmark of Monty's battles was that they were always very well prepared and resource-intensive ('tee-ed-up' was his phrase) and this meant that artillery and engineers featured prominently in Monty's plans. After much research and many interviews, an engineer-orientated battlefield tour was mounted in September 1946, and a book, *The Royal Engineers Battlefield Tour Normandy to the Seine*, was produced for all participants. The exercise was regarded as a great success and repeated when a companion volume, *The Seine to the Rhine*, was subsequently produced in February 1947. The battle sites chosen were highly selective, always Monty's and always portraying Monty in a favourable light. It is inconceivable that the tours would have happened without Monty's blessing.

Monty also ensured that 'his' men were well placed throughout the army, including Sir Richard 'Dick' McCreery, to whom Monty handed command of BAOR in 1946.[9] As soon as McCreery arrived at the Rhine Army, he mounted a major four-day battlefield tour to Normandy in June 1947. To accompany the study, participants were issued with four painstakingly researched battlefield guidebooks – one title for each day – for Operations Goodwood, Totalise, Bluecoat and Neptune. Beautifully produced in hardback, all four guidebooks made frequent reference to 'Field Marshal Viscount Montgomery's book *Normandy to the Baltic*', which was also quoted extensively. Monty's former subordinates were 'doing their job' of preserving the Monty legend for the next generation.

Another of Bernard's protégés generating pro-Monty history was Major General Richard Hull, Commandant of the Army Staff College at Camberley.[10] Hull was then just thirty-nine and very definitely owed his seniority to his old boss. In 1947 Hull initiated an annual Staff College battlefield tour over the ground where he and his wartime boss had campaigned in Normandy. This highlight of the Staff College year was repeated to much the same itinerary for the next twenty-two years and discontinued only in 1979. Monty was always there in the background, smiling benignly on these projects, even intervening to make funds available. He had started the ball rolling by using the vast resources (including financial) of BAOR to publish his own two memoirs, ostensibly as training aids, and the practice had continued.

We should ask why there was an obsession with examining the Normandy battles – particularly Goodwood (which can be considered a tactical defeat) – and why the American triumph of Operation Cobra was not studied. The answer lies in the partisan and nationalistic approach of the battlefield tours themselves and the accompanying guidebooks: they do not emphasise multinational operations as they might. This also reflects Monty's poor relationships with the US commanders; any battlefield study of US ground operations could be seen as an acknowledgement of American achievement and expertise, something that Monty was clearly reluctant to do. Not only did the guidebooks present a skewed history, emphasising some actions and ignoring or downplaying others, they were also land-centric: although 'Mary' Coningham's Second Tactical Air Force was invaluable to Monty in 1944–5, there was only grudging acknowledgment of their contribution.

So well produced were the BAOR and Royal Engineer battlefield guides (in an age of austerity) that there was clearly something else going on here. They were printed by the British Control Commission, whose mission (which was the administration of the British occupation zone) was certainly *not* to produce guidebooks, yet some of these hardback study guides run to over 200 pages with extensive colour mapping. These guidebooks aimed to safeguard the Monty-BAOR version of the land war, moulding an army, even after successive commanders had long departed; and in this they succeeded.

As early as 6 May 1946, Monty had written to the Principal Under Secretary at the War Office for permission to publish commercially his two campaign narratives in order to answer the 'books and articles containing much indiscreet matter and much misinformed criticism'. Accounts of the war had begun to surface, including Australian war correspondent Alan Moorhead's balanced, but by no means uncritical, biography *Montgomery* of 1946. The following year the recently demobbed Major Milton Shulman (an intelligence officer with the Canadian Army who interviewed many senior surviving German commanders) brought out *Defeat in the West*, and Freddie de Guingand published *Operation Victory*. In 1948 Volume 1 of Churchill's *The Second World War* was published; the BBC war correspondent Chester Wilmot was known to be working on what would become *The Struggle for Europe*; and Bradley and Eisenhower were preparing their recollections of the war.

Monty was obviously concerned that criticism of his abrasive style of high command could prove damaging and wanted to get his version of history in early. He would have guessed that a wave of American criticism was shortly on its way, and initially it came in 1946 from Ralph Ingersoll, a member of Bradley's staff (*Top Secret*), Captain Harry C. Butcher, Eisenhower's naval aide (*Three Years With Eisenhower*), and in 1949 Kay Summersby, Ike's British driver, followed with *Eisenhower Was My Boss*. We know Monty's relations with Patton, Eisenhower and Bradley were fraught. Whilst Patton died in a motor accident in December 1945, his posthumous memoirs, *War As I Knew It*, highly critical of Monty, appeared in 1947. Eisenhower's equally critical *Crusade in Europe* arrived in 1948 and Bradley's *A Soldier's Story* in 1951. None of the three cared for Monty, but it was the depth of Ike's criticism that really took the Field Marshal by surprise. When Monty read *Crusade in Europe*, he was wounded and furious and started writing behind Eisenhower's back to newspaper editors ('his book is a mistake . . . mediocre . . . he wasn't a great soldier'), but they still serialised the bestseller, the *Sunday Dispatch* running with the headline on 28 January 1949, 'How Montgomery

Upset the Americans'.[11] Apart from de Guingand, Brooke and Churchill, Monty must have wondered who else out there could do his reputation harm. He was extremely fortunate that the view of his closest confidant and subordinate, the discreet and enigmatic Miles Dempsey, was:

> I am one (of the very few I am sorry to say) who believes that what was private once is private still and will remain so always . . . I deplore many of the books that have been written, and that is why I have resisted the temptation myself.[12]

*

In January 1946, Monty reached the theoretical pinnacle of his career when he was appointed Chief of the Imperial General Staff, the senior military post in Britain, while on 3 December 1946 – along with Alanbrooke, Alexander, Mountbatten and Portal – he was created a Knight of the Garter. He was a senior public figure and an established household name, yet was finding himself under increasing criticism and at his most vulnerable. Out of his depth as a Whitehall warrior, he was at odds with Arthur Tedder, Chief of the Air Staff (their wartime relationship had been famously difficult, with the latter nominally Monty's superior as Eisenhower's deputy), and had alienated Andrew Cunningham at the Admiralty. Gerald Templer later claimed that while Slim (Monty's successor) was a good CIGS, Monty 'was a disaster . . . Slim cared about the army. Monty only cared about himself.'[13] However, there was more to Monty than this: he believed passionately that the British Army 'must not, as after the First World War, be allowed to drift aimlessly without a policy or a doctrine', as it had done between 1918–39.[14] In this, he was successful, and right up until the end of the Cold War in 1991, the British Army – particularly the BAOR – was still very much 'his' creation and a force he would have recognised. It remained well funded, modern, forward-thinking and highly trained.

At the end of his tenure, Monty had canvassed Miles Dempsey as his successor for CIGS, but the latter decided to retire from the army (and – ever the passionate horseman – subsequently ran the Racecourse Betting Control Board, later known as the Tote, most successfully).[15] Monty then prepared another candidate, Sir John Crocker, but Attlee as Prime Minister preferred Bill Slim, who had in fact already retired. Monty explained to his Prime Minister why he regarded Slim as unsuitable – and in any case he had already told Crocker he would be succeeding him.

'Well, untell him!' was Attlee's famous retort, and Monty found himself outgeneralled.[16] In January 1949, Attlee recalled Slim to the army and elevated him to field marshal, whence he began a very successful three years as CIGS.

At the war's end Monty had hoped for the gift of a decent house or grant of money from a grateful nation, in the manner of Marlborough, Wellington or Haig. Despite an appeal to the King's Private Secretary, no such magnanimous gesture was forthcoming from Attlee's austere and bankrupt government. In February 1947, Monty bought a decrepit old watermill at Isington, near Alton in Hampshire, which various colonial governments helped him refurbish to a habitable condition by donating building materials. He had very few possessions, most having been burned in the Portsmouth blitz, and he surrounded himself with his wartime trophies, including his three caravans, parked in a purpose-built shed; and here he would reside until his death.

Working for a Labour government had persuaded Montgomery, when taking an active part in the House of Lords, to become a Conservative. He did, however, develop an affection and admiration for Attlee, though found two of his three Secretaries of State for War difficult. The one he found most congenial was Manny Shinwell, an oddball and maverick like himself: Churchill once related how Shinwell had borrowed tuppence from him to call a friend from a House of Commons payphone – Churchill's quick-witted response? 'Here's fourpence, call them all.'

Whilst CIGS, and after, Monty spent much time out of the country, travelling the world as a self-appointed statesman, meeting Chairman Mao in China, Stalin (and later Kruschev) in Moscow, and touring South Africa, Israel, the USA and Canada. Here, again, he left controversy in his wake, as highlighted recently when the National Archives released documents showing that Bernard was inclined to look down on many Third World countries as incapable of governing themselves or developing their own resources (alas, not a terribly uncommon view for those times in many political quarters). Whilst there are racist overtones in his 1947 report, based on a two-month fact-finding tour of eleven African countries (with talk of 'African savages'), these seem to reflect more the paternalist era in which he was brought up rather than an extreme political viewpoint – he appears to have genuinely wanted to help, but could only frame his views within the Victorian and imperial values he knew, which admittedly were very old-fashioned even for the late 1940s. It was certainly more than a Labour government could stomach and Attlee had the reports suppressed.[17]

Since the war, Monty had, in the words of Miles Dempsey, 'been his own worst advertising agent'.[18] In his *Memoirs*, Monty reflected on this difficult post-war period:

> I have made it more than clear that by now there were plenty of people anxious to see the back of me. When I recall those days I often think that Whitehall was my least happy theatre of war. It did not provide *my sort of battle*. I have never minded making myself an infernal nuisance if it produced the desired result ... As a result of it all I was pretty unpopular when I left Whitehall.[19]

Monty had never made much of an attempt to avoid insulting people, and it would perhaps have been unrealistic to expect him to change for the sake of his *Memoirs*, published in September 1958, after he had retired, aged seventy, at the end of exactly fifty years of soldiering. The conclusion must be that Monty simply did not care, for he avoided showing the draft manuscript to the War Office until the last moment, knowing they were going to be controversial. Even his proposed title, *Sparks Fly Upwards,* reflected his determination to speak out. He refused to consider any changes to the main body of the text, but his publishers persuaded him to change his title to *The Memoirs of Field Marshal Montgomery*.

The book had the same level of success and impact as Desmond Young's earlier biography of Rommel,[20] but his comments on the Italians and Belgians caused telegrams of outrage via ambassadors from those countries, firm NATO allies, and Monty was obliged to return profuse apologies. Then Field Marshal Auchinleck instructed counsel and threatened to sue over Monty's allegations that he had drawn up plans and was ready to retreat when Monty assumed command in August 1942. The BBC had to caution Monty when he proposed to pursue his claims on air. Eventually Bill Williams, his former intelligence chief, felt obliged to compose a note which the publishers could insert in future editions of *Memoirs*, clarifying that Monty was grateful to Auchinleck for stabilising the front, thereby allowing Monty to conduct his successful offensive. This was acceptable to all parties and the matter was considered closed. Reviewing *Memoirs* for the *Observer* in 1958, Attlee drew attention to Auchinleck's cause, stating: 'Auchinleck ... gets far less than his due ... whereas previous generals had been working on a shoestring, he [Monty] took over when everything was being provided for a really big drive

forward. The fact is that Montgomery got everything he asked for ...
[he] says exceedingly little about the complete change in the scale of his
fighting power.'[21]

It was then President Eisenhower's turn to feel aggrieved – assaulted
as he was by sales of Monty's book as well as its serialisation in *Life
Magazine*. Monty would have seen this as retribution for Ike's own
Crusade in Europe, but a more generous view might have concluded
that any feud of this nature was pointless. Monty made the front cover
of *Life Magazine* on 13 October 1958, seated in his full field marshal's
uniform and cap in front of a map of Europe, under the headline
'Monty's Outspoken Story: Triumphs, Blunders of World War II'. Inside
there was no mincing of words, with the serialisation beginning under
the heading: 'The great general tells how he disobeyed orders, took over
a whipped army, changed strategy – and smashed the Germans.'

This was the image Monty liked to present to the world: that of the
outsider having to fight doubly hard to get to the top, overcoming his
sadistic mother, bureaucracy, nepotism and politicians; the 'bad boy of
the family' who had grown into the anti-establishment figure that
'disobeyed orders'.[22] In fact, Montgomery personified the establishment,
with his father a bishop, his family owning a 'country seat' – albeit a
modest one – private education, and a product of Sandhurst and
Camberley. He did, however, remain almost a 'professional' rebel and
provocateur. Attlee observed that 'as Prime Minister, I myself had to give
him more than one tick-off, and once I had to give him a really good
raspberry ... even at seventy, he is still something of a naughty boy.'[23]
True, he may also once have been an 'outsider', but that was in the distant
past; by the time of the *Memoirs*, he was no longer so.

Much of the anti-Eisenhower rhetoric was Monty's own exaggeration
or fabrication; Montgomery, the master of deception on the El Alamein
battlefield was also a master of deception in print, at home, after the
fighting had ended. True, there had been wartime differences of opinion,
but their relevance in 1958 was minimal; what Monty did was to resurrect
them, dust them off, exaggerate them and make the debate highly
personal. Just to make sure that *Life*'s readers bought the next edition,
Monty's piece ended with the caption:

> The most widely publicised military disagreements of the war, between
> Generals Eisenhower and Bradley on one side and Montgomery on
> the other, flared up following the Normandy invasion. In his second

instalment Montgomery writes bluntly of these disputes, blames Eisenhower's tragedy for 'the dismal and tragic story of events' after Normandy and tells why he feels the war was prolonged by US failure to heed his advice.[24]

This was really pushing in the knife, then twisting it; ever loyal, Freddie de Guingand tried to intervene and build bridges both with Auchinleck and Eisenhower but was rejected on both counts. However, the observation can be made that both authors – Ike and Monty – made a great deal of money precisely because their memoirs were so controversial. By January 1959, *Memoirs* had sold 250,000 copies in the UK alone and would soon top a million; in 1960 it would emerge in a Companion Book Club edition, the best-selling title ever published by that organisation; but Ike and Monty never exchanged another word. What Montgomery did not seem to consider was that there was a diplomatic angle to all he said and did, for his criticism was not just aimed at Eisenhower's strategy, but his qualities as a leader.

The late 1950s saw the Anglo-US partnership at its lowest ebb, initiated by a difference of opinion over Suez in 1956. Monty's spat came during Eisenhower's second term as President (1956–61), and whilst Ike would have appreciated the limited impact of Monty's remarks, it does not mean he was not hurt; whilst Monty was not the cause of poor Anglo-American relations at this time, he did nothing to speed their recovery, and this – in the depths of McCarthyism, the Space Race, an escalating involvement in south-east Asia, and the Cold War – was an unforgivable, strategic error, but way beyond Bernard's somewhat narrow perspective. Whereas he took steps to repair the damage with Auchinleck and apologised for insults levelled at the Belgians and Italians, he made no attempts to heal the wound with the United States and actively caused it to fester with subsequent statements and broadcasts. To a certain extent the front lines over interpretations of Montgomery remain today where he drew them in 1958, with Americans generally taking a more hostile view of him (led until recently by the late Stephen Ambrose), and the British community genuinely torn between Monty and Auchinleck.

Before the publication of Monty's *Memoirs* a curious little volume had appeared, penned by the officer-actor engaged by Lieutenant Colonel David Niven and MI5 to impersonate Monty as part of the D-Day deception campaigns. *I was Monty's Double* ostensibly told the story of Clifton James studying Monty and imitating him so successfully that Axis agents in Gibraltar and Algiers were taken in by the subterfuge. Whilst acting as Monty, Bernard had nobly insisted his double receive the pay of a general; James later encountered an old lady in a holiday resort:

> 'Excuse me, sir, but you were Monty's double, weren't you? When I admitted it she laid her hand on my arm. 'He is a wonderful man. My son was one of his soldiers. When he was killed I wrote to the General about him and he wrote back. Now he sends me a letter every Christmas. My boy would have been so proud to know this.' She shook my hand and slowly walked away. I saw that Monty had brought great consolation to that old lady and that it was all done in secret.[25]

The success of Monty's *Memoirs* no doubt helped James' escapades in *I Was Monty's Double* to become a successful movie of the same name in 1958, starring John Mills, but with James playing himself, playing Monty.

Other friends rushed to Monty's aid. It is no coincidence that the popular Welsh thinker and prolific author Goronwy Rees published *A Bundle of Sensations* in 1960, a collection of essays which included a lengthy hagiographic chapter on Monty, for whom Rees had worked in 1942. By his own admission, his feelings for Monty in the war 'did not fall far short of idolatry'.[26]

Whilst Monty seemingly went out of his way to needlessly annoy people, he also appeared very old-fashioned and out of touch by the 1960s – especially with the younger generation. His talks to the Camberley Staff College were not always regarded as a success, as he tended to tear a strip off keen young officers asking probing questions.[27] Monty also diminished, or completely excluded, the role of others in describing life's journey. Of the inter-war period, Monty's *Memoirs* did not dwell at all on his time at either Staff College, Camberley or Quetta, when – as we have seen – innovative ideas for the future were being hotly debated by staff and students, the towering figures of Brooke and Alexander being amongst his contemporaries. His wife or son feature hardly at all, perhaps because the memories of his private life were too painful, but possibly because his professional life was, in fact, all-absorbing. Likewise it seems strange that in *Memoirs* Monty does not make mention of his close friend Basil Liddell Hart, a very well-known name in military history by the late 1950s.

In later years Monty would not debate, but continue to attack, alternative views of history to his own: 1960, for instance, saw the publication of the first major revisionist work *The Desert Generals,* in which the young historian Correlli Barnett dared to question publicly Monty's contention that until his arrival in the Western Desert the situation was 'defeat, retreat and confusion' under Auchinleck, who was 'preparing to withdraw from Alamein if again assaulted by Rommel'; that Bernard 'instantly cancelled this defeatist strategy, drew up a brilliant defensive plan for the battle of Alam Halfa', followed by victory at Alamein, all according to his 'Master Plan'.[28]

Now regarded as brilliant and provocative, Barnett and his book were roundly condemned at the Alamein reunion in the Albert Hall later that year by Monty himself and Oliver Leese, to 3,000 Eighth Army veterans. Nigel Hamilton records Monty's frosty recommendation that interested newspapers 'have nothing to do with the book . . . It will do the author no good'.[29] This polemic was followed in 1967 by R. W. Thompson's *The Montgomery Legend,* which set out specifically to question Monty's military genius. This same year, Monty returned to Egypt on the twenty-fifth anniversary of Alamein at the invitation of the *Sunday Times,* and though he included Oliver Leese in his party, declined to take Freddie de Guingand, who would have liked to have gone. (De Guingand had already specifically not been invited to the May 1945 German surrender on Lüneburg Heath, the subsequent Victory Parade in London or Monty's eightieth birthday; only lobbying by a disgusted Miles Dempsey persuaded Monty to issue Freddie a belated invitation to the latter.) Maybe it was the case, as has been argued, that the younger, fitter de Guingand would have stolen the show and made it obvious that he had been the brains behind much of Monty's success.[30] Monty's brother Brian has described how Bernard had developed a fixation not to allow anyone else to share credit for achievements that he, Monty, attributed to himself.[31]

In 1968, Monty indirectly countered his critics with *A History of Warfare.* This was not so much a reply as a project that had been running since 1964 and which had been largely written by Anthony Brett-James, Alan Howarth and Anthony Wainwright, (Brett-James would later chronicle the book's progress in his delightful *Conversations With Montgomery*). The point is that it kept Monty, now eighty, in the public eye as an authority on all things military. The *Sunday Times Magazine* did him proud with a serialisation over several weeks; he was featured on the cover of the 22 September 1968 edition, seated in his garden, wearing his field marshal's uniform, surrounded by soldiers dressed in uniforms of the past – French and English Guardsmen from Waterloo, one of Frederick the Great's Prussian musketeers and a Highlander from the Crimea. The article began: 'The aim of the general must be not only to win wars, but also play his part in preventing them', thus reflecting Monty's preoccupation with the Cold War, and underscoring his important role in the establishment of NATO. Ever his own worst enemy in retirement, just as Monty could denigrate his liaison officers or Freddie de Guingand, he would say of *A History of Warfare*, with maybe a hint of self-mockery: 'Read the first two and last four chapters – I wrote those myself.'[32]

Monty's treatment of de Guingand echoed the disastrous relationship with his own mother, Lady Maud, whom he shunned after the war. In September 1945 he threatened to leave a banquet held in his honour at Monmouth on discovering she too had been invited; in the event her place was removed to the far corner of the dining room and she was physically kept away from him.[33] Nevertheless, Maud remained intensely proud of her son, but Bernard resented any reflected glory from himself that she acquired. Ever aware of the family ancestry, shortly after D-Day she had sent him a postcard which read, 'I believe you landed at the same spot from which William the Conqueror sailed in 1066. Our ancestor Roger Montgomery was his 2nd in Command – like you. The old Montgomery Castle was near Lisieux.'[34] He never replied.

From his retirement in 1958, Monty wrote, travelled and remained in the public eye; but from 1968 onwards he withdrew gradually from public life, hurt and perplexed by the reaction of some people to his pronouncements. He avoided funerals, including that of his mother in 1949 and Winston Churchill in 1965: though Monty had joined the Churchills in Nice for their 1958 golden wedding anniversary, when Lady Churchill specifically asked Bernard to be a pall-bearer for Winston, he refused, claiming that he could not as he was recovering from a prostate operation. Bernard's brother Brian, meanwhile, recorded his disgust that Monty failed to show up for his wedding, choosing to accept an invitation to a football match instead..

In his last years, Monty's reputation as master of the battlefield was further assaulted following the publication of Group Captain Fred Winterbotham's book *The Ultra Secret*. Winterbotham had been responsible for the distribution of Enigma-generated intelligence gleaned by the Ultra process at Bletchley Park. Whilst previous authors in the know had hinted at the Bletchley operation, his was the first book written, after the official ban on all mention of Ultra had been lifted, that gave an account of the uses to which Enigma-derived intelligence was put by the Allies, in the Mediterranean, the Battle of the Atlantic and perhaps most crucially in North Africa. The contribution to victory was extensive (it is usually measured by the significant number of months or years by which Ultra is judged to have shortened the Second World War) and is still being analysed and debated by historians today. Inevitably new literature started to reassess Monty's influence on the desert war; Correlli Barnett brought out a revised edition of his *Desert Generals* in 1983 and Alexander McKee produced *El Alamein: Ultra and the Three Battles* in 1991. These lessened Monty's perceived cleverness and revealed that he knew what to do because he was indeed reading Rommel's mind, via Erwin's Enigma-encoded correspondence to and from Berlin.

In 1966 and 1967, Monty carried the Great Sword of State during the Queen's State Opening of Parliament; however during the third occasion of performing this demanding ceremonial duty, in 1968, Monty began to sway, and feeling unwell, was obliged to turn the sword over to a neighbouring peer. That November, whilst celebrating his eightieth birthday in London, thieves broke into his Hampshire home at Isington Mill and stole his more valuable possessions, including medals and his gold field marshal's baton, a personal gift of King George VI (which remains missing to this day). The theft seemed to break his spirit (as a similar occurrence would do to most individuals of his age), and this really marked the end of his public life. He did, however, venture out in 1971 when his adored, widowed sister, Winsome, married Sir Godwin Michelmore, DSO, MC, in Exeter Cathedral. As the bride recalled:

> Bernard, the old warrior, had driven down from Bournemouth, where he was spending the winter. He was far too old and tired really to undertake such a journey, but he did. He came in a large black car with the Union Jack flying in front and the Chief Constable sent two motor cyclists to direct him into Exeter to the hotel and was there to greet him ... There were people all along the road and outside the church, television cameras and reporters ...[35]

But the 1968 burglary seemed to have knocked the stuffing out of him; he slowed right down and his health deteriorated. Nevertheless he turned up to the Staff College at Camberley in May 1972 for the unveiling of

Terence Cuneo's portrait of him. Of all the painted likenesses, this is the best, depicting a relaxed, confident-looking Monty, clad in casual military attire, with his trademark beret, binoculars at the ready, on a smoking desert battlefield. As good artists do, Cuneo could see beyond the sitter and realised that Monty's happiest moments were in North Africa, far distant from the criticism of Blighty's shores. Cuneo's masterpiece frequently catches my eye from its new home, on the first floor of the Joint Services Command and Staff College at Watchfield.

By contrast, the last photographs of Bernard betray the fact that he became gaunt rapidly, his high cheekbones protruding, his slight frame bowed and his piercing blue eyes dulled. By 1976 he had become a bedbound recluse. The deaths of those he committed to battle, particularly in the desert, began to weigh heavily on his conscience: 'I've got to go and meet God – and explain all those men I killed at Alamein,' was one of his last recorded thoughts.[36] A relatively peaceful end was not far off: as his brother Brian described it: 'His last day on earth was 24 March 1976. It was so quiet and peaceful, and only my wife and I were at his bedside when, at 1.30 a.m., he just gave a little sigh and "crossed over Jordan".'[37]

As a Knight of the Garter and field marshal, Monty's passing was marked by a state funeral at St George's Chapel in Windsor Castle on 1 April 1976*. Afterwards, the family accompanied Monty's hearse for thirty miles to his local church at Binstead, where the Field Marshal was laid to rest; Brian Montgomery captured the mood of that day:

> In every village and hamlet through which we passed the people, men, women and little children were lining the street as the funeral cortege went by. They stood quietly with heads bare, all the traffic had stopped and there was great stillness . . . cyclists dismounted and passing motorists halted . . . At every village school the pupils, boys and girls, had left their classes and stood outside in ranks, in silence, with their teachers. At Aldershot, the general in command had turned out the garrison to line the road along the route . . . No commands were heard, but officers saluted and men with rifles presented arms.[38]

---

* In the end, Mr Sherlock Holmes, who had first appeared in 1887, the year of Monty's birth, outlived him. He had featured on the Big Screen most frequently in the war years, when Monty was fighting Rommel. As Monty died in 1976, two feature films featuring the great detective were released, *The Seven-Per-Cent Solution* and *Sherlock Holmes in New York*.

Winsome remembered 'that coffin, with only the Union Jack and his famous beret on it . . . eight enormous Guardsmen carried him to the site of the grave which Bernard had chosen . . . and a thrush sang his heart out as the Vicar read the service'.[39]

Shortly after Monty's death, his wartime intelligence chief, Sir Edgar ('Bill' to Monty) Williams, penned the Montgomery entry in the *Dictionary of National Biography*, in which he reflected:

> It was easy to deride his strut, his preoccupation with personal publicity, his two-badged beret, his evangelical messages to his troops; but, whatever one's distaste for 'Montification', he was not 'just a PR general'. He was a very professional soldier . . . In the dread trade, to which his life was dedicated, of killing considerable numbers of enemy in the most economical way possible, he exhibited a clinical concentration on the essential with morals taking a major share in the calculations . . . he was at his best in the Western Desert commanding Eighth Army in whom he instilled his own special arrogance.[40]

24

The Desert Fox Reborn

Erwin Rommel died some seven months before the end of the European war. By 1945 his surname and old sobriquet, the Desert Fox, had drifted off the front pages and out of public consciousness. The circumstances of his death were not generally known; many were unaware even that he had died. In a war-torn Europe full of refugees, Rommel's was just another name that had disappeared. Yet keen readers of newspapers following the Nuremberg war crimes trials of November 1945 to October 1946 might have caught the evidence which emerged of Rommel's forced suicide. Any sense of rivalry between Montgomery and Rommel had also diminished in the wake of Monty's triumph in Europe, while, as the very real threat from Russia increased, thoughts turned to the possibility of another European war.

Within a generation, however, Rommel's name had returned to the public eye to an extent that surpassed the fame he knew in his lifetime. Today he is better known than Monty. In tracing the renaissance of Rommel's name and reputation it becomes clear that his image, carefully cultivated during his lifetime, was also carefully reworked after his death.

The first phase of this public relations voyage started with the young Erwin pushing his own fortune, in 1917, demanding to be given the *Pour le Mérite*, then winning it, sustained by a toughness and stamina that would put him in any nation's special forces today. *Infantry Attacks* published twenty years later marked an equally determined attempt to advance his career, which succeeded beyond his dreams. The second phase saw the Rommel image being promoted by Karl Hanke in France in 1940 and Alfred Berndt in the Western Desert, both media-aware professionals seconded by Josef Goebbels. This promotion went up a gear in November 1941, around the time of Rommel's fiftieth birthday, when the Reich Propaganda Minister wrote: 'I . . . would strongly advise that now, as soon as the battle for North Africa has been decided, Rommel be elevated to a kind of popular hero.'[1] Fed by Rommel's innate military ability, his public persona was developed by newsreels, postcards, magazine articles, radio and newspapers, and spread eventually to non-German-speaking countries such as Britain and America. It was aided by the fact that Rommel himself was a natural showman.

This phase was possible, or made easier, for two reasons. Rommel's advances over the winter of 1941–2 became a very useful distraction away from Germany's failure before Moscow. More importantly, Rommel's image was possible because the entire Third Reich was largely a sandcastle illusion built on propaganda. When Hitler came to power, he printed money and pretended to be a miracle worker, and after ten years of economic hardship, many Germans were ready to believe the Hitler dream. A glance at Hitler's gang is enough to realise the chasm between image and reality. Himmler – balding, bespectacled and paunchy – was anything but the ideal blond-haired, blue-eyed Aryan warrior he required his minions to emulate; Goebbels, master of propaganda, was short, afflicted with a club foot and was an outrageous womaniser – all carefully hushed up. Röhm, the SA brownshirt bully of the early years, was a notorious homosexual in a Nazi Germany where this was banned; Goering was a clinically obese morphine addict with a comically deranged liking for medals and uniforms; and Hitler himself, a Bavarian peasant with few social graces, no table manners, whose vegetarian diet and need of spectacles was kept secret. None of what the Nazis appeared to be was real – the hardy, Nordic, back-to-the-earth Wagnerian warrior *volk* was dreamy-eyed nonsense, out of place in the twentieth century – it actually put back aspects of modernity, industrialisation and women's rights.

Instead, the Nazis became experts in spin – the soundbite that caught a mood; the newsreel that galvanised a nation; the uniforms and rallies that promoted an illusion of world power. By the time Rommel had signed up to this machine it was running at full speed, at its most efficient. As the regime's crimes became increasingly horrible, bribes were used to eliminate opposition by corrupting senior officers and binding them further to it.[2] These secret payments and gifts – practices outlawed in the Third Reich – compromised each recipient morally and legally and bought silence and cooperation. We now know that all German field marshals from the summer of 1940 were paid 4,000 Reichsmarks per month tax-free from a secret Nazi party bank account, whilst a *Generaloberst* received RM 2,000[3] (at a time when a *Feldwebel* earned around RM 3,000 and a major some RM 8,400 per annum). Thus, from January 1942, on Rommel's promotion to *Generaloberst* and subsequently to *Generalfeldmarschall*, the Desert Fox was accepting a secret monthly bribe from the regime amounting to the equivalent of a sergeant's *annual* salary.[4]

Once part of the charade, it was difficult to part with it: the lies and half-truths were convenient to believe; the sinister forces beyond were equally inconvenient to ignore, or forget. Why else would Rommel forbid his son Manfred from joining the SS when the latter had expressed a desire to do so, motivated, as Manfred said, by that organisation's uniforms? So caught up in the propaganda machine was he that at this stage Rommel had probably come to believe his own rhetoric. Thus, in front of Hitler and Goebbels, on 30 September 1942, on receipt of his field marshal's baton, he promised victory in North Africa, which he *already knew* was impossible given the meagre resources he could deploy against Montgomery's growing might.

The third phase of Rommel's image management came after his untimely death. In a Cold War era of reconstruction and reconciliation Rommel was presented as the acceptable face of German militarism, the 'good' German who stood apart from the Nazi regime.

We will never know how much, if at all, Rommel was in league with the Stauffenberg plotters. As is now clear, Rommel had been very close to Hitler and the Third Reich, but had grown disillusioned, first at the time of Alamein and latterly during the Normandy campaign. In early 1944 we know that Rommel's optimism had revived, for during one long night-time drive from Le Mans to Paris Rommel and Blumentritt (Rundstedt's Chief of Staff) had a long private conversation. Blumentritt recalled: 'The interesting thing about these hours was in getting to know Rommel the man, and hearing of his bitterness because, owing to no fault of his own, the African campaign had not been successful'.

On this February night Rommel said nothing against Hitler personally. On the contrary. But he indeed found bitter words for the system as a whole. All the while he confined himself to purely military apprecia- tion, and the political side appeared not to interest him. He expressed the hope that the Allies would give him time up to May and ended these reflections with the words; 'Then they can come.'[5]

Thus, in February 1944, Rommel was still apologising for Hitler, and blaming his staff for the course of the war. The turning point seems to have come when Hans Speidel, one of the plotters, arrived as Rommel's Chief of Staff on 15 April.[6] In the secret world of nuances and nods, the plotters may – in their enthusiasm – have thought Rommel was more sympathetically disposed to them than he actually was; the most that Rommel may have ventured was a request to be 'kept informed'. The best evidence seems to suggest they made him vaguely aware they might launch a coup – which he was against; if anything he wanted Hitler put on trial.

Why did Rommel abhor the notion of violence against Hitler? Partly because he had been brought up in the army tradition of being non-political (regular officers were not even allowed to vote) he regarded himself as unable to assist in any form of political activity. Partly because he felt he owed Hitler something – a last chance to change things – in exchange for all the promotions and medals he had received from him. He knew he had bypassed the military system and got to the top by being close to Hitler. Partly too because he had lived through the tyranny of 1918–19, with soviets of soldiers roaming the streets looking for trouble, law and order having broken down, and parts of Germany in a state of civil war. Rommel would not want to have been associated with any activity that might have led to a similar breakdown of law and order: it would have been seen as another form of betrayal of the state.

Whilst Rommel was slowly coming round to the idea of a bloodless regime change, Stauffenberg and his fellow conspirators were planning something quite different. Rommel was out of the picture, recovering from his car crash, when the plotters struck on 20 July 1944. The end result was that Rommel realised too late that he was not immune from the regime, and bartered his family's life for his own. This devil's bargain was kept secret first by the state and then after 1945 by Manfred and Lucie. For at least five years after the war, Stauffenberg's coup was seen as disloyal to the German officers' code, however evil the regime they had been serving, and the family issued statements to the effect that the Field Marshal had nothing to do with the plot. Bizarrely, as one historian points out, in West Germany's early years, former officers were still divided between the 'oath-breakers', who sympathised with Stauffenberg, and those who sided with the regime – a division which had little to do with politics and everything to do with military pride and tradition.[7]

Rommel managed to destroy all the campaign narratives, thoughts and diary entries written about Normandy prior to 17 July, when convalescing after his car crash, so there is no firm evidence as to what he knew or did not know; consequently we are left with conflicting accounts from the survivors – all of whom are now dead – but whose public position on 20 July changed as Germany grew more democratic and was integrated into NATO. Gradually the Rommel family, probably under Hans Speidel's direction, shifted their stance about 20 July 1944. Desmond Young's biography of Rommel also assisted in developing a new version of 'the truth' – that Rommel had been an active (possibly a leading) plotter after all.

This interpretation gathered momentum in the 1950s as the Western Allies sought ways of rehabilitating West Germany within the new Europe and NATO, bringing respectability to some of her past military traditions. Immediately after the war Hans Speidel, who would eventually oversee the smooth integration of the *Bundeswehr* into NATO, started to talk up the Rommel legend and his own role in the Stauffenberg plot, thus strengthening his anti-Nazi credentials. He alleged he had deliberately engineered the late arrival in Normandy of two army panzer divisions (2nd and 116th, commanded by anti-Nazi colleagues) in order to support a change of regime on the Western Front, challenge any Waffen-SS backlash and perhaps hasten an Allied victory. They certainly arrived late, though in his later days as a NATO general he sought to bury these earlier claims. There has also been the suggestion that when Speidel was eventually arrested by the Gestapo on 7 September 1944, he 'sang like a canary', saying that yes, he knew of the plot and passed the information on to his superior; thus, an unwitting Rommel took the fall for Speidel's involvement in the July coup.[8]

The more senior Speidel became in NATO, the more this contention – that a blameless Rommel took the blame for Speidel's plotting activities – was first ignored, then revised and finally attacked. The line by the time of Speidel's death in 1984 was that he didn't talk, and the incriminating evidence came from elsewhere. Maybe. But amongst

Montgomery's papers at London's Imperial War Museum is an intriguing witnessed deposition made by one Manfred Rommel at Riedlingen, Württemberg, on 27 April 1945. This was before the war had ended and when the sixteen-year-old Manfred Rommel (serving in an anti-aircraft unit) was first captured, but long before the circumstances of Rommel's death were public knowledge. In it, Manfred recounted the circumstances of his father's death and suicide, specifically that:

> my father's former Chief of General Staff, Lieutenant General Speidel, who had been arrested a few weeks before, had stated that my father had taken a leading part in the 20 July 1944 plot and had only been prevented from direct cooperation by his injuries.[9]

Whatever Manfred's later views of Speidel, it is clear that in April 1945 he believed him to have been responsible for his father's death. As Riedlingen was first captured by the French, we do not know when Monty first came across this intriguing piece of paper; equally unclear is why this sworn affidavit did not come to light during Monty's lifetime. Monty seems nowhere to have discussed the circumstances of Rommel's death in public or in print, so it is all the more intriguing that Manfred's deposition sits in his archive.

Unlike his boss, Speidel appears to have been liked and admired for his efficiency throughout the German General Staff; consequently, when he appeared before a Court of Honour on 4 October 1944, von Rundstedt, Guderian and Keitel refused to expel him from the army, thus sparing him Freisler's People's Court and certain execution. Nevertheless, Speidel remained under suspicion and was imprisoned for seven months, then went into hiding and was eventually freed by French troops on 29 April 1945. Thus, Speidel was one of the very few senior conspirators to survive the July 1944 plot; the rest were executed or committed suicide (like Beck, Kluge and Henning von Tresckow).

Post-war, Speidel put forward his own version of the events surrounding D-Day and June–July 1944 in *Invasion 1944: Rommel and the Normandy Campaign*, published in German in 1949 and English the year after. He also supplied testimony and information to three English-speaking Rommel biographers, Charles Marshall, Desmond Young and Basil Liddell Hart. Yet some Britons remained suspicious of German generals, even 'good' ones, in the first decade after the war. When Speidel visited Britain in May 1953, *Time Magazine* noted disquiet among certain MPs:

'Is it wise, is it in the interest of security ... that this ex-Nazi should be permitted an inspection of modern British arms?' The problem cropped up at question time last week in the House of Commons. Labour members were objecting to the presence of Major General Hans Speidel, wartime Chief of Staff to Field Marshal Erwin Rommel, in a group of 100 Allied officers invited to inspect the British Army's latest infantry weapons. Speidel ... is now the top military adviser to Büro Blank, the embryonic German Defence Ministry, which will raise and train German recruits for the European Army.[10]

In response, Churchill repeated his wartime tribute to Rommel (pleased, no doubt, that his 1942 wartime assessment of Rommel's honourable character was holding up), as *Time Magazine* described:

Prime Minister Churchill's comment on Speidel's visit was brusque: both parties in Britain support the European Army project; Speidel is doing his best to make it reality. One Labour backbencher called Speidel's visit 'an affront to the men who fought the Rommel army'. Said Sir Winston sharply: 'In the height of the war, I paid my tribute to General Rommel's outstanding military gifts, and I am bound to say now, in time of peace, that I also regard his resistance to the Hitler tyranny, which cost him his life, as an additional distinction to his memory. This keeping alive of hatred is one of the worst injuries that can be done to the peace of the world, and any popularity gained thereby is shame to the member who attempts to gather it.'[11]

Eventually, attitudes to 20 July 1944 became the deciding factor in rehabilitation and credibility within West Germany – and by the late 1950s the Stauffenberg plot had acquired many more alleged sympathisers than can have been the case in 1944.

Other writers and historians also began to promote Rommel's story. The true circumstances of Rommel's death (revealed in 1950 with the publication of Desmond Young's biography) had in fact already been established before the war's end by one Captain Charles F. Marshall, a US VI Corps intelligence officer. Major General Edward H. Brooks's US VI Corps had advanced as far as the Vosges before crossing the Rhine in March and was heading deep into south-western Germany when Marshall visited Lucie Rommel on 25 April 1945 to request any useful documents she might have. Marshall, whose family had emigrated from

Hungary a generation earlier, could switch from educated German (*Hochdeutsche*) to the wildly different Bavarian-Swabian dialect, which through its guttural informality, had a knack of loosening tongues. However he often let his translator and colleague, Sergeant Greiner, ask the questions, giving him time to think and phrase follow-on queries. In April 1945, after a pleasant chat, Frau Rommel had broken down and revealed initially only to Greiner that

> her husband had not died of a heart attack as she had first told us, but had been poisoned by two generals sent to Herrlingen by Hitler for that purpose. Since she still had a son in the Army, she begged Greiner to say nothing until the boy was either killed or captured.[12]

After keeping this to himself for a couple of days, Greiner revealed this to Marshall on 28 April and the latter wrote an embargoed news story for the *Beachhead News*, the newspaper of the US VI Corps, under the headline, 'Exclusive! How Hitler Murdered Rommel'. The next day Marshall received news that Manfred Rommel had been captured by the French on 27 April. He now felt under no further obligation to keep the details secret according to Frau Rommel's wishes. The story was formally released and finally appeared in the 2 May 1945 edition of *Beachhead News*. However, Marshall's account was no longer the main news story: he had been scooped by another, more important announcement 'Nazis Proclaim Hitler's Death'. Although the Rommel murder was still front page, Hitler's demise masked the impact of the Rommel story and robbed Marshall of the scoop of a lifetime. As every correspondent in the world focused on Hitler's last days, few editors noticed what would have been a major story on any other day of the year. 'It was still a helluva story,' he recalled; 'I remember the editor telling me, "Don't feel bad. Hitler had to kill himself to top your story."'

At that initial meeting Marshall and Greiner took away a trunk-load of private letters between Erwin and Lucie, though all his official correspondence had been earlier reclaimed by Berlin. In reading the letters Marshall was captivated, and decided to stay in touch with the family. Using his privileged position as a senior interrogator at a corps headquarters, he began interviewing Rommel's friends, including *Hauptmann* Hermann Aldinger, 'a wiry Swabian of about forty-five'. He also tracked down other key figures in Rommel's life, including Speidel, then Professor of Modern History at Tübingen University, ('a well-built, freckled,

sandy-haired, bespectacled man in his early forties, of a forceful mind
. . . professorial in speech and manner, he was multilingual and a perfect
example of a Renaissance man') and Rommel's doctor, Dr Albrecht, also
of Tübingen University. Marshall translated and edited Rommel's
personal letters before sending them to the official US archives in
Washington DC.

Marshall put together all he had gleaned as a manuscript, which he
submitted to several US publishers, with photographs borrowed from
Lucie Rommel, after his return to America in May 1946. But this proved
too soon in America for a book seen to be glorifying a German general
and the project ground to a halt. 'The anti-German feeling was so
strong and so bitter at the time. My timing was just bad,' he later recalled.
The official US Army historian, S. L. A. Marshall, offered to help find
a publisher in 1950 but was posted to Korea before receiving the
manuscript and eventually Charles Marshall, then pursuing a successful
civilian career in a New York family business, gave up and assumed the
manuscript was lost. In retirement, he discovered it in the S. L. A. Marshall
Archives at the University of Texas, and eight years before his death
Marshall's book eventually saw the light of day as *Discovering the Rommel
Murder* (2002). It is an important work, as the original manuscript
predates Desmond Young's influential, adulatory biography of Rommel.

Rommel's post-war renaissance in popularity was boosted by a flood
of literature, which began in January 1950 with the publication of
Desmond Young's *Rommel: The Desert Fox*. The larger-than-life Young,
who in 1918 had served with the King's Royal Rifle Corps in
Montgomery's 47th Division, was a breezily confident socialite whose
father was a famous international maritime salvage expert. He seems to
have known everyone – Slim, Auchinleck – and was by nature one of
life's adventurer-explorers. Placed in charge of Indian Army public relations in 1941, he worked alongside war correspondents, and was captured
briefly by the Vichy French in Syria. The adventurous Young later again
roamed too far and was caught by the Afrika Korps. He briefly encountered Rommel in person; on recognising the Desert Fox he saluted him!
After a time in an Italian POW camp, Young escaped to Switzerland and
there ran a newspaper, *Marking Time*, for interned troops. At the war's
end, he returned to India and resumed work for the Indian Army as a
brigadier.

I expected to discover a burning desire in Desmond Young to write
about the German field marshal he had briefly encountered en route to

the POW cage. The truth was more mundane. Released from the army in the summer of 1948, Young found himself jobless, and was 'having an economy lunch in the Ritz Grill' with his wife in London when she came up with the idea for the book. Thereafter, he seems to have hit the same trail as Charles Marshall – Lucie and Manfred Rommel, Hans Speidel, Hermann Aldinger – and pushed his finished manuscript towards the London firm of Collins in June 1949. Three years can make all the difference and what fell on deaf ears in New York in 1946 was accepted with alacrity in the London of 1949.

A couple observations can be made here. America in the 1940s and 1950s – partly because of its large Jewish population – was less inclined to look on any German general favourably. Secondly, Rommel had always been a more popular in Britain than in the United States because of the unique nature of the twenty-seven-month duel between the Eighth Army and Afrika Korps, whilst Churchill had acknowledged his tactical genius. Wavell was quoted in *Time Magazine* as rating Rommel 'amongst the chosen few, among the very brave, the very true',[13] and Auchinleck, reviewing Young's biography, saluted him 'as a soldier and a man'.

It is perplexing how and why Britain managed to talk up such a ruthless opponent, yet Churchill even repeated his wartime tribute to Rommel in Volume Four of his history of the Second World War, published in 1951: 'we have a very daring and skilful opponent against us, and, may I say across the havoc of war, a great general'.[14] None of these were adjectives Churchill used of Montgomery. Andrew Roberts makes the shrewd observation that Britons have always had a soft spot for their brave opponents, whether Napoleon after his surrender, King Cetewayo of the Zulus, Jan Christian Smuts, leader of the Boers in 1901, Paul von Lettow-Vorbeck running rings around the Allies in East Africa during the First War – or Erwin Rommel.[15] Or it may be that raising the profile of the Desert Fox was akin to the very British tweed-and-grouse-moor habit of stalkers mounting their hunting trophies.[16]

Rommel was Collins's biggest seller of the year, and it made Young a fortune that enabled him to retire in some comfort to the island of Sark, where he died in 1966. The book made the sort of impact in January 1950 that *All Quiet on the Western Front* had in 1929 and caused many readers to begin to view the war differently; it also did well when translated into German, being enthusiastically received at the first post-war reunion of Deutsches Afrika Korps (DAK) veterans on 18 September 1951.

The 'clean desert war' cult enunciated by Young clearly centred on the personality of Rommel; there is none of the same admiration for the Italians, even though the British had fought them for much longer in the same theatre, and they had later become co-belligerents; on the contrary, they were frequently treated as objects of scorn or mirth. Young's argument that Rommel was a 'clean' German, killed by the Nazis, also brought respect to the Afrika Korps and their record as a whole.[17] Once the *Bundeswehr*, founded in 1955, was up and running, General Sir Richard Gale (who had commanded 6th Airborne Division in Normandy and was then commander of BAOR) felt able to speak at the DAK's 1956 veterans' meeting of his belief in the growing comradeship of British and German soldiers, all of which kept Rommel in the public eye and in a favourable light. It suited post-war politics, the NATO alliance and certain veterans to promote Rommel and maintain his profile abroad, as an example of how militarily professional the Germans really were – a good ally to have on one's side in the event of a war with Russia. Young even pushed this political line when promoting his book in the *Manchester Guardian*:

> Since you cannot remove this military spirit from the German race, you had better see if you can build on their soldierly virtue. Many of our leaders who take the soldierly view think the next best thing to a good friend is a good enemy.[18]

The ongoing debate about German integration into NATO was also bringing the former opponents closer; this had been prompted by the sight of Allied aircraft helping Berliners during the long months of the 1948–9 airlift and the realisation that there was a common enemy to the east. This was a trend also reflected by the French, who were making great strides to overcome an understandable Germanophobia, but who had also participated in the North African campaign, notably at Bir Hakeim (where a French contingent under General Koenig held out against the Afrika Korps in May–June 1942), and were equally enthralled by Rommel, on account of his 1944 sojourn in France (the events of May 1940 largely whitewashed from public memory).

The trend of sympathy towards Rommel was reinforced when the Hollywood movie mogul Darryl F. Zanuck of Twentieth Century Fox bought the film rights and created a box-office success out of Young's book. *The Desert Fox: The Story of Rommel* featured James Mason as

Rommel and Jessica Tandy as Lucie Rommel. Desmond Young, telling the story of his 1942 capture, meeting with Rommel and uncovering the murder, played himself. The film came as a shock to the world when released in October 1951, for Mason's Rommel was the first time in any film that a German officer had been portrayed with sympathy instead of the usual blond-haired, murderous fiend from Central Casting; James Mason spoke in perfect English, 'not affecting a guttural German accent, the standard alienating device of previous films'.[19] General F. W. von Mellenthin, who served on Rommel's staff in North Africa, thought Mason's Rommel was 'altogether too polite'.[20]

Rommel's early close association with Hitler was left out and little of the film was set on the battlefield, where Erwin spent most of his time (all the combat footage was actually taken from the documentary film *Desert Victory*). The Desert Fox was portrayed as camera shy, when the opposite was true; and his part in the 20 July plot was vastly exaggerated. Even von Rundstedt was more sympathetically portrayed than perhaps he deserved, possibly because he was employed as a technical adviser on the movie.[21] Zanuck would deploy Rommel again into cinemas with his 1962 blockbuster *The Longest Day*, building on Cornelius Ryan's bestselling book about D-Day, but despite the assistance of Lucie Rommel and General Günther Blumentritt as technical consultants, the Desert Fox's role was downgraded to a cameo, rather than the dominating influence originally envisaged by the studio.[22]

The Desert Fox encountered problems in America, where hostile critics poured scorn on the sympathetic treatment of the Germans and cinemas were picketed. In early 1952, Communists disrupted showings of *The Desert Fox* in Milan and Vienna sufficiently for it to be withdrawn, whilst the US High Commissioner in Germany tried to delay its release in Germany. It was screened eventually in August 1952, three months after Germany's entry into the European Defence Community, but with censored German dialogue dubbed over the English-language soundtrack.[23]

The success of *The Desert Fox* bred the less memorable *Desert Rats* in 1953, recounting the siege of Tobruk, starring Richard Burton, but with James Mason again playing Rommel. The North African battles came to inspire more than their fair share of war movies – perhaps because they were also perceived as worthy successors to the mediaeval Crusades in the Holy land and the 1917–18 exploits of Lawrence of Arabia. This fascination with desert warfare is a peculiarly British one (and can be traced back at least to P. C. Wren's novel *Beau Geste* of 1924, which inspired four movies), but has been a major factor in keeping Rommel and his Afrika Korps (particularly) and the Eighth Army more generally in the forefront of public attention ever since.[24]

Shortly after Young's biography, Rommel's own papers, notes and diaries also appeared in 1950, edited into a single volume by his wife, son and General Fritz Bayerlein, published in Germany as *Krieg Ohne Hass* (War Without Hate), a title Erwin had apparently chosen himself. Manfred stated:

> My father undoubtedly intended to publish another book on the military lessons to be derived from his experiences in the Second World War. From the moment he crossed the frontier on 10 May 1940 he began to keep a personal account of his operations, which he generally dictated daily to one of his aides . . . He preserved all his official orders, reports and returns . . . For the book he planned to write on the Second World War . . . he took literally thousands [of photographs], both in Europe and in Africa, including a large number in colour . . . Only a proportion of all this material survived the various vicissitudes which it underwent.[25]

After the first German edition of *Krieg Ohne Hass*, Basil Liddell Hart (who had lobbied the authorities in Washington DC for the return of the Erwin–Lucie correspondence Charles Marshall had requisitioned in April 1945) helped the family edit their book into a shorter English-language edition, *The Rommel Papers*, which appeared in 1953.[26] This volume, edited by a respectable British historian, proved another good seller and helped readers to understand more of the genius behind Rommel's military successes in France and the Western Desert.

Other memoirs of Rommel's former comrades in arms appeared at this time – significant amongst which were the delightful recollections of Heinz Werner Schmidt, Rommel's aide-de-camp for much of the desert campaign. *With Rommel in the Desert* appeared in English in 1951; within three years it had been reprinted nine times, and remains in print today. This publishing success is a testament not only to Schmidt's straightforward, voice-from-the-ranks style – no campaign history, no politics and no ideology – but to Rommel's fame in the 1950s. Schmidt was a South African-born German, to where he retired after the war, dying only in 2007, aged ninety, just before I was able to take up a kind invitation to visit him. He wrote only the one book, which records his time as Rommel's aide and the good fortune that ensured he was granted leave to get married in Germany days before the North African war ended with the fall of Tunis in May 1943. Schmidt wrote without adulation of Rommel, and his portrait of a distant and at times hard, obsessive commander was quite different to Young's portrayal.

Liddell Hart's interviews and Young's biography were key texts in evoking general sympathy for the 'clean' image of the *Wehrmacht* in the 1950s – the more so as the authors were British, not German apologists.[27] At the same time Chancellor Adenauer was exploring the form of the new German armed forces and the possible contenders for senior posts, who would all have risen to general rank in the *Wehrmacht*: the front runners were Speidel, von Manteuffel, von Schwerin and Geyr von Schweppenburg, all of whom had served with Rommel. Heinz Guderian (whose own memoirs, *Panzer Leader*, had also appeared in 1950) was considered too old and unpopular, but his son Heinz-Günther, who died in 2004, and had been Operations Chief in von Schwerin's 116th Panzer Division in 1942–5 eventually followed his father as Inspector of Armoured Troops in the *Bundeswehr*.

Most German memoirs which appeared in the 1950s focused on North Africa: the desert was, it seems, the only theatre that German veterans dared to write about. As other generals put pen to paper, often initially writing study papers for the US Army (Westphal, Guderian, Blumentritt, Kesselring, Mellenthin and Manstein, for example), their memoirs were collectively both exculpatory and lobbying for German participation in the new European Defence Community (EDC). They highlighted who had been 'good' Germans in the war and advertised German military prowess and the General Staff's efficiency (Manstein's thesis, for example, was that if the generals had been in charge of strategy instead of Hitler, the war on the Eastern Front could have been won[28]).

The Federal Republic entered NATO on 9 May 1955, by which time the idea of 'good' and 'bad' Germans had firmly lodged in West European consciousness; yet, in the early days of NATO, the Communist East German government relentlessly drew attention to the former 'Nazis' serving in the *Bundeswehr* as a means of trying to discredit the North Atlantic Alliance. If Rommel was at one end of the scale, personifying the 'good' German, the return from Russian captivity to East Germany of the ruthless Ferdinand Schörner, Rommel's old bête noire from 1917 and Dresden days, who had ordered the execution of many Germans without trial in the last month of the war, represented the 'bad'. Schörner entered West Germany in 1958, was arrested and charged with the illegal executions of 1945 – his trial was the media sensation of the year – and when found guilty, served a further four and a half years in a federal jail, before his release in 1963.

Military analysts and novelists have teased out some pretty enticing alternative histories for Erwin Rommel, on the basis that he somehow survived Hitler's wrath and the war. Often, they replace the fortune of Hans Speidel, Rommel's Chief of Staff, with that of the Desert Fox himself, and speculate how Erwin might have headed a new German army, based in Bonn. Speidel, who died in 1984, rose to four-star rank in the post-war *Bundeswehr*, and between 1957 and 1963 served as commander-in-chief of Allied ground forces in Central Europe (COMLANDCENT). But what the counterfactual historians overlook is that Rommel was not Speidel, and that had he survived the war by whatever miracle, he – like Montgomery – would not have had the patience to serve his own country as a desk warrior. Both their temperaments suited battlefield leadership; neither was equipped with the diplomacy or patience to walk the gilded corridors of power and suffer fools gladly. Rommel, with his lack of staff training and poor language skills, was a soldier's soldier (like George Patton) and, like Monty, would have found the post-war world difficult.

Although the Nuremberg trials exposed the *Wehrmacht* officers' code for the humbug it was, the blind eye that saw nothing it didn't wish to see, it was convenient in the 1950s for NATO countries to believe that most German war crimes had been perpetrated by the SS, and focus on the military achievements of 'good' German generals, particularly in fighting Russians. Today several scholars of the German military system have exposed the fact that Nazi excesses on the battlefield against opposing soldiers, civilians, ethnic minorities, commissars and Jews were not merely the preserve of the SS, as argued by *Wehrmacht* veterans and generals in the 1950s.[29] Nevertheless, under Speidel's guidance, a third phase in the management of Rommel's reputation was complete.

Young's version of Rommel was not seriously contested until David Irving's challenging biography of 1977.[30] Probably his best work, *The Trail of the Fox* was widely translated and sold well in Germany, and although *Der Spiegel* confidently announced 'Feldmarschall Rommel: Das Ende einer Legende' in its 21 August 1978 edition, perhaps because the Cold War remained very much a threat and Speidel was still alive, few were ready to shed the myth first constructed of Rommel in 1950. Throughout the 1970s Irving was able to gain the confidence of many of the Third Reich's 'old guard' and, as his many books imply, has acquired for future scholars a wealth of primary source material. His Rommel was closer to Hitler than most had suspected, whilst the soldiers amongst his readership were surprised at how dangerously dismissive of logistics Irving's Rommel appeared to be. Subsequently David Fraser's 1993 comprehensive biography *Knight's Cross* examined Rommel's life more from the aspect of a military commander – the author was a distinguished general – but, again, the Rommel legend remained intact. Fraser's contribution was to illustrate how Rommel's unique leadership translated into military success on the battlefield; he acknowledged Rommel was no staff officer, but argued he was the supreme pragmatist on the battlefield. He also suggested that Rommel was more strategically wise than previous historians claimed: if not in 1942, then certainly by the spring of 1944 he was far more aware of Germany's strategic situation than most other commanders, or Berlin.

In 2002, Munich-born Maurice Philip Remy produced a three-part documentary film, *Mythos Rommel* ('The Rommel Myth'), for German TV with a book of the same name, which chipped away at the Rommel legend dramatically. By then Germany was reunified and confident – *Bundeswehr* troops were deployed overseas on peacekeeping missions serving alongside Second World War adversaries – and the new nation was ready for a spot of iconoclasm and re-examination of sacred figures from its more recent post-war past. In the manner of Jeremy Isaacs' award-winning *World at War* series of 1973, Remy's exhaustive *Mythos Rommel*, later released with an English-language soundtrack, relied on much pre-war and wartime newsreel footage of Rommel, skilfully weaving in interviews with surviving members of the Field Marshal's staff including Heinz Werner Schmidt; his nurse in North Africa; soldiers who fought for and against him, including Field Marshal Lord Carver; one of Churchill's former secretaries; and the unrelated but intriguingly named Italian soldier Mario Rommel. By then, news of Rommel's illegitimate daughter had broken in 2001, and both his grandson and granddaughter Helen and Josef Pan, and Erwin's son Manfred also made important contributions.

Since 1945, then, the world's understanding of the Desert Fox has fluctuated wildly. Rommel has been portrayed variously as a model officer, folk hero, dashing combat leader, strong-willed Swabian, remarkable tactician, chivalrous knight, enthusiastic Hitler supporter, war criminal, resistance fighter, representative of a clean *Wehrmacht* and, finally, sacrificial victim of the Third Reich. There was also a simpler man underneath, as Rommel revealed to General Blumentritt in early 1944, when he confided that after the war was over 'his greatest desire was to lay out large fruit and vegetable gardens at his home, with glass and greenhouses, as he had admired them in Holland'.[31]

Britain was more responsible for many of these various interpretations than Germany, and on 1 January 1990 Her Majesty the Queen approved the award of an Honorary CBE to *Oberburgermeister* Manfred Rommel in recognition of his contribution to Anglo-German relations. Many also read into the decoration a belated British tribute to his father.

After 1945 many writers set out to show that contrary to most other generals Rommel was a black-and-white character, a transparently 'good' German; decades later it is clear that the Desert Fox was composed of infinite shades of grey – a complicated man of many contradictions.

Afterword

IN APRIL 1945, Captain Charles F. Marshall was leaving Frau Rommel's home when he remembered seeing a 'German translation of a booklet on the art of generalship by Rommel's first desert opponent, Field Marshal Archibald P. Wavell'.[1] Monty also admired the book, stating that Wavell

> was a tremendous thinker, with a clear brain and great strategic insight. He wrote well. In 1939 he gave the Lees Knowles Lectures at Cambridge, the subject being 'Generals and Generalship'. The lectures were the best treatise on that subject which I have ever read.[2]

Generals and Generalship – a slim volume incorporating three of Wavell's lectures – was, then, a work with which both commanders were familiar. The Desert Fox carried his own copy of Wavell's book through the desert campaign, and, post-war, Lucie Rommel presented her late husband's battered volume to Lady Wavell.[3] We know also that Erwin respected his British opponents; he had said to his aide-de-camp, Heinz Werner Schmidt, in a relaxed moment:

> We must beat the English. They are opponents fit for our own steel. They pretend they are not trained for war, and indeed one sees it is true. But they are good men at the art of war, or the game of war. Our task is clear. Every man in the desert must realise this simple fact. We are here to win.[4]

But to what extent did Monty and Rommel follow Wavell's advice and live up to the contemporary expectations of a good commander?

Firstly, Wavell suggested that a good general needed to be physically and mentally robust enough to stand the shock of modern war.

> Delicate mechanism is of little use in war; and this applies to the mind of the commander as well as to his body; to the spirit of an army as well as to the weapons and instruments with which it is equipped. All material of war, including the general, must have a certain solidity, a high margin over the normal strain.[5]

Both field marshals went about this by emphasising the importance of physical fitness in an era when this was uncommon and to a level not out of place in today's special forces. Monty espoused teetotalism, whilst the Desert Fox similarly drank very little.[6] Both remained very fit, exercising whenever they could. We might recall how Rommel out-skied all his fellow officers when he arrived to command the *Goslar-Jäger-Battalion* in October 1933. Such physical and mental fitness paid off for Monty in his extraordinary recovery from near death in October 1914 and for Rommel in surviving his car crash in July 1944. Though both suffered from medical complaints exacerbated by wounds sustained in the First World War, they drove themselves mercilessly, Rommel possibly too much so, leading to the 1942 collapse in his health, from which he nevertheless recovered.[7]

Wavell understood the ultimate value of the mind that still had 'the capacity to withstand unexpected shocks, and to put into execution bold and unorthodox designs'. He argued that the most vital quality a leader must possess is 'the fighting spirit, the will to win . . . the man who plays his best when things are going badly, who has the power to come back at you when apparently beaten, and who refuses to acknowledge defeat'.[8] Monty appears to have shown that grit, both in 1914 and when commanding his 3rd Division in 1940, where it out-performed other British formations in what was technically a defeat. He was equally pugnacious when he returned to England after the shocks of Dunkirk; he and his wartime boss Winston Churchill were similar in this respect – Winston experienced many political setbacks, yet always bounced back. Rommel, too, possessed fighting spirit in spades, particularly demonstrated during his time with the *Württemberger Gebirgsjäger-Batalion* in 1915–17, in France and the Desert.[9] Wavell observed that 'no battle was ever lost until the leader thought it so ... this is the first and true function of a leader, never to think the battle or cause lost'.[10] Monty and Rommel both possessed a deep well from which to draw mental sustenance; neither ever gave up.

In terms of personal courage, Wavell foresaw that 'in mechanised warfare we may again see the general leading his troops almost in the front of the fighting, or possibly reconnoitring and commanding from the air'.[11] This certainly was Rommel's hallmark, until 1944: always in front – though perhaps too far – and, in the Western Desert, in the air also. Siegfried Westphal, Chief of Staff to Rommel in North Africa, recalled:

> In thought and action he remained a front-line officer, even after he had become a senior commander. The front, the fighting, the longing to be with his soldiers in the struggle, it was these which always contrived to spirit him away from his battle headquarters into the fighting-line.[12]

Rommel spent his days during the Second World War chasing around the battlefield, meeting commanders; on the one hand obtaining an invaluable sense of events, his *fingerspitzengefühl*, but on the other hand exposing himself to danger and exhaustion. In May 1940, Rommel's aide Leutnant Most died standing next to him. Between 30 May and 1 June 1942 in the Western Desert both his Chief of Staff, Colonel Gause, and his deputy, Siegfried Westphal, were nearby when wounded; whilst twice, in 1941 and 1944, his staff car was bounced by an Allied fighter, both times the driver being killed, and latterly when Rommel himself was gravely wounded. Monty developed a different technique, of sending his personal liaison officers forward as his own eyes and ears. Yet, both generals saw the need for, and developed, a method of real-time and accurate knowledge of the battlefield. They both realised that timely and confident decision-making was the key to success, and that this was enabled by a good battlefield picture ('granularity' in modern military speak).

In his book, Wavell also thought a leader

> must have a spirit of adventure, a touch of the gambler in him. As Napoleon said: 'If the art of war consisted mainly in not taking risks, glory would be at the mercy of very mediocre talent.' Napoleon always asked if a general was lucky. What he really meant was, 'Was he bold?' A bold general may be lucky, but no general can be lucky unless he is bold.[13]

This is where the pair diverge. Monty certainly was no gambler and he is often accused of being excessively cautious to the extent of being risk averse. The one exception was the September 1944 Arnhem operation, which was certainly risky, if not a complete gamble; many see its motivation as a riposte to George Patton stealing all the glory before what Monty thought would be the war's end in late 1944.[14] When Monty took risks he did everything possible to minimise the chances of something going wrong – his risks were always calculated, measured. Alamein was a risk, so he used deception to lessen that risk and leant heavily on the illusory powers of Jasper Maskelyne to conceal his strength in the north and display a non-existent force to his south. During the run-up to D-Day, Monty employed the specialised armour of Hobart's Funnies to multiply the effectiveness of his Royal Engineers and mitigate any disaster. Rommel, too, was bold. But he was more than that; he took risks on the battlefield where the outcome could not be determined, frequently attacking before expected and giving the illusion of greater strength than he actually possessed. He was by nature a gambler: his successes in the First World War, against all odds, suggest this.

Wavell also felt that the knowledge of 'what is and what is not possible' was vital. By this he meant a profound understanding of logistics, of 'military movement and supply'. In this Montgomery was very well versed, but Rommel shared the same disadvantage as Hitler and Churchill, in not having studied logistics in depth, of not knowing enough. In the Western Desert, the Desert Fox came to rely on captured transport and fuel, which prolonged his success, but this was not a policy founded on certainties.[15] Monty, by contrast, undoubtedly had a grasp of the operational level of war, understanding the picture of the larger campaign formed from individual battles,[16] knowledge that he acquired through staff experience in the First World War and at Staff College. Rommel was less sympathetic to staff work and frequently became immersed in the tactical battle with little thought for the morrow. Time and time again he was the inspired combat commander on the battlefield, outperforming many of his opponents in combat but unable to thread his battles – and logistics – together into a viable campaign plan.[17]

Wavell also observed that a good general should 'never try to do his own staff work';[18] however, *Generalmajor* F. W. von Mellenthin[19] recalled:

Normally it is the staff officer who does the guiding, but Rommel had an incredible sense for direction and terrain and did his own navigation. During his visits to the front he saw everything and nothing escaped him. When a gun was inadequately camouflaged, when mines were laid in insufficient number ... Rommel would see to it. Everywhere he convinced himself personally that his orders were being carried out.[20]

None of this should be a commander-in-chief's function. Rommel's presence may have been inspirational on the battlefield, a 'force multiplier', but much of this 'eyes forward' work should have been done by trained staff officers (as required by Monty), whilst attending to the camouflage of guns hints at an unnecessary obsession with detail that bedevils some generals, such as Gort in 1940. Contrast that with Monty on his second day commanding Eighth Army:

I was woken up soon after dawn by an officer with the morning situation report. I was extremely angry and told him no one was ever to come near me with situation reports; I did not want to be bothered with details of patrol actions and things of that sort ... if anything was wrong the Chief of Staff would tell me; if nothing was wrong I didn't want to be told. The offending officer was very upset; so we had an early morning cup of tea together and a good talk, and he went away comforted.[21]

Here was Monty stating explicitly that he did not want to be overwhelmed with detail (unlike his predecessor); it also offers us an insight into his management style. He tore a strip off the officer who'd woken him, but then took pains to reassure the young messenger that he was not at fault. This provides us with an important difference in the style of Monty and Rommel: the former had the 'detail' summarised and packaged for him; Rommel enjoyed being in the midst of the fight, from which he formed his own impression of the action, then made his decisions accordingly; inspirational, yes, but efficient – no. One must return to the observation that a staff college education would, no doubt, have ironed this particular habit out of him. Instead of Rommel using his staff to do the jobs they were trained for, he often undertook those functions himself. In this sense, his command was very self-centred, or self-indulgent. Mellenthin noted:

> Rommel had some strange ideas on the principles of staff work. A particularly irksome characteristic was his interference in details that should have been the responsibility of the Chief of Staff. As a rule Rommel expected his Chief of Staff to accompany him on his visits to the front – which frequently meant into the very forefront of the battle. This was contrary to the accepted general staff principle, that the Chief of Staff is the deputy of the commander-in-chief during the latter's absence. But Rommel liked to have his principal adviser always at his elbow, and if he became a casualty, well – he could always be replaced. During critical periods the absence of Rommel and his Chief of Staff sometimes lasted not only for a day, but for several days. This threw a heavy responsibility on the junior staff . . . we accepted it gladly.[22]

Monty painstakingly selected his own staff, building what he considered, and what proved to be, an outstanding team of many talents. He provided the direction and they enacted his wishes. Rommel was inclined to impose his own tempo and methods on his staff, much to their consternation. Unfortunately for the efficiency of the Afrika Korps, Rommel's own desire to be at the front was also reflected in his requirement that his subordinates follow his example. A telling statistic is that from February 1941 to January 1943, of the thirty-seven generals who held major commands in Libya and Egypt under Rommel, five were killed, six wounded and three taken prisoner. Without counting those Rommel sacked, fourteen battle casualties out of thirty-seven equates to thirty-eight percent of Rommel's senior leadership pool over those two years, probably unequalled in any other campaign – far too high, and ultimately unnecessary. Many would have commanded equally well, and more safely, from an astute distance back.

Monty was a proscriptive commander: he demanded freedom of action himself but was reluctant to grant this to his subordinates. In choosing his commanders (unusual in itself that he acquired the ability to pick his own team) Monty generally opted for compliant, competent leaders, often those whom he had taught at Staff College or had commanded in 1940. He was disinclined to launch subordinates on independent missions. One of the observations made by Michael Howard of Miles Dempsey, British Second Army commander, was that he had been selected because Monty knew he would do as he was told. Yet the good direction from above also applied to Montgomery himself: he was probably at his best (and happiest) as an army commander in an independent theatre like North Africa. Whilst running 21st Army Group (for which he received no formal training), Monty appears to have treated his army commanders at times like corps commanders and was demonstrably out of his depth in his relations with Eisenhower and SHAEF, the American forces in general, and even with Churchill.[23]

Rommel, on the other hand, was born of a system that encouraged independent initiative; he personally had benefited from it in the First World War with the enlightened example of Major Theodor Sprösser, and expected the same of others. Unfortunately he never confided in a mentor or a deputy in the way Sprösser had confided in him. Rommel was perplexed when he found OKW and Hitler more proscriptive as the war ground on. The very command style that had given the *Wehrmacht* so many victories in 1939–42 – trusting the initiative of subordinates (today 'mission command', *Auftragstaktik* in German) at whatever level – was replaced by ever-increasing centralised control from Berlin, in ignorance of the real situation on the ground. Hitler's 'no retreat' order received by Rommel during Alamein was just one instance of this, but one that profoundly shocked the Desert Fox. Yet Rommel himself was always viewed as a difficult subordinate.

Writing in 1939, Wavell argued prophetically that:

> It seems to me immaterial whether he is a soldier who has really studied the air or an airman who has really studied land forces. It is the combination of the two, never the action of one alone, that will bring success for a future war.[24]

Sadly, there is no evidence that Wavell encountered such cooperation in North Africa – this is a reflection of the systemic failure of both services to mesh together. The RAF was a young organisation, founded only in 1918, and keen to preserve its independence. It had already fought and lost an extensive battle with the Royal Navy over control of maritime aircraft and viewed excessive cooperation with the land force as the ultimate path back to reintegration into the army – a view some accuse it of possessing to this day. Monty understood the capabilities that air could deliver on the modern battlefield, having witnessed the German use of close air support on the 1940 battlegrounds of France, yet could not get on with his air advisers, Tedder and 'Mary' Coningham. Johnny Henderson, in Monty's TAC, recalled that Coningham was known to them as the 'snake in the grass', for his perceived disloyalty, and that

Monty would only deal with his deputy, Harry Broadhurst.[25] While Monty understood air power, it appears he never respected the RAF as equals; he was fortunate that he took command in the desert just as the Allies achieved air superiority.

Rommel not only understood air power, but spent much time, as a divisional commander in 1940, aloft in his personal Fieseler Storch. He was even awarded a diamond-encrusted pilot's badge, one of several honorary awards made by Goering.[26] Yet in 1940 he was criticised by the German Fourth Army commander, Günther von Kluge, for failing to acknowledge the contribution of the Luftwaffe in his accounts of 7th Panzer Division's dazzling advance through France – exactly the criticism the RAF's Desert Air Force had of Monty, that he stole their glory. In 1941–42, despite his best efforts, Rommel had no direct Luftwaffe support, though this was not his fault but an institutional flaw typical of the Third Reich, which saw the German air force run its own chain of command, via the OKL in Berlin, down to field commanders.[27] In Normandy, Rommel had even less interaction with *Luftflotte* 3, commanded by his fellow Württemberger, the portly fifty-nine-year-old Hugo Sperrle, whose hands were in any case tied, having few fighters or ground attack aircraft.[28] Ironically, of all the German generals, it was Rommel who grasped the immense potential of close air support, even if he was denied the tools to deploy it. Although tactical airpower worked well for Monty's armies in Europe, it might have been even more efficient. Consequently, the lack of inter-service cooperation eroded the efficiency of both Monty and Rommel as battlefield commanders.

Wavell went on to say that the 'general who sees that the soldier is well fed and looked after . . . will naturally have his confidence. Whether he will have his affection is another story. Wellington was most meticulous about his administrative arrangements . . . but he was certainly not popular.'[29] Both commanders fought, with varying degrees of success, for the logistics and resources they felt their armies deserved. Whilst Monty was popular with his men, it seems that Rommel was less so. He wasn't necessarily liked by his men in North Africa, as he drove them so hard and appears to have had little interest in preventative medicine (significantly lower British sickness rates contributed to the manpower imbalance in 1941–42[30]). Mellenthin recalled of his boss that he:

was not an easy man to serve; he spared those around him as little as he spared himself. An iron constitution and nerves of steel were needed to work with Rommel, but I must emphasise that although Rommel was sometimes embarrassingly outspoken with senior commanders, once he was convinced of the efficiency and loyalty of those in his immediate entourage, he never had a harsh word for them.[31]

Heinz Werner Schmidt, Rommel's aide-de-camp for much of the desert campaign, found his chief aloof.

It was months before he called me anything but the formal *Leutnant*, and only after he had decided that I was more than a necessary additional limb did he bother to address me by name, bother to find out my age, whether I was married, whether I was happy, or, indeed, even to think of me other than as something that filled a uniform and answered a command. It was, in fact, almost strange, after a long acquaintance, to find the impersonal General actually human.[32]

Another of the Desert Fox's staff officers, Siegfried Westphal, observed:

His strongly developed spirit of contrariness and his occasional pig-headedness often made it difficult for both superiors and subordinates to get along with him . . . he was no diplomat. He said what he thought bluntly.[33]

To this day, the name Bernard Montgomery excites a polarised reaction in conversation with Britons: undiminished antipathy or unrestrained praise, there is no middle ground. Monty's long-suffering staff were devoted to him, but to many he was a distant figure. His (transient) wartime popularity stemmed from the fact that he was the first British general to make a point of addressing everyone serving under his command, and thereafter to keep them informed. Bill Deedes observed of him: 'Montgomery was not loved by his contemporaries in the British services, being all too prone to argue . . . but he found a path into the hearts and minds of the soldiers he commanded and planted confidence there. That was one reason for his success.'[34] Even some of Monty's most ardent critics would acknowledge his talent to inspire: Sir Brian Urquhart, future Under Secretary General of the UN and an outspoken opponent of Operation Market Garden, remembered: 'he made a tremendous impression ... Whatever else he may have been, Montgomery was a genius at morale-building and training ... and we left with a feeling of at last having a leader in full charge.'[35]

Wavell also explored what motivated a soldier to risk his life; today we might ask 'Why should anyone follow you?' In essence Wavell was alluding to the ability to raise and sustain the morale of an army. Monty was a master at raising morale and sustaining it. Wavell (unwittingly foreshadowing another desert campaign) cited the fact that when General Edmund Allenby took over a demoralised Egyptian Expeditionary Force in 1917 he dashed about visiting every unit, where his 'bearing radiated an impression of tremendous resolution, quick decision, and steely discipline. Within a week of his arrival Allenby had stamped his personality on the mind of every trooper ... and every infantryman.'[36] Monty's major contribution in the desert war was taking over a demoralised Eighth Army under remarkably similar circumstances, and reviving its morale and fighting spirit within a few weeks. He did this by personal appearances and the delivery of simple, clear messages that radiated confidence; in the weeks preceding D-Day these reached exhaustive levels. Monty supported his forces with logistics aplenty; he used troops' newspapers (*Eighth Army News* and *Crusader*) and photographers to raise morale and on 19 March 1945 launched the British Army's *Soldier* magazine – still published today. Rommel also understood what made soldiers tick: he had emerged from relatively humble origins and achieved success (against the norm of the Prussian military caste) through sheer determination; he trained his men hard, and rewarded his soldiers with victories, medals and loot, though his ability to do so diminished through the war. The Goebbels propaganda machine and the favours bestowed personally by Hitler did the job of popular image-making on Rommel's behalf.

Professionally, Monty and Rommel may have been ahead of their time in many respects but subordinates close to them have observed that it was not easy to get to know either. There was a sense that neither really opened up; they remained guarded and reluctant to confide in others, a result possibly of having watched earlier friendships destroyed by combat in the trenches. One of Montgomery's aides, Johnny Henderson, later recalled that Monty would 'never let you get too close to him. I mean, this is the extraordinary thing, that you couldn't.'[37]

Biographer Nigel Hamilton has suggested another dimension to the British field marshal's character in latent (if not actual practising) homosexuality. He based his argument on Monty's preference for the company of handsome and dashing young liaison officers.[38] This was resolutely refuted by Henderson, who mused shortly before his own death in 2005: 'I can truthfully say that such a thought never crossed my mind in the four years I was with him.' He added that if rumours had circulated to that effect (and the upper echelons of the army were a very small club) then Monty's reputation and credibility would have been totally undermined, which rings true. Rommel's headquarters was also staffed with attractive young examples of Aryan manhood, and yet the same observation has never been made of him. Youth has always had a prominent place on the battlefield and is often rather better at war than the older, more fearful and sensitive, generations. War is a young man's business (as any walk through a war cemetery will confirm) and to master absolute proficiency in almost any walk of life requires monk-like dedication. Certainly both field marshals incited adulation from the young men who fought under them, and both basked in the reception they received; but that is not homosexuality. We have seen earlier how Goronwy Rees was frank enough to admit to idolising Monty. Across the divide, Major Hans-Joachim Schraepler's letters to Lucie, written on Erwin's behalf in France and North Africa were full of fawning phrases.[39] The idea of non-homosexual love and loyalty, especially at war in the far-off deserts of North Africa, is best captured in a phrase used by the British military historian Richard Holmes – 'the bondship of mates'.

Monty – and perhaps Rommel – liked to surround themselves with young men because they posed no threat to their careers. Confiding in contemporaries was difficult for either in the competitive world of senior command: threatening in a democracy and positively dangerous in a dictatorship. Taking very junior, bright and able officers (such as Monty's team of liaison officers), and giving them status and promotion beyond their expectations would have been one way of creating an utterly loyal court in whom to confide; this is the principle behind a general's aide-de-camp in some armies. Without a mentor or respected contemporary to provide the odd sanity check, coteries of worshipping junior officers run the risk of becoming a gaggle of sycophantic yes-men (as surrounded Hitler), but Monty and Rommel's headquarters were notable for containing a generation of talent – wise minds, young and old.

Sexual promiscuity is the downfall of many a talented leader. Rommel learned this lesson very early in his career, with Walburga Stemmer, but the incident was successfully buried for the rest of his life, with Lucie's assistance. Although Monty and Rommel both fathered sons in 1928, for most of their lives they distanced themselves from women; even their wives seem to have been on the periphery of their lives. Yet both had strong marriages and were romantics at heart (Rommel was up early on 6 June 1944 wrapping birthday presents for his wife). Monty and Rommel shared a monk-like devotion to their trade and an ability to shut out emotion, which allowed them to perform to their utmost in the most demanding years of their careers. Monty seems to have been equally abrasive to men and women – it may be that he just did not understand people very well. He seems to have suffered from an inferiority complex when it came to women, whom he perceived as a threat to which he had no effective defence, except creating distance. Similarly, Rommel's friend Kurt Hesse wrote of him:

> I only recall him talking about them [women] once. He had been to visit a town in East Prussia, and told me he had noticed that it was full of pretty girls. What he liked were horses and dogs, but he never spent any money on them either.[40]

In place of human affection, Monty and Rommel lavished attention on animals. Both marshals were useful horsemen and in 1945 Monty was pictured in the *Illustrated London News* riding a white Arab stallion, Rommel, that had once been owned by his namesake.[41] The human Rommel recorded on 21 January 1944 being presented with a pair of dachshunds: 'The younger one was very affectionate immediately, but the older one is not so forthcoming. The two of them are now lying beneath my writing desk.'[42] A photograph also exists of Erwin cradling his pet fox – how appropriate! – in a First World War dugout. In Normandy, Monty acquired a King Charles spaniel whom he named Rommel and a Jack Russell terrier improbably called Hitler, and others followed.[43] Visiting Monty's TAC HQ in July 1944, A. P. Herbert noted, 'the absurd dogs, Rommel, Hitler, Keitel and company romped on the camouflage netting; and in the General's caravan a canary sang its head off'.[44]

The personal reserve evident in Monty and Rommel may, in turn, have failed to provide either with the crucial ability to judge others; such a quality is important in order to get the best out of peers, contemporaries and subordinates. Montgomery was notoriously difficult as a commander. It was not that he was unprepared to suffer fools, but he went out of his way to actively denigrate some talented contemporaries, such as Auchinleck, as though he had to put distance between himself and competition. He was not a team player, and constantly looked down on the Americans, Ike, Patton and Bradley. Rommel badly misjudged Hitler, despite having commanded his personal bodyguard for a while, and excited animosity amongst his fellow generals, who consequentially omitted to help him in the aftermath of the July 1944 plot, when his fate (and life) hung in the balance. He was no team player either, and his squabbles with Kesselring, the Mediterranean C-in-C, von Arnim in Tunisia and later Geyr von Schweppenberg in Normandy badly corroded his efficiency.

Wavell suggested that whilst 'political generals are anathema to the British military tradition',[45] successful commanders should be politically aware, and cited Cromwell, Marlborough and Wellington as appropriate examples. He went on to refer to the relationship of President Abraham Lincoln with his generals in the American Civil War. Lincoln had tested an extraordinary number of Union commanders in battle before he settled on Ulysses S. Grant, promoting him in March 1864 to lieutenant general, and appointing him General in Chief of the Union armies.[46] Total war requires a degree of ruthlessness in the man at the top, and Churchill echoed Lincoln precisely in his purging of commanders in the Western Desert until in Monty he found the right instrument that suited the moment. Lincoln and Churchill, presiding over democracies, found they had to be as ruthless as a dictator like Hitler, in their hiring and firing of commanders. Monty said to the journalist Bill Deedes over dinner in 1947: 'Politicians bully. You must put decisions squarely to them.'[47]

Although both Monty and Rommel were propelled ultimately to stardom by the specific and personal decisions of their political masters, Churchill and Hitler, they were also incredibly naive in their dealing with politicians. Both thought they could make strategy on their own: Rommel in 1942 wanted to reach the Middle Eastern oilfields and Monty in 1944 was determined on a narrow thrust to Berlin when neither had been adopted as formal strategies. Both were headstrong (and bold) enough to question and disobey their superiors. When this happened occasionally they were forgiven, but frequency wore down any sense of gratitude; Hitler had Rommel killed, while Churchill said of Monty: 'In defeat, unbeatable; in victory, unbearable; indomitable in retreat, invincible in advance, insufferable in victory.'

It is a tribute to the army which served a democracy that someone like Monty was allowed to rise and flourish; someone as overtly abrasive would have surely run foul of the regime in Germany – as happened to other *Reichsheer* and *Wehrmacht* generals who disliked or disagreed with Hitler. In fact, the Nazis would have surely taken the opportunity after the failed Stauffenberg coup of 20 July 1944 to dispose of someone as obstreperous as Montgomery, as happened to other vocal critics not even connected to the plot.

Dictatorships, more than democracies, also have a habit of appointing the incompetent over the able: safe politically, but suspect militarily. Rommel encountered these throughout his career in the Third Reich, who had the ear of the SS or Gestapo. Yes-men are highly dangerous and will always constrain the options of the able in a dictatorship. The lazy *Generaloberst* Friedrich Dollmann, commanding Seventh Army in Normandy, or his notoriously inept subordinate, *Generalleutnant* Edgar Feuchtinger – commanding the only armoured formation within reach of the D-Day beaches on 6 June 1944 – held their appointments chiefly because they were overt Nazi sympathisers, not because of any military skill. Thus Monty is as unlikely to have flourished in the German army as Rommel in the British: they were creatures, and the products, of their respective societies.

If fate had enabled Rommel to pursue his field sketching into later life, his battlefield technique suggests a modernist painter, applying big, bold splashes of unremitting primary colour by instinct onto large canvases (whether or not he had been commissioned to paint large). Inevitably eye-catching, demanding one's attention, there would have been an equally bold signature beginning with a flourishing 'R' in one corner. Had Montgomery acquired any of his wife's considerable painting skills he would have been, by contrast, a careful (yet equally vain) crafts-man – more of the seventeenth-century miniaturist, with meticulous attention to detail.[48] Examination of his work would have revealed many preparatory sketches, doubtless done by subordinates, before he committed himself to paint, the careful mixing of which he himself would have overseen. Monty, an accomplished court painter who believed his influence was greater than it was, renowned for his range of personal idiosyncrasies, would have maintained a studio full of apprentices, whereas Rommel might have been regarded for what he was – a careless, inspirational, often brilliant, loner.

Notes

Prologue

1 Basil Liddell Hart (ed.), *The Rommel Papers*, p. 521.
2 George Lane, MC, Obituary, *Daily Telegraph*, 26 March 2010; *The Times*, 7 April 2010; David Irving, *Trail of the Fox*, p. 3.
3 Antony Brett-James, *Conversations With Montgomery*, p. 100.

1. The Irishman

1 Monty's youngest brother, Brian, reminds us that these also included Thomas Hughes, Matthew Arnold (the headmaster's son), Burne-Jones, Holman Hunt and many other literary and artistic celebrities of the day. See Brian Montgomery, *Monty: A Life in Photographs*, p. 25.
2 Bernard's siblings were: Sibyl; Harold Robert, CMG, Senior Provincial Commissioner, Kenya (1884–1958); Donald Stanley, KC, (1886–1970); Una (1889–1936), who married Andrew Holden; Maud Winifred (1895–1987), who married firstly Lieutenant Colonel William Holderness, and secondly Major General Sir Godwin Michelmore, KBE, CB, DSO, MC, TD, JP, DL; Desmond; the Revd Colin Roger, sometime Rector of Vryburg, South Africa (1901–59) and Lieutenant Colonel Brian Frederick (1903–89).
3 Tom Carver, *Where the Hell Have You Been?*, p. 37.
4 *The Memoirs of Field Marshal the Viscount Montgomery of Alamein, K.G.*, p. 14.
5 St Paul's was one of the original nine great schools included in the Public Schools Act of 1868, designed to reform and regulate England's leading boys' schools. The others were Charterhouse, Eton, Harrow, Merchant Taylors', Rugby, Shrewsbury, Westminster and Winchester. Most of Bernard's regular army colleagues attended one of these.
6 Carver, *Where the Hell*, p. 37.

7   Montgomery, *Memoirs*, p. 25.

8   The British Army is back on the North West Frontier, albeit Afghanistan, as I write, and undertaking the sort of operations that Monty trained for whilst there, but never actually undertook. Today's army is utterly professional – of which Monty would thoroughly approve – and an excellent, literary picture of its current activities was recently published in Patrick Hennessey's *The Junior Officer's Reading Club*.

2. The Swabian

1   At this stage, 7 million Germans were calculated to be living in the USA, including 2 million who emigrated between 1871–90. See *Whitaker's Almanack*, 1892.

2   Erwin's father died in 1913, his mother in 1940. His siblings were: Helene Rommel (1887–1973); Karl Rommel (1897–1968); Gerhardt Rommel (1905–70).

3   *Guardian*, 9 May 2001.

4   There is a lively debate amongst historians as to whether the Prusso–German militarism that trained Rommel sprang from a long, aggressive tradition of which the Wehrmacht, led by its General Staff, was merely the inheritor. Sir John Wheeler-Bennett thought it did and in his 1953 study of the German Army and Politics, *Nemesis of Power*, he identified three quotations to support his view; he quoted Tacitus: 'The Germans have no taste for peace'; the Comte de Mirabeau (1788): '*La Prusse n'est pas un pays qui a une armée, c'est une armée qui a un pays… La guerre est l'industrie nationale de la Prusse*'; and Julius Fröbel (1859): 'The German nation is sick of principles and doctrines, literary existence and theoretical greatness. What it wants is Power, Power, Power! And whoever gives it Power, to him will it give honour, more honour than he can ever imagine.' Other revisionist historians question whether there is any actual continuity in German foreign and military policy and suggest that the Third Reich period was a 'break' with the past, rather than continuity.

5   Lucie Mollin, born 6 June 1894 in Danzig, died 26 September 1971 in Stuttgart.

6   *The Sunday Times*, 22 July 2000. Walberga Stemmer (1892–1928); the daughter Gertrud Stemmer (1913–2000) married Joseph Pan and had two children, Joseph and Helga.

7   Ibid.

8   Adolf Hitler, *Mein Kampf*, p. 6.

9   Charles Carrington, *Soldier From the Wars Returning*, p. 256.

10 Desmond Young, *Rommel*, p. 32.

3. First Combats

1   Werner Maser, *Hitler*, p. 81.

2   Hitler, *Mein Kampf*, pp. 147–8.

3   Ian Kershaw, *Nemesis*, p. 73.

4   Joachim C. Fest, *Hitler*, pp. 105–6. This is borne out by the so-called Tandy incident which has puzzled many historians. On 15 September 1938, whilst

discussing Munich at Berchtesgaden, Chamberlain's attention was drawn by Hitler to a print of a British soldier from 2/Green Howards winning a VC. Hitler alleged that 'this English soldier spared my life in 1914'. This may have been part of a negotiating ploy to demonstrate Hitler's affection for Britain; however the individual in the print was the much-decorated Sergeant Henry Tandy VC, DCM and MM, who had apparently spared the life of a limping German during the storming of the Nieuwe Kruisecke crossroads on the Menin Road, on 29 October 1914. There is absolutely no supporting evidence to suggest that Tandy and Hitler ever met in battle (though both the relevant units may have been in the vicinity on that day) but Hitler seems to have believed that he had been spared by a 'Tommy' at some stage, and had convinced himself that it was Tandy. Certainly, the Green Howards have a thank-you note from the German embassy for supplying a copy of their painting. The point is that *Hitler believed it to be true*, and evidence of destiny sparing him for 'great things'. (It has to be said that Tandy had no recollection of the incident but went to his grave in 1977 as not only the most decorated 'Tommy' of the Great War, but – as the media reported – 'the man who spared Hitler's life'.)

5   Kershaw, *Nemesis*, p. 91.

6   *Statistics of the Military Effort of the British Empire*, pp. 30, 156–59 and 363–64.

7   Erwin Rommel, *Infantry Attacks*, p. 1.

8   Ibid., p. 8.

9   Ibid., p. 9.

10 Ibid.

11 Ibid., pp. 10–11.

12 Rommel, *Infantry Attacks*, p. 12.

13 Ibid., p. 13.

14 Ibid., p. 14.

15 *War Diary*, August 1914. There are two copies of the 1/Warwicks War Diary; one in the National Archives at WO 95/1484 and a second at the Royal Regiment of Fusiliers (Royal Warwickshire Regiment) Museum, Warwick.

16 Frank Richards, *Old Soldiers Never Die*, pp. 9–11.

17 Ibid., pp. 16–17.

18 Nigel Cave & Jack Sheldon, *Le Cateau*, Chapter 3, pp. 141–96.

19 Montgomery, *Memoirs*, p. 27.

20 Brian Montgomery, *A Field Marshal*, p. 159.

21 Now in the B. L. Montgomery Archive at the Imperial War Museum, the letters are an incomplete set, and came to Brian Montgomery in a trunk the latter discovered when clearing out their mother's estate.

22 Tim Carew, *The Vanished Army*, p. 64.

23 Montgomery, *Memoirs*, p. 28.

24 IWM, BLM 1/23.

25 John Ashby, *Seek Glory, Now Keep Glory*, pp. 74–6.

26 Brigadier Sir James Edmonds, *Official History, Volume One*: 1914, p. 148.

4. Cheating Death

1 Rommel, *Infantry Attacks*, p. 47.

2 Ibid., p. 49.

3 Ibid.

4 It also underlines the shortcomings of the five-round magazine of the German 7.92mm Mauser over the ten rounds carried within the French 8mm Lebel rifle (and, indeed the ten-round British Lee Enfield). That said, the bolt-action Lebel, which dated back to 1886 – the first rifle to use a rimmed, smokeless cartridge with an aerodynamically tapering (the technical term is 'boat-tailed') bullet head – was slow to reload (and was about to be replaced had not the war erupted), so 'rapid fire' was a relative term.

5 The 'one-year volunteer' served a year as a private at his own expense; after twelve months, he transferred to the Reserve for officer training. This was a quick way for an upwardly mobile, ambitious young man to raise his social standing and join the officer class.

6 Rommel, *Infantry Attacks*, pp. 49–50.

7 Ibid., p. 50.

8 Exactly four years later, III Corps of Pershing's First US Army attacked the Varennes area from the south in the opening phase of their Meuse-Argonne offensive. By a bizarre twist of fate, also just a few metres north of the D38, but heading east out of Varennes, Rommel's American nemesis, Lt. Col. George S. Patton, was seriously wounded – also in the upper left thigh – on 26 September 1918, leading a tank attack from Vauquois to the south, via the ground between Varennes and Cheppy. If the helpful sketch supplied in *A Genius for War* – Carlo D'Este's excellent life of Patton – is accurate, then I suspect I am the first to observe that the distance between the two identical woundings cannot be more than about 3,000 feet. Dennis Showalter has written a thought-provoking joint biography of the pair but missed the stunning Varennes connection. The little town also has a museum which contains a First World War section, and an imposing memorial dedicated to the Pennsylvania men of US 28th Division who attacked the area alongside Patton's Renault tanks – one of Patton's tankers won a Medal of Honor in the vicinity – and nearby Cheppy has a memorial to those from Missouri who also took part, as well as a German cemetery containing some of Rommel's comrades. One of Missouri's sons who survived the 1918 battle was a thirty-four-year-old National Guard lieutenant colonel named Harry Truman whose artillery battalion supported Patton's attack, and his HQ was located south of Varennes atop the obvious hill just west of Neuvilly. Bringing history full circle, in the aftermath of the Great War, Pershing began a series of battlefield tours of the late European battlefields for his officers, which evolved into *A Guide to the American Battlefields in Europe*, published by the American Battle Monuments Commission (ABMC) in 1927, to commemorate the tenth anniversary of America's entry into the war. This was designed for US Army officers and was a comprehensive work (the 1992 reprint of the 1938 edition runs to 547 pages). Pershing's chief staff officer assigned to the project in 1926 was a Major Dwight D. Eisenhower. Absent for his War College course (September

1927–June 1928), Ike then returned to the ABMC for a further year of duty in France gathering information for a revised edition, when Varennes and the surrounding area was one of the key stops on his itinerary. At some stage, Rommel returned to Varennes – as he did to everywhere where he had fought, taking photographs and making some of the sketches which would appear eventually in *Infantry Attacks*. One can imagine him on a day in the 1920s scrambling around the west side of Varennes, nodding to an anonymous Patton, Ike or Truman who were about to explore the old battleground to the east. For Eisenhower, this in some ways made up for his lack of combat service in France during 1917–8 and gave him valuable insights for the European war to come.

9   BLM Archives, Imperial War Museum, BLM1/15.
10 Ashby, *Seek Glory*, p. 112.
11 IWM, BLM1/18.
12 Ibid.
13 Ibid.
14 IWM, BLM 1/22.
15 Ibid.
16 Brian Montgomery, *Monty A Life in Photographs*, 1985.
17 Ibid.
18 IWM, BLM 1/25.
19 Ibid.
20 Ashby, *Seek Glory*, p. 113.
21 Montgomery, *Memoirs*, p. 28.
22 Ibid., pp. 28–9.
23 Ashby, *Seek Glory*, p. 113.
24 Ironically, today close-quarter encounters in Afghanistan are decided by bayonet, and its skilful use remains on the curriculum for officer cadets at Sandhurst.
25 IWM, BLM 1/30.
26 E. S. Turner, *Dear Old Blighty*, p. 132.
27 IWM, BLM 1/31.
28 IWM Jackson MSS, 78/59/1.
29 Montgomery, *Memoirs*, p. 29.
30 The twelve senior sniper victims were Brigadiers Bull (1916), Bulkeley-Johnson (1917), Clifford (1916), Fitton (1916), Hepburn-Stuart-Forbes-Trefusis (1915), Heyworth (1916), Hodson (1916), Johnston (1917), Lumsden (1918), Maclachlan (1917), Maxwell (1917) and Riddell (1915).
31 IWM, BLM 1/36.
32 Brian Montgomery, *A Field Marshal*, p. 163.
33 Letter to Trumbull Warren, 17 November 1969, B. L. Montgomery Collection, IWM.
34 IWM, BLM 1/28.

5. *Learning the Trade*

1   Winston Churchill, *The World Crisis*, Vol. 1, p. 236.
2   The essential figures are that, excluding the Regulars, Reserves and Territorials,

4,907,902 men enlisted in the army from August 1914, a figure equivalent to 22.11 percent of the entire male population, of all ages, then in the United Kingdom. Of these, the overwhelming majority served in France: 60 of Britain's 101 divisions were on the Western Front in November 1918. Naval strength peaked at 400,000 in 1918, and when the RAF was founded in April 1918, its personnel all came from the Army and Royal Navy. Statistics from John Turner (ed.), *Britain and the First World War*, pp. 43, 68, 100–01.

3   Harold Macmillan, *Winds of Change*, p. 59.

4   Joining by professions as many of the Pals invariably did, the informal titles of their battalions indicate their differing backgrounds and social origins: 10/Royal Fusiliers were the Stockbrokers' Battalion; 16/Middlesex formed as the Public Schools Battalion; 1st and 2nd Football Battalions were officially the 17th and 23/Middlesex; the Lord Mayor and Corporation of London formed the Banker's Battalion (26/Royal Fusiliers); Glasgow raised the 15th, 16th and 17/Highland Light Infantry, otherwise known as the Tramways, Boys' Brigade and Commercials Battalions; 14/Royal Irish Rifles were the Belfast Young Citizens; formally, two Newcastle battalions – the Commercials and Railway Pals – were 16th and 17/Northumberland Fusiliers; in Hull there were four Pals battalions of the East Yorkshire Regiment, called the Hull Commercials, Tradesmen, Sportsmen – the fourth battalion were such a diverse assortment of volunteers, so called themselves the Hull T'Others.

5   See Ian F.W. Beckett, Chapter 1 in Beckett and Keith Simpson (eds.), *A Nation in Arms*.

6   We can take Montgomery's own Royal Warwickshire Regiment as a yardstick to illustrate the development of these various armies, but every other regiment grew in a similar way except the cavalry and Guards. The Royal Warwicks had two Regular battalions, which grew out of the old 6th of Foot, a line infantry regiment dating back to 1675; as the 1881 Cardwell reforms obliged all regiments to drop their numerical designations and adopt county titles, the 6th became the Royal Warwickshire Regiment, with its regimental training depot in the county town, Warwick; the organisation adopted an antelope as its cap badge (the emblem coming from a Moorish battle flag, captured by the regiment in 1710). Also sporting the antelope badge were two locally recruited Rifle Volunteer battalions; it was this latter pair that became the nucleus for two of the Royal Warwicks Territorial battalions when that force was established on 1 April 1908. Exploiting the large pool of enthusiastic manpower in nearby Birmingham, four Territorial battalions were created, which formed the Warwickshire Infantry Brigade. The County Territorial Associations also 'owned' their local Officer Training Corps (OTC), in the Royal Warwicks' case at Birmingham. The OTCs were designed to train the better-educated as potential officers and divided into two 'divisions', Junior (which recruited at public schools) and Senior (based at universities); their existence was soon justified – in the first seven months of the war, nationally they produced a staggering 29,500 commissioned officers. One twenty-one-year-old teacher who enrolled as a cadet in Birmingham OTC in 1912 was commissioned in August 1914, and

joined 9/Royal Warwicks, a New Army battalion. Despite being wounded at Gallipoli, he did rather well and finished his career as Field Marshal Lord Slim of Burma; it is thus a strange coincidence that the two key Army commanders of 1945 – Slim and Montgomery – had both started their army careers wearing the Royal Warwicks' antelope. The regiment also 'owned' eight New Army battalions, mostly recruited in Birmingham, of which three were known as the Birmingham Pals..

7   IWM, BLM 1/28.

8   Desmond Young, *Try Anything Twice*, p. 47.

9   Sidney Allinson, *The Bantams*, p. 69.

10 Ray Westlake, *Kitchener's Army*, p. 138.

11 Modern soldiers found themselves in not dissimilar conditions when fighting in Bosnia in the mid-1990s. Often the opposing Bosnian, Croat or Serb lines would consist of trenches dug through forests – studded with bunkers made of logs for protection – and one realised that some of the basics of soldiering have not changed at all. With generous coils of barbed wire arrayed in front, the trenches I inspected in 1996 were revetted, fire-stepped and sandbagged almost as per the *Field Service Pocket Book* of 1915–16 and – though recently abandoned – reeked of unwashed bodies and wet clothing, as must have done the original article. By a sniper's position – the spent cartridge cases told the story – notches in a wooden post suggested 'kills'. It all gave any visiting student of military history a vivid impression of how the Western Front must have seemed.

12 Rommel, *Infantry Attacks*, pp. 55–6.

13 Ibid., pp. 58–9.

14 Dennis Showalter, *Patton and Rommel*, p. 46.

15 IWM, BLM 1/39.

16 Ibid.

17 IWM, BLM 1/54.

18 Born in the same year as Winston Churchill (1874), James Walter Sandilands was commissioned into the Manchester Regiment in 1895 and transferred to the Queen's Own Cameron Highlanders in 1897. He fought in the Sudan campaign at Omdurman in September 1898 (where Churchill charged with the 21st Lancers), and was Mentioned in Despatches. Sandilands then served in the Second Boer War, when he was wounded and again Mentioned in Despatches: he was evacuated under fire from the battlefield by Sergeant Donald Farmer who was awarded the Victoria Cross for this act of bravery. Sandilands commanded 7/Cameron Highlanders (whose history he wrote in 1922) before leading 104th Brigade, 1916–18. After the war he became Military Attaché in Berlin and was subsequently GOC of British Troops in China and Hong Kong, 1929–32, and retired in 1933. He retired with a CB, CMG and DSO dying in 1959.

19 Brian Montgomery, *A Field Marshal*, p. 168.

20 IWM, BLM 1/55.

21 Allinson, *The Bantams*, pp. 204–05.

22 IWM, BLM 1/55.

23   Ibid.

24   Ernst Jünger, *The Storm of Steel*, p. 99. Jünger (1895–1998) was a German novelist, man of letters, scientist and thinker. In September 1918 he was awarded the Pour le Mérite, the youngest ever to receive the decoration. At his death, aged 102, he was the last surviving bearer of the award. His autobiography of the Great War, *Storm of Steel*, brought him to Hitler's attention and favour, but he refused to join the Nazi Party. An inspiration to anti-Nazi conservatives in the German Army, he served on the staff of the Military Governor in Paris (von Stulpnagel), 1940–44, knew Hans Speidel well and was on the fringes of the Stauffenberg bomb plot. In 1984 he joined Chancellor Helmut Kohl and President François Mitterrand at a 1984 Franco-German ceremony at Verdun, He was incredibly lucid and active when I met him briefly at a dinner for winners of the Pour le Mérite and Knight's Cross, in Germany in 1995.

25   Montgomery, *Memoirs*, p. 30.

26   Allinson, *Bantams*, p. 213.

27   Ibid., pp. 214–15.

28   IWM, BLM 1/57.

29   Montgomery to his father, IWM, BLM1/39 and 61.

30   Allinson, *Bantams*, p. 225.

31   These senior Nazi figures included: Commander of the Bavarian Guard Regiment, Oberstleutnant Franz von Epp, who became the senior SA (Brownshirt) leader in Bavaria during the Third Reich. Leutnant Ferdinand Schörner, of the same regiment, became a rival of Erwin's and the last field marshal appointed by Hitler in 1945. In three other Alpenkorps battalions were Walter Bronsart von Schellendorf who won a *Pour le Mérite* in 1918; Heinrich Kirchheim, who would win a Pour le Mérite in the Great War and a Knight's Cross in the Second (Kirchheim would go on to serve in the DAK under Rommel, who eventually sacked him in 1941; he would turn up again in unusual circumstances in 1944 – sitting on the 4 October 1944 Army Court of Honour hearing to determine the complicity of Rommel's Chief of Staff, Hans Speidel, in the failed Stauffenberg bomb plot); and Hans Kreysing, commander of 3rd Mountain Division during the Second World War.

32   Major Theodor Sprösser entered the *Reichsheer* after the Great War, was promoted *Oberstleutnant* on 1 March 1920, and *Oberst* on 1 February 1922. He subsequently became Kommandant of Glatz, in Lower Silesia (now in Poland) and retired as *Generalmajor* on 31 March 1925. He died in 1932.

33   Henri Barbusse, *Under Fire*, p. 212.

34   Jünger, *Storm of Steel*, p. 81.

35   David Irving, *Trail of the Fox*, pp. 12–13.

36   Turner (ed.), *Britain and the First World War*, p. 43.

37   Amongst the troops hastily despatched to shore-up this south-eastern corner of the Central Powers' empire was the 3rd Bavarian Infantry Regiment, which included a Leutnant Wilhelm von Thoma (1891–1948), who had won a high Bavarian decoration for bravery on the Eastern Front. The bestowal of the Bavarian Max Josef Order also included a title, Ritter (Knight), and as Ritter von Thoma,

he commanded German tanks in the Spanish Civil War and would one day act as Rommel's deputy in the Afrika Korps until his capture at Alamein.

38   Rommel, *Infantry Attacks*, pp. 82–5.

6. Bloody Red Tabs

1   Scrapbook annotation, IWM, BLM 131.
2   Simon Robins has calculated that of the 447 Staff College graduates from Camberley and Quetta (the Indian Army's equivalent) in the August 1914 *Army List*, nearly fifty percent – 219 – were killed in the war, mostly in the opening months. Robins, *British Generalship on the Western Front*, p. 37.
3   Jünger, *Storm of Steel*, p. 109.
4   IWM, BLM 1/61.
5   Ibid.
6   Captain J. C. Dunn, *The War the Infantry Knew*, p. 342.
7   Brian Montgomery, *A Field Marshal*, p. 169 He states it was an XV Corps Instruction, but this cannot be correct; 33rd Division was serving under VII Corps all through the period and XV were technically inactive.
8   Montgomery to Captain Cyril Falls, 8 October 1938, CAB45/116, National Archives, cited in Robins, *British Generalship*, p. 65.
9   Richards, *Old Soldiers Never Die*, p. 217.
10 Keith Simpson, Introduction to Dunn's *The War the Infantry Knew*, p. xli.
11 Siegfried Sassoon, 'The General', No. 12 in *Counter-Attack and Other Poems* (1918).

7. The Mountain Lion

1   Rommel, *Infantry Attacks*, p. 108.
2   Cited in John Wilkes and Eileen Wilkes, *Rommel and Caporetto*, p. 20.
3   Irving, *Trail of the Fox*, p. 13.
4   Rommel, *Infantry Attacks*, p. 118.
5   Ibid., p. 125.
6   Ibid., p. 135.
7   Ibid., p. 167.

8. Mud and Mountains

1   'The God of War and the Vale of the Passion: BEF Artillery Development 1914–17' lecture by Rob Thompson to Yorkshire branch of the Western Front Association, 14 February 2009.
2   John Terraine, *The Smoke and the Fire. Myths and Anti-Myths of War 1861–1945*, p. 132.
3   Montgomery, *History of Warfare*, p. 478.
4   IWM, BLM 1/61.
5   Ibid.
6   Ibid., 1/64.

7   See John Lee's Introduction to Chris McCarthy, *The Third Ypres Passchendaele Day-by-Day Account.*

8   Rommel, *Infantry Attacks*, p. 173.

9   Ibid., p. 177.

10 Ibid., p. 189.

11 Ibid., p. 191.

12 Ibid., pp. 201–02.

13 Ibid., p. 221.

14 Rommel, *Infantry Attacks*, p. 223.

15 Ibid., p. 224.

16 Kenneth Macksey, *Rommel. Battles and Campaigns*, p. 19.

17 Irving, *Trail of the Fox*, p. 14.

18 Showalter, *Patton and Rommel*, p. 74.

19 Young, *Try Anything Twice*, p. 119.

20 Ralf Georg Reuth, *Rommel: End of a Legend*, p. 15.

21 Rommel, *Infantry Attacks*, p. 229.

22 Ibid., p. 243.

23 Ibid., p. 187.

24 Barbara Tuchman, *August 1914*, p. 95.

25 For this reason, armies then – and even more emphatically today – send their chosen, talented leaders (the future generals) to command and staff colleges where they are admitted to the dark arts of staff work and logistics. Naturally this kind of work is not to everyone's taste and – in the midst of a total war – it is a huge mental leap from leading men to glory over the Italian Alps to completing forms and filing returns. This incompatibility works in the opposite direction also, which is potentially far more dangerous: examples of brilliant staff officers miscast as combat commanders abound, of which two examples will suffice. Keith Simpson said of the brave but over-promoted Arthur Percival (Monty's exact contemporary in age), who lost Malaya and surrendered Singapore to the Japanese in February 1942, that he was '…an efficient and intelligent staff officer who was an indefatigable worker… There was always something of the perpetual Staff College instructor about Percival even when he was supposed to be exercising operational command…' See Simpson in John Keegan (ed.) *Churchill's Generals.* Friedrich Paulus, whom we met in the Alpine Corps and in January 1943 surrendered the German Sixth Army in Stalingrad, had spent his military career as a distinguished and talented staff officer rising to the post of Chief of Staff, Tenth Army. However in late 1941 he was promoted over the heads of colleagues to command Sixth Army on the recommendation of its C-in-C, Field Marshal von Reichenau (his former chief), who was promoted to Army Group command. Paulus (who had only held brief, peacetime commands of a rifle company and a battalion) was a natural subordinate, but lacked moral courage, initiative and drive, which contributed towards Sixth Army's hopeless position – along with Hitler's insane orders. As gifted staff officers, both Percival and Paulus found themselves out of their depth and overtaken by events when having to lead, and would have worked better responding to more bullish, proactive chiefs. Rommel

– the ultimate combat commander – (whom Paulus would encounter in North Africa) seems to have found himself in 1918 similarly – and hopelessly – miscast as a staff officer. See Paulus in Correlli Barnet (ed.) *Hitler's Generals*.

9. The Last Year

1   Montgomery to Captain Cyril Falls, 8 October 1938, CAB45/116, National Archives, cited in Simon Robins, *British Generalship*, p. 65.

2   IWM, BLM 1/59.

3   Letter, *Sir Miles Dempsey to Ronald Lewin*, 15 November 1968, p. 3, paragraph 10, RLEW 7/7, Churchill College Archives, Cambridge.

4   Chalfont, *Montgomery*, p. 64.

5   See www.firstworldwar.bham.ac.uk/nicknames/gorringe.htm

6   The medical historian Nick Bosanquet points out that nearly 1,200 young American doctors were seconded to the BEF by mid-1918, usually employed as battalion MOs. Bosanquet, 'Health Systems in Khaki: the British and American Experience', Chapter 33 in Hugh Cecil and Peter Liddle (eds), *Facing Armageddon*, p. 457.

7   IWM, BLM 1/63.

8   As early as 28–30 March, with the Michael offensives hammering towards Amiens, engineers of the 3rd US Infantry Division attached to the British were involved in the successful defence of Warfusée-Abancourt, a hamlet astride the Amiens–St Quentin road, a couple of miles east of Villers-Brettoneux. The first offensive action had been the capture of the village of Cantigny, about twenty miles southeast of Amiens in the French sector, where on 28 May troops of the 1st US Infantry Division's 28th Infantry Regiment seized the settlement and held it against all German counterattacks (Pershing made sure a monument was erected postwar to record this). Monty's IX Corps had, back in May and June, already good reason to be grateful to the 2nd and 3rd American Divisions, who had hurried to the Marne to provide a backstop against the Blücher-Yorck steamroller. The motorised machine-gun battalion of 3rd Division had shot up German troops trying to cross the Marne at Chateau-Thierry on 31 May, whilst 2nd Division – which included a US Marine Corps brigade – deployed and dug in between the village of Vaux and Belleau Wood on 1 June. The fighting sucked in both divisions and lasted until 25 June, when Belleau Wood was finally cleared; Vaux was retaken by 2nd Division on 1 July, by which time the engagement had cost 5,000 casualties. This was a significant contribution, but possibly Montgomery was so distracted by the carnage inflicted on IX Corps that he was unaware of this key assistance not far away. As IX Corps then left the area, other US divisions helped French formations counterattack German units astride the Marne throughout July. On Independence Day 1918, about a thousand soldiers of the US 33rd Division assisted 4th Australian Division in retaking the village of Hamel, on the south bank of the Somme and close to the scene of the March fighting at Warfusée-Abancourt.

9   Edmunds, *Official History of the War, 1918: Volume IV*, p. 199.

10 Jill Knight, *The Civil Service Rifles*, pp. 134–5.

11 IWM, BLM 1/65.

12 Ibid.

13 Edmonds, *Official History*, p. 210.

14 The others were: Blomberg, Bock, Brauchitsch, Busch, Keitel, Kleist, Kluge, Kuchler, Leeb, List, Manstein, Model, Paulus, Reichenau, Rundstedt, Schorner, Weichs and Witzleben.

15 Rowland Fielding, *War Letters to a Wife*, p. 290.

16 Ibid., p. 116.

17 Fielding, *War Letters*, pp. 305–06.

18 Alan H. Maude, *History of 47th London Division*, p. 194.

19 IWM, BLM 1/65.

20 IWM, BLM 1/66.

21 M.L. Walkinton, *Twice in a Lifetime*, pp. 161–3.

22 Martin Gilbert, *Winston Churchill, Volume IV*, pp. 154–9; Major John (Jack) Churchill died in 1947; his daughter Clarissa married Anthony Eden in 1952.

23 Chalfont, *Montgomery*, pp. 64–5.

24 Carrington, *Soldier From the Wars*, pp. 277–8.

25 Montgomery, *Memoirs*, pp. 31–2.

26 Letter Montgomery to Col. Henry Abel Smith, 7 August 1969, quoted in Gary Mead, *The Good Soldier*, p. 6.

27 IWM, BLM 1/35.

28 Gary Sheffield and John Bourne, *Douglas Haig*, p. 282.

29 Montgomery, *Memoirs*, p. 30.

30 Sir John Smyth, *Leadership in Battle*, p. 12.

31 Modern historians have effectively refuted the popular allegations that Haig and his Chief of Staff were negligently out of touch, an argument first advanced by Basil Liddell Hart in *The Real War* (1930) and repeated by Lloyd George in his scandalously inaccurate *War Memoirs* of 1933. In 2005, Douglas Haig's 1914–1918 diaries were published by Gary Sheffield and John Bourne and give an overwhelming impression of a man on the road a great deal, visiting army, corps, divisional and even brigade headquarters – not the activities of a château-loving recluse. During the late-1918 battles, Douglas Haig actually left the cosy environs of Montreuil-sur-Mer and commanded with his advance HQ staff from a train – not unlike Monty with his 1944–45 caravans. These allegations of the physical and mental distance of Great War generals from the front can be neatly summed up under the accusation of 'château generalship'. Reinforced by Stephen Fry's caricature general in the BBC's *Blackadder* television series, it is a neat phrase that nevertheless misleads, for all higher headquarters necessarily (even today) tend to have large staffs that need accommodating together and require big rooms from which to direct their wars. Châteaux, be they castles, manor houses or large old farm complexes, were ideal for this purpose, though were often damp and devoid of luxury or warmth. Even big châteaux are too small for modern, higher formation headquarters: in Bosnia in the 1990s, NATO used a selection of redundant factories, former barracks and hotels to house its various headquarters; in

Iraq it was barracks and airports – around the world today, industrial ware-housing is probably most suitable – though was rarely available and certainly not in the right places in 1914–18. The château trend continued into 1939–45, when all the armies adopted the practice of commanding from mansions and other large buildings. 'Château generalship', then, ought to be more a statement of prac-ticality than derision.

32   I am grateful to Simon May of the Old Paulines for his diligent research into the statistics. Monty's school also suffered greatly in the Second World War, when 243 old boys were killed.

10. Coping with Peace

1   Robertson enlisted aged seventeen as a private in the 16th Lancers in November 1877, was commissioned in 1888 into the 3rd Dragoon Guards, and entered Staff College in January 1895. Having been promoted colonel in 1904, Robertson was Quartermaster General (QMG) to Sir John French's British Expeditionary Force, then Chief of the General Staff (CGS) until 1918; when the post was renamed in December 1915 as Chief of the Imperial General Staff, he received promotion to Lieutenant General. A staunch supporter of Sir Douglas Haig, Robertson blocked Lloyd George's attempts to divert effort from the Western to the Eastern Front; unlike Lloyd George, Robertson was a keen 'Westerner', believing that the war could only be won on the Western Front. Robertson resigned in February 1918, and commanded the British forces on the Rhine from 1919–20.

2   As a result, Alun Chalfont wittily suggests: 'the Battle of El Alamein may have been won on the tennis courts of Cologne'. The unlucky may grumble about the 'old school tie' in operation; far from it: this is an instance of the able and ambi-tious making their own luck. In all large organisations there are formal chains of command and informal ones. These latter can include everything from shared schools and universities, common sporting interests, social and family ties, even (most unfashionable these days) freemasonry has played a big role in the past. Arguably this operates as a natural system of checks and balances that often elevates on merit those overlooked for other reasons – and can also alert to the unsuitable, the sycophant and the charlatan. Such are the ways many a fortunate general, politician or magnate reaches the top, with an opportunity exploited at the right moment.

3   IWM, BLM 1/68.

4   Author's calculations from Owl Pie 1920, list of students.

5   The Staff College historian noted that the 1919 and 1920 courses together fielded twenty officers who had served as brigadier generals, and their ranks included five VCs, thirty-eight CBs and CMGs, plus 170 DSOs (three of whom had bars). Godwin-Austin, The Staff and the Staff College, p. 272.

6   Hamilton, Monty, The Making, p. 151.

7   Mitcham, Rommel's Lieutenants, pp. 9–11.

8   Charles Richardson, Send for Freddie, p. 21.

9   Ibid.

10   The many exchanges between the pair are preserved in their respective archives, at the Imperial War Museum (Monty) and King's College, London (Liddell Hart).

11   Hamilton, *Monty, The Making*, p. 163.

12   The reserve soldiers of Bernard's understanding remain a key component within the British Army as I write; the aspiration is still a 'one army concept' and during recent operations in Bosnia, Iraq and Afghanistan, the Territorials (who celebrated their centenary in 2008) have provided an indispensable ten percent of the ground force; as importantly, they invariably deploy with professional skills their regular counterparts do not possess: insurgents cannot tell them apart.

13   Tom Carver, *Where the Hell Have You Been*, p. 33.

14   Charles Richardson, *Send for Freddie*, p. 21.

15   Ibid., p. 36.

16   Later General Sir Bernard Paget, 1887–1961, GCS Home Forces 1940, C-in-C SE Command 1941, C-in-C Home Forces 1941–43, C-in-C 21st Army Group until sidelined by Monty in Dec 1943, thereafter C-in-C Middle East Force 1944–45; Lieutenant General Sir Henry Pownall, 1887–1961, DMO 1938–39, CGS to BEF 1939–40, C-in-C Far East 1941; COS to Mountbatten 1943–44.

17   See Dominick Graham, *The Price of Command: A Biography of General Guy Simonds*, p. 27.

18   Report of No. 6 Syndicate (Major E. G. W. W. Harrison, Captains F. W. Messervy, R. H. Campbell and M. B. Burrows), 'The Battle of the Marne Sept 6th–9th 1914, Foreign Tour', BTC 016, JSCSC Archives, 'Annex "B", France's Attitude Towards Great Britain and the United States of America'.

19   'Foreign Tour Report on Operations of the BEF, Mons 1914', BTC008, JSCSC Archives.

20   'Report on the Operations of the French Armies, 4–9th September 1914', BTC015, JSCSC Archives.

21   Later Field Marshal the Earl Alexander of Tunis, 1891–1969, C-in-C Middle East 1942, GOC 18th Army Group, 1943–45, Governor General of Canada, 1946–52, Minister of Defence, 1952–54.

22   Later Major General D. N. Wimberley, 1896–1983, GOC 51st Highland Division (1941–43), wounded in Sicily, Commandant Staff College 1943–44, Director of Infantry, 1944–45.

23   The syndicate was Lieutenant Colonel the Hon H. R. L. G. Alexander DSO, MC, Irish Guards; Major E. S. D. Martin DSO, MC, 5th Inniskilling DG; Major GHB De Chair, OBE, MC, Royal Sussex; Captain J. A. C. Whitaker, Coldstream Guards.

24   'Foreword to syndicate report', BTC024, JSCSC Archives.

25   Ibid., Part IV, pp.1–2.

26   Ibid., pp. 2–3.

27   Ibid.

28   Ibid., p. 8.

29   'Report on the Retreat from Mons', BTC018, JSCSC Archives.

30   Ibid., Annex B.

31   Later General the Lord Robertson of Oakridge, Bt., GCB, GBE, KCMG, KCVO, DSO, MC, Chief Admin Officer, 15th Army Group, Italy 1943–45; Deputy Military Governor, Germany Control Commission 1945–47; C-in-C Occupation Zone 1947–49; C-in-C Mid East 1951–53; Chairman of British Transport Commission 1953–61; Life Peer 1961.

32   David G. Williamson, *A Most Diplomatic General,* p. 22.

33   Rowland Ryder, *Oliver Leese,* p. 31.

34   Later General Sir E. H. Barker, 1894–1983, KCB, KBE, DSO, MC, GOC 49th Division 1943–44, VIII Corps (after O'Connor) 1944–45.

35   Later the Viscount Bridgeman, 1896–1982, KBE, DG Home Guard and TA 1941 and 44, then DAG.

36   Later Major General E. E. Dorman-Smith (subsequently, from 1949, O'Gowan), 1895–1969, later an instructor at Camberley (1936–37). COS to Auchinleck at 8th Army in 1942 but purged with his boss and after brigade command at Anzio was dismissed from the army in 1944. The brilliant 'Chink's' spat with Monty ultimately wrecked his subsequent career. See the useful biography by Lavinia Greacen, *Chink: a Biography* (London, 1990).

37   Later Field Marshal Lord Harding of Petherton.

38   Later Major General Sir Leslie Hawkesworth, KBE, GOC 46th Div in Italy, X Corps Nov 1944, he died of a heart attack in May 1945.

39   Later Lieutenant General Sir Oliver Leese, 1894–1978, GOC XXX Corps at Alamein, in Tunisia and Italy, GOC 8th Army, 1944–45.

40   Later Major General Sir Ronald Penney, GOC 1st Division at Anzio, where he was wounded.

41   Later Major General Philip Gregson-Ellis, 1898–1956; GOC of 5th Division at Anzio; later Commandant Staff College 1945–46; retired 1950

42   Later Marshal of the RAF Sir Arthur 'Bomber' Harris, 1892–1984, AOC Bomber Command 1942–45.

43   Later General Sir Richard McCreery, 1898–1967, GOC 8th Army 1944–45.

44   Later General Sir Gordon MacMillan of MacMillan, 1897–1986, KCB, KCVO, GOC 15th Scottish Division 1944, then 49th Div, then 51st Div, subsequently GOC Gibraltar 1955.

45   Later Field Marshall Sir Gerald Templer, 1898–1979.

46   Henry Probert, *Bomber Harris His Life and Times,* pp. 59–61.

47   Ibid., p. 61.

11. Preparing for War

1   The *Vossische Zeitung* was the principal German newspaper published in Berlin (1721–1934); it was generally regarded as the national newspaper of record, like *The Times* and *Le Monde,* but was discontinued by the Nazi party (perhaps on account of its liberal views, and not least because it had serialised *All Quiet*) and replaced by the NSDAP's newspaper, *Der Vôlkische Beobachter.*

2   Milestone starred in and directed the film, winning one of the two Academy Awards given to *All Quiet.* It also had five other nominations and, according to

Halliwell's Film Guide it was 'a landmark of American cinema and Universal's biggest and most serious undertaking until the sixties; this highly emotive war film with its occasional outbursts of bravura direction fixed in millions of minds the popular image of what it was like in the trenches, even more so than *Journey's End* which had shown the Allied viewpoint. Despite dated moments, it retains its overall power and remains a great pacifist work.'

3 Robert Cedric Sherriff (1896–1975)'s *Journey' End* was first published by Samuel French in 1928 as Remarque was finishing *All Quiet*. The play opened at the end of 1928, starring the unknown Lawrence Olivier (then twenty-one) and soon transferred to the Savoy Theatre in January 1929 and opened on Broadway in 1930. By the autumn of 1929 it was playing by fourteen companies in English and seventeen in other languages; it, too, was made into a film version in 1930. It was one of BBC television's first plays, broadcast on Armistice Day 1937. Sherriff later wrote the screenplays to *Goodbye, Mr Chips* (1933) and *The Dambusters* (1955), both of which were nominated for awards.

4 Hamilton, *Monty, The Making*, pp. 190–91.

5 Ibid.

6 Montgomery, *Memoirs*, p. 36.

7 Hamilton, *Monty, the Making*, p. 213.

8 David Fraser, *Knight's Cross*, p. 97.

9 Irving, *Trail of the Fox*, pp. 20–21.

10 Ibid., pp. 21–22.

11 List commanded the 14th Army in Poland (1939), 12th Army in France (1940), Greece and Yugoslavia (1941) and Army Group South in Russia (1942). His career hit the buffers after dismissal on Hitler's whim on 9 September 1942 and he was never re-employed. He died in 1971.

12 Mitcham, *Hitler's Field Marshals*, p. 339.

13 Irving, *The Trail of the Fox*, pp. 21–2.

14 Von Luck, *Panzer Commander*, p. 9.

15 Charles Richardson, *Send for Freddie*, p. 31.

16 Hamilton, *Monty, the Making*, p. 217.

17 Charles Richardson, *Send for Freddie*, p. 33.

18 Carver, *Where the Hell*, p. 43.

19 Ibid., p. 43.

20 Siegfried Knappe and Ted Brusaw, *Soldat*, p. 104.

21 Ibid., pp. 104–09.

22 In the United States today this is taught by the Marine Corps as the '70 Percent Solution' – the imperfect answer that can nevertheless be made immediately, based on the majority of the required data and good situational awareness. See David H. Freedman, *Corps Business*, pp. 7–10.

23 We might observe that Rommel's volume is very similar in appearance to, and was almost certainly inspired by, *Infantry in Battle*, a military handbook published in 1934 by the US Army. In April 1945, Captain Charles F. Marshall, an intelligence officer at US VI Corps headquarters, found a translation of *Infantry in Battle* in Rommel's study when he called to interrogate Frau Rommel.

24   Rommel, *Infantry Attacks*, p. 265.

25   Barker, 'Monty and the Mandate', *History Today*, March 2009.

26   Montgomery, *Memoirs*, p. 34.

27   After Winsome's first husband, Lieutenant Colonel John Holdeness, died of cancer, she married Sir Godwin Michelmore, in 1971. As Winsome Michelmore she self-published the charming memoir, *Thursday Child*, in 1985.

28   Lord Douglas of Kirtleside, *Autobiography, Volume Two, Years of Command*, pp. 20–21.

29   Later Admiral of the Fleet Sir James Somerville, 1882–1949.

30   Admiral of the Fleet Viscount Cunningham of Hyndhope, *A Sailor's Odyssey*, p. 139.

31   Happily today, British senior officers after their Staff College year also benefit – if fortunate to be selected – from a Higher Command and Staff Course *and* the IDC, now renamed as the Royal College of Defence Studies.

32   Irving, *Trail of the Fox*, p. 29.

33   *Against All Odds*, p. 12.

34   Ibid., p. 6.

35   'Gun Buster' (pseudonym), *Return via Dunkirk*, pp. 35–6. Although a wartime novel, it was conditioned by the author's own experiences of 1939–40.

36   Brigadier B. L. Montgomery, DSO, 'The Problem of the Encounter Battle as Affected by Modern British War Establishments', *Royal Engineers Journal*, September 1937, pp. 339–54.

37   Sydney Jary, *18 Platoon*, pp. 19–20; Professor Gary Sheffield has made the same point in his chapter, 'The Shadow of the Somme: the Influence of the First World War on British Soldiers' Perceptions and Behaviour in the Second World War', in Paul Addison and Angus Calder, (eds), *Time to Kill: the Soldier's Experience of War in the West 1939–1945*.

38. Jean Paul Pallud, 'Hitler on the Western Front', article in *After the Battle* magazine, Number 117, pp. 3–13. See also Kershaw, p. 299. Hitler's two comrades from the First World War were Max Amann (head of the Nazi party's publishing concerns) and Ernst Schmidt.

39   Donald Lindsay, *Forgotten General*, p. 106.

40   Rhodes James, *Anthony Eden*, p. 136.

12. Phoney War

1   The school, founded in 1751, was reopened in 1958, and remains Austria's military academy.

2   The train generally consisted of a locomotive, armoured anti-aircraft wagon with quadruple 20mm guns and a twenty-six-man crew, baggage and generator wagons, the Führer's Pullman (in which his Chief *Wehrmacht* Adjutant, his chief Personal Adjutant and a valet were also quartered), conference wagon with communications centre including teletype machines, escort wagon for Rommel's team, dining car, two sleeping cars (for entourage and guests), bath car, second dining car, two cars for personnel (secretaries, cooks, etc.), press car with communications centre

including a 700-watt short-wave transmitter, baggage and generator car, and a final anti-aircraft wagon. David Fraser argues that the train was named after Amerika, a small hamlet near Gheluvelt, where Hitler had fought in 1914. (See Fraser, *Knight's Cross*, p. 143.)

3   Ralf Georg Reuth, *Rommel: The End of a Legend*, p. 37.

4   In April 1945 the Zossen complex fell into Russian hands intact. Throughout the Cold War huge numbers of Soviet soldiers manned the sprawling site of surface and underground bunkers, barracks, living quarters and communication centres, and abandoned it only in 1994. The Zossen base is now a museum.

5   Both the OKW and OKH were housed in different wings of the Bendlerblock, the old War Ministry building, bordered on one side by the Bendlerstrasse and on another by the Tirpitzufer.

6   Alexander Stahlberg, *Bounden Duty*, p. 119.

7   Samuel W. Mitcham, *Rommel's Lieutenants*, p. 22 and Yves Buffetaut, *Rommel*, pp. 14–16.

8   Martin Windrow, *The Panzer Divisions*, p. 5.

9   Sources for the strength of 7th Panzer on 10 May 1940 seem to conflict, but the most authoritative seems to be in George F. Nafziger, *German Order of Battle: Panzers and Artillery*, pp. 64–66.

10   Heinz Guderian, *Panzer Leader*, p. 73.

11   Preface to Mitcham, *Rommel's Lieutenants*.

12   Montgomery, *Memoirs*, p. 43.

13   Field Marshal Lord Carver, *Seven Ages of the British Army*, p. 203.

14   Concluding paragraph in *Notes on Certain Lessons of the Great War*, War Office, April 1934.

15   Alex Danchev and Dan Todman (eds.) *War Diaries of F. M. Lord Alanbrooke*, p. 11.

16   Robin McNish, *Iron Division*, p. 73.

17   Lieutenant General Sir Brian Horrocks, *A Full Life*, pp. 98–9.

18   The full text of the 15 November 1939 Divisional Order is cited in Brian Montgomery, *A Field Marshal in the Family*, pp. 253–54.

19   Ibid.

20   Montgomery, *The Path to Leadership*, p. 126.

21   C. L. Potts, *Gordon and Michael: A Memoir*, p. 34. Lieutenant G. F. A. Potts (8th Royal Warwicks) was killed in action on 21 May 1940.

22   Danchev and Todman (eds.) *F. M. Lord Alanbrooke*, p. 20.

13. Blitzkrieg

1   Even the origin of the word 'blitzkrieg' has made for a lively debate amongst modern historians, for it seems that the Germans had no specific word for the 'lightning blow' that blitzkrieg implies. English language media occasionally used the term (for example in *The Times* on 14 June 1939), but in the context of discussing German air strategy. *Time Magazine* was the first recorded media outlet to use the 'B-word' to describe the ground war in Poland. It had appeared in

German military documents earlier but was employed in a general sense to mean a surprise, knockout attack – which in the late 1930s meant an assault with air forces (as at Guernica in April 1937 or as would happen at Pearl Harbor in December 1941). As there was no specific word for what we now assume was an established doctrine, it is difficult to argue that blitzkrieg was then a fully thought-out concept either in Poland or even France.

2   Karl-Heinz Frieser, *The Blitzkrieg Legend*, p. 56.

3   Ibid., p. 33.

4   Liddell Hart (ed.), *Rommel Papers*, p. 22.

5   John Keegan, *A History of Warfare*, p. 71.

6   Desmond Young, *Rommel*, p. 42.

7   Liddell Hart (ed.), *Rommel Papers*, p. 39.

8   Horrocks, *A Full Life*, p. 86.

9   Author's interview with Brigadier Peter Vaux, April 2008.

10 Liddell Hart (ed.) *Rommel Papers*, pp. 30–32.

11 I've explored the Arras area many times over the years, studying both world wars fought over the same terrain but most recently on a British Army staff ride with Brigadier Miles Wade and his Chief of Staff Major Andy Leighton. Their insights proved to me than one never stops learning. With the eye of a modern tank commander, the Brigadier immediately identified that in 1940 Rommel had moved around Arras clockwise, keeping to the reverse slope of a rim of high ground, so that he never revealed his strength to a potential opponent. Major Leighton then started identifying from the map where he would station armoured infantry in an advance; in every case he correctly identified the deployments of 7th Panzer Division, ordered instinctively after just a few minutes consideration by Rommel back in May 1940 – some things never change.

12 In fact, King Leopold III capitulated without consulting his own cabinet or the allies on 28 May, whilst the British garrison in Calais had already surrendered at 4 p.m. on 26 May.

13 McNish, *Iron Division*, p. 87.

14 Montgomery, *The Path to Leadership*, p. 128.

15 Horrocks, *A Full Life*, p. 85.

16 National Army Museum, *Against All Odds*, p. 73.

17 Bredin Obituary, *The Times*, 9 March 2005.

18 Liddell Hart (ed.), *The Rommel Papers*, pp. 59–60.

19 Mitcham, *Rommel's Lieutenants*, p. xi and Young, *Rommel*, pp. 76–7.

20 The victory in the West also silenced the anti-Nazi plotters in the German High Command until a string of defeats encouraged a renewal of assassination plans in 1943–44. In fact, the *Wehrmacht*'s victory was more of a French military collapse than a victory for blitzkrieg. France's politicians and many senior generals, with no clear picture of what was actually happening, and paralysed with inde-cision, seemed to have lost the will to carry on the fight early on, and this was reflected further down the chain of command by the surrenders of intact forma-tions. Immediately the Battle for France had ended, its citizens attempted to come to terms with the reasons for the catastrophic collapse. *L'Etrange defaite (Strange*

Defeat) by Marc Bloch, published after the Liberation in 1944, was just the first of many works reflecting the puzzlement many patriots felt at how their country had fallen apart within just six weeks of the German invasion. In 1995 I was dining alone in the town of Albert. A nearby party of three – also foreigners – kindly invited me to join them at their table after dinner. They turned out to be a German father, his son and son-in-law, who were retracing another ancestor's war on the Somme in 1916. Over cognac, the old man spoke up and told me he had driven this way in 1940; my ears pricked up as his own tale unfolded. He had been an NCO in charge of an armoured car detachment in von Senger's cavalry brigade (under Rommel's command), probing ahead looking for road-blocks, broken bridges and clear routes. At one town he had driven his eight-wheeler into the market place and came across a French brigade drawn up on parade, being addressed by their general. Momentarily paralysed with the shock of encountering around 5,000 armed French troops, he expected at the very least to find his own soldiering days over. There was one of those few-second pauses that seemed to last for hours, before the French brigade commander turned on his heel, saluted the armoured car – and surrendered his brigade. For that the German NCO was promoted to sergeant major and awarded the Iron Cross. 'Do you want to see it?' the old man asked. I thought I had misheard him, until he pushed the worn and scratched decoration, still attached to a grubby piece of ribbon, across the table. The grime, I later discovered, was collected later on the Eastern Front, along with the wound that had made him deaf in one ear and hampered our conversation. In a flash, I understood the nature of the German victory in France that summer.

21   Montgomery, *Memoirs*, p. 61.
22   Author's interview with Norman Field, December 2008.
23   Horrocks, *A Full Life*, p. 100.
24   Ibid., p. 100.
25   Carlo d-Este. *Bitter Victory*, p. 106.
26   Colonel John Hughes-Wilson, *Military Intelligence Blunders and Cover-Ups*, pp. 133–64.

14. Duel in the Desert I

1   IWM, BLM 12/6.
2   Despite all the preparations made to withstand a siege (the rock of Gibraltar more resembles a giant slice of mousetrap cheese, so riddled is it with a warren of tunnels and caverns built for every conceivable purpose, storage of supplies not least amongst them) the base would not have lasted long, and without its airstrip, still the lifeline today, would have fallen easily to a Spanish–German assault, had they bothered to attempt it – and plans for Operation *Felix* were, indeed, drawn up.
3   MacGregor Knox, *Hitler's Italian Allies*.
4   Martin Kitchen, *Rommel's Desert War*, p. 24.
5   Liddell Hart (ed.) *Rommel Papers*, p. 261.

6   Knox, *Hitler's Italian Allies*.

7   Richard Holmes, *The World at War: The Landmark Oral History from the Previously Unpublished Archives*, p. 265.

8   Hans-Albrecht Schraepler (ed.), *At Rommel's Side: The Lost Letters of Hans-Joachim Schraepler*, p. 96.

9   Liddell Hart (ed.) *Rommel Papers*, p. 104.

10 Heinz Werner Schmidt, *With Rommel in the Desert*, p. 12.

11 Schraepler, *At Rommel's Side*, p. 67.

12 Liddell Hart (ed.) *Rommel Papers*, p. 111.

13 The judges at the Nuremberg trials of 1946 found the pair guilty on all counts and sentenced them to death by hanging.

14 Liddell Hart (ed.) *Rommel Papers*, p. 118.

15 Schraepler, *At Rommel's Side*, p. 124.

16 Ibid., pp. 124–25.

17 Irving, *Trail of the Fox*, pp. 21–2.

18 Schraepler, *At Rommel's Side*, p. 165.

19 Ibid., p. 89. This incident of 20 April 1941 anticipated exactly the circumstances of Rommel's serious injury by air attack in Normandy on 17 July 1944, when his driver was also killed.

20 Schraepler, *At Rommel's Side*, p. 97.

21 O'Connor, a Cameronian who was two years younger than Monty, but his contemporary both as a student and instructor at Staff College and divisional commander in Palestine just before the war, would later escape captivity and emerge as VIII Corps commander in Normandy under Monty. Neame, a Cheltenham-educated Royal Engineer and an all-round Boy's Own hero, was five months younger than Monty and had uniquely won a VC in 1914 and an Olympic Gold Medal in 1924. Aged fifty-five, he would eventually make a successful escape attempt from his Italian POW camp in the company of Dick O'Connor.

22 Colin Smith reminds us in *England's Last War Against France* that this little-known conflict witnessed combat between opposing detachments of the Foreign Legion and saw the future Israeli leader Moshe Dyan sacrifice an eye in fighting for the British against the French.

23 Schraepler, *At Rommel's Side*, p. 81.

24 Lawrence Durrell, the future author and a wartime press officer in Cairo, recalled of an interview with Auchinleck and Wavell, 'Auchinleck was absolutely charming but an extremely sedentary man. He had no real personality, he was handsome and personable and quiet in a Scots way but he hadn't any, you know, for a critical moment you need a bit of rhetoric, a bit of panache and a sort of "Charge, boys, charge" thing – and he hadn't got that precisely. The people who had it, like General Harding, were not in evidence and were commanding detachments in the desert and being shot up. It's sad about Wavell because he had that. His orders for the day were inspiring and you needed that inspiration.' Churchill later wrote that he had been 'much impressed with his personal qualities, his presence, and high character . . . [and] had the conviction that in Auchinleck I should bring a new, fresh figure to bear the multiple strains of the Middle East'. See

Richard Holmes, *The World at War*, p. 261 and Winston S. Churchill, *Second World War, Vol. III, The Grand Alliance*, p. 309.

25 Montgomery, *Memoirs*, p. 62.

26 Nick Smart, *Biographical Dictionary of British Generals of the Second World War*, p. 15. Monty regarded Auchinleck as a 'poor picker of men', which is an unfair observation, for in 1942–44 Monty would have had the pick of men who had been his students at Staff College and had already seen action. Back in 1941, when the Auk was appointed from India, there was no such pool of talent in which to fish for the brightest and the best. After the war Monty unfortunately polarised opinion about his own generalship: you were either pro-Monty, and if you weren't, you were an opponent: there was no room for middle ground. Monty's view of the war, decidedly anti-Auchinleck, prevailed: thus he was able not merely to disagree with the Auk, but later to denigrate his achievements.

27 Josef Goebbels, *The Goebbels Diaries*, pp. 5–6.

28 John Harvey (ed.), *The War Diaries of Oliver Harvey, 1941–1945*, p. 88.

29 Goebbels, *Diaries*, p. 8.

30 Desmond Young found this in Rommel's papers.

31 Schmidt, *With Rommel in the Desert*, pp. 32–3.

32 Ibid., p. 27.

33 Liddell Hart (ed.) *The Rommel Papers*, p. 176.

34 Schraepler, *At Rommel's Side*, p. 165.

35 Irving, *Trail of the Fox*, p. 95.

36 Schmidt, *With Rommel in the Desert*, p. 56.

37 Goebbels, *Diaries*, entry for 16 December 1942, p. 184.

38 One of the principal colonies, from where many of the first Afrika Korps was recruited, was German South West Africa (modern-day Namibia), whose first Governor was Heinrich Goering, father of Hermann. There is a *Göring Strasse* in its German-speaking capital, Windhoek, to this day.

39 See *Rommel's Year of Victory: The Wartime Illustrations of the Afrika Korps* by Kurt Caesar (Greenhill, 1998).

40 Schmidt, *With Rommel in the Desert*, pp. 57–8.

41 Ibid.

42 Schraepler, *At Rommel's Side*, p. 116.

43 Young, *Rommel*, pp. 40–42.

44 Lieutenant Colonel P. Richards, 'Rommel – *Fingerspitzengefuhl* Or Common Eavesdropper?', article in *British Army Review*, No. 131, p. 63.

15. Duel in the Desert II: The Battles of El Alamein

1 H. W. McIntosh, 'Heat Exhaustion and Heat Stroke', article in *Canadian Medical Association Journal*, Vol. 76, p. 873.

2 Holmes, *The World at War*, p. 271.

3 Ibid.

4 Hamilton, *Monty Vol.1, The Making of a General*, p. 797.

5 Liddell Hart (ed.) *Rommel Papers*, p. 150.

6   Schmidt, *With Rommel in the Desert*, p. 72.

7   *Daily Telegraph*, Obituary, 28 August 1997.

8   Rudolf Schneider, interview, *Independent*, 19 June 2009.

9   'Chivalrous Rommel Wanted to Bring Holocaust to Middle East', *Independent*, 25 May, 2007 and 'New Research Taints Image of Desert Fox Rommel', *Der Spiegel*, 23 May 2007.

10 Danchev and Todman (eds), *F. M. Lord Alanbrooke*, p. 235.

11 Esebeck wrote for *Signal* whilst recuperating from a wound sustained in Russia, but he had been wounded leading 15th Panzer in Libya also; arrested after the 20 July 1944 plot and sent to a concentration camp, Esebeck survived the war to become a leading light of the DAK Old Comrades Association.

12 Churchill, *Second World War*, Vol. IV, pp. 343–4.

13 Author's interview with Herr Otto Carius, 26 May 2005.

14 A decision point is the latest moment at which a predetermined course of action is (or must be) initiated, and modern staff officers are taught to plan a campaign by decision points – plotting them schematically on maps and chronologically on wall planners/calendars – to ensure that all necessary decisions are taken in time, and not too late. Logistics often drive decision points, and in a remote theatre they are all the more important.

15 See James Holland's *Fortress Malta: An Island Under Siege, 1940–1943*.

16 Danchev and Todman (eds), *F. M. Lord Alanbrooke*, pp. 292–93.

17 Churchill to Deputy PM, National Archives, CAB/65/31/11, WM42/106 Reflex Telegram No. 27.

18 Harvey (ed.), *Oliver Harvey, 1941–45*, p. 135.

19 *The Times*, 22 August 1942.

20 Smart, *Biographical Dictionary of British Generals*, pp. 125–6.

21 For a deeper analysis of Brooke, see Andrew Roberts, *Masters and Commanders*.

22 Earl of Avon, *The Eden Memoirs: The Reckoning*, pp. 86 and 159.

23 Harvey, whose edited diaries appeared in 1978, went to great pains to assure his readers that these entries were genuine immediate impressions, not the product of reflection and hindsight.

24 Montgomery, *Memoirs*, pp. 83–4.

25 Danchev and Todman (eds), *F. M. Lord Alanbrooke*, p. 292.

26 See John Grainger's recent work, *The Battle for Palestine 1917*.

27 James Holland's interview is contained in his book *Heroes*, and also at www.secondworldwarforum.com.

28 Harvey (ed.), *Oliver Harvey, 1941–45*, p. 148

29 The fact that Monty backed down when Auchinleck threatened a lawsuit over the retreat claim in Monty's 1958 *Memoirs* maybe suggests some merit in the Auck's recollection of events.

30 The full text was copied down in shorthand and resides amongst the Cabinet Papers in the National Archives at Kew (CAB 106/654).

31 Liddell Hart (ed.), *Rommel Papers*, p. 270.

32 Ibid., pp. 270–72.

33 Goldensohn, *The Nuremberg Interviews*, p. 321.

34 Goebbels, *Diary*, 31 March 1942; Reuth, *Rommel*, p. 54.

35 It was the publication of *The Ultra Secret* by Group Captain Fred Winterbotham in 1974 that revealed the massive impact of Ultra on the course of the war. Winterbotham was in charge of the Special Liaison Units that distributed Ultra intelligence to field commanders.

36 This cycle still holds good for combat operations today, and is called an Air Tasking Order.

37 Field Marshal B. L. Montgomery, *Some Notes on the Use of Air Power in Support of Land Operations* (21st Army Group, December 1944), p. 5.

38 See David Fisher, *The War Magician: the Man Who Conjured Victory in the Desert*.

39 Hermione, Countess of Ranfurly, *To War with Whitaker*, pp. 168–9.

40 Peter Hoffmann, *Stauffenberg: A Family History*, pp. 162–80.

41 Basil Liddell Hart, *The Other Side of the Hill*.

42 Monty's slouch hat resides in the Australian War Memorial in Canberra, whilst his Grant command tank (since refurbished) and one of his berets can be seen at the Imperial War Museum, London.

43 'I was Monty's Hat-man', contribution to the BBC's Second World War People's War Website by BBC Radio Norfolk Action Desk on behalf of Jim Fraser, Article ID: A6080924, contributed 10 October 2005. Accessed at www.bbc.co.uk/ww2peopleswar/stories/24/a6080924.shtml.

44 Ibid.

45 Ibid.

46 Johnny Henderson with Jamie Douglas-Home, *Watching Monty*, p. 29

47 Jack Swaab, *Field of Fire*, p. 148.

48 D'Este, *Bitter Victory*, p. 106.

49 Sir Edgar ('Bill') Williams, Monty's Intelligence Chief, entry for B. L. Montgomery in the *Dictionary of National Biography 1971–80*.

50 Montgomery, *Memoirs*, p. 154.

51 D'Este, *Bitter Victory*, p.106

52 Martin van Creveld, *Supplying War: Logistics from Wallenstein to Patton*, pp. 200–01.

53 Albert Kesselring to Leon Goldensohn, 4 February 1946. Goldensohn was an American psychiatrist charged with caring for the mental health of the twenty-one Nazi defendants awaiting trial at Nuremberg in 1946. See Leon Goldensohn and Robert Gellately, (eds), *The Nuremberg Interviews*, pp. 317–28.

16. Two Return to France

1 Although he was by far the oldest commander in any theatre, it is perhaps significant that the Allies' oldest general in Normandy was George Patton, a good ten years younger, and possessing the energy of someone half his age.

2 Difficult to determine, the average age of the Afrika Korps has been taken from the recorded ages of German troops in US and British POW camps, who prior to mid–1943 were Afrika Korps veterans.

3 Hamilton, *Monty: Master of the Battlefield*, p. 459.

4   Alanbrooke, *War Diaries* (ed. Danchev and Todman), p. 496.

5   Ibid., p. 499.

6   Bradley, *A Soldier's War*, pp. 207–08.

7   Roy Jenkins, *Churchill*, p. 727–8.

8   Captain Noël Chavasse, MC, MBE, Middlesex Regiment, was a nephew of Captain Noël Chavasse (after whom he was named), RAMC, who had won two VCs during the Great War. Young Noël's father, Christopher (younger twin of the double VC-winner) was Bishop of Rochester; his grandfather was Bishop of Liverpool and thus a friend and contemporary of Bishop Montgomery, Monty's father. Monty (b. 1887) and Christopher Chavasse (b. 1884) were near contemporaries, and met during 1917, when the latter was Senior Chaplain of 62nd Division, 'I recognised him at once though I hadn't seen him since they came to Ireland one summer a long time ago.' IWM, BLM 1/60.

9   John Colville, *The Fringes of Power*, p. 442.

10   Churchill, *The Second World War, Vol. V, Closing the Ring*, p. 393.

11   I am indebted to Gary Giumarra (unpublished paper *Operation Sledgehammer*) and Major-General Graham Hollands for this reference.

12   Theodore A. Wilson (ed.), *D-Day 1944*, p. 224.

13   The floating harbour idea can be traced back to 1917, when Churchill had drafted detailed plans for the capture of two islands, Borkum and Sylt, which lay off the Dutch and Danish coasts. He envisaged using a number of flat-bottomed barges or caissons which would form the basis of an artificial harbour when lowered to the seabed and filled with sand. Events moved on and Churchill's proposal was filed away and was never published. General Lord Ismay, *Memoirs*, p. 346.

14   Eisenhower, *Crusade in Europe*, pp. 270–71.

15   Warren F. Kimball (ed.), *Churchill and Roosevelt: The Complete Correspondence*: Vol. I, p. 280; also Vol. II, pp. 491–92. Churchill had clear ideas about what constituted appropriate names and after coming across several that he considered inappropriate he demanded that all future code names be submitted to him for approval; he dropped this demand when he learned the magnitude of the task, but he did take the precaution of writing down some commonsense principles to guide his subordinates.

16   Churchill, *Closing the Ring*, p. 662. Churchill would no doubt have been apoplectic at the drab code names chosen for Britain's major operations today, reserving particular venom, I feel sure, for the anodyne *Banner* (Ireland), *Corporate* (Falklands), *Granby* (First Gulf), *Grapple* (Bosnia), *Telic* (Iraq operations), *Fingal* and *Herrick* (both Afghanistan).

17   Michael Howard, *History of the Second World War: United Kingdom Military Series: Vol. IV, Grand Strategy, August 1942–September 1943*, p. 430

18   W. G. F. Jackson, *Overlord: Normandy 1944*, p. 89; Omar N. Bradley, *A Soldier's Story*, p. 172; Forrest C. Pogue, *George C. Marshall, Vol. II, Organizer of Victory*, p. 413.

19   Historian Mark A. Stoler tantalises us with the argument that Churchill may have deliberately recast *Anvil* as *Dragoon* to represent his feelings of being 'dragooned'

into the operation.

20 Sir Alexander Stanier, *Sammy's Wars, Recollections of War in Northern France and Other Occasions*, pp. 37–8. See also *Daily Telegraph*, Obituary, 11 January 1995.

21 A view first advanced by Ronald Lewin in his excellent 1971 study, *Montgomery as a Military Commander*.

22 Bradley, A Soldier's Story, p. 208.

23 Alanbrooke, *War Diaries* (ed. Danchev and Todman), p. 531.

24 James, *I Was Monty's Double*, p. 107.

25 Bradley, A Soldier's Story, p. 209.

26 James, *I Was Monty's Double*, p. 106.

27 Author's interview with Langdon, Army Junior Division (AJD) Study Day, Staff College, 3 November 2000.

28 Stephen Badsey, 'Faction in the British Army: Its Impact on 21st Army Group Operations in Autumn 1944', *War Studies Journal*, 1/1, Autumn 1995, pp. 13–28.

29 Ibid.

30 Eversley Belfield and Hubert Essame, *The Battle for Normandy*, p. 47.

31 Keith Simpson, 'A close run thing? D-Day, 6 June 1944: The German perspective', article in *RUSI Journal*, Vol. 139/3 (June 1994), pp. 60–71.

32 Facsimile in Wilson (ed.), *D-Day 1944*, p. 201.

33 There is a link between Normandy and the 1991 Gulf War when one considers the enormous sea transport and logistics planning that made both possible, the crushing weight of air power, and the fact that cruise missiles were fired from the refurbished Second World War battleships *Missouri* and *Wisconsin*. Their sister ships, *Nevada*, *Texas* and *Arkansas*, had provided fire support on D-Day.

34 Adrian R. Lewis, *Omaha Beach: A Flawed Victory*, p. 37.

35 Bradley, *A Soldier's Story*, p. 138.

36 Rear Admiral A. G. Kirk USN, *Rough Draft* [After-]*Action Report on OVERLORD*, 25 July 1944, Naval Historical Centre, Washington DC.

37 James Jay Carafano, *After D-Day: Operation Cobra and the Normandy Breakout*, p. 99.

17. Defending Normandy

1   Karl Friedrich, Freiherr von dem Knesebeck, and Gottlieb, Graf von Haesler.

2   David Fraser, *Wars and Shadows*, pp. 11–14.

3   Author's analysis taken from muster roll of 3rd Kompanie/SS-*PanzerGrenadier* Regt. 25, 13 July 1944.

4   Samuel W. Mitcham, *The Desert Fox in Normandy*, p. 81.

5   Luck, *Panzer Commander. The Memoirs of Hans von Luck*, p. 132.

6   Blumentritt, *von Rundstedt*, pp. 207–08.

7   Major-General F. W. von Mellenthin, *Panzer Battles*, p. 56.

8   Schmidt, *With Rommel in the Desert*, p. 56.

9   Department of the US Navy, *Manuscripts A-982 and B-282*, Naval Historical Center, Washington DC.

10 Blumentritt, *von Rundstedt*, pp. 205–06.

11   Department of the US Navy, *Manuscripts A-982 and B-282*.

12   Ibid.

13   By contrast there were 4,120 Luftwaffe aircraft on the Eastern Front at this time.

14   Ibid.

15   Department of the US Navy, *Manuscripts B-352 and B-353*.

16   Richard Collier, *Ten Thousand Eyes*, pp. 182–7.

17   James, *I Was Monty's Double*, pp. 73–7.

18   Stephen A. Hart, *Colossal Cracks*, p. 25.

19   Montgomery, *Memoirs*, p. 100.

20   Ibid., pp. 208–09.

21   Author's interview with Sir John Archibald Ford, KC MG, MC, artillery officer with 50th Division, May 2000.

22   Moorhead, *Montgomery*, p. 188.

23   James, *I Was Monty's Double*, pp. 100–01.

24   Montgomery, *Memoirs*, pp. 209–10.

25   Bradley, *A Soldier's Story*, pp. 239–41.

26   Major General Sir John Kennedy, *The Business of War*, pp. 326–7.

27   Eisenhower, *Crusade in Europe*, p. 269.

28   Sir John Wheeler Bennett, *King George VI*, pp. 599–600.

29   Kennedy, *Business of War*, p. 328.

30   Ismay, *Memoirs*, p. 352.

31   Montgomery, *Memoirs*, p. 221.

32   Ibid.

33   Tom Pocock, *Norfolk*, p. 37.

34   Stanier, *Sammy's Wars*, p. 29.

35   The quotation came from James Graham, 1st Marquis of Montrose, Scottish Covenanter, who later supported the Stuarts and was hanged by order of the Scottish Parliament in Edinburgh in 1650.

18. Britain's Last Hurrah!

1   Figures from Anthony Kemp, D-Day. *The Normandy Landing and the Liberation of Europe*, pp. 52 and 81, *The European* newspaper, *Élan* supplement, 3–9 June 1994 and Michael Glover, *Battlefields of Northern France and the Low Countries*, pp. 106–07.

2   In addition, German forces lost another 200,000 as prisoners of war.

3   The daily rates are my own, calculated by dividing the total days of the campaign by the known casualties of both sides, but beware, for Great War casualty figures are notoriously inaccurate. I have taken these from Richard Holmes, *The Western Front*, pp. 110, 139 and 174. Total casualties at Verdun are cited as 714,231; on the Somme at 1,015,000, and at Passchendaele as 520,000. The Somme figures are most open to question, and may be higher still.

4   Eddy Florentin, *The Battle of Normandy*; and 'Did We Fight in France?', article by Giles Coren in *The Times*, 3 June 1994.

5   3rd Canadian Infantry Division, Operation Order No. 1, 'Overlord', 13 May 1944,

Appendix G.

6   Blumentritt, *von Rundstedt*, p. 221.

7   I first heard John Howard's story from his own lips at Pegasus Bridge in July 1975. He and Hans von Luck formed an unlikely friendship and performed many a double act, relating their 1944 adventures to dozens of British Army battlefield tours. In the 1962 film *The Longest Day*, the actor Richard Todd played John Howard with knowing realism; Todd himself, as adjutant of 6/Parachute Battalion, had also landed on D-Day and had met Howard at Pegasus Bridge. Though the bridge is a modern replacement, it looks not dissimilar to its predecessor, and a visit to the Café Gondrée is a must. In 1944 it was run by Georges and Thérèse and hosted first the German bridge guards, then John Howard's wounded. Today it remains in the family and is presided over by their daughter, the hospitable Arlette, who still recalls the sound of the gliders landing and the blackened faces of her liberators. Before his death in 2009, Richard Todd returned to the Café Gondrée and donated the officers' webbing he had worn on D-Day to Arlette's small museum, where one fights one's way past her collection of treasured mementos to order at the bar.

8   In 2008, I accompanied Herb Fussell, formerly a sergeant with 22nd Independent Parachute Company, and a group of his friends around the churchyard at Ranville. Herb regaled us with his minute-by-minute account of battling with the German defenders amongst the tombstones of the churchyard; he accounted for a row of obvious bullet holes chipped into the stone by the church door, as having come from 'a German NCO – he missed me, of course, firing too high, so we got him'. Herb, one of that seemingly bullet-proofed race of heroes, went to join the Great Muster Beyond shortly before I finished this book.

9   Author's interview with Lieutenant Colonel Kingston Adams, 1991; and *The Warwick Courier*, June 2004.

10 On 6 June 2008, I found scattered .303-inch cartridges and a NAAFI fork, dated 1944, dropped by a careless Tommy in the fields around Hillman, testifying to the bitter struggle and was pleased to be able to take Krug's son and grandson, both *Bundeswehr* colonels, to the scene of their forebear's textbook defence.

11 Was an alteration to the Sword Beach plan possible, and if so, where would the extra troops have landed? The main objection to widening 3rd Division's front may have lain with the offshore shoals; but other landing sites would have been possible – Ouistreham itself, or further east, to the obscure reserve beach, code-named Band, which incorporated the seafront towns of Cabourg and Houlgate, either side of the Dives river mouth. Though this would have been on the far side of two significant water obstacles – the Caen Canal and River Orne – the terrain offered no real challenges and good road access to the hinterland, and led directly onto the high ground of the Bréville ridge overlooking Caen. In their book *Monty, The Lonely Leader*, Bernard's son David and Alistair Horne never-theless observed that: 'Had the additional landing craft only been available, the beaches from the Orne eastwards to Cabourg, half a mile wide at low tide, would have offered as good landing as anywhere else. The hinterland was flat and open ... given the immense Allied air and naval superiority, the argument that the

approach would have come under fire from the Le Havre heavy coastal guns hardly holds water.' p. 125.

12 Amongst the many Canadian troops who landed on D-Day was a twenty-four-year-old captain in the Royal Canadian Artillery, who was struck by eight bullets and evacuated that evening. After the war James Doohan became an actor and in 1966 found himself cast as Scotty, the chief engineer of the USS *Enterprise* in the original *Star Trek* series, battling Klingons rather than Germans. Obituary, *Daily Telegraph*, 21 July 2005.

13 Capa, *Slightly Out of Focus*, pp. 139–40.

14 Ibid, p. 152.

15 Author's interview with George Laws, December 2001.

19. Where is Rommel?

1 He was the son of a distinguished historian, Professor Erich Marcks, who had written biographies of Bismarck and Wilhelm I. Young Erich grew up in Potsdam, in an educated, liberal, middle-class family that had known Bismarck intimately; his father had been an exchange professor of history in the USA, broadening the mind of father and son. Erich was commissioned into the artillery in 1911 and served in the First World War, winning both classes of Iron Cross; by 1935 he had been promoted to colonel and was Chief of Staff of VIII Corps.

2 Which included 243rd Division (Western Cotentin), 709th Division (Cherbourg), 352nd Division (between Bayeux and Isigny), 716th Division (Caen area), 319th Division (Channel Islands) and 30th Quick Reaction Brigade (between Coutances and Avranches).

3 Liddell Hart (ed.), *The Rommel Papers*, pp. 467–8.

4 Much knowledge of Feuchtinger was furnished by Hans von Luck, either in his memoirs, or during innumerable battlefield tours and staff rides I shared with him. He did not rate his divisional commander highly: 'General Edgar Feuchtinger, an artilleryman, had no combat experience, and none at all of panzer units. He had become known in Germany as the organiser of the military part of the so-called *Reichsparteitage*, the national party rallies, and through that was very familiar with Hitler and the party apparatus... Feuchtinger was a live-and-let-live person. He was fond of all the good things of life, for which Paris was a natural attraction. Knowing that he had no combat experience or knowledge of tank warfare, Feuchtinger had to delegate most things, that is, leave the execution of orders to us experienced commanders.'

5 In March 1945, Nazi friends lobbied successfully for his pardon and he was transferred to 20th Panzer Grenadier Division, remaining as *Kanonier*. Feuchtinger then fled to Celle, where he eventually surrendered to the British. Hans von Luck recalled his reaction to Feuchtinger's court martial in his memoirs: 'Although I, too, hold to the saying de *mortuis nil nisi bene* (of the dead say nothing but good), when I think of our brave men, who fought so brilliantly, and of the thousands of dead, wounded, and missing... I cannot help reproaching Feuchtinger with having done us all poor service.' Feuchtinger spent two years, until 1947, as a

British POW, and was under investigation for corruption when he died in obscurity in West Berlin in 1960.

6 Monty's title of overall Ground Commander would be termed 'land component commander' today.

7 Omar N. Bradley and Clair Blair, *A General's Life: An Autobiography by General of the Army Omar N. Bradley*, p. 252.

8 In am indebted to General Scott-Bowden's son, Brigadier Jamie, with whom I explored Normandy in 2008, for making available interview transcripts and articles by his father.

9 Three more were landed dry; and of those five available tanks, three were destroyed within minutes, acting as magnets for German fire. Recent marine archaeology has established that the tanks sank beam-on to the waves – they had been told to aim for the church spire at Colleville, which meant they swam in at a forty-five degree angle to the coast, presenting an easier target for the waves to swamp. Strong winds and adverse currents likewise pushed a significant number of landing craft off course, into areas where commanding officers' maps were useless.

10 Lewis, *Flawed Victory*, p. 296.

11 Ronald J. Drez (ed.) *Voices of D-Day: The Story of the Allied Invasion Told by Those Who Were There*, p. 233.

12 www.3cbluering.com/Wehrmacht/HTML20Pages/unit20history.html, accessed 2 August 2001.

13 S. L. A. Marshall, *The Soldier's Load and the Mobility of a Nation*, p. 38.

14 Played by Robert Mitchum in the film *The Longest Day*.

20. Exploiting the Beachhead

1 Horne and Montgomery, *Monty, The Lonely Leader 1944–45*, p. 121.

2 Eisenhower, *Crusade in Europe*, pp. 278–9.

3 Horne and Montgomery, *Lonely Leader*, p. 123.

4 Interview with Albert 'Bert' Williams, *Western Daily Press*, 6 June 2009.

5 Churchill, *The Second World War, Volume VI, Triumph and Tragedy*, p. 11.

6 The aged Marquis de Druval never knew the identity of the mysterious guest with a hundred vehicles who occupied his estate whilst he remained in his house until the TAC had moved on, but his descendants live there still; a plaque on the gatepost records that HM King George VI visited Creullet on 16 June 1944 and it was a matter of some pride for me to take one of His Majesty's grandsons and his wife there in 2008.

7 In 2003, Andy and Sue Parlour wrote the first history of this secretive and little-known regiment: *Phantom at War*.

8 John Buckley has observed that most engagements with German tanks took place at very short ranges, where even the standard Allied 75mm gun could damage most German armour, though not necessarily a Tiger.

9 Buckley, *British Armour in the Normandy Campaign 1944*, and 'British armoured operations in Normandy, Chapter 6 in Buckley (ed.), *The Normandy Campaign 1944: Sixty Years On*.

10 Belfield and Essame, *The Battle for Normandy*, p. 127.

11 Major Johnny Langdon MC of 3/Royal Tank Regiment told me in 2002 that in the spirit of the friendly rivalry between infantry and armour, in training they used to let their tank telephones trail on the ground 'like snakes' and get great amusement from watching the infantry scurry after them. He had served in the Middle East, but in 21st Army Group felt there was no evidence of any cross-theatre doctrine.

12 Wittmann's remains were discovered only in March 1983 by the *Volksbund Deutsche Kriegsgräberfürsorge* (German People's War Graves Association) in an unmarked battlefield grave alongside the four other members of his tank crew; they were re-interred in the German Military Cemetery at La Cambe, south of Omaha Beach, where their plot is now much visited.

13 This has prompted the roguish academic Martin Van Creveld to question whether all the resources poured into making the Mulberries and towing them over to France would have been better employed elsewhere, observing that the prefabricated harbours themselves were an example of an 'over-engineered luxury' the Allies didn't really need.

14 Anthony Powell, *The Military Philosophers*, pp. 173–4.

15 Interview with Herr Erich Sommer, broadcast on ABC (Australia), 17 May 2004.

16 Robin Neillands, *The Conquest of the Reich*, p. 5.

17 *Michelin Guide to Normandy*, 1972 edition, p. 25.

18 *Guardian*, 7 June 1994.

19 Montgomery, *Memoirs*, pp. 233–4.

20 Staff College Battlefield Tour Archives 1956, Watchfield JSCSC/BTC 155.

21 Nigel Hamilton has cited Basil Liddell Hart's post-war correspondence with Dempsey of March 1952, when the former Second Army commander stated that: 'To obtain this [large-scale bomber support] from the Air Staff and Bomber Command, who disliked being diverted to aid ground operations, Monty felt it was necessary to overstate the aims of the operation,' Hamilton, *Monty: Master of the Battlefield, Volume II*, p. 719.

22 Subsequently, on 8 September, the first V-2 rockets arrived over England; 1,115 were launched in total, of which 517 fell on London, killing another 2,754 people and injuring 6,523. The V-weapons blitz continued into 1945, when the last V-1 was destroyed over Sittingbourne on 27 March 1945 and two days later the last V-2 fell on Orpington in Kent.

23 Brooke (ed. Danchev and Todman), *War Diaries*, p. 571.

24 Luck, *Panzer Commander*, p. 158.

25 Ibid. , pp. 153–4.

26 John Keegan referred to the incident in his excellent 1982 account of the campaign, *Six Armies in Normandy*. In the years before his death in 1997, Luck escorted dozens of British army battlefield tours to Normandy, recounting the famous story of himself and the Luftwaffe captain. I last enjoyed his stories and sparkling company on his old battlefields in September 1992. It was in 2005 that Ian Daglish produced his revolutionary book *Goodwood: Over the Battlefield*, which analysed the story of Operation Goodwood using the extensive aerial photography archive provided by the RAF's Photographic Reconnaissance Unit (PRU), who overflew

the battlefield almost hourly. It is an innovative and forensically foolproof way of studying the events and, to the consternation of many, Daglish has failed to find any trace of the battery of anti-aircraft guns in Luck's portion of the battlefield. Hans von Luck was undoubtedly a brave and inspirational warrior and one desperately wants his tale of drawing his pistol on the Luftwaffe captain and the 88s to be true, but so far the jury is out.

27   Montgomery, *Memoirs*, p. 235.

28   David Holbrook, *Flesh Wounds*, pp. 175–6. Holbrook's highly autobiographical novel first appeared in 1966. His description of the death of a Sherman tank (here) was set a few days after D-Day, but holds true for Goodwood, by which time Holbrook had already been wounded.

21. Plots and Breakouts

1   The huge Margival site is today set in beautiful parkland and contains a network of over twenty vast concrete bunkers of different shapes and sizes, in pristine condition, but betrays some deep feral instinct of Hitler's for life underground. He, of course, died underground in his Berlin bunker.

2   The ADC was Alexander Stahlberg, deeply involved in the Stauffenberg plot, and one of its few survivors. His account (which cannot be corroborated) of the three field marshals can be found in Alexander Stahlberg, *Bounden Duty*, pp. 306–10.

3   Cited in many works, including Mitcham, *The Desert Fox in Normandy*, p. 184.

4   Von Mellenthin, *Panzer Battles*, p. 55.

5   Young, *Rommel*, pp. 210–12.

6   Alain Roudeix initially had no idea who the injured were, but in 1970 retraced the route of Rommel's last car journey in Normandy with Helmuth Lang, who was researching his own wartime adventures. The road is now the busy N579, but it is still possible to follow Rommel's route, coming from the north, and see the minor road to Lisores on the left (Lang incorrectly stated it was on the right) that Daniel had been trying to reach to get under cover.

7   Charles Messenger, *Hitler's Gladiator*, p. 132.

8   Every year on 20 July, successive German governments have commemorated the coup attempt with sobering ceremonies, both in the courtyard of the War Ministry Bendlerblock, on Stauffenbergstraße, where Stauffenberg and three co-conspirators were shot, and at the Plötzensee Prison, now a Memorial Centre, where most were executed. Remer, who later won promotion to *Generalmajor* and commanded a division, was active from 1950 until his death in neo-Nazi movements; in 1994 he fled to Spain to escape a jail term imposed in connection with Holocaust denial, where he died in 1997.

9   Like General David Fraser in his biography of the Desert Fox.

10   Siegfried Knappe and Ted Brusaw, *Soldat*, pp. 254–5.

11   Milton Shulman, *Defeat in the West*, pp. 156–66.

12   James Jay Carafino, *After D-Day: Operation Cobra and the Normandy Breakout*, pp. 40–41, 59 and 99.

13   These were the world's first armoured, fully tracked armoured personnel carriers

(APCs) and adapted from the Canadian artillery's 105mm self-propelled gun carriers – the Priests. The guns were removed and the gap in the armour screen replaced with two steel plates separated by a layer of ballast. As the Priests had been constructed on a Sherman chassis, there was perfect interoperability between the two – tank and APC; after conversion, each Kangaroo carried twenty fully equipped infantrymen with armoured protection.

14   Kurt Meyer, *Grenadiers: The Story Of Waffen SS General Kurt 'Panzer' Meyer*, p. 278.

15   Jim Dan Hill, 'Mighty Both the Sword and Pencil', review of *Montgomery's Memoirs*, in *Saturday Review*, 1 November 1958.

16   The events are related in the 1965 book *Is Paris Burning?* by Larry Collins and Dominique Lapierre, made into a film the following year.

17   Belfield and Essame, *The Battle for Normandy*, p. 227.

18   Commander Czarnecki, *Boisjos Hill 262*, p. 22.

19   Stan Z. Biernacik, 'The D-Day Invasion of Normandy: Poles Played A Crucial Role in Bringing Nazi Germany to Its Knees', article in the *Polish American Journal*, June 1994.

20   Professor John Buckley, University of Wolverhampton Normandy Field Trip Notes, April 2001.

21   Recent analysis of the destruction recorded in Normandy has, however, indicated that much German armour and some soft-skinned transport were abandoned intact. Whilst some vehicles had undoubtedly run out of fuel, many had been left by their crews at the very sight of Allied aircraft, and the actual damage caused by rocket or cannon attack from the air was proven to be negligible. Thus, Allied air superiority was as much a psychological weapon as an actual one. See Ian Gooderson, 'Allied fighter-bombers versus German Armour in North-West Europe 1944–1945: Myths and realities', *Journal of Strategic Studies*, Vol. 14/2 (June 1991), pp. 210–31.

22   Norman Kirby, *1100 Miles With Monty*, p. 61.

23   Reuth, *Rommel*, pp. 192–4.

22. Beyond the Bocage

1   Horne and Montgomery, *Lonely Leader*, pp. 240–42. I have hunted boar and stag here and the huge, silent forest remains delightful.

2   Lieutenant Colonel Christopher 'Kit' Dawnay, CBE, MVO, (1909–89), Coldstream Guards, Monty's Personal Military Assistant. Monty had two British ADCs, Captains Noël Chavasse and Johnny Henderson, an American ADC, Captain Ray BonDurant, and a Canadian Personal Assistant, Lieutenant Colonel Trumbull Warren.

3   John Harvey (ed.), *The War Diaries of Oliver Harvey*, pp. 352–3.

4   Ibid.

5   Powell, *Military Philosophers*, pp. 189–90.

6   The manpower situation had been less critical during the First World War, when the British raised eighty-four divisions and when twenty-two percent of the adult

male population served in the army. But during the Second World War British labour – male and female – also had to man the factories, farms and mines, as well as the Royal and Merchant Navies, and the RAF.

7   Chapter 14 in Wilson (ed.) *D-Day 1944*. The exception was 'Pip' Roberts of 11th Armoured, born in 1906.

8   Montgomery, *Memoirs*, p. 276.

9   Bradley, *A Soldier's Story*, p. 416.

10 Horne and Montgomery, *Monty the Lonely Leader*, pp. 291–2.

11 Harvey, *The War Diaries of Oliver Harvey*, p. 355.

12 In Dinant a commemorative stone marks the spot where on 23 December 1944 a jeep-load of Germans wearing American uniforms were blown up on mines laid by a section of British soldiers guarding the bridge across the Meuse. In nearby Celles stands a Panther, the lead tank of an armoured column of 2nd Panzer Division, immobilised by a mine on 24 December 1944.

13 The full text of Monty's address to the 7 January 1945 press conference (though not of the subsequent question and answer session with the journalists!) is contained in Montgomery's *Memoirs*, pp. 287–91.

14 Ibid.

15 Ibid.

16 21 Army Group Log, 8 May 1945; US National Archives, microfilm: 331.10, *Records of HQ 21st Army Group (SHAEF) 1944–45*.

23. How Will History Judge Me?

1   Showalter, *Patton and Rommel*, p. 417.

2   Annette Kuhn, '*Desert Victory* and the People's War, article in *Screen*, Vol. 22/2 (1981), pp. 45–68.

3   S. P. Mackenzie, *British War Films:1939–1945: the Cinema and the Services*, pp. 109–11.

4   The sequel publication, *Tunisia*, which went on sale in 1944, concluded the story of the Desert Campaign, whilst the Air Ministry (though perhaps not to Monty's liking) related their contribution in *RAF Middle East*, which dealt with the air war of February 1942 to January 1943.

5   After the war, when the medals were issued, the '1' and '8' were represented on the full-size medal ribbon by a bar reading 'First' or 'Eighth Army'.

6   *Illustrated Magazine* devoted a five-page feature to the litter of war at Alamein in the 25 October 1947 edition, to coincide with the 1947 reunion; *Illustrated Magazine*, 25 October 1947, pp. 7–11.

7   See Monty's correspondence relating to the publication of this and *From Normandy to the Baltic* in the Liddell Hart Collection, King's College Archives, ref BLM 173–4. Major General David Belchem's papers also reside in the KCL Archives (ref 84/2/1).

8   Monty's successors at BAOR were: Dick McCreery (1946–7), Brian Robertson (1947–8), Brian Horrocks (1948–9), Charles Keightley (1949–51), John Harding (1951–2) and Richard (Windy) Gale (1952–7).

9   McCreery (1898–1967) was a shy old Etonian who had been a 12th Lancer, winning an MC in 1918, became Chief of Staff to Alexander in 1942, commanded X Corps under Monty in Sicily and Italy and had succeeded Oliver Leese at Eighth Army; then becoming GOC British Troops in Austria – the postwar name for Eighth Army.

10 Richard Hull (1907–89, the last CIGS) was a light cavalryman who had – after Charterhouse and Trinity, Cambridge – commanded 17/21 Lancers, 26th Armoured Brigade, 1st Armoured Division and 5th Division in Italy and north-west Europe, before arriving at the Camberley Staff College as Commandant in May 1946.

11 Hamilton, *Monty Volume 3, The Field Marshal*, p. 736.

12 Peter Rostron, *The Military Life and Times of General Sir Miles Dempsey*, p. 191.

13 Hamilton, *Monty Volume 3*, p. 727.

14 Montgomery, *Memoirs*, p. 382.

15 Rostron, *Dempsey*, pp. 173–89.

16 Hamilton, *Monty Volume 3*, p. 725.

17 'Secret Papers Reveal Monty's Racist Masterplan', *Guardian*, 7 January 1999.

18 Rostron, *Dempsey*, p. 192.

19 Montgomery, *Memoirs*, p. 462.

20 Young later claimed that serialisation of *Rommel* added 92,000 to the *Sunday Express's* circulation in January 1950, whilst Nigel Hamilton observed the *Sunday Times* claimed a boost of 100,000 copies because of *Memoirs*.

21 Frank Field (ed.), *Attlee's Great Contemporaries*, p. 50.

22 In terms of his mother, the more one reads about Monty the more it appears to have been he who was odd, not she. She was administering the usual discipline expected in a Victorian/Edwardian household full of children. Whilst not excusing her behaviour as recorded in Montgomery's memoirs, it never seems to have occurred to Bernard that *he* might have been the one at fault, not she: her admonition to his siblings, cited in *Memoirs*, 'Go and find out what Bernard is doing and tell him to stop it', now rings true – for Bernard in later life still needed to be told to 'stop it', and usually disobeyed, like the naughty schoolboy he still was at heart. See *Memoirs*, p. 13.

23 Frank Field (ed.), *Attlee's Great Contemporaries*, pp. 45–57.

24 *Life Magazine*, 13 October 1958, p. 82.

25 James, *I Was Monty's Double*, p. 92.

26 Goronwy Rees, *A Bundle of Sensations*, p. 156.

27 Interview, Lieutenant Colonel John Stevens, Royal Signals, in June 2008. The specific Monty 'incident' happened to a member of the Directing Staff at Camberley in 1967.

28 Barnett, 'The Desert Generals Revisited', Address to the British Commission for Military History (BCMH), 5 February 2005.

29 Hamilton, *Monty Volume 3*, p. 925.

30 Ibid., pp. 929–30.

31 Brian Montgomery, *A Field Marshal*, pp. 320–21.

32 Ibid.

33 Hamilton, *Monty Volume 3*, pp. 589–90; Brian Montgomery, *A Field Marshal*, p. 321; Chalfont, *Montgomery*, p. 305.

34 Horne and Montgomery, *Monty: the Lonely Leader*, p. 126.

35 Winsome Michelmore, *Thursday's Child*, p. 90.

36 Hamilton, *Monty Volume 3*, p. 941.

37 Brian Montgomery, *Monty: A Life in Pictures*, p. 152.

38 Ibid., p. 155.

39 Winsome Michelmore, *Thursday's Child*, pp. 97–8.

40 Entry for 'B. L. Montgomery', *Dictionary of National Biography 1971–1980*.

24. The Desert Fox Reborn

1 Reuth, *Rommel*, p. 137.

2 Norman J. W. Goda, 'Black Marks: Hitler's Bribery of His Senior Officers during World War II', article in *The Journal of Modern History*, No. 72 (June 2000), pp. 413–52.

3 Ibid. This was in addition to the taxed basic salary of a *Generalfeldmarschall* (of RM 26,500) or a *Generaloberst* of RM 24,000, on top of which tax-free wartime supplements were paid for housing, food, clothing and medical care, which could amount to another RM 10,000. These bribes applied to the senior ranks of the *Kriegsmarine* and Luftwaffe also, and followed a practice whereby senior civilian officials had been receiving secret monthly payments since 1936.

4 Ibid. Some generals such as Keitel, von Leeb and Guderian received country estates, and many benefited from a tax-free gift of RM 250,000 (effectively ten years' salary) on a landmark birthday – fiftieth, sixtieth, sixty-fifth. Goda stated that Frau Rommel 'made the distinction after the war between the illegitimacy of the larger gifts . . . and the [supposed legitimacy of the] monthly gifts that all field marshals received'.

5 Blumentritt, *Von Rundstedt*, pp. 208–09.

6 Bright, bespectacled and crafty, Speidel, like Rommel, was a Württemberger, born in Metzingen, but six years younger. His father, Dr Emil Speidel, was a senior forestry official, not a soldier at all, and it was the First World War that intervened and put young Hans on a military career path. Enrolled as a *Fahnenjunker* (officer cadet) in November 1914, he would spend most of the next forty-nine years in military uniform, serving three German armies. He was commissioned in 1915 into the Württemberg infantry and fought in Flanders, on the Somme and at Cambrai, which won him both classes of Iron Cross and the Württemberg Service Medal in gold. As we have seen, amongst the many fellow officers he ran into during 1915 was a Leutnant Rommel, in the Argonne Forest. After the war, Speidel was retained in the 100,000-man *Reichswehr*, but his father's scholarly influence ensured that he continued his academic studies which had been cut short in 1914. In 1923–4 he studied history and economics at the Universities of Tübingen, Berlin and at the Technische Hochschule, Stuttgart, and was one of the few serving interwar German officers to gain a PhD in February 1925. Unlike their British counterparts, it was by no means uncommon for German officers

to take university degrees (though doctorates were unusual); by 1932 he was *Hauptmann* Dr Speidel. He attended the two-year General Staff course over 1930–33 and was posted in the early months of the Third Reich (October 1933) to Paris, as assistant military attaché. Three years later, he was back in Berlin as Director of Foreign Armies West at OKW; 1937 saw *Oberstleutnant* Dr Speidel assigned as Chief of Staff to the 33rd Infantry Division, with whom he fought in the 1940 French campaign; with the obvious asset of his fluency in French, he was soon appointed Chief of Staff of the Military Governor in Paris and promoted *Oberst*. In 1942 Speidel was sent to the Eastern Front, where he served as Chief of Staff of V Corps, then of Army Group South in 1943, by which time he had reached the rank of *Generalmajor*. He was promoted *Generalleutnant* in January 1944 and on 15 April, Speidel, aged forty-seven, was appointed to Army Group B under Rommel; they had also served together in the same regiment briefly in the interwar period. By this time, Speidel had been awarded the German Cross in Gold and wore at his neck the coveted Knight's Cross, awarded for his service in Russia and advertising his special qualities as an extremely able and trustworthy Chief of Staff.

7   See Searle, *Wehrmacht Generals, West German Society and the Debate on Rearmament, 1949–59*.

8   See Irving, *Trail of the Fox*, pp. 386–405.

9   IWM, BLM 1/91.

10   'Friend or Foe?', article in *Time Magazine*, 11 May 1953.

11   Ibid.

12   Marshall, *Discovering the Rommel Murder*, pp. 183–201.

13   'Armoured Knight', *Time Magazine*, 22 January 1951.

14   Churchill, *The Hinge of Fate*, p. 59.

15   Andrew Roberts, *Masters and Commanders*, p. 264.

16   Patrick Major, 'Our Friend Rommel: The *Wehrmacht* as "Worthy Enemy" in Postwar British Popular Culture', article in *German History*, Vol. 26/4, 2008, pp. 529.

17   Lockenour, *Soldiers as Citizens*, p. 59. General Ludwig Crüwell, the DAK veterans' first chairman, stated their vision was to 'build bridges with our former enemies and to see that the old wounds slowly form scars'. Both groups (Eighth Army and DAK) wielded huge influence within their respective nations' veterans lobby because they were nationally, rather than regionally, recruited. The Eighth Army Veterans Association, which held a big rally in London's Albert Hall on the anniversary of El Alamein each year, and the Deutsches Afrika Korps veterans, through reunions and their newsletter *Die Oase* (the *Oasis*), promoted this vision. The DAK veterans even sponsored a soccer match against a representative unit of the Eighth Army during its 1953 annual meeting in Hanover. Frau Rommel was very closely associated with the DAK's meetings, featuring in nearly every photo, brochure or publication. Just as the British would reference their desert campaign to Lawrence of Arabia, so the Afrika Korps would build on the reputation of an earlier DAK, led by General Paul von Lettow-Vorbeck, who fought the British in East Africa throughout 1914–18, and at reunions adopted their battle cry *Heia Safari!* The phrase was a Swahili war cry, meaning 'Let's go get 'em', or 'Tallyho!'

18   *Manchester Guardian*, 25 January 1950, and see Patrick Major, 'Our Friend Rommel', p. 522.

19   Major, 'Our Friend Rommel', p. 521. Mr Zanuck actually knew the campaign, for as Colonel Zanuck he had served in Tunisia as a film-maker with the US Signal Corps, making the 1943 War Department movie *At The Front* and narrating his adventures in the book *Tunis Expedition*.

20   Von Mellenthin, *Panzer Battles*, fn, p. 54.

21   Patrick Major, 'Rehabilitating the *Wehrmacht*: The Desert Fox and West German Rearmament', p. 6.

22   Directly attributable to the influence of the film *The Desert Fox*, the formula of the 'good German' was ventured again in 1956 with the *Battle of the River Plate*, where the (real-life) Captain Langsdorff of the German battleship *Graf Spee* was revealed as a kindly anti-Nazi, and in the 1957 movie *The One That Got Away*, where Hardy Kruger played the (real-life) young Luftwaffe fighter pilot Franz von Werra escaping from a British POW camp. In both these films the central character was a German officer with whom the audience was encouraged to side. The success of sympathetically portrayed Germans also encouraged the inclusion of the Anthony Quayle character in the 1958 film *Ice Cold In Alex*; James Mason had pioneered the type – courting admiration and some hostility – but by the time of the 1969 movie *The Bridge at Remagen*, also portraying real events, it was no longer controversial for Robert Vaughan to portray an anti-Nazi German officer trying to do his duty, who is eventually shot by the SS. The Anglo-German commonality of being at war with the desert as well as each other was the central theme of the fictional *Ice Cold in Alex*; in the film Quayle played a German disguised as a South African, but who was key to helping John Mills and his party escape to Alexandria after the 1942 fall of Tobruk. Along the way he saved their lives, and they returned the favour and prevented his execution as a spy. As we have noted before, the Italians had by then already fallen from public memory as even present, let alone combatants: for Germany's Italian allies there was no real hatred, but rather an affectionate and amused contempt.

23   Patrick Major has noted that there was a wide discrepancy between the original US version and the diluted message of the German-language edition, which attempted to play down the impression that the Third Reich might have won with generals like Rommel, were it not for the bungling of Hitler.

24   The desert campaign also inspired the supremely enthralling *Ice Cold in Alex* of 1958, which starred John Mills leading an escape from Tobruk in an ambulance. The equally unmemorable *Tobruk*, a 1967 epic with Rock Hudson, though having only a passing acquaintanceship with truth, nevertheless portrayed many aspects of desert war; as did the distinctly mediocre *Raid on Rommel* of 1971, starring Richard Burton again, which blatantly reused many of the sequences from *Tobruk*. During 1966–8, a total of fifty-eight episodes of the US-made *Rat Patrol* were aired on television, with story lines based on the LRDG's wartime adventures in North Africa. At the time, the series offended many English audiences because of the dominance of American characters, who took no part in this aspect of the

North African campaign; in consequence, the series was pulled from British television. Back on the big screen, in 1970 the hugely successful *Patton*, with George C. Scott in the title role (which won him an Oscar in 1971), had told the story of America's outstanding armour commander, with much of the film set in North Africa; whilst the 1973 TV movie *Death Race* set a crippled Allied P-40 fighter crawling across the North African desert pursued by a panzer commanded by Lloyd Bridges playing a German general. In 1981 the Afrika Korps also formed the opposition team in the factually irrelevant but highly entertaining Spielberg-directed *Raiders of the Lost Ark* (the first of the Indiana Jones trio of moves), and much of *The English Patient* of 1996 wove a fictional narrative around the real-life Hungarian noble Laszlo de Almásy, who was – in reality – Rommel's desert adviser in the war. Other movies set in the North African desert war included *Five Graves to Cairo* (1943), *The Fighting Rats of Tobruk* (1945), the disappointing *Play Dirty* of 1969, and *The Key to Rebecca* (1985). Even during the war, the exotic setting of North Africa had never been far from the cinema-going public's mind (anywhere far away to take one's mind off the Blitz), with Humphrey Bogart swanning around the palm trees in *Casablanca* (1942) and *Sahara* of 1943. *Sahara* coincided with the release of the Ealing Studios' *Nine Men*, essentially an updated *Beau Geste*, where – for once – the British (including a young Gordon Jackson) were fighting the Italians; and in the same year, Henry Fonda starred in *Immortal Sergeant*, as a young Canadian who enlists and finds himself leading a patrol behind enemy lines in Libya after his sergeant is killed.

25 Manfred Rommel, Introduction to *The Rommel Papers*, pp. xxiii–xxiv.

26 In 1948, Liddell Hart himself had published *The Other Side of the Hill*, based on his interviews with senior German generals; it sold well on both sides of the Atlantic and was known in the USA as *The German Generals Talk*; it was translated into German in 1949 as *Jetzt düren sir reden* ('Now They Are Allowed to Speak') – and was extremely popularly received. In 1951 he brought out an expanded edition with fresh material, including a chapter 'The Soldier in the Sun', whom of course he had not interviewed. John Mearsheimer has, however, shown conclusively that Basil Liddell Hart contrived to demonstrate that Rommel had been a pre-war pupil of his (when there was no evidence of this) by writing explanatory footnotes for later editions of Desmond Young's biography and for Fritz Bayerlein to insert in *The Rommel Papers*. He was seeking to raise his profile and had already successfully induced Heinz Guderian to include favourable references that effectively exaggerated his influence on German tactical thinking before the Second World War, which would appear in the 1952 English-language edition of Guderian's *Panzer Leader*.

27 Alaric Searle, *Wehrmacht Generals, West German Society and the Debate on Rearmament, 1949–1959*, especially Chapter 6, pp. 185–230.

28 Other commanders' memoirs which also appeared at this time include General Siegfried Westphal's 1950 volume *Heer in Fesseln* ('Army in Chains'), translated as *The German Army in the West* in 1951); Heinz Guderian's *Erinnerungen eines Soldaten* ('Memories of a Soldier') in 1950, translated into English as *Panzer Leader* in 1952; General Günther Blumentritt's biography of his wartime boss,

Von Rundstedt, the Soldier and the Man, 1952; Field Marshal Albert Kesselring's *Soldat bis zum letzten Tag* ('A Soldier to the Last Day') in 1953, (published in English as *A Soldier's Record* in 1954); Major General F. W. Mellenthin's *Panzer Battles: A Study of the Employment of Armour in the Second World War* (1956); and Field Marshal Erich von Manstein's *Verlorebe Siege* in 1955 (published as *Lost Victories* in English in 1958).

29 In 2006 Wolfram Wette charted the construction of this particular myth in his study *The Wehrmacht: History, Myth, Reality*, whilst Christopher Browning in 1998 examined the way 101st Reserve Police Battalion (i.e. non-SS) murdered Jews in Poland, in *Ordinary Men*; Philip Blood in *Hitler's Bandit Hunters* assessed the extraordinary brutality of German anti-partisan operations conducted by army, airforce and even naval personnel, concluding that it was easy for ordinary citizens to get carried away doing what they perceived to be their patriotic duty – and suggested that such events might not be confined to history in the post-11 September 2001 world.

30 Whilst Irving was later discredited as a result of the Deborah Lipstadt libel trial of 2000, which concluded that he was guilty of Holocaust denial in distorting the historical record of the Second World War, this should not discount his earlier scholarship, which was profound.

31 Blumentritt, *Von Rundstedt*, pp. 209–10.

Afterword

1   Marshall, *Discovering the Rommel Murder*, p. 6.

2   Monty thought that Wavell had 'failed' on two scores: because he was 'not sufficiently ruthless with those below him whom he reckoned were not fit for their jobs', and because he 'never seemed able to establish good relations with his political masters'. (Montgomery, *The Path to Leadership*, p. 45.) Whilst Monty may certainly have remedied the former issue in his own career, he was a far worse offender than ever Wavell was in terms of relations with political masters. I chuckled when I first read this for it seems beyond comprehension that Monty could have written that sentence without realising his own extensive faults.

3   Lewin, *Rommel As Military Commander*, p. 238.

4   Heinz Werner Schmidt, *With Rommel in the Desert*, p. 69.

5   General (later Field Marshal) Sir Archibald Wavell, *Generals and Generalship*, p. 17.

6   This teetotalism even continued when Monty was visited by his monarch, King George VI. In his October 1944 visit to Monty's TAC, a royal equerry was obliged to ask Freddie de Guingand for a bottle of whisky, as the King was longing for a drink. 'Freddie was delighted to hand over a bottle of the hard stuff' and assured the King there would be wine at dinner. Johnny Henderson, *Watching Monty*, p. 126.

7   In terms of age, Monty was fixated by the need for youth on the battlefield and equated youth with mental robustness, having seen many elderly commanders

fail in 1940; he ruthlessly sacked those he thought were too old, unfit or found wanting in other ways; Rommel did so too in North Africa. For political reasons, Rommel was unable to replace the poor commanders he found in Army Group B during 1944, such as Dollmann or Feuchtinger; von Rundstedt, as his superior or contemporary (the ambiguity was never satisfactorily settled) at OB West, was far too old a field commander at sixty-eight.

8   Wavell, *Generals and Generalship*, p. 22.

9   Kesselring and the SS commander Sepp Dietrich claimed in later years that Rommel was inclined to depression when faced with setbacks.

10 Wavell, *Generals and Generalship*, p. 62.

11 Ibid., p. 19.

12 Westphal, *The German Army in the West*, pp. 126–7.

13 Wavell, *Generals and Generalship*, pp. 23–4.

14 For example, Brian Urquhart *in A Life in Peace and War*, pp. 69–72.

15 Rommel fought a campaign, initially against orders, that was never logistically sustainable. In achieving success against his opponents he came to believe in his own genius, but was always going to risk failure the moment the Allies outweighed him logistically. This happened under Montgomery, but might equally have occurred under Wavell or Auchinleck.

16 This also implies a comprehension of timescales – time spent in planning, logistic preparation and fighting sequential battles. A previous commander of British forces, the Duke of Wellington, understood this well when fighting the French in the Peninsular Campaign; he was obliged to plan all his battles at least six months in advance because of the time lapse between requesting resources or men from England, and receiving them. In the Western Desert, whose campaign was Rommel fighting? Initially his job was to stabilise the military situation of Hitler's Italian allies in Libya – a tactical mission; but Rommel immediately commenced an operational-level campaign (without telling Berlin) aimed ultimately at reaching the Suez Canal and Middle Eastern oilfields. Unaware of the impending invasion of Russia, accordingly he demanded operational-level logistics for a tactical mission and was surprised when turned down by Berlin: in their eyes, Rommel was advocating a disproportionate use of resources, and his fighting spirit got the better of him.

17 Much of Montgomery's achievement we can now attribute to the intelligence gained from cracking German signals sent by Enigma enciphering machines and deciphered at Bletchley Park. Neither was averse to deploying every capability at their disposal, from the kinetic end of the spectrum (war-fighting) to the 'soft' – in their use of deception, the media and psychological operations, towards minimising casualties and achieving victory. As with all successful commanders, their approaches echoed the teachings of the ancient Chinese military philosopher Sun Tzu, who wrote, 'he who wins without fighting demonstrates the epitome of skill'.

18 Wavell, *Generals and Generalship* , second lecture, 'The General and his Troops'.

19 Mellenthin was Rommel's Chief Intelligence Staff Officer and then acting Chief of Operations Staff from June 1941–September 1942.

20 Mellenthin, *Panzer Battles*, p. 55.

21 Montgomery, *Memoirs*, p. 92.

22 Mellenthin, *Panzer Battles*, pp. 54–5.

23 Wavell's vision was more in keeping with a 'band of brothers', a command team almost of equals, who knew and could anticipate each other's wishes and actions. This is how Nelson commanded his battle fleets, and how some German formations who served together for extended periods on the Eastern Front came to a high pitch of efficiency: it is the fusion of sheer professionalism with character. This was not Monty's style: to him a subordinate was a lesser mortal.

24 Wavell, *Generals and Generalship*, p. 30.

25 Henderson, *Watching Monty*, p. 105.

26 The *Flugzeugführer und Beobachterabzeichen*, an equivalent of a set of pilot's wings and visible on many portrait photographs of him.

27 In North Africa, this was achieved through a *Fliegerführer Afrika*, surbordinated to *Luftflotte* 2 in Italy, commanded by Albert Kesselring until June 1943. Luftwaffe air sorties should have been in support of Rommel's land forces, but were poorly directed; their targets should have been British bombers and airfields; instead, of 653 reported Allied aircraft losses between 10 February and 23 October 1942, only fifty-four were twin- or four-engined bombers and the remainder fighters. (It is an interesting reflection on the Luftwaffe and *Regia Aeronautica Italiana* that their claims over the same period totalled 1,200, almost double.) The Messerschmitt 109 aces – Hans-Joachim Marseille and his chums – were content to shoot down Hurricanes and Kittyhawks by the score, yet it was the RAF's bombers that were harming Rommel, sinking his ships in harbour and attacking his supply columns. See Christopher Shores and Hans Ring, *Fighters Over the Desert*.

28 In contrast to Coningham or Broadhurst living cheek-by-jowl with Monty like Gypsies in a circle of caravans, the monocled Sperrle emulated his boss, Hermann Goering, in enjoying a notoriously sybaritic lifestyle in the Palais du Luxembourg, Paris; he was dismissed (one feels, partly for this) on 23 August 1944 and never re-employed.

29 Wavell, *Generals and Generalship*, p. 39.

30 See Mark Harrison, *Medicine and Victory: British Military Medicine in the Second World War*.

31 Von Mellenthin, *Panzer Battles*, p. 54.

32 Schmidt, *With Rommel in the Desert*, p. 70.

33 Westphal. *The German Army in the West*, pp. 127–8.

34 Bill Deedes, *Daily Telegraph*, 30 June 2004.

35 Brian Urquhart, *A Life in Peace and War*, p. 43.

36 Wavell, *Generals and Generalship*, p. 42.

37 Cited in Hamilton, *Monty, Volume 3*, p. 590.

38 The volume concerned is Nigel Hamilton's *The Full Monty* of 2001; its central theme is an analysis of Monty's 'confused sexuality' and that his emotional interest after he was widowed in 1937 was exclusively in men, an argument that failed to convince many reviewers.

39   Letters to Lucie, signed Schraepler, 27 May 1940 and 22 April 1941, *The Rommel Papers*, (ed. Liddell Hart), pp. 39 and 131. See also Hans-Albrecht Schraepler (ed.), *At Rommel's Side: The Lost Letters of Hans-Joachim Schraepler* (2009).

40   David Irving, *The Trail of the Fox*, pp. 54–5.

41   *The Illustrated London News*, 18 August 1945. There is surely a psychological angle to acquiring something once owned by your rival. Andrew Roberts has observed how the Duke of Wellington after Waterloo not only lived in Napoleon's house, but cavorted with at least one of his mistresses.

42   Liddell Hart (ed.), *The Rommel Papers*, p. 462.

43   Henderson, *Watching Monty*, pp. 102–03

44   A. P. Herbert, *Independent Member*, p. 321.

45   Wavell, *Generals and Generalship*, pp. 53–4.

46   Generals George B. McClellan, Henry W. Halleck, William S. Rosecrans, Ambrose Burnside, Joseph Hooker, George G. Meade.

47   Bill Deedes, *Daily Telegraph*, 30 June 2004.

48   Monty was indeed known as 'the Master' at TAC.

Bibliography

Archives

British Library Newspaper Collection, Colindale, London NW9

Camberley/JSCSC Archives, Watchfield: Staff Ride Collection and other papers

Churchill Archives Centre, Cambridge: papers of Ronald Lewin, GBR/0014/RLEW

Hauptstaatsarchiv, Konrad-Adenauer Str., Stuttgart: Württemberg military records

Haus der Geschichte, Baden-Württemberg: records relating to 2008–09 Rommel exhibition

Imperial War Museum, London: BL Montgomery Archive and other papers

Liddell Hart Archive, King's College, London: Alexander, Liddell Hart and other correspondence

The National Archives, Kew: War Diaries (WO series) and Cabinet Office (CAB series) papers

National Army Museum, London: Photograph collection of Lt-Col Brian Montgomery, NAM 2000–06–19 series

Naval Historical Center, Washington DC: Commentaries on Normandy campaign, Department of the US Navy

Rommel Archiv, Herrlingen

Royal Regiment of Fusiliers (Royal Warwickshire Regiment) Museum, Warwick

US National Archives: National Archives and Records Administration (NARA)

Unpublished Sources

Ben-Reuven, Colonel Eyal, *Command and Control in North African Battles, World War II, 1941–1943* (MA Thesis, US Army War College, Carlisle, Pennsylvania, 1989)

Cockerham, Colonel William C., *Clausewitz's Concept of CPV* [Culminating Point of Victory] *in the North African Campaigns of Rommel and Montgomery* (MA Thesis, US Army War College, Carlisle, Pennsylvania, 1987)

Davenport, Major Brian W., *Operational Sustainment, Defining the Realm of the Possible* (MA Thesis, US Army Command and General Staff College, Fort Leavenworth, 1986)

Everett, Major Michael W., *Tactical Generalship, A View From the Past and a Look Toward the 21st Century* (MA Thesis, US Army Command and General Staff College, Fort Leavenworth, 1986)

Gibson, Commander Charles M., *Operational Leadership as Practised by Field Marshal Erwin Rommel During the German Campaign in North Africa, 1941–1942, Success or Failure?* (MA Thesis, US Naval War College, Newport, RI, 2001)

Kirkland, Major Donald E., *Rommel's Desert Campaigns, February 1941–September 1942, A Study in Operational Level Weakness* (MA Thesis, US Army Command and General Staff College, Fort Leavenworth, 1986)

Kitchin, Major M. J., Australian Army, *Field Marshal B.L. Montgomery, Twenty-First Century Leadership in the Industrial Era* (Research Paper)

Knightly, Major William S., *Campaigning in the Secondary Theater, Challenges for the Operational Commander* (MA Thesis, US Army Command and General Staff College, Fort Leavenworth, 1987)

LeFace, Major Geoffrey L., *Tactical Victory Leading to Operational Failure, Rommel in North Africa* (MA Thesis, US Command and General Staff College, Fort Leavenworth, 2001)

Machin, Commander Mark A., *Rommel, Operational Art and the Battle of El Alamein* (MA Thesis, Naval War College, Newport, R.I., 1994)

McFetridge, Charles D., *In Pursuit, Montgomery after El Alamein* (MA Thesis, US Army War College, Carlisle, Pennsylvania, 1991)

McMahon, Major T. L., *Operational Principles, The Operational Art of Erwin Rommel and Bernard Montgomery* (MA Thesis, US Army Command and General Staff College, Fort Leavenworth, 1985)

Major, Patrick, *Rehabilitating the Wehrmacht, The Desert Fox and West German Rearmament* (Research Paper, University of Reading)

Meisner, Lieutenant Colonel Bruce L., *The Culminating Point – A Viable Operational Concept or Some Theoretical Nonsense?* (MA Thesis, US Army Command and General Staff College, Fort Leavenworth, 1986)

Price, Major Michael J., USAF, *Field Marshal Erwin Rommel, A Great Captain* (MA Thesis, Air Command and Staff Course, Air University, Maxwell Air Force Base, Alabama, 2006)

Ritter, Lieutenant Colonel George P., *Leadership of the Operational Commander, Combat Multiplier or Myth?* (MA Thesis, US Army Command and General Staff College, Fort Leavenworth, 1993)

Schreiber, Major Paul K., USMC, *Rommel's Desert War, the Impact of Logistics on Operational Art* (MA Thesis, US Naval War College, Newport, RI, 1998)

Starbuck, Colonel F. Randall, *Air Power in North Africa, 1942–43, An Additional Perspective* (MA Thesis, US Army War College, Carlisle, Pennsylvania, 1992)

Tillson, Lieutenant Commander Patrick A., *Operational Planning and Logistical Sustainment, The Axis Experience in Two North African Offensives* (MA Thesis, Naval War College, Newport, R.I., 1995)

Tosch, Major David F., *German Operations in North Africa, A Case Study of the Link Between Operational Design and Sustainment* (MA Thesis, US Army Command and General Staff College, Fort Leavenworth, 1987)

Waddington, Major CC, MBE, PARA, *Rommel and the Art of Manoeuvre Warfare* (MA Thesis, Cranfield University, July 1995)

Wiegert, Lieutenant Commander Robert N., *An Analysis of the Operational Leadership of Field Marshal Erwin Rommel in the Afrika Korps* (MA Thesis, US Naval War College, Newport, R.I., 1997)

Williams, Lieutenant Commander Ethan R., *50 Div in Normandy, A Critical Analysis of the British 50th (Northumbrian) Division on D-Day and in the Battle of Normandy* (MA Thesis, US Naval Academy, Annapolis, Maryland, 1997)

Willmuth, Major Thomas J., *The Study of Military History Through Commercial War Games, A Look at Operation Crusader with the Operational Art of War* (MA Thesis, US Army Command and General Staff College, Fort Leavenworth, 2001)

Wolff, Major Terry A., *The Operational Commander and Dealing with Uncertainty* (MA Thesis, US Army Command and General Staff College, Fort Leavenworth, 1991)

Published Sources

Abenheim, Donald, *Reforging the Iron Cross, The Search for Tradition in the West German Armed Forces* (Princeton University Press, 1988)

Addison, Paul and Calder, Angus (eds), *Time To Kill, The Soldier's Experience of War in the West, 1939–45* (Pimlico, 1997)

Addington, Larry H., 'Operation Sunflower, Rommel Versus the General Staff', *Military Affairs*, Vol. 31/3 (Autumn 1967), pp. 120–30

Aldrich, Richard J., *Witness to War: Diaries of the Second World War in Europe and the Middle East* (Doubleday, 2004)

Alexander, Martin, 'After Dunkirk, The French Army's Performance against "Case Red", 25 May to 25 June 1940', *War in History*, Vol. 14/2 (April 2007)

Allinson, Sidney, *The Bantams* (Howard Baker, 1981)

Almasy, Laszlo and Kubassek, Janos, *With Rommel's Army in Libya* (Author House, 2001)

Ambrose, Stephen E., *The Supreme Commander, the War Years of General Dwight D. Eisenhower* (Doubleday, 1970)

——'Eisenhower, the Intelligence Community and the D-Day Invasion', *Wisconsin Magazine of History*, Vol. 64/4 (Summer 1981)

——*D-Day, 6 June 1944 – The Climactic Battle of World War II* (Simon & Schuster, 1994)

American Battlefield Monuments Commission, *American Armies and Battlefields in Europe* (US Government Printing Office, 1938)

Arsenian, Seth, 'Wartime Propaganda in the Middle East', *Middle East Journal*, Vol. 2/4 (October 1948)

Ashby, John, *Seek Glory, Now Keep Glory, The Story of the 1st Battalion, Royal Warwickshire Regiment 1914–1918* (Helion & Company, 2000)

Atkinson, Rick, *An Army at Dawn, the War in North Africa, 1942–1943* (Henry Holt, 2002)

—— 'Speech to the US Commission on Military History, Washington, 1 November 2003', *Journal of Military History*, Vol. 68/2 (April 2004), pp. 527–33

—— *The Day of Battle, The War in Sicily and Italy, 1943–1944* (Henry Holt, 2007)

Badsey, Stephen, *Normandy 1944*, (Osprey, 1990)

Badsey, Stephen and Bean, Tim, *Battle Zone Normandy, Omaha Beach* (Sutton, 2004)

Bahnemann, Günther, *I Deserted Rommel* (Jarrolds, 1961)

Bagnold, Brigadier R. A., 'Early Days of the Long Range Desert Group', *The Geographical Journal*, Vol. 107, 1945

Barbusse, Henri, *Under Fire* (French edition, 1916, Everyman English edition, 1926)

Barker, Christine R. and Last, R. W., *Erich Maria Remarque, Author of All Quiet on the Western Front* (Oswald Wolff, 1979)

Barker, James, 'Monty and the Mandate', *History Today* (March 2009), pp. 30–34

Barr, Niall, *Pendulum of War, The Three Battles of El Alamein* (Jonathan Cape, 2004)

Barnett, Correlli, *The Desert Generals* (William Kimber, 1960)

—— 'The Education of Military Elites', *Journal of Contemporary History*, Vol. 2/3 (July 1967)

—— (ed.), *Hitler's Generals, Authoritative Portraits of the Men Who Waged Hitler's War* (Weidenfeld and Nicolson, 1989)

Bartlett, Captain Sir Basil, *My First War, An Army Officer's Journal for May 1940 Through Belgium to Dunkirk* (Chatto and Windus, 1940)

Bassford, Christopher, *Clausewitz in English, The Reception of Clausewitz in Britain and America, 1815–1945* (Oxford University Press, 1994)

Battistelli, Pier Paolo, *Rommel's Afrika Korps, Tobruk to El Alamein* (Osprey, 2006)

—— *Panzer Divisions, The Blitzkrieg Years 1939–40* (Osprey, 2007)

The Battle of Egypt, the Official Record in Pictures and Maps (Government Information Booklet, HMSO, 1943)

Battlefield Tour 1st Day – Op. GOODWOOD, (BAOR, 1947/ MLRS, 2005)

Baxter, Colin F., (comp.), *The Normandy Campaign, 1944, A Selected Bibliography* (Greenwood Press, 1992)

—— 'Did Nazis Fight Better Than Democrats? Historical Writing on the Combat Performance of the Allied Soldier in Normandy', *Parameters* (Autumn 1995)

—— *Field Marshal Bernard Law Montgomery 1887–1976, A Selected Bibliography* (Greenwood Press, 1999)

Bayerlein, Fritz, 'El Alamein', chapter in Cyril Falls, *Fatal Decisions* (Michael Joseph, 1956)

Baynes, Sir John, *Morale, A Study of Men and Courage* (Cassell, 1967)

—— *Forgotten Victor, General Sir Richard O'Connor* (Brassey's, 1989)

Beaumont, Roger, 'The Bomber Offensive as a Second Front', *Journal of Contemporary History*, Vol. 22/1 (January, 1987), pp. 3–19

Bechthold, B. Michael, 'A Question of Success, Tactical Air Doctrine and Practice in North Africa, 1942–43', *Journal of Military History*, Vol. 68/3 (July 2004), pp. 821–51

Beckett, Ian F. W., and Simpson, Keith (eds), *A Nation in Arms: A Social Study of the British Army in the First World War* (Tom Donovan, 1990)

———(Beckett), *The Great War, 1914–18* (Longman, 2001)

———(Beckett), *Ypres: The First Battle, 1914* (Longman, 2004)

Beevor, Antony, *D-Day. The Battle for Normandy* (Viking, 2009)

Begg, Dr Richard Campbell and Liddle, Dr Peter, *For Five Shillings A Day. Personal Histories of World War II* (HarperCollins, 2000)

Behrendt, Hans-Otto, *Rommel's Intelligence in the Desert Campaign* (William Kimber, 1985)

Belchem, Major-General David, *Victory in Normandy* (Chatto & Windus, 1981)

Belfield, Eversley and Essame, Hubert, *The Battle for Normandy* (BT Batsford, 1965)

Bellamy, Ronald F. and Llewellyn, Craig H., 'Preventable Casualties, Rommel's Flaw, Slim's Edge', [US] *Army Magazine*, Vol. 40 (May 1990)

Below, Nicolaus von, *At Hitler's Side. The Memoirs of Hitler's Luftwaffe Adjutant 1937–1945* (Greenhill, 2001)

Bennett, Ralph, *Ultra in the West, The Normandy Campaign of 1944–45* (Hutchinson, 1979)

———*Ultra and Mediterranean Strategy, 1941–45* (Hamish Hamilton, 1989)

———(ed.), *Intelligence Investigations, How Ultra Changed History* (Cass, 1996)

Bergot, Erwan, *The Afrika Korps* (Allan Wingate, 1976)

Bessel, Richard, *Germany After the First World War* (Oxford University Press, 1993)

———*Germany 1945: From War to Peace* (Simon & Schuster, 2009)

Bierman, John and Smith, Colin, *The Battle of Alamein, Turning Point of World War II* (Viking, 2002)

Biernacik, Stan Z., 'The D-Day Invasion of Normandy, Poles Played A Crucial Role in Bringing Nazi Germany to Its Knees', *Polish American Journal*, June 1994

De la Billière, Major James, 'The Political–Military Interface, Friction in the Conduct of British Army Operations in North Africa 1940–1942', *Defence Studies*, Vol. 5/2 (June 2005)

Bird, Antony, *Gentlemen, We Will Stand and Fight, Le Cateau 1914* (Crowood Press, 2008)

Black, Jeremy (ed.), *European Warfare 1815–2000* (Palgrave/Macmillan, 2002)

Blaxland, Gregory, *Amiens, 1918* (Frederick Muller, 1968)

——— *Destination Dunkirk* (William Kimber, 1973)

——— *The Plain Cook and the Great Showman. The First and Eighth Armies in North Africa* (William Kimber, 1977)

——— *Alexander's Generals, The Italian Campaign, 1944–45* (William Kimber, 1979)

Bloch, Marc, *Strange Defeat, a Statement of Evidence Written in 1940* (Oxford University Press, 1949)

Bloem, Walter, *The Advance From Mons* (Peter Davies, 1930)

Blood, Philip W., *Hitler's Bandit Hunters: The S.S. and the Nazi Occupation of Europe* (Potomac Books, 2006)

Blumenson, Martin, *Rommel's Last Victory. The Battle of Kasserine Pass* (Allen and Unwin, 1969)

——— *The Battle of the Generals: The Untold Story of the Falaise Pocket – the Campaign That Should Have Won World War II* (William Morrow, 1993)

Blumentritt, Günther, *Von Rundstedt, The Soldier and the Man* (Odhams Press, 1952)

Bond, Brian, (ed.), *Chief of Staff, The Diaries of Lt-Gen Sir Henry Pownall*, Volume One, 1933–40 (Leo Cooper, 1972)

—— *France and Belgium 1940* (Clarendon Press, 1979)

—— *British Military Policy Between The Two World Wars* (Clarendon Press, 1980)

——'Judgement in Military History', *RUSI Journal*, Vol. 134/1 (Spring 1989), pp. 69–72

Bond, Brian and Cave, Nigel, *Haig, A Re-appraisal 70 Years On* (Pen and Sword, 2009)

Bond, Brian and Murray, Williamson, 'The British Armed Forces, 1918–39', in Millett, Allan R., and Murray, Williamson (eds.), *Military Effectiveness, Volume II, The Interwar Period* (Allen and Unwin, 1980)

Bond, Brian and Robbins, Simon (eds), *Staff Officer. The Diaries of Walter Guinness (First Lord Moyne) 1914–1918* (Leo Cooper, 1987)

Bond, Brian and Taylor, Michael (eds), *The Battle of France & Flanders Sixty Years On* (Leo Cooper, 2001)

Böselager, Philipp von, *Valkyrie, The Plot To Kill Hitler* (Weidenfeld and Nicolson, 2009)

Bourke, Joanna, 'Remembering War', *Journal of Contemporary History*, Vol. 39/4 (October 2004)

Bradford, George and Morgan, Len, *50 Famous Tanks* (Ian Allan, 1968)

Bradford, George R., *Rommel's Africa Korps, El Agheila to El Alamein* (Stackpole, 2008)

Bradley, Dermot, 'The *Bundeswehr* and German Unification 1955–91', *Irish Studies in International Affairs*, Vol. 3/4 (1992), pp. 53–66

Bradley, General Omar N., *A Soldier's Story, Tunis to the Elbe* (Eyre and Spotteswoode, 1951)

Bradley, General Omar N., and Blair, Clay, *A General's Life, An Autobiography by General of the Army Omar N. Bradley* (Simon & Schuster, 1983)

Brett-James, Antony, *Conversations With Montgomery* (William Kimber, 1984)

Brett-Smith, Richard, *Hitler's Generals* (Presidio, 1977)

Brighton, Terry, *Masters of Battle, Monty, Patton and Rommel at War* (Viking, 2008)

Brooks, Stephen, 'Montgomery and the Preparations for Overlord', *History Today*, Vol. 34/6 (June 1984)

——(ed.) *Montgomery and the Eighth Army* (Bodley Head, 1991)

Brown, Malcolm, *The Imperial War Museum Book of the Somme* (Sidgwick and Jackson, 1996)

——*Imperial War Museum Book of 1918, Year of Victory* (Sidgwick and Jackson Ltd, 1998)

——*Imperial War Museum Book of 1914, The Year of Lost Illusions* (Sidgwick and Jackson, 2004)

——*Imperial War Museum Book of the Western Front* (Macmillan, 2004)

Browning, Christopher, *Ordinary Men, Reserve Police Battalion 101 and the Final Solution in Poland* (Harper, 1998)

Bryant, Arthur, *Turn of the Tide 1939–1943: A Study Based on the Diaries and Autobiographical Notes of Field Marshal The Viscount Alanbrooke* (Collins, 1957)

Buckley, John, *British Armour in the Normandy Campaign 1944* (Frank Cass, 2004)

————*The Normandy Campaign 1944, Sixty Years On* (Frank Cass, 2006)

Buffetaut, Yves, *Rommel, France 1940* (Editions Heimdal, 1985)

Bungay, Stephen, *Alamein* (Aurum Press, 2002)

Burleigh, Michael, *The Third Reich, A New History* (Macmillan, 2000)

Butcher, Captain H. C., *My Three Years With Eisenhower* (Heinemann, 1946)

Caddick-Adams, Peter, *By God They Can Fight! A History of 143rd Infantry Brigade, 1908–1995* (Shrewsbury, 1995)

————*Lessons From Normandy, A Study of the 1944 Normandy Campaign* (Cranfield University Press, 2007)

————'General Sir Miles Christopher Dempsey (1896–1969) – Not a Popular Leader', *RUSI Journal*, Vol. 150/5 (October 2005)

Calder, Ritchie, *Men Against the Desert* (Country Book Club, 1952)

Campbell, John P., *Dieppe Revisited, A Documentary Investigation* (Frank Cass, 1993)

Capa, Robert, *Slightly Out of Focus* (Henry Holt, 1947)

Carafano, Lieutenant Colonel James J., *After D-Day, Cobra and the Normandy Breakout* (Lynne Rienner, 2000)

Carew, Tim, *The Vanished Army* (William Kimber, 1964)

Carrell, Paul (Paul Karl Schmidt), *The Foxes of the Desert* (Macdonald and Co., 1960; originally published as *Die Wüstenfüschse*)

————*Invasion – They're Coming,* (George Harrap, 1962/Corgi Paperback, 1974)

Carrington, Charles, *Soldier From The Wars Returning* (Hutchinson, 1965)

Carruthers, Bob and Trew, Simon, *The Normandy Battles* (Cassell, 2000)

————(Carruthers), *Servants of Evil* (André Deutsch/Carlton, 2001)

Carver, Field Marshal (Lord) Michael, *El Alamein* (BT Batsford, 1962)

————*Tobruk* (BT Batsford, 1964)

————*The War Lords. Military Commanders of the Twentieth Century* (Weidenfeld and Nicolson, 1976)

————*Harding of Petherton* (Weidenfeld and Nicholson, 1978)

————*Seven Ages of the British Army* (HarperCollins, 1984)

————*Dilemmas of the Desert War, A New Look at the Libyan Campaign 1940–1942* (BT Batsford, 1986)

————*Out of Step* (Hutchinson, 1989)

————'El Alamein to Desert Storm, Fifty Years from Desert to Desert', *RUSI Journal*, Vol. 137/3 (June, 1992), pp. 52–6

———— *Britain's Army in the Twentieth Century* (Macmillan, 1998)

———— *Imperial War Museum Book of the War in Italy, A Vital Contribution to Victory in Europe 1943–1945* (Sidgwick and Jackson, 2001)

Carver, Tom, *Where The Hell Have You Been?* (Short Books, 2009)

Cave, Nigel and Sheldon, Jack, *Le Cateau* (Pen and Sword Battleground, 2008)

Cecil, Hugh and Liddle, Peter, *Facing Armageddon, The First World War Experienced* (Pen and Sword, 1996)

Chalfont, Alun, *Montgomery of Alamein* (Weidenfeld and Nicolson, 1976)

Chandler, David G. and Collins, J. Lawton, *The D-Day Encyclopædia* (Simon & Schuster, 1994)

Chant, Christopher, Humble, Richard, Fowler, William and Shaw, Jenny, *Hitler's Generals and their Battles* (Purnell/Salamander Books, 1976)

Chasseaud, Peter, *Topography of Armageddon. A British Trench Map Atlas of the Western Front 1914–1918* (Mapbooks, 1991)

Churchill, Rt. Hon. Winston S., CH, MP, *The World Crisis, Vol. One, 1911–1914* (Thornton Butterworth, 1923)

———*Marlborough, His Life and Times*, 4 volumes (Odhams, 1933–38)

———*The Second World War*, 6 volumes (Cassell, 1948–1953)

———*A History of the English Speaking Peoples, Volume IV The Great Democracies* (Cassell, 1958)

Clarke, I. F., *Voices Prophesying War 1763–1984* (Oxford University Press, 1966)

Clayton, Tim and Craig, Phil, *Finest Hour* (Hodder and Stoughton, 1999)

———*The End of the Beginning* (Hodder and Stoughton, 2002)

Clemente, Steven E., *For King and Kaiser! The Making of a Prussian Army Officer 1869–1914* (Greenwood, 1985)

Coakley, Robert W. and Leighton, Richard M., *Global Logistics and Strategy 1944–5* (Office of the Chief of Military History of the US Army, Washington DC, 1968)

Collier, Richard, *Ten Thousand Eyes* (Collins Publishing, 1958)

——— *Fighting Words, The Correspondents of World War II* (St Martin's Press, 1989)

Collins, J. Lawton, *Lightning Joe, An Autobiography* (Baton Rouge, 1979)

Collins, Major-General R. J., *Lord Wavell, A Military Biography* (Hodder and Stoughton, 1947)

Colville, Sir John, *The Fringes of Power, Downing Street Diaries 1939–55* (Hodder and Stoughton, 1985)

Connell, John, *Wavell, Scholar and Soldier* (Collins, 1964)

The Conquest of North Africa. Prepared from Official Records and First-Hand Accounts (Burke Publishing, c.1944)

Coombes, Rose E. B., MBE, *Before Endeavours Fade, A Guide to Battlefields of the First World War* (After the Battle, 1976)

Cooper, John St John, *Invasion! The D-Day Story* (Daily Express/Beaverbrook Newspapers Limited 1954)

Cooper, Matthew and Lucas, James, *Panzer, The Armoured Force of The Third Reich* (Macdonald and Jane's, 1976)

Copp, Terry, *Fields of Fire, The Canadians in Normandy* (University of Toronto Press, 2004)

Corrigan, Gordon, *Mud, Blood and Poppycock, Britain and the First World War* (Cassell, 2003)

Cortesi, Lawrence, *Rommel's Last Stand* (Zebra Books, 1984)

Cowley, Robert, 'WWI, Massacre of the Innocents', *The Quarterly Journal of Military History*, Spring 1998

Crang, Jeremy A., *The British Army and the People's War 1939–1945* (Manchester University Press, 2000)

Creveld, Martin van, 'Rommel's Supply Problem', *RUSI Journal*, Vol. 119/3 (1974)

———*Supplying War: Logistics from Wallenstein to Patton* (Cambridge University Press, 1977)

——— 'Thoughts on Military History', *Journal of Contemporary History*, Vol. 18 (1983)

——— *Command in War* (Harvard University Press, 1985)

——— *The Training of Officers, From Military Professionalism to Irrelevance* (Free Press/Macmillan, 1990)

——— *The Art of War: War and Military Thought* (Cassell, 2000)

Crimp, R. L., (ed. Alex Bowlby), *The Diary of a Desert Rat* (Leo Cooper, 1971)

Crisp, Robert, *Brazen Chariots, An Account of Tank Warfare in the Western Desert, November–December 1941* (Norton, 2005)

Crozier, Brigadier General Frank Percy, *A Brass Hat in No Man's Land* (Jonathan Cape, 1930)

Cruickshank, C., *Deception in World War II* (Oxford University Press, 1979)

Cunningham of Hyndhope, Admiral of the Fleet the Viscount, *A Sailor's Odyssey* (Hutchinson, 1951)

Czarnecki, Commander, Lieutenant *Boisjos, Hill 262* (Mont-Ormel Memorial, 1995)

Daddow, Oliver J., 'Facing the Future, History in the Writing of British Military Doctrine', *Defence Studies*, Vol. 2/1 (Spring 2002)

Daglish, Ian, *Goodwood, Over the Battlefield* (Leo Cooper, 2005)

Dalton, Lieutenant E. H. J. H., *With British Guns in Italy, A Tribute to Italian Achievement* (Methuen, 1919)

Danchev, Alex, *Alchemist of War, The Life of Basil Liddell Hart* (Weidenfeld and Nicolson, 1998)

———and Todman, Dan (eds), *War Diaries 1939–45, FM Lord Alanbrooke* (Weidenfeld and Nicolson, 2001)

Dancocks, Daniel G., *The D-Day Dodgers, the Canadians in Italy, 1943–1945* (McClelland & Stewart, 1991)

David, Saul, *Churchill's Sacrifice of the Highland Division, France 1940* (Brassey's, 1994)

Davies, Frank and Maddocks, Graham, *Bloody Red Tabs: General Officer Casualties of the Great War, 1914–1918* (Leo Cooper, 1995)

Davson, Lieutenant Coloneo H. M, *The History of the 35th Division in the Great War* (Sifton Praed, 1926)

Dear, I. C. B and Foot, M. R. D., *The Oxford Companion to the Second World War* (Oxford University Press, 1995)

Deighton, Len, *Blitzkrieg, From the Rise of Hitler to the Fall of Dunkirk* (Jonathan Cape, 1979)

Delaney, John, *Fighting the Desert Fox, Rommel's Campaigns in North Africa, April 1941–August 1942* (Cassell, 1998)

Desquesnes, Remy, *Gold Beach, Asnelles 6th June 1944* (Edns. Ouest-France, 1990)

D'Este, Carlo, *Decision in Normandy* (William Collins, 1983)

———*Bitter Victory. The Battle for Sicily July–August 1943* (Collins, 1988)

———*A Genius for War, A Life of George S. Patton* (HarperCollins, 1995)

———*Eisenhower, Allied Supreme Commander* (Weidenfeld and Nicolson, 2002)

———*Warlord, Churchill at War, 1874–1945* (Allen Lane, 2009)

Dick, Charles, 'The Goodwood Concept – Situating the Appreciation', *RUSI Journal*, Vol. 127/1 (March 1982)

DiNardo, R. L. and Bay, Austin, 'Horse-Drawn Transport in the German Army', *Journal of Contemporary History*, Vol. 23/1 (January, 1988), pp. 129–42

———*Germany's Panzer Arm in World War II* (Stackpole, 2006)

———'Guderian, Panzer Pioneer or Myth Maker?', *Journal of Military History*, Vol. 71/2 (April 2007)

———*Mechanised Juggernaut or Military Anachronism, Horses and the German Army of World War II* (Stackpole, 2008)

Dixon, Norman, *On the Psychology of Military Incompetence* (Jonathan Cape, 1976)

Doherty, Richard, *A Noble Crusade, A History of Eighth Army 1941–1945* (Spellmount, 1999)

———*Normandy 1944, The Road to Victory* (Spellmount, 2004)

———*Ireland's Generals in The Second World War* (Four Courts Press Ltd, 2004)

———*None Bolder, The History of the 51st (Highland) Division in the Second World War* (Spellmount, 2006)

———*Eighth Army in Italy 1943–45, The Long Hard Slog* (Pen & Sword, 2007)

Dorling, Captain Taprell, *Ribbons & Medals, Naval, Military, Air Force and Civil* (George Philip & Son, 1963 edition)

Dornan, Peter, *The Last Man Standing, Herb Ashby and the Battle of El Alamein* (Allen & Unwin, 2006)

Doughty, Robert A., *The Breaking Point, Sedan and the Fall of France, 1940* (Archon, USA, 1990)

Douglas, Keith, (ed. Dominic Graham) *Alamein to Zem-Zem* (Faber and Faber, 1992)

Douglas-Home, Charles, *Rommel* (Weidenfeld and Nicolson, 1973)

Douglas of Kirtleside, the Lord, *Autobiography, Volume Two, Years of Command* (Collins, 1966)

Downing, David, *The Devil's Virtuosos, German Generals at War 1940–45* (New English Library, 1977)

Doyle, Peter and Bennett, Matthew R., *Fields of Battle. Terrain in Military History* (Kluwer Academic, NL, 2002)

Drez, Ronald J., (ed.), *Voices of D-Day, The Story of the Allied Invasion Told by Those Who Were There* (Louisiana State University Press, 1994)

Duffy, Christopher, *Frederick the Great, A Military Life* (David and Charles, 1974)

———*Through German Eyes, The British and the Somme 1916* (Phoenix Press, 2007)

Dunn, Captain J. C., *The War the Infantry Knew 1914–1919* (P. S. King, 1938)

Dunphie, Christopher, *The Pendulum of Battle* (Pen and Sword, 2004)

Dupuy, Colonel Trevor, *A Genius for War, The German Army and General Staff, 1807–1945* (Macdonald and Jane's, 1977)

Durschmied, Erik, *The Hinges of Battle, How Chance and Incompetence Have Changed the Face of History* (Hodder and Stoughton, 2002)

Edmonds, Brigadier General Sir James, *Official History of the Great War, Military Operations, France and Belgium, 1914, Volume One* (Macmillan, 1922); *1914, Volume Two* (Macmillan, 1925); *1915, Volume Two* (Macmillan, 1928); *1916, Volume One* (Macmillan, 1932); *1917, Volume Two* (HMSO, 1948); *1918, Volume One* (Macmillan, 1935); *1918, Volume Two* (Macmillan, 1937); *1918, Volume Three*

(Macmillan, 1939); *1918, Volume Four* (HMSO, 1947); *1918, Volume Five* (HMSO, 1947); *The Occupation of the Rhineland, 1918–1929* (HMSO, 1944 and 1987)

Edmonds Brigadier General Sir James E. and Wynne, Captain G. C., *Official History of the Great War, Military Operations, France and Belgium, 1915, Volume One* (Macmillan, 1927)

Edwards, Jill (ed.), *Al-Alamein Revisited, Proceedings of a Symposium held on 2 May 1998 at the American University in Cairo, the Battle of Al-Alamein and its Historical Implications* (American University of Cairo Press, 2001)

Eighth Army, September 1941–January 1943 (Government Information Booklet, HMSO, 1944)

Eisenhower, General Dwight D., *Report by the Supreme Commander to the Combined Chiefs of Staff on the Operations in Europe of the Allied Expeditionary Force, 6 June 1944–8 May 1945* (HMSO/US Government Printing Office Pamphlet, 1946)

———*Crusade in Europe* (Doubleday, 1948)

———*At Ease. Stories I Tell to Friends* (Robert Hale, 1968)

Ellis, Chris, *21st Panzer Division, Rommel's Afrika Korps Spearhead* (Ian Allan, 2002)

Ellis, John, *World War II, The Sharp End* (Windrow and Greene, 1980)

Ellis, Major L. F., *Official History of the Second World War, The War in France and Flanders, 1939–1940* (HMSO, 1953)

———*Official History of the Second World War, Victory in the West, Vol. 1, The Battle of Normandy* (HMSO, 1962)

———*Official History of the Second World War, Victory in the West, Vol. 11, The Defeat of Germany* (HMSO, 1968)

English, John A., *The Canadian Army and the Normandy Campaign, A Study of Failure in High Command* (Praeger Publishers, 1991)

Esebeck, H. G. von, *Rommel et L'Afrika-Korps 1941–1943* (transl. René Jouan, Paris, Editions Payot, 1950)

Essame, Hubert, *Normandy Bridgehead* (Ballantine, 1970)

Evans, Richard J., *The Third Reich at War, How the Nazis Led Germany from Conquest to Disaster* (Allen Lane, 2008)

Falls, Captain Cyril, *Official History of the Great War, Military Operations, France and Belgium, 1917, Volume One* (Macmillan, 1940)

——— (ed.), *The Fatal Decisions, 6 Decisive Battles of the 2nd World War from the Viewpoint of the Vanquished* (Michael Joseph, 1956)

Fanning, William J. Jnr., 'The Origin of the Term "Blitzkrieg", Another View', *Journal of Military History*, Vol. 61 (1997)

Ferris, John, 'The "Usual Source", Signals Intelligence and Planning for the Eighth Army Crusader Offensive, 1941', *Intelligence and National Security*, Vol. 14/1 (Spring 1999)

———*Intelligence and Strategy, Selected Essays* (Routledge, 2005)

Fest, Joachim C., *The Face of the Third Reich, Portraits of the Nazi Leadership* (Weidenfeld and Nicolson, 1970)

———*Hitler* (Weidenfeld and Nicolson, 1974)

———*Plotting Hitler's Death, The German Resistance to Hitler, 1933–45* (Weidenfeld & Nicolson, 1996)

Field, Frank, MP, *Atlee's Great Contemporaries, The Politics of Character* (Continuum, 2009)

Fielding, Lieutenant Colonel Rowland, *War Letters to a Wife* (Medici Society, 1929)

Fisher, David, *The War Magician. The Man Who Conjured Victory in the Desert* (Weidenfeld and Nicolson, 2004)

Fitzgerald, Professor Michael, 'Did Field Marshal Bernard Montgomery Have Asperger's Syndrome', *Indian Journal of Psychiatry*, 2000, Vol. 42/1

Florentin, Eddy, *The Battle of the Falaise Gap; an Army Dies in Normandy* (Hawthorn Books, 1967) – Translated from the French, *Stalingrad en Normandie*
————*The Battle of Normandy* (Editions Ouest-France, 1994)

Foley, Robert T., *Alfred von Schlieffen's Military Writings* (London, Frank Cass, 2003)

Folkestad, William B., *The View From the Turret, The 743rd Tank Battalion During World War II* (Burd Street Press, 1996)

Ford, Ken, *D-Day 1944, Sword Beach & the British Airborne Landings* (Osprey, 2002)
————*D-Day 1944, Gold and Juno Beaches* (Osprey, 2002)
————*Caen 1944, Montgomery's Breakout Attempt* (Osprey, 2004)
————*Falaise, 1944, Death of an Army* (Osprey, 2005)
————*El Alamein, 1942, The Turning of the Tide* (Osprey, 2005)
————*Assault on Sicily, Monty and Patton at War* (Sutton, 2007)
————*Gazala 1942, Rommel's Greatest Victory* (Osprey, 2008)

Forney, Nathan A., 'Rommel in the Boardroom', *Business Horizons*, July–August 1999

Forty, George, *The Armies of Rommel* (Arms and Armour Press, 1998)
————*The Fall of France. Disaster in the West 1939–1940* (Guild Publishing, 1990)

Fox, Colin, 'The Myths of Langemarck', *Imperial War Museum Review*, No. 10 (1995)

Fraser, David, *Alanbrooke* (Collins, 1982)
————*Knight's Cross, A Life of F. M. Erwin Rommel* (HarperCollins, 1993)
————*And We Shall Shock Them, The British Army in the Second World War* (Cassell, 1999)
————*Wars and Shadows* (Allen Lane, 2002)

Freedman, David H., *Corps Business: The 30 Management Principles of the US Marines* (Harper Business, 2000)

French, David, *The British Way in War 1688–2000* (Unwin, 1990)
————'Colonel Blimp and the British Army, British Divisional Commanders in the War Against Germany 1939–45', *English Historical Review*, Vol. 111/444 (November 1996)
————'Tommy Is No Soldier, The Morale of the Second British Army in Normandy', *Journal of Strategic Studies*, Vol. 19/4, December 1996
————*Raising Churchill's Army. The British Army and the War Against Germany 1919–1945* (Oxford University Press, 2000)
————'You Cannot Hate the Bastard Who is Trying to Kill You, Combat and Ideology in the British Army in the War Against Germany, 1939–45', *Twentieth Century British History*, Vol. 11/1 (2000)
————'Invading Europe, The British Army and its preparations for the Normandy campaign, 1942–4', *Diplomacy and Statecraft*, Vol. 14/2 (June 2003)

Frieser, Lieutenant Colonel Karl-Heinz, *The Blitzkrieg Legend, The 1940 Campaign in the West* (Naval Institute Press, 2005)

Fuller, Major-General J. F. C., *Generalship, its Diseases and Their Cure* (Skeffington, 1933)

———*Memoirs of an Unconventional Soldier* (Ivor Nicholson and Watson, 1936)

———*The Conduct of War, 1789–1961. A Study of the Impact of the French, Industrial and Russian Revolutions on War and its Conduct* (Eyre & Spotteswoode, 1961)

Galland, Adolf, *The First and the Last* (Henry Holt, 1954)

Galloway, Colonel Strome, *With The Irish Against Rommel – A Diary of 1943* (Battleline Books, 1978)

Gat, Azar, *British Armour Theory and the Rise of the Panzer Arm, Revising the Revisionists* (Palgrave Macmillan, 2000)

Gause, Alfred, 'Command Techniques Employed by Field Marshal Rommel in Africa', [US] *Armor Magazine*, Vol. 67 (July–August 1958)

Gavaghan, Michael, *An Illustrated Pocket Guide to the Battle of Le Cateau 1914* (M & L Publications, 2000)

Gavin, General James, *On to Berlin, Battles of an Airborne Commander 1943–1946* (Viking, 1978)

Gelb, Norman, *Ike and Monty, Generals at War* (Constable, 1994)

German Army Handbook, April 1918, (Arms and Armour Press, 1977)

Germany 1944: The British Soldier's Pocketbook (Reprinted by the National Archives, 2006)

GHQ Middle East Training Pamphlet No.10, Lessons of the Cyrenaica Campaign December 1940–February 1941 (GHQ Middle East, 1942)

Gilbert, Adrian and Bramall, FM Lord, *The Imperial War Museum of the Desert War* (Sidgwick and Jackson, 1992)

——— (Gilbert), *Germany's Lightning War* (David & Charles, 2000)

Gilbert, Martin, *Winston S. Churchill 1941–45 Volume VII, Road to Victory* (Heinemann, 1986)

———*Second World War* (Weidenfeld and Nicolson, 1989)

———*Somme, The Heroism and Horror of War* (John Murray, 2007)

Gill, Anton, *An Honourable Defeat, A History of German Resistance to Hitler, 1933–1945* (William Heinemann, 1994)

Gleichen, Lord Edward, *Infantry Brigade, 1914 – The Diary of a Commander of the 15th Infantry Brigade, 5th Division, British Army, During the Retreat from Mons* (Leonaur Ltd, 2007)

Gliddon, Gerald, *The Battle of the Somme 1916, A Topographical History* (Sutton, 1987)

Glover, Michael, *Battlefields of Northern France and the Low Countries* (Michael Joseph, 1987)

Goda, Norman J. W., 'Black Marks, Hitler's Bribery of His Senior Officers During World War II', *Journal of Modern History*, Vol. 72 (June, 2000), pp. 413–52

Goebbels, Josef, *The Goebbels Diaries, 1942–43*, ed. & trans. Louis P. Lochner (Hamish Hamilton, 1948)

Goldensohn, Leon N., and Gellately, Robert, (eds), *The Nuremberg Interviews. Conversations with the Defendants and Witnesses* (Alfred A. Knopf, 2004)

Golembiewski, Robert T., 'Toward the New Organization Theories, Some Notes on Staff', *Midwest Journal of Political Science*, Vol. 5/3 (August 1961)

Gooch, John (ed.), *Decisive Campaigns of the Second World War* (Frank Cass, 1990)

Gooderson, Dr Ian, 'Allied Fighter-Bombers Versus German Armour, Myths and Realities', *Journal of Strategic Studies* Vol. 14/2 (June 1991)

———*Air Power at the Front, Allied Close Air Support in Europe 1943–1945* (Routledge, 1998)

Gordon, Harold J., *The Reichswehr and the German Republic* (Princeton University Press, 1957)

Görlitz, Walter, *History of the German General Staff 1657–1945* (Praeger, 1953)

Graham, Dominick and Bidwell, Shelford, *Tug of War, The Battle for Italy, 1943–45* (Hodder and Stoughton, 1986)

———(Graham), *The Price of Command. A Biography of General Guy Simonds* (Stoddart, 1993)

Grainger, John, *The Battle for Palestine, 1917* (Boydell Press, 2006)

Gray, Colin S., *Another Bloody Century, Future Warfare* (Weidenfeld & Nicolson, 2005)

Greacen, Lavinia, *Chink, A Biography* (Macmillan 1989)

Greene, Jack and Massignani, Alessandro, *Rommel's North African Campaign September 1940–November 1942* (Combined Books, 1994)

Greenwood, Alexander, *Field Marshal Auchinleck* (Pentland Press, 1990)

Griffith, Paddy, *World War II Desert Tactics* (Osprey, 2008)

Grigg, John, *1943, The Victory That Never Was* (Eyre Methuen, 1980)

Grint, Keith, *Leadership, Management and Command, Rethinking D-Day* (Palgrave Macmillan, 2008)

Guderian, General Heinz, *Panzer Leader* (Michael Joseph, 1952)

———*Achtung-Panzer! The Development of Tank Warfare* (Arms and Armour Press, 1992)

de Guingand, Major-General. Sir Francis, *Operation Victory* (Hodder and Stoughton, 1947)

———*Generals at War* (Hodder and Stoughton, 1964)

Hackett, General Sir John, *The Profession of Arms* (Macmillan, 1983)

Haldane, General Sir Aylmer, *A Brigade of the Old Army* (Wm Blackwood and Sons, 1920)

———*A Soldier's Saga* (Wm Blackwood and Sons, 1948)

Hall, David Ian, *The Birth of the Tactical Air Force, British Theory and Practice of Air Support in the West, 1939–1943* (DPhil Thesis, University of Oxford, 1996)

Hamilton, Nigel, *Monty, Making of a General, 1887–1942* (Hamish Hamilton, 1981)

———*Monty, Master of the Battlefield, 1942–1944* (Hamish Hamilton, 1983)

———*Monty, The Field Marshal, 1944–1976* (Hamish Hamilton, 1986)

———*The Full Monty, Montgomery of Alamein 1887–1942* (Allen Lane, 2001)

Handel, Michael I. (ed.), *Leaders and Intelligence* (Frank Cass, 1989)

——— *Intelligence and Military Operations* (Frank Cass, 1990)

Harding, William, *A Cockney Soldier, Duty Before Pleasure* (Merlin Books, 1989)

Hargreaves, Richard, *The Germans in Normandy* (Pen and Sword, 2006)

Harman, Nicholas, *Dunkirk – The Necessary Myth* (Hodder and Stoughton, 1980/Coronet, 1981); earlier title, *Dunkirk, the Patriotic Myth*

Harpur, Glyn and Hayward, Joel, *Born to Lead? Portraits of New Zealand Commanders* (Exisle Publishing 2003)

Harris, J. P., *Douglas Haig and the First World War* (Cambridge, 2008)

Harrison, Frank, *Tobruk, the Great Siege Revisited* (Arms and Armour Press, 1997)

Harrison Gordon A., *Cross-Channel Attack* (Official History, Office of the Chief of Military History of the US Army, Washington DC, 1951)

Harrison, Mark, *Medicine & Victory: British Military Medicine in the Second World War* (Oxford University Press, 2004)

Hart, Captain B. H. Liddell, *When Britain Goes To War, Adaptability and Mobility* (Faber and Faber, 1935)

———*Thoughts on War* (Faber and Faber, 1944)

———*The Other Side Of The Hill* (Cassell, 1948; published in USA as *The German Generals Talk*)

——— *Memoirs*, Two Volumes (Cassell, 1965)

Hart, Russell A., *Clash of Arms, How the Allies Won in Normandy* (University of Oklahoma Press, 2004)

———'Feeding Mars, The Role of Logistics in the German Defeat in Normandy 1944', *War in History*, Vol. 3/4 (November 1996)

Hart, Stephen A., *Colossal Cracks, Montgomery's 21st Army Group in Northwest Europe, 1944–45* (Stackpole, 2007)

Harvey, A. D., 'The Royal Air Force and Close Air Support, 1918–1940', *War in History*, Vol. 15/4 (November, 2008), pp. 462–86

Harvey, John (ed.), *The War Diaries of Oliver Harvey* (Collins, 1978)

Hastings, Max, *Overlord. D-Day & the Battle for Normandy* (Michael Joseph, 1984)

———*Armageddon, The Battle for Germany 1944–45* (Macmillan, 2004)

———*Finest Years, Churchill as Warlord 1940–45* (HarperPress, 2009)

Haswell, Jock, *The Intelligence and Deception of the D-Day Landings* (Batsford, 1979)

Hautecler, Georges, *Rommel and Guderian against the Belgian Chasseurs Ardennais, The Combats at Chabrehez and Bodange, 10 May 1940* (Belgian staff study, transl. George F. Nafziger, 2003. Orig. Belgian edns pub. by Ministere de la Defense Nationale, 1955 and 1957)

Hawkins, Desmond (comp. & ed.), *War Report, BBC Radio Reports From the Western Front 1944–45, D-Day to VE-Day* (BBC Books, 1985)

Heckmann, Wolf, *Rommel's War in Africa* (Granada, 1981)

Heffer, Simon, *Like the Roman, The Life of Enoch Powell* (Weidenfeld and Nicolson, 1998)

Henderson, Johnny, with Douglas-Home, Jamie, *Watching Monty* (Sutton, 2005)

Herbert, A. P., *Independent Member* (Methuen, 1952)

Herrera, Geoffrey L., 'Inventing the Railroad and Rifle Revolution, Information, Military Innovation and the Rise of Germany', *Journal of Strategic Studies*, Vol. 27/2 (June 2004)

Hesketh, Roger, *Fortitude. The D-Day Deception Campaign* (St Ermin's Press, 1999)

Hill, Gordon L., Jnr, 'Rommel, Fox or Fake?' *Marine Corps Gazette*, Vol. 45 (June 1961)

Hillgruber, Andreas, *Germany and the Two World Wars* (Harvard University Press, 1981)

Hinsley, F. H., *British Intelligence in the Second World War, its Influence on Strategy and Operations*, Volumes I and II (HMSO, 1979 and 1981)

———and Stripp, Alan (eds), *Codebreakers, The Inside Story of Bletchley Park* (Oxford University Press, 1993)

Hitler, Adolf, *Mein Kampf* (first English edition, Hurst and Blackett, 1939)

Hoffmann, Karl, *Erwin Rommel* (Brassey's, 2004)

Hoffmann, Professor Heinrich, *Hitler in Polen* (Zeitgeschichte-Verlag, 1939)

Hoffmann, Peter, *History of the German Resistance, 1933–1945* (Macdonald and Jane's, 1977)

———'Ludwig Beck, Loyalty and Resistance', *Central European History*, Vol. 14/4 (1981), pp. 332–50

———*Stauffenberg, A Family History, 1905–1944* (Cambridge, 1995)

Holbrook, David, *Flesh Wounds* (Methuen,1966)

Holland, James, *Together We Stand, North Africa 1942–1943, Turning the Tide in the West* (HarperCollins, 2005)

———*Heroes: The Greatest Generation and The Second World War* (HarperCollins, 2007)

Holmes, Richard, *Bir Hacheim. Desert Citadel* (Ballantine Books, 1971)

———*The Little Field Marshal, A Life of Sir John French* (Weidenfeld and Nicolson, 1981)

———*Fatal Avenue* (Jonathan Cape, 1992)

———*Army Battlefield Guide, Belgium and Northern France* (HMSO, 1995)

———*Riding the Retreat, Mons to the Marne 1914 Revisited* (Jonathan Cape, 1995)

———*War Walks From Agincourt to Normandy* (BBC Books, 1996)

———*The Western Front* (BBC Books, 2000)

———(ed.) *The Oxford Companion to Military History* (Oxford University Press, 2001)

———*Battlefields of the Second World War* (BBC Books, 2001)

———*Tommy, The British Soldier on the Western Front* (HarperCollins, 2004)

———*In the Footsteps of Churchill* (BBC Books, 2005)

———*The World at War: The Landmark Oral History from Previously Unpublished Archives* (Ebury Press, 2007)

Holt, Toni and Valmai, *Maj & Mrs Holt's Battlefield Guide to the Normandy Landing Beaches* (Pen & Sword, 1994)

———*Maj & Mrs Holt's Battlefield Guide to the Somme* (Pen & Sword 1996)

———*Maj & Mrs Holt's Guide to The Ypres Salient* (Pen & Sword, 1997)

———*Maj & Mrs Holt's Concise Guide to The Western Front – North* (Pen & Sword, 2003)

———*Maj & Mrs Holt's Concise Guide to The Western Front – South* (Pen & Sword, 2005)

Horne, Alistair, *To Lose A Battle, France 1940* (Macmillan, 1969)

———with David Montgomery, *Monty, The Lonely Leader, 1944–1945* (HarperCollins, 1994)

Horrocks, General Sir Brian, *A Full Life* (Collins, 1960)

————*Corps Commander* (Sidgwick and Jackson, 1977)

Howard, Michael (ed.), *The Theory and Practice of War: Essays Presented to Captain B. H. Liddell Hart* (Cassell, 1965)

————*The Mediterranean Strategy in the Second World War* (Cambridge University Press, 1966)

————*History of the Second World War, United Kingdom Military Series, Vol. IV, Grand Strategy, August 1942–September 1943* (HMSO 1972)

————*British Intelligence in the Second World War, Volume V, Strategic Deception* (HMSO, 1990)

————*The First World War* (Oxford University Press, 2002)

Howarth, T. E. B. (ed.), *Monty at Close Quarters, Recollections of the Man* (Leo Cooper, 1985)

Hudson, Miles and Stanier, John, *War and the Media. A Random Searchlight* (Sutton, 1997)

Hughes, Colin, *Mametz, Lloyd George's Welsh Army at the Battle of the Somme* (Orion Books, 1982)

Humble, Richard, *Hitler's Generals* (Arthur Barker, 1973)

Hurst, Sidney C., *The Silent Cities, An Illustrated Guide to the War Cemeteries & Memorials to the 'Missing' in France and Flanders* (Methuen, 1929)

Ingersoll, Ralph, *The Battle Is the Pay-Off* (New York, Harcourt Brace & Co., 1943)

Instructions for British Servicemen in France 1944 (Reprinted by the Bodleian Library, Oxford, 2005)

Irving, David, *The Rise and Fall of the Luftwaffe: The Life of Field Marshal Erhard Milch* (Weidenfeld and Nicolson, 1973)

————*The Trail of the Fox, The Life of Field-Marshal Erwin Rommel* (Weidenfeld and Nicolson, 1977)

————*Hitler's War* (Hodder and Stoughton, 1977)

————*The War Path* (Michael Joseph, 1979)

————*The War Between the Generals* (Allen Lane, 1981)

————'The Trail of the Desert Fox, Rommel Revised', *Journal for Historical Review*, Vol. 10/4 (Winter 1990)

Isby, David C., (ed.), *Fighting in Normandy, the German Army from D-Day to Villers Bocage* (German Debrief – Greenhill, 2002)

———— *Fighting the Breakout, The German Army in Normandy from 'Cobra' to the Falaise Gap* (German Debrief – Greenhill, 2004)

————*Fighting the Invasion, The German Army at D-Day* (German Debrief – Greenhill, 2006)

Ismay, Lord, *The Memoirs of General The Lord Ismay* (Heinemann, 1960)

Jackson, Ashley, *The British Empire and the Second World War* (Hambledon Continuum, 2006)

Jackson, W. G. F., *The Battle for Italy* (Harper and Row, 1967)

————*Alexander of Tunis as Military Commander* (BT Batsford, 1970)

————*The North African Campaign, 1940–43* (BT Batsford, 1975)

————*Overlord, Normandy 1944* (Davis-Poynter Publishers, 1978)

Jacobsen, H. A. and Rohwer, J. (eds), *Decisive Battles of World War II, The German View* (André Deutsch, 1965)

James, Anthony, *Informing the People. How the Government Won Hearts and Minds to Win WW2* (HMSO, 1996)

James, M. E. Clifton, *I Was Monty's Double* (Rider & Co./Popular Book Club, 1957)

James, Robert Rhodes, *Anthony Eden* (Weidenfeld and Nicolson, 1986)

Jary, Sydney, *18 Platoon* (Sydney Jary Ltd, 1987)

Jenkins, Roy, *Churchill* (Macmillan, 2001)

Jentz, Thomas L., *Tank Combat in North Africa, The Opening Rounds, Operations Sonnenblume, Brevity, Skorpion and Battleaxe* (Schiffer, 1998)

Jewell, Derek (ed.), *Alamein and the Desert War* (Sphere Books/Sunday Times, 1967)

Jörgensen, Christer, *Rommel's Panzers: Rommel, Blitzkrieg and the Triumph of the Panzer Arm* (Zenith Press, 2003)

Jowett, Philip S., *The Italian Army 1940–45, Volume 2, Africa 1940–43* (Osprey, 2001)

Jünger, Ernst, *The Storm of Steel. From the Diary of a German Storm-Troop Officer on the Western Front* (Chatto and Windus, 1929)

Kahn, David, *The Codebreakers, The Story of Secret Writing* (Macmillan 1967)

Keegan, John, *The Face of Battle* (Jonathan Cape, 1976)

———*Six Armies in Normandy* (Jonathan Cape, 1982)

———(ed.), *Churchill's Generals* (Weidenfeld and Nicolson, 1991)

———*A History of Warfare* (Hutchinson, 1993)

Kelly, Orr, *Meeting the Fox, The Allied Invasion of Africa, from Operation Torch to Kasserine Pass to Victory in Tunisia* (Wiley, 2002)

Keisling, Eugenia C., 'Illuminating Strange Defeat and Pyrrhic Victory, The Historian Robert A. Doughty', *Journal of Military History*, Volume 71/3 (July 2007), pp. 875–88

Keitel, Wilhelm, *The Memoirs of Field Marshal Wilhelm Keitel*, trans. David Irving and ed. Walter Görlitz (William Kimber, 1965)

Kemp, Anthony, *D-Day: The Normandy Landing and the Liberation of Europe* (Thames & Hudson, 1994)

Kennedy, Gregory C. and Neilson, Keith (eds), *Military Education, Past, Present and Future* (Praeger, 2002)

Kennedy, Major-General Sir John, *The Business of War* (Hutchinson, 1957)

Kershaw, Ian, *Hitler, 1889–1936, Hubris* (Allen Lane, 1998)

———*Hitler, 1936–1945, Nemesis* (Allen Lane, 2000)

———*Luck of the Devil, The Story of Operation Valkyrie* (Penguin, 2009)

Kesselring, Albrecht, (ed. Kenneth Macksey), *The Memoirs of Field-Marshal Kesselring* (William Kimber, 1953)

Khan, David, 'Codebreaking in World Wars I and II, The Major Successes and Failures, Their Causes and Their Effects', *The Historical Journal*, Vol. 23/3 (1980)

Kimball, Warren F. (ed.), *Churchill and Roosevelt, The Complete Correspondence*, Vols I and II (Princeton University Press, 1984)

Kinvig, Clifford, *Scapegoat. General Percival of Singapore* (Brassey's, 1996)

Kippenberger, Sir Howard, *Infantry Brigadier* (Oxford University Press, 1949)

Kitchen, Martin, *The German Officer Corps 1890–1914* (Clarendon Press, 1968)

————*A Military History of Germany, From the Eighteenth Century to the Present Day* (Weidenfeld and Nicholson, 1975)

————*The Silent Dictatorship, The Politics of the German High Command, 1916–1918* (Croom Helm, 1976)

————'The Traditions of German Strategic Thought', *International History Review*, Vol. 1/2 (April 1979)

————*The German Offensives of 1918* (Tempus, 2001)

————*Rommel's Desert War, Waging War in North Africa, 1941–1943* (Cambridge University Press, 2009)

Knapp, Andrew, 'The Destruction and Liberation of Le Havre in Modern Memory', *War in History*, Vol. 14/4 (November 2007)

Knappe, Siegfried and Bursaw, Ted, *Soldat, Reflections of a German Soldier 1936–1949* (Airlife Publications, 1992)

Knight, Jill, *The Civil Service Rifles in the Great War, 'All Bloody Gentlemen'* (Pen and Sword, 2004)

Knightley, Phillip, *The First Casualty. The War Correspondent as Hero, Propagandist and Myth Maker* (André Deutsch, 1975)

Knopp, Guido, *Hitler's Warriors* (Sutton, 2005)

Knox, MacGregor, *Hitler's Italian Allies* (Cambridge University Press, 2000)

Koch, Lutz, *Erwin Rommel* (Editions Correa, Paris, 1950)

Krammer, Arnold, 'German Prisoners of War in the United States', *Military Affairs*, Vol. 40/2 (April 1976)

————'American Treatment of German Generals during World War II', *Journal of Military History*, Vol. 54/1 (January, 1990), pp. 27–46

Krause, Michael D., and Phillips, R. Cody, *Historical Perspectives of the Operational Art* (Center of Military History, US Army, 2005)

Kriebel, Colonel Rainier (author) and Gudmundsson, Bruce L. (ed.), *Inside the Afrika Korps, The Crusader Battles 1941–1942* (German Debrief – Greenhill, 1999)

Kuhn, Annette, 'Desert Victory and the People's War', *Screen*, Vol. 22/2 (1981)

Kühn, Volkmar, *Mit Rommel in der Wüste* (Motorbuchverlag Stuttgart, 1992)

Kurowski, Franz, *Das AfrikaKorps: Der Kampf der Wüstenfüchse* (Heyne Verlag, 1978)

————*Erwin Rommel: Der Mensch, Der Soldat, Der Generalfeldmarschall* (Heinrich Poppinghaus, 1978)

Kurtz, Robert, *Rommel's AfrikaKorps Groupings, Awards and Ephemera* (Schiffer, 2004)

Lamb, Richard, *Montgomery in Europe 1943–45, Success or Failure?* (Buchan & Enright, 1983)

————*War In Italy, 1943–1945, A Brutal Story* (John Murray, 1993)

Lande, D. A., *Rommel in North Africa* (MBI Publishing, 1999)

Langmaid, Rowland, *'The Med', The Royal Navy in the Mediterranean 1939–45* (Batchworth Press, 1948)

Latimer, Jon, *Alamein* (John Murray, 2001)

————*Tobruk 1941* (Osprey, 2001)

————*Tobruk 1941, Rommel's Opening Move* (Praeger, 2004)

Ledwoch, Janusz, *Tank Power Volume LXV, Rommel's Tanks* (Wydawnictwo Militaria, 2008)

Lefèvre, Eric, *Panzers in Normandy, Then and Now* (After the Battle, 1983)

Levine, Alan J., *The War Against Rommel's Supply Lines 1942–1943* (Praeger, 1999)

Lewin, Ronald, *Rommel as Military Commander* (BT Batsford, 1968)

———*Montgomery as Military Commander* (BT Batsford, 1971)

———*The Life and Death of the Afrika Korps* (BT Batsford, 1977)

———*ULTRA Goes to War* (Hutchinson, 1978)

———'A Signal-Intelligence War', *Journal of Contemporary History*, Vol. 16/3 (July, 1981), pp. 501–12

———*Hitler's Mistakes* (Leo Cooper, 1984)

Lewis, Adrian R., *Omaha Beach, A Flawed Victory* (University of North Carolina Press, 2001)

Lichem, Heinz von, *Rommel 1917 Der Wüstenfuchs als Gebirgssoldat* (Hornung-Verlag Viktor Lang München, 1975)

Liddle, Peter (ed.), *Aspects of Conflict 1916* (Michael Russell, 1985)

———*Home Fires and Foreign Fields. British Social and Military Experience in the First World War* (Brassey's, 1985)

———*Passchendaele in Perspective, 3rd Battle of Ypres* (Pen & Sword, 1997)

———*D-Day, By Those Who Were There* (Leo Cooper, 2004)

Lindsay, David, *Forgotten General. A Life of Andrew Thorne* (Michael Russell, 1987)

Lloyd, David W., *Battlefield Tourism, Pilgrimage and Commemoration of the Great War in Britain, Australia and Canada 1919–1939* (Berg, 1998)

Lockenour, Jay, *Soldiers As Citizens, Former Wehrmacht Officers in the Federal Republic of Germany 1945–1955* (University of Nebraska Press, 2001)

Lodge, Brett, *Lavarack, Rival General* (Sydney, Allen and Unwin, 1998)

Loose, Gerhard, 'The German High Command and the Invasion of France', *Military Affairs*, Vol. 11/3 (Autumn 1947)

Loring-Villa, Brian, *Unauthorized Action, Mountbatten and the Dieppe Raid 1942* (Oxford University Press, 1989)

———'Mountbatten, the British Chiefs of Staff, and Approval of the Dieppe Raid', article in the *Journal of Military History*, Vol. 54/2 (April, 1990), pp. 201–26

Lucas, James, *Panzer Army Africa* (Macdonald and Jane's, 1977)

———Lucas, James and Barker, James, *The Battle of Normandy, The Falaise Gap,* (Holmes and Meier, 1978)

———*War in the Desert, Eighth Army at Alamein* (Arms & Armour Press, 1982)

———*Rommel's Year of Victory. The Wartime Illustrations of the Afrika Korps by Kurt Caesar* (Greenhill, 1998)

Luck, Oberst Hans von, *Panzer Commander* (Praeger, 1989)

———'The End in North Africa', *MHQ, The Quarterly Journal of Military History*, Vol. 1/4 (1989), pp. 118–27

Luvaas, Jay, *The Education of an Army, British Military Thought 1815–1940* (University of Chicago Press/Cassell and Co., 1964)

Lyman, Robert, *The Longest Siege, Tobruk – the Battle That Saved North Africa* (Macmillan, 2009)

MacDonogh, Giles, *After the Reich: From the Liberation of Vienna to the Berlin Airlift* (John Murray, 2007)

Macdonald, Lyn, *1914: The Days of Hope* (Michael Joseph, 1987)

MacKenzie, Major-General J. J. G., and Reid, Brian Holden, *The British Army and the Operational Level of War* (Tri-Service Press, 1989)

Macksey, Kenneth, *Armoured Crusader. The Biography of Maj-Gen Sir Percy 'Hobo' Hobart* (Hutchinson, 1967)

———*Panzer Division, The Mailed Fist* (Macdonald and Jane's, 1969)

———*Crucible of Power. The Fight for Tunisia 1942–1943* (Hutchinson, 1969)

———*Afrika Korps* (Purnell/Pan Ballantine, 1970)

———*Guderian, Panzer General* (Macdonald and Jane's, 1975)

———*Kesselring, The Making of the Luftwaffe* (BT Batsford, 1978)

———*Rommel. Battles and Campaigns* (Arms and Armour Press 1979)

Major, Patrick, '"Our Friend Rommel", The Wehrmacht as "Worthy Enemy" in Postwar British Popular Culture', *German History*, Vol. 26/4 (2008), pp. 520–35

———*Rummel um Rommel, The Desert Fox Myth in the Early Federal Republic*, Conference Paper, 2007

Mallaby, George, *Each In His Office, Studies of Men in Power* (Leo Cooper, 1972)

Man, John, *The Penguin Atlas of D-Day & The Normandy Campaign* (Penguin, 1994)

Manstein, Field Marshal Erich von, *Lost Victories* (Methuen & Co, 1958; originally published as *Verlorene Siege*, 1955)

Manteuffel, General Hasso von, *The 7th Panzer Division, An Illustrated History of Rommel's Ghost Division 1938–45* (Schiffer, 2000)

Marshall, Charles F., *Discovering the Rommel Murder, The Life and Death of the Desert Fox* (Stackpole, 2002)

Marshall, S. L. A., *Men Against Fire. The Problem of Battle Command in Future War* (William Morrow, 1947)

———*The Soldier's Load and the Mobility of a Nation* (Marine Corps Association, Quantico, 1980)

Marshall-Cornwall, James, *Wars and Rumours of Wars* (Leo Cooper, 1984)

Martin, Charlie, *Battle Diary, From D-Day and Normandy to the Zuider Zee* (Dundurn Press, 1994)

Maser, Werner, *Hitler* (Allen Lane, 1973)

Maude, Alan H. (ed.), *The 47th (London) Division 1914–1919. By Some Who Served With It In The Great War* (Amalgamated Press, 1922)

Maule, Henry, *Caen, The Brutal Battle and Breakout from Normandy* (David and Charles, 1976)

Mayer, S. L. (Intro), *Signal, Hitler's Wartime Picture Magazine* (Bison Books, 1976)

McCallum, Neil, *Journey With a Pistol. A Chronicle of War (from Alamein to the Invasion of Sicily) from a Point of View Other than That of Montgomery of Alamein's Memoirs* (Victor Gollancz Ltd, 1959)

McCarthy, Chris, *The Third Ypres Passchendaele Day-by-Day Account* (Arms and Armour Press, 1995)

McGuirk, Dan, *Rommel's Army in Africa* (Century Hutchinson, 1987)

McIntosh, H. W., 'Heat Exhaustion and Heat Stroke', *Canadian Medical Association Journal*, Vol. 76 (15 May 1957), p. 873

McKee, Alexander, *Caen, Anvil of Victory* (Souvenir Press, 1964)

———*El Alamein, Ultra and the Three Battles* (Hutchinson, 1981)

McManus, John C., *The Americans at Normandy, The Summer of 1944 – The American War from the Normandy Beaches to Falaise* (Forge Books, 2004)

———*The Americans at D-Day, The American Experience at the Normandy Invasion* (Forge Books, 2005)

McNish, Robin, *Iron Division. The History of 3rd Division* (Ian Allen, 1978)

Mead, Gary, *The Good Soldier, The Biography of Douglas Haig* (Atlantic Books, 2007)

Mearsheimer, John J., *Liddell Hart and the Weight of History* (Brassey's, 1988)

Mechanised and Armoured Formations, Instructions for Guidance When Considering Their Action (War Office, 1929)

Megargee Geoffrey P., *Inside Hitler's High Command* (University Press of Kansas, 2002)

Mellenthin, Major-General F. W. von, *Panzer Battles 1939–1945. A Study in the Employment of Armour in the Second World War* (Cassell, 1955)

de Mendelssohn, Peter, 'The German Military Mind. A Survey of Recent German Military Writing', *The Political Quarterly*, Vol. 23/2 (1952), pp. 182–90

Messenger, Charles, *The Tunisian Campaign* (Ian Allen, 1982)

———*Hitler's Gladiator: The Life and Times of Oberstgruppenführer Sepp Dietrich* (Brassey's, 1988)

———*The Last Prussian, A Biography of F. M. Gerd von Rundstedt 1875–1953* (Brassey's, 1991)

———*Rommel. Leadership Lessons From the Desert Fox* (Palgrave Macmillan, 2009)

Meyer, Kurt, *Grenadiers* (JJ Fedcrowicz, 2001)

Michelin Guide to Normandy, (Michelin 1972)

Middlebrook, Martin and Mary, *The Somme Battlefields* (Viking, 1991)

Miksche, F. O., *Blitzkrieg* (Faber and Faber, 1941)

Miles, Captain W., *Official History of the Great War, Military Operations, France and Belgium, 1916, Volume Two* (Macmillan, 1938); *1917, Volume Three* (Macmillan, 1949)

Milano, James M., 'How Rommel Applied Lessons Learned in WWI to his Afrika Korps Operations in WWII', [US] *Armor Magazine*, Vol. 100 (Sept–Oct 1991)

Militärgeschichtliches Forschungsamt (Research Institute for Military History – Boog, Horst, Rahn, Werner, Stumpf, Reinhard and Wegner, Bernd), *Germany and the Second World War, Vol. VI, The Global War 1941–1943* (Clarendon Press, 2001)

Miller, Robert A., *August 1944, The Campaign for France* (Presidio, 1988)

Miller, Russell, *Nothing Less Than Victory, an Oral History of D-Day* (Michael Joseph, 1993)

Mitcham, Samuel W., Jnr, *Rommel's Desert War. The Life and Death of the Afrika Korps* (Stein and Day, 1982)

———*The Desert Fox In Normandy, Rommel's Defence of Fortress Europe* (Stein and Day, 1983)

———*Triumphant Fox, Erwin Rommel and The Rise of the Afrika Korps* (Stein and Day, 1984)

———*Hitler's Field Marshals and Their Battles* (William Heinemann, 1988)

———*Rommel's Greatest Victory. The Desert Fox and the Fall of Tobruk, Spring 1942* (Presidio, 1998)

————*Retreat to the Reich; The German Defeat in France 1944* (Presidio, 2000)

————*The Panzer Legions, A Guide to the German Army Tank Divisions of WWII and Their Commanders* (Praeger, 2000)

————*Rommel's Lieutenants, The Men Who Served the Desert Fox, France, 1940* (Praeger, 2007)

————*Rommel's Desert Commanders, The Men Who Served the Desert Fox, North Africa, 1941–1942* (Praeger, 2007)

Michelmore, Winsome, *Thursday's Child* (privately published, 1985)

Model, Hans Georg, *Der Deutsche Generalstabsoffizier, Seine Auswahl und Ausbildung in Reichswehr, Wehrmacht und Bundeswehr* [*The German general staff officer, his selection and education in the Reichswehr, Wehrmacht and Bundeswehr*](Bernard & Graefe Verlag, 1968)

Molinari, Andrea, *Desert Raiders, Axis and Allied Special Forces 1940–43* (Osprey, 2007)

Molony, Brigadier C. J. C., *Official History of the Second World War, The Mediterranean and Middle East Volume V* (HMSO, 1973)

Montgomery, Field Marshal (later Viscount) Sir Bernard, 'The problem of the Encounter Battle as Affected by Modern British War Establishments, *Royal Engineers* Journal, September 1937, pp. 339–54

————'Major Tactics in the Encounter Battle', *The Army Quarterly*, July 1938

————*Eighth Army. Some Notes on High Command in War* (Eighth Army, Tripoli, January 1943)

————'21st Army Group in the Campaign in North-West Europe 1944–45', Lecture given to the Royal United Services Institution in London (BAOR, 1945)

————*Forward To Victory* (Hutchinson, 1945) – Facsimiles of Monty's messages to his troops

————*El Alamein to the River Sangro* (BAOR, 1946/Hutchinson, 1947)

————*From Normandy to the Baltic* (BAOR/21st Army Group, 1946/Hutchinson, 1947)

————*Forward From Victory* (Hutchinson, 1948)

————*Memoirs of FM Montgomery* (Collins, 1958)

————*An Approach to Sanity* (Collins, 1959)

————*The Path to Leadership* (Collins, 1961)

————*Three Continents* (Collins, 1962)

————*A History of Warfare* (Collins, 1968)

————'A Voice From the Past (Notes for the First Senior War Staff Course, 1940)', *British Army Review*, No. 59 (August 1978), pp. 34–6

————*Notes on War* (Notes on the Infantry Division in Battle; Use of Air Power in Support; The Armoured Division in Battle; all 1944, reprinted Military Library Research Service, 2006)

Montgomery, Brian, *A Field Marshal in the Family* (Constable, 1973)

————*Monty, a Life in Photographs, Family, Life and Times of FM the Viscount Montgomery of Alamein, 1887–1976* (Blandford Press, 1985)

Moore, Bob, 'Unwanted Guests in Troubled Times, German Prisoners of War in the Union of South Africa, 1942–1943', *Journal of Military History*, Vol. 70/1 (January, 2006), pp.63–89

Moore, John Hammond, 'Hitler's Wehrmacht in Virginia, 1943–1946', *The Virginia Magazine of History and Biography*, Vol. 85/3 (July, 1977)

Moorhead, Alan, *African Trilogy, the Desert War up to the Surrender of the Axis Armies in 1943* (Hamish Hamilton, 1944)

———*Montgomery, A Biography* (Hamish Hamilton, 1946)

Moorhouse, Roger, *Killing Hitler, The Third Reich and the Plots Against the Führer* (Jonathan Cape, 2006)

Moran, Lord (Charles Moran Wilson), *The Anatomy of Courage* (Constable, 1945)

Moreman, Tim, *Desert Rats, British 8th Army in North Africa 1941–43* (Osprey, 2007)

Morgan, Lieutenant General Frederick E., *Overture to Overlord* (Hodder and Stoughton, 1950)

Morris, Eric, *Circles of Hell, The War in Italy 1943–1945* (Hutchinson, 1993)

Mosse, George, *Fallen Soldiers, Reshaping the Memory of the World Wars* (Oxford University Press, 1990)

Moynihan, Michael, *War Correspondent* (Leo Cooper, 1994)

Murray, Patrick, *Eisenhower vs. Montgomery, The Continuing Debate* (Praeger 1996)

Murray, Williamson, 'ULTRA, Some Thoughts on its Impact on the Second World War', *Air University Review* (July–August 1984)

———and Sinnreich, Richard Hart (eds.), *The Past as Prologue, The Importance of History to the Military Profession* (Cambridge University Press, 2006)

Mythos Rommel, Katalog zur Sonderaussrellung 18.12.2008 – 30.8.2009 (Haus der Geschichte Baden-Württemberg, 2008)

Nafziger, George F., *The German Order of Battle. Panzers and Artillery in World War II* (Greenhill, 1999)

Neaman, Elliott, 'Ernst Jünger's Legacy', *South Central Review*, Vol. 16–2/3 (1999), pp. 54–67

Neillands, Robin, *The Desert Rats, 7th Armoured Division, 1940–1945* (Weidenfeld and Nicolson, 1991)

———and de Normann, Roderick, *D-Day 1944, Voices From Normandy* (Weidenfeld and Nicolson, 1993)

———*The Conquest of the Reich. D-Day to VE-Day – A Soldier's History* (Weidenfeld and Nicolson, 1995)

——— *The Great War Generals on the Western Front, 1914–18* (Robinson, 1998)

——— *The Battle of Normandy 1944* (Cassell, 2002)

——— *Eighth Army. From the Western Desert to the Alps, 1939–45* (John Murray, 2004)

——— *The Dieppe Raid, The Story of the Disastrous 1942 Mission* (Aurum, 2006)

Nevin, Thomas, *Ernst Jünger and Germany: Into the Abyss 1914–1945* (Constable, 1997)

Nicholls, Jonathan, *Cheerful Sacrifice. The Battle of Arras 1917* (Leo Cooper, 2005)

Nicolson, Nigel, *Alex, the Life of F. M. Earl Alexander of Tunis* (Weidenfeld and Nicolson, 1973)

Norling, Bernard, 'The Generals of World War I', *The History Teacher*, Vol. 2/4 (1969), pp. 14–26

Norman, Albert, *Operation OVERLORD, Design and Reality, The Allied Invasion of Western Europe* (Greenwood Press, 1970)

North, John, *North-West Europe 1944–45: The Achievement of 21st Army Group* (HMSO, 1953)

Ommert, Stefan, *Recon For Rommel, The 2.Staffel Nahaufklarungsgruppe (Heer)/14 – Air Recon Flyers in Africa* (Military Machine, 2007)

O'Connor, Damian P., 'The Twenty Year Armistice, RUSI Between the Wars', *RUSI Journal*, Vol. 154/1 (February 2009), pp. 84–90

O'Neill, Robert J., *The German Army and the Nazi Party, 1933–1939* (Cassell, 1966)

Orange, Vincent, *Coningham, A Biography of Air Marshal Sir Arthur Coningham* (Methuen, 1990)

————*Tedder, Quietly in Command* (Frank Cass, 2004)

Overy, Richard, *Why the Allies Won* (Jonathan Cape, 1995)

Owen, Edward, *1914, Glory Departing* (Buchan and Enright, 1986)

Owen, Roderic, *The Desert Air Force* (Hutchinson, 1948)

Pack, S. W. C., *Operation Husky: The Allied Invasion of Sicily* (David & Charles, 1977)

Pallud, Jean-Paul, *Blitzkrieg in the West, Then and Now* (After the Battle, 1991)

Paret, Peter, Craig, Gordon A. and Gilbert, Felix (eds), *Makers of Modern Strategy from Machiavelli to the Nuclear Age* (Princeton & Oxford University Press, 1986)

Parkinson, Roger, *Clausewitz, A Biography* (Wayland Press, 1970)

————*The Auk, Auchinleck, Victor at Alamein* (Granada, 1977)

Parlour, Andy and Sue, *Phantom at War, The British Army's Secret Intelligence & Communication Regiment of WWII* (Cerberus Publishing, 2003)

Paterson, Michael, *Winston Churchill: His Military Life 1895–1945* (David & Charles, 2005)

Patton, General George S., *War As I Knew It* (Houghton Mifflin, 1947)

Peitz, Bernd and Wilkins, Gary, *AfrikaKorps: Rommel's Tropical Army in Color* (Schiffer, 2004)

Penrose, Jane (ed.), *The D-Day Companion* (Osprey Books, 2004)

Phillips, Brigadier C. E. Lucas., *Alamein* (Heinemann, 1962)

Philpott, William James, *Bloody Victory: The Sacrifice on the Somme and the Making of the Twentieth Century, The Battle, the Myth, the Legacy* (Little, Brown, 2009)

Piekalkiewicz, Janusz, *Rommel and the Secret War in North Africa, 1941–1943, Secret Intelligence in the North African Campaign* (Schiffer, 1992)

Pitt, Barrie, *The Crucible of War, Vol. 1, Wavell's Command, 1941* (Jonathan Cape, 1980)

————*The Crucible of War, Vol. 2, Auchinleck's Command, 1942* (Jonathan Cape, 1982)

————*The Crucible of War, Vol. 3, Montgomery and Alamein* (Jonathan Cape, 1986)

Place, Timothy Harrison, *Military Training in the British Army, 1940–1944, From Dunkirk to D-Day* (Routledge, 2000)

Pocock, Tom, *Norfolk* (Pimloco Press, 1995)

Poett, General Sir Nigel, *Pure Poett* (Leo Cooper, 1991)

Pogue, Forrest C., *George C. Marshall, Vol. II, Organizer of Victory* (Viking Press, 1973)

Porch, Douglas, 'Military Culture and the Fall of France in 1940', *International Security*, Vol. 24/4 (Spring 2000)

————*Hitler's Mediterranean Gamble, The North African and Mediterranean Campaigns in World War 2* (Weidenfeld and Nicolson, 2004)

Potts, C. L., *Gordon & Michael, A Memoir* (Privately printed, 1940)

Powell, Anthony, *The Military Philosophers* (William Heinemann, 1968)

Powers, Stephen T., 'The Battle of Normandy, The Lingering Controversy', *Journal of Military History*, Vol. 56/3 (July 1992), pp. 455–71

Prior, Robin and Wilson, Trevor, *Command on the Western Front, Military Career of Sir Henry Rawlinson, 1914–18* (Blackwell, 1992)

————'The Great War in History, Debates and Controversies, 1914 to the Present', *English Historical Review*, Vol. CXXI, (2006)

Probert, Air Commander Henry, *Bomber Harris. His Life and Times* (Greenhill Books, 2001)

Quarrie, Bruce, *Panzers in the Desert* (Patrick Stephens, 1978)

————*Encyclopaedia of the German Army in the 20th Century* (Patrick Stephens Ltd, 1989)

Ramsay, Admiral Sir Bertram, *The Year of D-Day. The 1944 Diary of Admiral Sir Bertram Ramsay*, ed. by Love, Robert W., and Major, John (University of Hull, 1994)

Ramsden, John, *Don't Mention the War: The British and the Germans Since 1890* (Little, Brown, 2006)

Ranfurly, Countess of, *To War With Whitaker. The Wartime Diaries of the Countess of Ranfurly 1939–45* (William Heinemann, 1944)

Rankin, Nicholas, *Churchill's Wizards, The British Genius for Deception, 1914–1945* (Faber and Faber, 2009)

Reed, Paul, *Walking Arras* (Pen and Sword Battleground, 2007)

Reid, Brian Holden, *J. F. C. Fuller, Military Thinker* (Macmillan Press, 1987)

Reichsarchiv/Kraft von Dellmensingen, *Schlachten des Welkrieges: Der Durchbruch am Isonzo, Band 12a and 12b* (Gerhard Salling, Berlin, 1926)

Reiss, Matthias, 'Bronzed Bodies behind Barbed Wire, Masculinity and the Treatment of German Prisoners of War in the United States during World War II', *Journal of Military History*, Vol. 69/2 (April 2005), pp. 475–504

Remarque, Erich Maria, *All Quiet on the Western Front* (GP Putnam's Sons, 1929)

Remy, Maurice Philip, *Mythos Rommel* (Munich, List, 2002)

Renz, Irina; Krumeich, Gerd and Hirschfeld, Gerhard, *Scorched Earth, The Germans on the Somme 1914–18* (Pen and Sword Military, 2009)

Reuth, Ralf Georg, *Rommel. The End of a Legend* (Haus Publishing, 2005)

Reynolds, Major-General Michael, *Steel Inferno, 1st SS Panzer Corps in Normandy* (Spellmount 1997)

Rhodes, Anthony, *Sword of Bone, Phoney War and Dunkirk, 1940* (Faber & Faber, 1942; much reprinted).

Richards, Frank, *Old Soldiers Never Die* (Faber and Faber, 1933)

Richards, Lieutenant Colonel P., 'Rommel – *Fingerspitzengefuhl* Or Common Eavesdropper?', *British Army Review*, No. 131, pp. 61–9

Richardson, General Sir Charles, *Send For Freddie. The Story of Montgomery's Chief of Staff, Major-General Sir Francis de Guingand DSO.* (William Kimber, 1987)

Roberts, Andrew, *Napoleon and Wellington* (Weidenfeld and Nicolson, 2001)
——*Hitler & Churchill, Secrets of Leadership* (Weidenfeld and Nicolson, 2003)
——*Masters and Commanders, The Military Geniuses Who Led the West to Victory in World War II* (Allen Lane, 2008)
—— (ed.) *The Art of War, Great Commanders of the Modern World* (Quercus, 2008)
——*The Storm of War, A New History of the Second World War* (Allen Lane, 2009)
Robins, Simon, *British Generalship on the Western Front, 1914–1918, Defeat into Victory* (Frank Cass, 2005)
Robinson, Colonel James R., 'The Rommel Myth', *Military Review*, Vol. 77/5 (September–October 1997)
Rolf, David, *The Bloody Road to Tunis, Destruction of the Axis Forces in North Africa, November 1942–May 1943* (Greenhill, 2001)
Rommel, Erwin, *Infantry Attacks*, with a New Introduction by Manfred Rommel (Greenhill Books, 1990); originally published as *Infanterie greift an, Erlebnisse und Erfahrungen* (Ludwig Voggenreiter Verlag, 1937. Also translated and published by the US Army Infantry Journal (1944)
Rommel, Erwin, (ed. B. H. Liddell Hart, with Lucie-Maria and Manfred Rommel and General Fritz Bayerlein), *The Rommel Papers* (Collins 1953); originally published in a longer edition as *Krieg ohne Hass. Afrikanische Memorien* ('War Without Hate, African Memoirs') by Heidenheimer Druck und Verlags GmbH, 1950, published in France as *Rommel* by Jacques Mordal (1952)
Rommel, Erwin, (ed. Dr John Pimlott), *Rommel and His Art of War* (Greenhill, 2003)
Rostron, Peter, *The Military Life and Times of General Sir Miles Dempsey* (Pen & Sword, 2010)
Royle, Trevor, *War Report: The War Correspondent's View of Battle from the Crimea to the Falklands* (Mainstream, 1987)
Ruge, Admiral Friedrich, *Rommel in Normandy* (Macdonald and Jane's, 1979)
——'The Invasion of Normandy', Chapter IX in *Decisive Battles of World War II, the German View* (André Deutsch, 1965)
Russell, Douglas S., *Winston Churchill – Soldier, The Military Life of a Gentleman at War* (Brassey's UK, 2005)
Ryan, Cornelius, *The Longest Day* (Victor Gollancz Ltd, 1960)
Ryan, Professor Chris (ed.), *Battlefield Tourism: History, Place and Interpretation* (Elsevier Science, 2007)
Ryder, Rowland, *Oliver Leese* (Hamish Hamilton, 1987)
Sadkovich, James J., 'Understanding Defeat, Reappraising Italy's Role in World War II', *Journal of Contemporary History*, Vol. 24/1 (1989), pp. 27–61
——'German Military Incompetence through Italian Eyes', *War in History*, Vol. 1/1 (March, 1994), pp. 39–62
Scarfe, Norman, *Assault Division, A History of the 3rd Division From the Invasion of Normandy to the Surrender of Germany* (Collins, 1947)
Schellenberg, SS-General Walter, *Invasion 1940. The Nazi Invasion Plan for Britain* (St Ermin's Press, 2000)
Schellendorf, General Bronsart von, *Duties of the General Staff* [*Der Dienst des Generalstabes*] (Berlin, E. S. Mittler; translated War Office Intelligence Division, 1893)

Schivelbusch, Wolfgang, *The Culture of Defeat, On National Trauma, Mourning and Recovery* (Granta Books, 2003)

Schmidt, Heinz Werner, *With Rommel in the Desert* (George G. Harrap, 1951)

Schofield, Victoria, *Wavell, Soldier & Statesman* (John Murray, 2006)

Schramm, Percy Ernst (ed.), *Kriegstagebuch des Oberkommando der Wehrmacht, 1939–45 (Wehrmachtführungsstab)*, 8 volumes [High Command of the German Armed Forces Daily War Dairy] (München, 1982)

Schraepler, Hans-Albrecht (ed.), *At Rommel's Side, The Lost Letters of Hans-Joachim Schraepler* (Frontline, 2009)

Schweppenburg, General Leo Geyr von, 'Reflections on the Invasion', *Military Review*, Part 1, February 1961 and Part 2, March 1961

Searle, Alaric, *Wehrmacht Generals, West German Society, and the Debate on Rearmament, 1949–1959* (Praeger, 2003)

Seaton, Albert, *The German Army 1933–45* (Weidenfeld and Nicolson, 1982)

Sebag-Montefiore, Hugh, *Enigma, The Battle for the Code* (Weidenfeld and Nicolson, 2000)

———*Dunkirk, Fight to the Last Man* (Viking, 2006)

Senger und Etterlin, General Frido von, *Neither Fear Nor Hope* (Macdonald, 1964)

Seth, Ronald, *Caporetto: The Scapegoat Battle* (Macdonald and Co., 1965)

Shea, William L. and Pritchett, Merrill R., 'The Afrika Korps in Arkansas, 1943–1946', *Arkansas Historical Quarterly*, Vol. 37/1 (1978)

———'The Wehrmacht in Louisiana', *Louisiana History*, Vol. 23/1 (1982)

———'A German Prisoner of War in the South, the Memoir of Edwin Pelz', *Arkansas Historical Quarterly*, Vol. 44/1 (1985)

Sheffield, Gary, *Forgotten Victory, The First World War – Myths and Realities* (Headline, 2001)

———*The Somme* (Cassell, 2003)

———and Bourne, John, *The Haig Diaries, War Diaries and Letters – 1914–1918, The Diaries of FM Sir Douglas Haig* (Weidenfeld and Nicolson, 2005)

———and Todman, Dan, *Command and Control of the Western Front, The British Army's Experience 1914–18* (The History Press, 2007)

Sheldon, Jack, *The German Army on the Somme 1914–1916* (Pen & Sword Military; 2007)

———*The German Army at Passchendaele* (Pen & Sword Military, 2007)

———*The German Army on Vimy Ridge 1914–1917* (Pen & Sword Military, 2008)

———*The German Army at Cambrai* (Pen & Sword Military, 2009)

Shepherd, Alan, *France 1940, Blitzkrieg in the West* (Praeger, 2004)

Shores, Christopher and Ring, Hans, *Fighters Over the Desert* (Neville Spearman, 1969)

Showalter, Dennis E., 'Army and Society in Imperial Germany, The Pains of Modernization', *Journal of Contemporary History*, Vol. 18/4 (October 1983)

———*Patton And Rommel, Men of War in the Twentieth Century* (Berkley, 2005)

Shulman, Milton, *Defeat in the West* (Cassell and Co., 1947)

Siegler, Fritz, Freiherr von, *Die Höheren Dienstellen Der Deutschen Wehrmacht 1933–1945* (Institut für Zeitgeschichte 1953)

Simkins, Peter, *Kitchener's Army, The Raising of the New Armies 1914–1916* (Manchester University Press, 1988)

Simpson, Andy, *Directing Operations, British Corps Command on the Western Front 1914–18* (Spellmount, 2005)

Simpson, Keith, 'A Close Run Thing? D-Day, 6 June 1944, The German Perspective', *RUSI Journal*, Vol. 139/3 (June 1994)

Sixsmith, Major-General E. K. G., *British Generalship in the Twentieth Century* (Arms and Armour Press, 1970)

———*Eisenhower as Military Commander* (BT Batsford, 1973)

Slessor, Sir John, *The Central Blue, Recollections and Reflections* (Cassell and Co, 1956)

———(as Wing Commander J.C.), *Air Power and Armies* (Oxford University Press, 1936)

Smart, Nick, *Biographical Dictionary of British Generals of the Second World War* (Pen & Sword, 2005)

Smith, Colin, *England's Last War Against France, Fighting Vichy 1940–42* (Weidenfeld and Nicolson, 2009)

Smith, Michael, *Station X: The Codebreakers of Bletchley Park* (Macmillan, 1998)

Smith, Peter C., *Massacre at Tobruk, The British Assault on Rommel 1942* (Stackpole, 2008)

Smith, Rupert, *The Utility of Force, The Art of War in the Modern World* (Allen Lane, 2005)

Smith, W. H. B., *Small Arms of the World* (A&W Visual Library, 10th edition, 1973)

Smyth, Sir John, VC, *Leadership in War 1939–1945: The Generals in Victory and Defeat* (David & Charles, 1974)

———*Leadership in Battle 1914–1918: Commanders in Action* (David & Charles, 1975)

———*Milestones* (Sidgwick and Jackson, 1979)

Smurthwaite, David, Nicholls, Mark and Washington, Linda, *'Against All Odds': The British Army of 1940* (National Army Museum, 1989)

Sparrow, Lieutenant Colonel. J. H. A., *Morale* (The War Office, 1949)

Spayd, P. A., *From Afrikakorps to Panzer Lehr, The Life of Rommel's Chief of Staff – Generalleutnant Fritz Bayerlein* (Schiffer, 2003)

———and Wilkins, Gary, *The After Action Reports of the Panzer Lehr Division Commander from D-Day to the Rhur* (Schiffer, 2004)

Spears, Brigadier General Edward Louis, *Liaison 1914, A Narrative of the Great Retreat* (Heinemann, 1930)

———*Assignment to Catastrophe, 2 Volumes, (1) Prelude to Dunkirk (2) The Fall of France* (Heinemann, 1954)

———*The Picnic Basket* (Secker and Warburg, 1967)

Speidel, Lieutenant General Hans, *Invasion 1944. Rommel and the Normandy Campaign* (Herbert Jenkins, 1951) Alternative English title, *We Defended Normandy*

Spiller, Roger, 'Some Implications of ULTRA', *Military Affairs*, April 1976

Spires, David N., *Image and Reality, The Making of a German Officer 1921–1933* (Greenwood Press, 1984)

Sprösser, Theodor, *Die Geschichte der Württembergischen Gebirgsschützen* (Stuttgart, 1933)

Stahlberg, Alexander, *Bounden Duty: The Memoirs of a German Officer 1932–45* (Brasseys, 1990)

Stanier, Brigadier Sir Alexander, Bt., DSO, *Sammy's Wars, Recollections of War in Northern France and Other Occasions* (Welsh Guards OCA, 1998)

Statistics of the Military Effort of the British Empire during the Great War (HMSO, 1922)

Steinhoff, Johannes, Pechel, Peter and Showalter, Denis (eds), *Voices From the Third Reich, An Oral History* (Grafton/HarperCollins, 1991)

Stephenson, Michael, (ed.), *Battlegrounds, Geography and the History of Warfare* (National Geographic, 2003)

Stephenson, Jill, *Hitler's Home Front: Württemberg under the Nazis* (Hambledon Continuum, 2006)

Stolfi, R. H. S., 'Equipment for Victory in France in 1940', *History*, Vol. 55 (1970), pp. 1–20

Stone, David, *Fighting for the Fatherland, The Story of the German Soldier from 1648 to the Present Day* (Conway Press, 2006)

Strachan, Hew, *The First World War, A New History* (Simon & Schuster, 2003)

Strange, Joseph L., 'The British Rejection of Sledgehammer, An Alternative Motive', *Military Affairs*, Vol. 46/1 (1982)

Strawson, Major-General John, *The Battle for North Africa* (BT Batsford, 1969)

———*Hitler as Military Commander* (BT Batsford, 1971)

———*The Italian Campaign* (Secker and Warburg, 1987)

Strobl, Gerwin, *The Germanic Isle, Nazi Perceptions of Britain* (Cambridge, 2000)

Sundberg, Ulf E., and Harvey, A. D., 'The Siege of Tobruk. The Struggle to Maintain the Coastal Supply Line, 1941', *RUSI Journal*, Vol. 154/1 (February 2009), pp. 78–82

Szygowski, Ludwik Józef, *Seven Days and Seven Nights in Normandy* (White Eagle Press, 1976)

Tedder, Arthur, *With Prejudice, The War Memoirs of Marshal of the Royal Air Force Lord Tedder* (Cassell, 1966)

Terraine, John, *Douglas Haig, the Educated Soldier* (Hutchinson, 1963)

———*The Smoke and the Fire, Myths and Anti-Myths of War 1861–1945* (Sidgwick and Jackson, 1980)

———*The Right of the Line, The Royal Air Force in the European War 1939–1945* (Wordsworth, 1997)

Thompson, Major-General Julian, *Dunkirk, Retreat to Victory* (Sidgwick & Jackson, 2008)

Thompson, R. W., *The Montgomery Legend* (Allen and Unwin, 1967)

———*Montgomery The Field Marshal, the Campaign in North-West Europe 1944–45* (Allen and Unwin, 1969)

Todman, Dan, *The Great War, Myth and Memory* (Hambledon, 2005)

Toland, John, *Adolf Hitler* (Doubleday, 1976)

Tooley, Hunt, *The Western Front. Battle Ground and Home Front in the First World War* (Palgrave Macmillan, 2003)

Tout, Ken, *Tank!* (Robert Hale, 1985)

Trauschweizer, Ingo Wolfgang, 'Learning with an Ally: The US Army and the Bundeswehr in the Cold War', *Journal of Military History*, Vol. 72/2 (April 2008), pp. 477–508

Travers, T. H. E., *How the War Was Won, Command and Technology in the British Army on the Western Front, 1917–18* (Routledge, 1992)

Trevor-Roper, H. R., *Hitler's War Directives, 1939–1945* (Sidgwick & Jackson, 1964)

Trummer, Peter, *Die Bruder Stauffenberg und der deutsche Widerstand* (Landeszentrale für politische Bildung, Baden-Württemberg, 2009)

Trythall, A. J., *'Boney' Fuller, The Intellectual General* (Cassell, 1977)

Tsouras, Peter, *Disaster at D-Day, The Germans Defeat the Allies, June 1944* (Greenhill Books, 1994)

Tuchman, Barbara W., *The Guns of August – August 1914* (Constable, 1962)

Turner, E. S., *Dear Old Blighty* (Michael Joseph, 1980)

Turner, John (ed.), *Britain and the First World War* (Routledge, 1988)

Tute, Warren, *D-Day* (Sidgwick and Jackson, 1974)

——*The North African War* (Sidgwick and Jackson, 1976)

Ulin, Lieutenant Colonel R. R., 'The Role of the Middle East in Allied Strategy during World War II', *RUSI Journal*, Vol. 129/4 (December 1984), pp. 54–9

Urquhart, Brian, *A Life in Peace and War* (Weidenfeld and Nicolson, 1987)

Various authors, *D-Day Then And Now*, 2 volumes (London, Battle of Britain Prints/After the Battle, 1995)

Vaughan-Thomas, Wynford (introduction), *Great Front Pages. D-Day to Victory 1944–1945* (Collins, 1984)

Walker, Ian, *Iron Hulls, Iron Hearts, Mussolini's Elite Armoured Divisions in North Africa* (Crowood, 2006)

Walkinton, L. M., *Twice In A Lifetime* (Samson Books, 1980)

Wallach, Jehuda, *The Dogma of the Battle of Annihilation, the Theories of Clausewitz and Schlieffen and Their Impact on the German Conduct of the Two World Wars* (Greenwood Press, 1986)

Warlimont, Walter, *Inside Hitler's Headquarters* (Weidenfeld and Nicolson, 1964)

Warner, Philip, *Auchinleck, The Lonely Leader* (Buchan and Enright, 1981)

——*Horrocks: The General Who Led From The Front* (Hamish Hamilton, 1984)

Watson, Alexander, 'Junior Officership in the German Army during the Great War, 1914–1918', *War in History*, Vol. 14/4 (November 2007), pp. 429–53

Watson, Bruce Allen, *Exit Rommel. The Tunisian Campaign, 1942–43* (Praeger, 1999)

Wavell, General Sir Archibald, *Generals and Generalship. The Lees Knowles Lectures Delivered at Trinity College, Cambridge in 1939* (The Times/Penguin, 1941)

Wedemeyer, General A. C., *Wedemeyer Reports!* (Henry Holt and Co., 1958)

Weigley, Russell F. (ed.), *New Dimensions in Military History* (Presidio, 1975)

——*Eisenhower's Lieutenants, The Campaigns of France and Germany 1944–45* (Indiana University Press, 1981)

The Western Front, Then and Now (C. Arthur Pearson Ltd, 1938)

Westlake, Ray, *Kitchener's Army* (Nutshell Publishing, 1989)

Westphal, Gen. Siegfried, *The German Army in the West* (Cassell, 1951)

——'Notes on the North African Campaign, 1941–1943', *RUSI Journal*, Vol. 105/1 (February 1960), pp. 70–81.

Wette, Wolfram, *The Wehrmacht, History, Myth, Reality* (Harvard University Press, 2006)

Wheeler-Bennett, Sir John, *The Nemesis of Power, The German Army in Politics 1918–1945* (Macmillan, 1954)

——*King George VI: His Life and Reign* (Macmillan, 1958)

Whicker, Alan, *Whicker's War* (HarperCollins, 2005)

Whitaker, Brigadier Denis & Shelagh, *Victory at Falaise, the Soldiers' Story* (HarperCollins, 2000)

——and Copp, Terry, *Normandy, The Real Story of How Ordinary Soldiers Defeated Hitler* (Presidio Press, 2004)

White, B. T., *Tanks and Other AFVs of the Blitzkrieg Era 1939–41* (Blandford Press, 1972)

Whiting, Charles, *The Field Marshal's Revenge, The Breakdown of a Special Relationship* (Spellmount, 2004)

Wilkes, John and Eileen, *The British Army in Italy 1917–1918* (Leo Cooper, 1998)

——*Rommel and Caporetto* (Leo Cooper, 2001)

Williams, John F., *Corporal Hitler and the Great War 1914–1918* (Frank Cass, 2005)

Williamson, David, *A Most Diplomatic General. The Life of Lord Robertson of Oakridge 1896–1974* (Brassey's, 1996),

Williamson, Gordon, *Afrikakorps 1941–43* (Osprey, 1991)

Wilmot, Chester, *The Struggle for Europe* (Collins, 1952)

Wilson, Major H. R. G., 'Monty, A Review Article', *British Army Review*, No. 77 (August 1984), pp. 5–11

Wilson, Theodore A (ed.), *D-Day 1944* (University Press of Kansas, 1994)

Wilt, Alan F., 'An Addendum to the Rommel–Rundstedt Controversy', *Military Affairs*, Vol. 39/4 (December 1975)

Windrow, Martin, *The Panzer Divisions* (Osprey, 1973)

Winterbotham, F. W., *The Ultra Secret* (Weidenfeld and Nicolson, 1974)

Wood, James A., 'Captive Historians, Captivated Audience, The German Military History Program, 1945–1961', *Journal of Military History*, Vol. 69/1 (January 2005), pp. 123–47.

——(ed.), *Army of the West. The Weekly Reports of German Army Group 'B' from Normandy to the West Wall* (Stackpole, 2007)

Young, Brigadier Desmond, *Rommel* (Collins, 1950)

——*Try Anything Twice. The Autobiography of Desmond Young* (Hamish Hamilton, 1963) US title, *All The Best Years*

Younger, Tony, *Blowing Our Bridges, A Memoir from Dunkirk to Korea Via Normandy* (Leo Cooper, 2004)

Zabecki, Major-General David T. (ed.), *Chief of Staff, The Principal Officers Behind History's Great Commanders, Vol. 2* (Naval Institute Press, 2008)

Zaloga, Stephen J., *Operation Cobra 1944, Breakout from Normandy* (Osprey, 2001)

——*Poland 1939; The Birth of Blitzkrieg* (Osprey, 2002)

——*D-Day 1944, Omaha Beach, Omaha Beach* (Osprey, 2003)

——*D-Day 1944, Utah Beach and US Airborne Landings* (Osprey, 2004)

——*Kasserine Pass 1943. Rommel's Last Victory* (Osprey, 2005)

——*D-Day Fortifications in Normandy* (Osprey, 2005)

Zetterling, Niklas, *Normandy 1944, German Military Organisation, Combat Power & Organisational Effectiveness* (Fedorwicz, Winnipeg, 2000)

Zuehlke, Mark, *Operation Husky, The Canadian Invasion of Sicily, July 10–August 7, 1943* (Douglas & McIntyre, 2009)

——*Ortona, Canada's Epic World War II Battle* (Douglas & McIntyre, 2004)

Acknowledgements

Like a case of treasured claret, this book has been the result of several years' maturity and many fine ingredients, supplied willingly and with enthusiasm by numerous friends and accomplices. I have been extremely fortunate to work under two outstanding military historians, who not only allowed me time to write, but accompanied me on numerous battlefield tours and were often willing to drop everything and discuss my ideas. Professors Richard Holmes and Chris Bellamy were much more to me than my commanding officers, but became firm friends and mentors to whom I owe much. I must also tip my hat to the encouragement provided to me by the historian James Holland, who has proffered advice and read endless drafts, well beyond the call of duty. I am profoundly grateful to other academics for their patience and incisiveness over the years, particularly Professors Brian Bond, John Buckley and Gary Sheffield, to whom, collectively, I owe an enormous debt.

Help has arrived from many quarters, and I wish to acknowledge the assistance (in no particular order) of Field Marshal the Lord Bramall; General Sir Reddy Watt; Major Generals John Drewienkiewicz, Graham Hollands, John Moore-Bick, Richard Sherriff and Dave Zabecki; Brigadiers David Allfrey, Ian Copeland, Charles Grant, Chris Murray, Hal Nelson, John Smales, Jim Tanner, Miles Wade and Sam Weller; Colonels Tom Bowers, Alistair Bruce of Crionaich, Tom Giesbors in Holland, Tom Huggan at the British Embassy in Rome, Ivar Hellberg, Peter Herrley in Paris, David Hopley, John Hughes-Wilson, Nick Reynolds USMC, John Rouse and Angus Taverner; Oberst Karl-Heinz Frieser and Hans-Joachim Krüg; Lieutenant Colonels John Cargill, Ron Laden, Alan Miller, Ingram Murray, Chris Pugsley, Thomas Ryan USMC and David Stone; Majors Peter Cottrell and Gordon Corrigan; Hauptmann Otto Carius, Josef Klein and Heinz Volker; Professors Martin Alexander, Ian Beckett, Paul Cornish, A. C. Grayling and Huw Strachan; Doctors Philip Blood, John Bourne, Roger Cirillo, Lloyd Clark, Ed Flint, David Hall, Chris Hobson, Charles Kirke, Steve Prince, Alan Ryan, Keith Simpson

MP, Jim Storr, and Simon Trew; and Diego Cancelli in Italy, Dale Clark, David Cross, Hap Halloran, Sydney Jary MC, Dave Kenyon, Peter Macdonald, Toby Macleod, Patrick Mercer MP, Andy Robertshaw, Graham Ross, Peter Rostron, John Stevens, Mike Taylor, Peter Trummer in Heidelberg, Avril Williams, the amazing proprietress of Tea Rooms in the Somme hamlet of Auchvillers, and many other soldiers, students and academics who read drafts, have passed me ammunition when my magazine was running low, or encouraged me to attack from an unguarded flank.

Then there are the legions of older soldiers – over one hundred whom I have interviewed – who served under the marshals and have since joined them in the great muster beyond, including Field Marshals Lord Carver and Sir John Stanier, but who on this earth enthralled me with their tales of military life, which I am privileged to pass down to others.

This is not to forget the long-suffering research staff of the Cranfield University Library at Shrivenham, led by Iain McKay, and Chris Hobson and his team at the JSCSC Library, Watchfield; the staff of the Aerial Reconnaissance Archive (TARA) in Edinburgh; *After the Battle* Magazine; the Airborne Museum, Oosterbeek; Bastogne Historical Center; Bletchley Park; fellow members of the British Commission for Military History; the British Library; British Library Newspaper Collection, Colindale Avenue, NW9; Churchill Archives Centre, Churchill College, Cambridge; Commonwealth War Graves Commission; D-Day Museum, Portsmouth; La Gleize Military Museum, Belgium; fellow members of the Guild of Battlefield Guides; Hauptstaatsarchiv, Stuttgart; Haus *der* Geschichte Baden-Württemberg, Stuttgart; L'Historial de la Grande Guerre, Peronne; Nigel Steel, Simon Robbins and staff at the Imperial War Museum, London; In Flanders Fields Museum, Ypres; Brigadier Tom Longland and staff at the Joint Lessons Cell (formerly the Tactical Doctrine Retrieval Centre), Doctrine and Concepts Development Centre, UK Defence Academy; Taff Gillingham and his colleagues of the Khaki Chums; Guy Leaning and his British Overseas Chums (BOCs); Liddell Hart Centre for Military Archives at King's College, London; The Liddle Collection at the University of Leeds Brotherton Library; Le Mémorial de Caen; the Memorial Museum Passchendaele 1917 (Zonnebeke); Militärgeschichtliches Forschungsamt (MGFA) in Potsdam; Militärhistorische Museum der Bundeswehr, Dresden; *the Musée du Débarquement,* Arromanches; the National Archives, Kew; Michael Ball and colleagues at the National Army Museum, Chelsea; the National Liberation Museum, Groesbeek; Simon May of the Old Pauline Club; Rob and Jacqueline de Ruyter at the Poteau-44 Museum; Prince Consort's Library, Aldershot; Rommelarchiv, Herrlingen; RMAS Library, Sandhurst; RUSI Library, London; Second World War Experience Centre, Wetherby; the Small Arms Weapons Collection at Shrivenham under the watchful (and tolerant) eye of Lt.-Col (retd.) John Starling, who taught me so much about the weapons that Monty and Rommel knew; Southwick House (formerly HMS *Dryad*); the Tank Museum, Bovington, Dorset; Thiepval Visitor's Centre; Volksbund Deutsche Kriegsgräberfürsorge (VDK) in Kassel; the Royal Regiment of Fusiliers (Royal Warwickshire Regiment); Soldiers of Gloucestershire and Staffordshire Regimental Museums, whose pencils I have snapped, coffee cups I have upset inadvertently, dusty shelves I have disordered and whose photocopiers I have overheated.

I am particularly grateful to the guidance of my agent, Patrick Walsh, whilst at

Preface Publishing, Trevor Dolby – especially – and Nicola Taplin, have, when needed, blown the whistle, summoning me 'over the top' through the barbed wire of deadlines and on to the objective; most importantly, organising me like a good adjutant, Kate Johnson, my long-suffering editor, has put in more hours than I deserve and crafted my literary meanderings into coherent form. In the background have been my indispensable secretaries, Steph Muir, Bella Platt, and latterly Anne Harbour.

Try as I might I can never put myself in the shoes of the rest of my family when writing. As the research progresses, the dining table is invaded by platoons of disorderly documents: I hog the home computer at strange hours, scattering more papers, while battalions of books start encroaching into every room in the house. No one sees them arrive, they infiltrate by stealth and suddenly they're there, standing in silent piles. Several book-skyscrapers are flanking me now, impressively tall and equally unsafe, which the cat and dog stalking each other have realised to their cost. I would like to take this opportunity to acknowledge the help and support of my wife Stefania and daughter Emmanuelle for their unfailing help, tolerance and fortitude during long periods of my self-enforced isolation whilst writing, or absence when exploring Monty's and Rommel's battlefields (and, in their eyes at least, beaches and bars). I simply say 'thank you' for putting up with me and this project. I promise to demobilise the books and papers and return them to their shelves soon and give you back your home.

Chronology

MONTGOMERY
(1887–1976)

1887	Nov 17	Born at St Mark's Vicarage, Kennington, London
1889–1901		At Bishopscourt, Hobart, Tasmania
1907	Jan 30	Officer Cadet, Royal Military College Sandhurst
1908	Sep 19	Commissioned 2nd Lieutenant, sails to India
1910	Apr 1	Promoted Lieutenant
1913	Jan	1/Royal Warwicks return to Shorncliffe, England
1914	Aug 5	Britain declares war on Germany: mobilisation
	Sep 26	Battle of Le Cateau
	Oct 13	Wounded at Méteren
	Dec 7	Promoted Captain; gazetted with DSO
1915	Feb 12	Appointed Brigade Major, 112th Brigade

ROMMEL
(1891–1944)

1891	Nov 15	Born in Heidenheim, near Ulm
1910	Jul 19	*Fähnrich* (Ensign, or Cadet Officer), 124th Regiment
1911	Mar 15	Assigned to Kriegsschule, Danzig
1912	Jan 27	Commissioned *Leutnant* at Kriegsschule, Danzig
1913	Dec 5	Death of father, Erwin Rommel
	Dec 8	Daughter Gertrud born to Walburga Stemmer
1914	Aug 4	Belgium invaded
	Sep 24	Wounded, Iron Cross, Second Class
1915	Jan 29	Argonne battles, Iron Cross, First Class
	Sep 18	*Oberleutnant*
	Oct 4	Assigned Wurttemberg Mountain Battalion

1916	Jan	Lands in France with 104th Brigade
	Jul 1	Battle of the Somme commences
1917	Jan 22	Appointed GSO2, HQ 33rd Division
	Feb–Mar	On Staff Course, Hesdin
	Apr 9	Battle of Arras: Vimy Ridge seized
	Jul 6	Appointed GSO2, IX Corps
	Jun 7	Messines Ridge seized
	Jul 31	Third Ypres (Passchendaele) campaign begins
	Nov 20–Dec 3	Battle of Cambrai
1918	Mar 21	German 'Michael' offensives begin
	Mar	Promoted Brevet Major
	Aug 8	Battle of Amiens; Allied counter-offensive begins
	Jul 16	Appointed GSO1, 47th Division, Temporary Lieutenant-Colonel
	Oct 27	Montgomery and Churchill at Lille Victory Parade
	Nov 11	Armistice; GSO2, British Army of the Rhine
1919	Sep–Dec	CO, 17/Royal Fusiliers
1920	Jan 22	At Staff College Camberley
1921	Jan 5	Brigade Major, 17th Infantry Brigade, Cork
1922	May 24	Brigade Major, 8th Infantry Brigade, Plymouth
1923	May	GSO2, HQ 49th Division (TA), York
1925	Mar	'A' Company Commander, 1/Royal Warwicks
	Jul 26	Substantive Major
1926	Jan 23	Lieutenant Colonel, DS, Staff College, Camberley
1927	Jul 27	Marries Betty Carver (née Hobart)
1928	Jan 1	Brevet Lieutenant Colonel
	Aug 18	Son David born
1929	Oct–Apr 1930	Revises *Infantry Training Manual*, Volume II
1931	Jan	CO, 1/Royal Warwicks, Palestine and Egypt
1932	Nov 25	Death of father, Bishop Sir Henry Montgomery, KCMG
1934	Jun 29	Colonel, DS, Staff College, Quetta

1916	Aug 8–20	Battle for Mount Cosna, Romania
	Nov 27	Marries Lucie Mollin in Danzig
1917	Oct 24	Caporetto offensive begins
	Oct 26	Captures Mount Matajur
	Nov 9–10	Battle of Longarone
	Dec 13	Awarded *Pour le Mérite* with Major Sprösser
1918	Jan 11–Dec 21	Assigned to Staff, LXIV Army Corps
	Oct 18	Promoted *Hauptmann*
	Nov 11	Armistice
1919	Jun 25–Dec 31 1920	25th Internal Security Regiment
1921	Jan 1	Company Commander, 13th Infantry Regiment
1928	Oct	Death of Walburga Stemmer
	Dec 24	Son Manfred born
1929	Oct 1–Sep 30 1933	Instructor, Infantry School, Dresden
1932	Apr 1	Promoted Major
1933	Jan 30	Hitler Chancellor of Germany
	Oct 1	Commander III Battalion, 17th Regiment, Goslar
1934		Publishes *Problems for Platoon and Company*
	Jun 30	Night of the Long Knives, threat to army removed
	Aug 2	Death of Hindenburg: army oath to Hitler
	Sep 30	First meeting with Hitler, Goslar
1935	Mar 1	*Oberstleutnant*
	Mar 16	German conscription announced
1935	Oct 15–Nov 9 1938	Instructor, *Kriegsschule*, Potsdam
1936	Mar 7	Rhineland demilitarised
	Sep 8–14	Führer's Military Escort, 8th Nuremberg Rally

1937	Aug 5	Brigadier, 9th Infantry Brigade, Plymouth
	Oct 19	Death of Betty Montgomery
1938	Dec	Major General, 8th Division, Palestine
1939	Mar 15	Germany occupies Prague
	Apr 27	Conscription introduced in UK
	Aug 28	Commander, 3rd Division, BEF
	Sep 3	War declared: BEF to France; Phoney War period
1940	May 10	Invasion of France and Belgium; Churchill PM
	May 26–Jun 4	Dunkirk evacuation
	Jul 10–Oct 31	Battle of Britain
	Jul 22	Lieutenant General, Commander, V Corps, South-east England
1941	Apr 27	Commander, XII Corps, England
	Dec 11	Germany declares war on USA
	Dec 25	Alan Brooke appointed CIGS
1942	Aug 4–10	Churchill and Brooke to Egypt
	Aug 7	Death of Gott; Monty C-in-C, Eighth Army
	Aug 19	Dieppe Raid
	Oct 23–Nov 4	Battle of El Alamein
	Nov 8	Operation Torch: Allies land in French North Africa
	Nov 11	General; awarded KCB
1943	Jan 14–24	Casablanca Conference (QUADRANT)
	Mar 6	Operation Capri, Battle of Medenine
	Mar 20–27	Battle of the Mareth Line
	May 13	Germans surrender in Tunisia
	Jul 9	Operation Husky: Invasion of Sicily
	Sep 3	Eighth Army lands in Italy
	Nov 28–Dec 1	Tehran Conference (EUREKA)

1937	Jan	Publishes *Infantry Attacks*. 400,000 copies sold by 1945
	Oct 1	Promoted *Oberst*
1938	Mar 12	*Anchluss* with Austria
	Oct 1–10	Commands Hitler's escort into Sudetenland
	Nov 10	*Kommandant, Kriegsschule*, Wiener Neustadt
1939	Mar 12–24	Commands Hitler's escort into Prague
	Aug 1	Promoted *Generalmajor*
	Aug 23	Assigned to Staff, *Führerhauptquartier*
	Sep 1–30	Poland invaded and overrun
1940	Feb 15	Commander, 7th Panzer Division
	May 10	Campaign in the West begins
	May 21	Arras counterattack
	May 26	Awarded Knight's Cross
	June	Death of mother, Helene Rommel
	Jun 22	Franco-German Armistice
1941	Feb 1	Promoted *Generalleutnant*
	Feb 12	Arrives in Tripoli to command Afrika Korps
	Mar 19	Awarded Oakleaves to Knight's Cross
	May 20	Crete invaded
	Jun 15	Operation Battleaxe launched
	Jun 22	Invasion of Russia begins
	Jul 1	Promoted *General der Panzertruppen*
	Nov 18	Operation Crusader launched
1942	Jan 21	Awarded Swords to Knight's Cross
	Jan 22	*Generaloberst*, Commander *Panzerarmee Afrika*
	Jun 22	Tobruk falls; promoted *Generalfeldmarshall*
	Aug 30–Sep 2	Battle of Alam Halfa
	Sep 23	Rommel home to Germany on sick leave
	Oct 23–Nov 4	Battle of El Alamein
	Nov 8	Allies launch Operation Torch
1943	Feb 2	Stalingrad surrenders
	February 19–21	Battle of Kasserine Pass
	Feb 23	Appointed C-in-C *Armeegruppe Afrika* in Tunisia
	Mar 9	Leaves North Africa
	Mar 11	Awarded Diamonds to Knight's Cross
	Jul 9	Allies invade Sicily
	Jul 15	Army Group B activated, Munich
	Aug 15	Army Group B in Northern Italy
	Sep 3	Montgomery lands in Southern Italy
	Nov 21	Army Group B moves to Fontainebleau, France

1944	Jan 3	C-in-C, 21st Army Group, England
	May 15	Final D-Day briefing at St Paul's School, Chiswick
	Jun 6	Invasion of Normandy
	Jul 18	Operation Goodwood
	Aug 21	Falaise Pocket closed
	Sep 1	Promoted Field Marshal
	Sep 17–26	Operation Market Garden: Arnhem
	Dec 16–Jan 25 1945	Battle of the Bulge
1945	Mar 23–24	Rhine Crossings, Operations Plunder and Varsity
	Apr 12	Death of Roosevelt; Truman US President
	May 4	Takes German Surrender at Lüneburg Heath
	Jul 6	General Election; result on 26th: Attlee PM
1946	Jan 31	Created Viscount Montgomery of El Alamein
	Jun 26–Dec 31 1949	Chief of the Imperial General Staff
	Dec 3	Created Knight of the Garter (KG)
1948–1951		Chairman of the Western European Union's Commander in Chiefs Commission
1949		Death of mother, Lady Maud Montgomery
	Apr 4	Formation of North Atlantic Treaty Organisation (NATO)
1950	Dec 19	Eisenhower Supreme Allied Commander Europe, Monty as his Deputy at SHAPE (Supreme Headquarters Allied Powers Europe)
1955	May 14	Warsaw Pact founded
1958	Sep 18	Retires after fifty years in British Army. Publishes *Memoirs*
1976	Mar 24	Dies at Isington Mill, Alton, Hampshire
	Apr 1	State funeral

1944	Mar 9	Army Group B arrives La Roche-Guyon, France
	Apr 15	Hans Speidel appointed to Army Group B
	Jun 6	Normandy invasion; Frau Rommel's birthday
	Jun 17	Conference with Hitler at Margival, France
	Jun 29	Last Meeting with Hitler, Berchtesgaden
	Jul 15	Final telegram to Hitler
	Jul 17	Injured in car crash, St Foy de Montgommery
	Jul 20	Stauffenberg bomb plot attempted and fails
	Oct 14	Generals Burgdorf and Maisel visit; dies after taking cyanide
	Oct 18	State funeral at Ulm
1945	May 8	VE Day; European war over

1950	Jan	Publication of Desmond Young's biography
1951	Oct 17	Film *Rommel the Desert Fox* released
1953		*The Rommel Papers* published in English

Guide to Ranks

US/BRITISH ARMY	GERMAN ARMY	WAFFEN-SS	RAF
General	*Generaloberst*	*Oberstgruppenführer*	Air Chief Marshal
Lieutenant General	*General*	*Obergruppenführer*	Air Marshal
Major General	*Generalleutnant*	*Gruppenführer*	Air Vice Marshal
Brigadier/Brig. Gen.	*Generalmajor*	*Brigadeführer*	Air Commodore
Senior Colonel	*Oberführer*		
Colonel	*Oberst*	*Standartenführer*	Group Captain
Lieutenant Colonel	*Oberstleutnant*	*Obersturmbannführer*	Wing Commander
Major	Major	*Sturmbannführer*	Squadron Leader
Captain	*Hauptmann*	*Hauptsturmführer*	Flight Lieutenant
Lieutenant	*Oberleutnant*	*Obersturmführer*	Flying Officer
Second Lieutenant	*Leutnant*	*Untersturmführer*	Pilot Officer

Note that the German rank of *Generalmajor* does not equate to a British/US Major General, but is more akin to Brigadier. The British ceased using the term Brigadier General in 1922.

List of Illustrations

While every effort has been made to contact copyright holders, the author and publisher would be grateful for information about any material where they have been unable to trace the source, and would be glad to make amendments in further editions.

Index